ADVANCED SENIOR MATHEMATICS 12

MATHEMATICS

PRINCIPLES

PROCESS

Authors

FRANK EBOS, Senior Author
Faculty of Education
University of Toronto

BOB TUCK
Mathematics Consultant
Nipissing Board of Education

WALKER SCHOFIELD
Department Head of Mathematics
Banting Secondary School
London, Ontario

Consultants

Roger Johnston
Assistant Head of Mathematics
South Grenville District High School
Prescott, Ontario

Ronald Reid
Department Head of Mathematics
York Memorial C.I.
Toronto, Ontario

Jim Turcott
Department Head of Mathematics
Levack District High School
Levack, Ontario

NELSON CANADA

W9-AOC-826

Published in 1986 by
Nelson Canada
A Division of International Thomson Limited
1120 Birchmount Road
Scarborough, Ontario

ISBN 0-17-602524-3

Canadian Cataloguing in Publication Data

Ebos, Frank
Mathematics: Principles and Process

Includes index.
For use in grade 12.
ISBN 0-17-602524-3

1. Mathematics — 1961–   I. Tuck, Bob,
1941–   II. Title.

QA107.E52 1986   510   C86-

Project Editor
Sheila Basset

Editors
Peter Gardiner
Anthony Luengo

Technical
Frank Zsigo

The symbol for year is a. For the sake of clarity, the word year has been used in place of a.

Printed and bound in Canada.

34567890   JD 893210987

Photo Credits
p. 19 Athlete Information Bureau; Province of Manitoba Department of Research; CITY-TV, Toronto; p. 92 Camerique/Miller Services Ltd.; p. 141 Ontario Hydro; p. 165 National Film Board; p. 170 Tourism New Brunswick; p. 189 Standard Oil Co. (N.J.); p. 202 Photography and Dominion Communications, Ottawa; p. 203 Harold M. Lambert/Miller Services; p. 220 Las Vegas News Bureau; p. 312 Cara Operations; Athlete Information Bureau; p. 357 British Columbia Government Photographic Services; H. Armstrong Roberts; p. 370 F. Ebos; p. 371 Hewlett Packard Ltd.; H. Armstrong Roberts; Cara Caswell/Wardair Canada; p. 409 Lufthansa Airlines; p. 421 National Film Board of Canada; p. 469 Transport Canada; p. 497 Canapress; p. 509 George Hunter

Using Mathematics: Principles and Process, Book 2

These pages explain how the text is organized. They tell you what to look for in each lesson and in every chapter.

Lesson Features

▶ Look for the lesson number and title.

Teaching

▶ The lesson begins with the information you need to learn. Look for pictures and photos that illustrate uses of mathematics. New words are printed in **bold type**.

▶ Examples and Solutions guide you step-by-step through new material.

▶ Always read the hints and helps printed in red type.

Exercise Features

▶ **Each lesson gives you lots of practice:**

A These questions let you practise the skills and concepts of the lesson. Many of these questions can be done with your teacher and the class.

B These questions give you practice with what you have learned. There are also lots of problems to solve.

C These questions provide an extra challenge, or may involve another approach.

Applications

These sections show how mathematics is a part of the everyday world. You will solve some problems and learn some interesting facts.

11.3 Parabolas and Their Applications

The method used to find the equation of a circle is applied in a similar way to investigate other loci. For example:

What type of curve is obtained if a point P is equidistant from a fixed point, and a fixed line? In other words, what is the locus of P if PF = PD?

To find the equation, you need to sketch the given information on a diagram as shown in the following example.

Example 1   Find the equation of the locus of all points equidistant from a fixed point, F(1, 2), and a fixed line defined by $y = -4$.

Solution   Let P(x, y) be any point on the locus. Sketch the information on a diagram.
The distances from P to F and P to D are equal.

Thus,
$$PF = PD,$$
$$\sqrt{(x-1)^2 + (y-2)^2} = \sqrt{(x-x)^2 + (y+4)^2}$$
$$(x-1)^2 + (y-2)^2 = (y+4)^2$$
$$(x-1)^2 + y^2 - 4y + 4 = y^2 + 8y + 16$$
$$(x-1)^2 - 12 = 12y$$

2.4 Exercise

A 1 Arrange each polynomial in descending powers of x.
(a) $-2x + 4x^2 + 9$ (b) $11 + 2x^2 - x - 5x^3$
(c) $x^4 - 2x + 12x^2 + 5 + 7x^3$ (d) $x^2 + 1 - x - \dfrac{1}{x^2} + \dfrac{2}{x}$

2 Fill in any missing terms of each polynomial.
(a) $4p^2 - 5$ (b) $3x^3 + x - 2$ (c) $y^5 + y + 7$ (d) $4w^4 + 3w^3 - 75$

B 3 Use your factoring skills to divide.
(a) $\dfrac{q^2 - q - 6}{q + 2}$ (b) $\dfrac{2a^2 - 25a + 63}{a - 9}$ (c) $\dfrac{x^2 + 4xy + 3y^2}{x + y}$
(d) $\dfrac{2ac - ad + 2bc - bd}{a + b}$ (e) $\dfrac{b^2 - c^2}{b + c}$ (f) $\dfrac{18x^2 - 2y^2}{3x + y}$

Applications: Atmospheric Pressure

The approximate distance above sea level, d, in kilometres, is given by
$$d = \frac{500(\log_{10} P - 2)}{27}$$
where P is the atmospheric pressure, in kiloPascals.
The reading on a barometer at sea level is 100 kPa of mercury. Use this value in the formula.
$$d = \frac{500(\log_{10} P - 2)}{27}$$
$$= \frac{500(\log_{10} 100 - 2)}{27}$$
$$= 0$$
Thus, the formula shows the distance above the earth at sea level is 0 km.
In Questions 12–16, round answers to 4 significant digits.

If the pressure inside a jet liner changes drastically, the passengers receive a supply of oxygen that automatically is made available in the cabin of the jet.

Reviews and Tests

These sections review or test skills and concepts *after* every chapter:

▶ **Practice and Problems: A Chapter Review**
▶ **Test for Practice**

These sections help you review and practise skills from *earlier* chapters:

▶ **Maintaining Skills**
▶ **Cumulative Review**

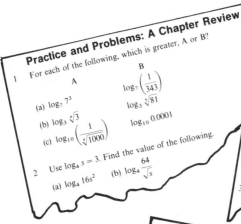

Practice and Problems: A Chapter Review

1 For each of the following, which is greater, A or B?

A	B
(a) $\log_7 7^3$	$\log_7\left(\dfrac{1}{343}\right)$
(b) $\log_3 \sqrt[4]{3}$	$\log_3 \sqrt[3]{81}$
(c) $\log_{10}\left(\dfrac{1}{\sqrt[4]{1000}}\right)$	$\log_{10} 0.0001$

2 Use $\log_4 s = 3$. Find the value of the following.

(a) $\log_4 16s^2$ (b) $\log_4 \dfrac{64}{\sqrt{s}}$

Looking Back: A Cumulative Review

1 Simplify.
(a) $(3\sqrt{2} - 3\sqrt{3})(2\sqrt{2} - 5\sqrt{3})$ (b) $(2\sqrt{5} - 3)^2$
(c) $(5\sqrt{3} - 2\sqrt{2})^2 - 3(\sqrt{3} - \sqrt{2})(\sqrt{3} + \sqrt{2})$

2 Solve for x: (a) $x^{\frac{1}{2}} = 5$ (b) $x^{\frac{1}{5}} = 8$ (c) $x^{\frac{1}{3}} = \dfrac{32}{\sqrt{x}}$

3 Solve $(x^2 + 2x)^2 - 2(x^2 + 2x) - 3 = 0$

Math Tip

The particular skills of graphing have their roots in the work of Rene Descartes (1596–1650). His contribution to the study of mathematics was honored by having the Cartesian plane named after him. The honoring of a mathematician by naming a law or curve or something mathematical after him or her is very common in mathematics.

Problem-Solving Features

There are lots of opportunities to learn and practise problem-solving skills — not just in the lessons, but also in special sections based on particular aspects of problem solving.

Solving Problems

Lessons in most chapters give you new problem-solving skills like *Looking for Clues*, *Working Backwards*, and other strategies.

▶ **Problem Solving** in every chapter give you a chance to do different types of problems and introduce you to interesting aspects of mathematics. Some show new ways to practice your skills.

Computer Tip

To calculate the mean, M, of a set of data, a computer program written in BASIC language is used, as follows.

```
10  INPUT N
15  DIM X(100)
20  LET S = 0
30  FOR I = 1 TO N
40  INPUT X(I)
50  LET S = S + X(I)
60  NEXT I
70  LET M = S/N
80  PRINT "THE MEAN IS", M
90  END
```

Problem-Solving: Show That . . .

Very often in mathematics you are asked to show that a fact is true. Often, there is more than one strategy for doing so. For example:

Show that $2\sqrt{2} + 3$ is a root of the quadratic equation $x^2 - 6x + 1 = 0$.

Strategy A
You could use your skills and find the roots of this quadratic equation and show that one of the roots is $2\sqrt{2} + 3$. This strategy would meet the requirements, but, would also involve the additional work of finding a second unasked-for root.

Strategy B
You could show directly that $2\sqrt{2} + 3$ satisfies the equation and is thus a root. For example:

▶ **Calculator Tips** give you practice with your calculator.

▶ **Computer Tips** will help you learn about micro computers.

Extension Features
Math Tips

Do you like learning shortcuts? Are you interested in who "invented" mathematics? *Math Tips* are for you! They are in every chapter.

8

Inventory of Essential Skills

In the study of mathematics, each skill is used to develop further skills.
The following questions provide a brief overview of some of the skills from
your previous studies. For each exercise
  • review the meaning of any important words that occur in the exercise.
  • do the exercise.
  • check your answers at the back of the book.

Skills with Integers

Calculations with integers and rationals occur constantly in mathematics.

1 Simplify.
   (a) $-4 - 3[2(5 - 3)]$ (b) $-2(-3^2 - 2^3)$
   (c) $300 - 2[5(-3^2 + 5) - 5^2]$ (d) $(2^2 - 3^2 - 1)(3^3 - 2^2 - 2)$
   (e) $(-2 - (-3)^2 - 4^2)(-2^2 + (-3)^3 - 5^2)$

2 Simplify.
   (a) $\left(\dfrac{-4}{3}\right) \div \left(\dfrac{2}{-3}\right)$ (b) $\left(\dfrac{3}{8}\right)\left(\dfrac{-5}{4}\right)$ (c) $\left(\dfrac{-2}{3}\right) \div \left(\dfrac{-3}{8}\right)$
   (d) $\left(\dfrac{-3}{-2}\right)\left(\dfrac{-2}{3}\right)$ (e) $-12 \div \left(-\dfrac{2}{3}\right)^2$ (f) $\left(\dfrac{-2}{3}\right)\left(-\dfrac{3}{4}\right)^2$

3 Simplify.
   (a) $\dfrac{1}{2} - \dfrac{-2}{3}$ (b) $\dfrac{-2}{5} + \dfrac{-1}{2}$ (c) $\dfrac{7}{12} - \dfrac{-3}{4}$
   (d) $-\dfrac{1}{8}\left(\dfrac{3}{-8} - \dfrac{-1}{2}\right)$ (e) $-\dfrac{2}{3}\left(\dfrac{1}{-4} - \dfrac{-1}{3}\right)$ (f) $-\dfrac{3}{4}\left(\dfrac{-3}{4} - \dfrac{1}{-4}\right)$

4 If $x = -3$ and $y = -2$, find the value of each expression.
   (a) $-x + 3y$ (b) $-4x - y$ (c) $3x - 8y$
   (d) $3x^2 - y$ (e) $-2x - 3y^2$ (f) $2y^2 - 3x^2$
   (g) $\dfrac{xy + 2y^2}{y}$ (h) $\dfrac{x^2 - y^2}{x + y}$ (i) $\dfrac{x^2 - 2xy + y^2}{x - y}$

5 (a) How much more is $-3(-3) + (-4)^2$ than $-9^2 + 6^2$?
   (b) By how much does $-4 - (-3)^2$ exceed $-9^2 - (-2)^3$?

Polynomials

1   Simplify each of the following.
  (a) $2y - 5 - 3[-2(y - 2) + 3]$ (b) $4[(3x - 6) - 5(-2x - 1)]$
  (c) $2y(y - 1) - 4(y^2 - 3y) + 3(y^2 - 2y)$
  (d) $(x - 3)(x + 9) - 3(x - 3)(x + 3)$
  (e) $(3x - y)(2x + y) - 2(x - y)^2$ (f) $-3(a - 4b)^2 - 2(a + 5b)^2$

2   (a) Simplify. Write the answer in descending powers of x.
$$2(x + 6)(x - 2) - 3(x - 4)(x + 6)$$
  (b) Simplify. Write the answer in ascending powers of x.
$$3(x^2 - 1) - 2x(x - 5) - 3x(x - 2)$$

3   (a) Subtract $3(x^2 - 2) - 3(x + 5)$ from $2(x^2 - 6) - 3(x + 5)$.
  (b) By how much does $3(y - 2)^2 - 3(y + 1)^2$ exceed $2(y - 4)^2 - 3(y + 5)^2$?
  (c) How much less is the product of $3y$ and $(2y + 5)^2$ than the sum of $(3y - 2)^2$ and $2(y + 5)$?

4   Factor each of the following.
  (a) $4x^3 + x^2$ (b) $9x^4 - 16y^2$ (c) $x^2 + 9x + 20$
  (d) $y^2 - 13y + 42$ (e) $5 + 6x + x^2$ (f) $x^4 - 64$
  (g) $x^4 - 3x^2 - 4$ (h) $y^4 + 6y^2 + 9$ (i) $x^4 + 2x^2 - 15$
  (j) $x^4 - 225y^2$ (k) $4(x - y)^2 - (x + y)^2$

5   Replace each (?) symbol with the correct sign (positive or negative) for the expression.
  (a) $\dfrac{x + y}{x - y} = (?) \dfrac{y + x}{y - x}$ (b) $\dfrac{x + y}{x - y} = (?) \dfrac{-x - y}{x - y}$ (c) $\dfrac{x + y}{x - y} = (?) \dfrac{-x - y}{y - x}$

  (d) $-\dfrac{x - y}{y - x} = (?) \dfrac{y - x}{x - y}$ (e) $\dfrac{-x + y}{y - x} = (?) \dfrac{y - x}{y - x}$ (f) $-\dfrac{x - y}{y - x} = (?) \dfrac{x - y}{x - y}$

6   Simplify each of the following.
  (a) $\left(\dfrac{x^2 - 4}{2x^2 + 11x + 5}\right)\left(\dfrac{x^2 + 2x - 15}{x^2 - x - 6}\right)$ (b) $\left(\dfrac{a + 1}{a - 1}\right)\left(\dfrac{a + 3}{1 - a^2}\right) \div \dfrac{(a + 3)^2}{1 - a}$

  (c) $\dfrac{x - 4}{4} + \dfrac{4}{x - 4} \div \dfrac{1}{4x - 16}$ (d) $\dfrac{3a - 1}{a^2 - 9} - \dfrac{5}{a - 3}$

7   If $x = -3$, find the value of each of the following.
  (a) $\left(\dfrac{x^2 + x - 2}{x^2 - x - 6}\right)\left(\dfrac{x^2 - 9}{x^2 + 2x - 3}\right)$ (b) $\dfrac{3x - 1}{x^2 - 1} - \dfrac{2x + 5}{x^2 - x - 2}$

Solving Equations

1   Solve and verify.
    (a) $15 + 5(x - 20) = 3(x - 1)$ (b) $4(m + 7) - 2(m - 5) = 3(m - 2)$

2   Solve.
    (a) $3 - (9y - 15) = 4(y - 2)$ (b) $8(2p + 4) - 3p = 6(3p + 7)$
    (c) $3(k - 7) + 2(k + 3) = 4(k - 1)$ (d) $2(m - 3) + 3(m - 5) = 4$
    (e) $\dfrac{x}{2} - \dfrac{1}{7}(x + 1) = \dfrac{1}{3}(2x - 6)$

3   Find the solution set.
    (a) $2(y + 1) - 3 = 4(2y + 1) - 11$ (b) $1 + 3x(x - 6) = x(3x - 1) + 1$

4   (a) For what value of y will $2(3y - 2)$ equal $(4y + 1)$?
    (b) What value of m will make $6(m - 4)$ exceed $-(m - 1)$ by 38?

5   Solve for the variable indicated.
    (a) $2x + y = 1$, y (b) $3y - 2x = 1$, y (c) $x + 2y = 3$, x
    (d) $2x - 5y = 6$, x (e) $t = 4(p + s)$, s (f) $PV = nRT$, n
    (g) $u^2 = v^2 - 2vt$, t (h) $A = \left(\dfrac{a + b}{2}\right) t$, a

Inequalities and Inequations

1   Which of the following are true? Which are false? $a, b, c \in R$
    (a) If $a < b$, then $b - a < 0$.   (b) If $b > a$, then $b - a > 0$.
    (c) If $b > a$, and $c < 0$, then $b - c > a$.
    (d) If $b > a$, then $-b > -a$.

2   Which of the following are true for all $a, b, c \in R$? Use numerical examples
    to test each one.
    (a) If $a < b$ and $b < c$ then $a < c$.
    (b) If $a \leqq b$ and $b \leqq c$ then $a \leqq c$.
    (c) If $a > b$ and $c < 0$ then $a + c < b + c$.
    (d) If $a > b$ and $c < 0$ then $ac < bc$.
    (e) If $a < b$ and $c > 0$ then $a + c < b + c$.
    (f) If $a < b$ and $c < 0$ then $ac > bc$.

3   Draw the graph of each of the following. The first one has been done for you.

(a) $\{x \mid x > 3 \text{ or } x \leq -2, x \in R\}$

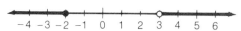

(b) $\{x \mid x < -4 \text{ or } x \geq 1, x \in R\}$

(c) $\{x \mid -6 < x < -3, x \in R\}$

(d) $\{x \mid x \geq 7 \text{ or } x < 5, x \in R\}$

(e) $\{x \mid -2 \leq x \leq 3, x \in R\}$

(f) $\{x \mid -4 \leq x < 0, x \in R\}$

(g) $\{x \mid x < 0 \text{ or } x \geq 4, x \in R\}$

4   Solve the inequations.

(a) $3y + 9 < 5y + 11$

(b) $4(2 - 3x) + 6 < -x - 8$

(c) $6(2 - x) \leq -4(x + 2)$

(d) $6m^2 + 3(m^2 + 1) > (3m + 1)^2 + 2$

(e) $2(x - 5) + x^2 \leq (x + 2)(x - 1)$

Ratio and Proportion

1   Find the value of each variable.

(a) $m:3 = 24:18$ (b) $2:3 = k:45$

(c) $45:25 = 9:x$ (d) $15:m = 3:2$

(e) $2:3:m = 14:n:49$ (f) $20:x:y = 4:3:7$

2   Solve for a and b.

(a) $4:2:1 = (a + 3):(b + 2):1$ (b) $(a + b):3:8 = 2:6:a$

3   Find the ratio $x:y$ from each of the following.

(a) $4x = 6y$ (b) $3y - 2x = 5y$

(c) $\dfrac{3y}{2} = \dfrac{4x}{5}$ (d) $\dfrac{3x - 2y}{5x + y} = \dfrac{4}{3}$

4   If $x:y = 5:3$, find the value of

(a) $\dfrac{2x + 3y}{x - y}$. (b) $\dfrac{4x - y}{x + y}$.

Analytic Geometry

1   The distance, d, between the points $P(x_1, y_1)$ and $Q(x_2, y_2)$ is given by
$$d = \sqrt{(x_2 - x_1)^2 + (y_2 - y_1)^2}.$$
Find the distance between each pair of points.

(a) $(-2, 1)$ and $(9, -4)$ (b) $(-3, -3)$ and $(9, 1)$

(c) $(9, -7)$ and $(5, 0)$ (d) $(7, 9)$ and $(-5, -7)$

2   Find the distance from each point to $(2, 5)$.

(a) $(0, 8)$ (b) $(-1, 7)$ (c) $(5, 7)$

(d) What do you notice about your answers?

3   Find the perimeter of $\triangle PQR$ for co-ordinates $P(-2, 3)$, $Q(1, 7)$, and $R(-5, 5)$.

4   The slope of the line segment PQ with co-ordinates $P(x_1, y_1)$ and $Q(x_2, y_2)$ is given by
$$m = \frac{y_2 - y_1}{x_2 - x_1}.$$
Find the slope of each line segment.

(a) $P(3, -4)$ and $Q(9, 2)$ (b) $R(14, -3)$ and $S(8, -9)$

(c) $A(-5, 1)$ and $B(5, 3)$ (d) $C(2, -1)$ and $D(8, -3)$

5   Three points are given by $P(-4, -2)$, $Q(4, 2)$, and $R(8, 4)$.

(a) Find the slopes of PQ, QR, PR. What do you notice about your answers?

(b) Why are the points P, Q, and R said to be collinear (to lie on the same line)?

(c) The following points are collinear.
$$(0, -3), \quad (k + 1, -1), \quad (3, 3)$$
What is the value of k?

6   (a) Draw a graph of the relation given by $2x + y = 6$. Describe its graph.

(b) Why is it appropriate to refer to the graph above as a linear relation?

(c) Which of the following are on the graph?
$$A(0, 6) \qquad B(3, 0) \qquad C(-1, 4)$$

7   For the line given by $x + 2y = 6$, the
    x-intercept is 6 and the y-intercept is 3.
    Find the x- and y-intercepts for each of the
    following lines.

    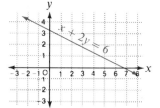

    (a) $2x - 6y = 6$ (b) $y - 3x + 6 = 0$

    (c) $\frac{1}{2}x - 2y = 8$ (d) $2(x - y) = 6$

    (e) The lines $2x - y = 1$ and $4x - 2y + k = 0$ have equal y-intercepts.
        Find the value of k.

8   (a) Parallel lines have equal slopes. Which of the following are parallel lines?
        A: $2x - y = 3$ B: $2y - x = 3$
        C: $3y = 2(9 - x)$ D: $2x - 3y - 15 = 0$
        E: $2y + x = 12$ F: $x - 2y = 5$
        G: $y - 2x - 5 = 0$ H: $y + 2x = 5$

    (b) The lines given by $x - 2y = 2$ and $4 + 2y = kx$ are parallel. Find the
        value of k.

9   The slope of a line given in the form $y = mx + b$ is m. A variable such as
    m is called a **parameter**. Find the slope of each equation.

    (a) $x + y = 8$ (b) $x = y + 6$ (c) $2x + y = 3$
    (d) $2y = x - 6$ (e) $3x + 2y - 6 = 0$ (f) $y - 2x = 3$

    (g) $3y - 2x - 6 = 0$ (h) $\frac{x}{3} - y = 4$

10 The y-intercept of a line given in the form $y = mx + b$ is b, where b is a
    parameter. Find the y-intercept of each equation.

    (a) $2x - y = 3$ (b) $2y - x = 3$
    (c) $2x - 3y - 15 = 0$ (d) $3y = 2(9 - x)$

11 (a) A parameter is used to write the family of lines given by $y = mx + 3$.
        What property do the lines have in common?

    (b) A parameter is used to write the family of lines given by $y = 3x + b$.
        What property do the lines have in common?

12 (a) The defining equation of a family of lines with y-intercept 6 is given by
        $y = mx + 6$ where m is the parameter. Find the member of the family
        with slope $\frac{2}{3}$.

(b) The defining equation of a family of lines with slope -2 is given by $y = -2x + b$, where b is a parameter. Find the member of the family with y-intercept 3.

(c) Write the defining equation of the family of lines parallel to the x-axis. What is the equation of the member of this family passing through $(-3, 8)$?

(d) Write a defining equation of the family of lines passing through the origin. What is the equation of the member of this family passing through $(-2, 5)$?

13 The slopes of perpendicular lines are **negative reciprocals**. What is the slope of a line perpendicular to each of the following?

(a) $y = 4x - 3$ (b) $2x + y = 8$ (c) $x - 3y = 6$ (d) $y = 6$

14 (a) If the lines given by $x - 4y = 24$ and $y - 4 = 2kx$ are parallel, find the value of k.

(b) If the lines given by $y - 8 = 2kx$ and $3y + x = 9$ are perpendicular, find the value of k.

15 The vertices of triangles are shown. Which triangles have a right angle?
(a) $\triangle ABC$: $A(2, -2)$, $B(5, 2)$, and $C(8, -2)$
(b) $\triangle DEF$: $D(-1, 4)$, $E(3, 2)$, and $F(-3, 0)$
(c) $\triangle GHI$: $G(-4, 5)$, $H(1, 7)$, and $I(3, 2)$

16 The area of $\triangle PQR$, with $P(x_1, y_1)$, $Q(x_2, y_2)$, and $R(x_3, y_3)$ is given by

$$\triangle PQR = \frac{1}{2} |x_1 y_2 + x_2 y_3 + x_3 y_1 - (x_1 y_3 + x_3 y_2 + x_2 y_1)|.$$

(a) Find the area of $\triangle PQR$: $P(5, 1)$, $Q(-4, 0)$, and $R(10, -4)$
(b) Find the area of $\triangle ABC$: $A(-2, 4)$, $B(-5, -4)$, and $C(0, -5)$

17 To find the distance, d, from a point $P_1(x_1, y_1)$ to a line give by $Ax + By + C = 0$, use the formula,

$$d = \frac{|Ax_1 + By_1 + C|}{\sqrt{A^2 + B^2}}.$$

(a) Calculate the distance from $(-3, 2)$ to the line $3x - 2y = 8$.
(b) What is the length of the perpendicular from $(-2, 3)$ to $5x - y - 8 = 0$?
(c) Find the distance from the origin to the line $2(x - y) = 9$.
(d) What is the distance from $(3, -2)$ to the line $2x - 3y = 12$? How would you interpret your answer?

Solving Systems of Equations

1   To solve systems of equations, the **method of substitution** may be used.

$$3x - y = -9 \qquad ①$$
$$y - 2x = 7 \qquad ②$$

From ②, $y = 2x + 7$.
Use $y = 2x + 7$ in ① to obtain

$$3x - (2x + 7) = -9. \qquad ③$$

(a) Solve equation ③ to find x.

(b) Use the value of x in (a) and substitute in ① to find y.

2   To solve systems of equations, the **addition-subtraction method** may be used to *eliminate one of the variables.*

$$3x + 4y = -11 \qquad ①$$
$$2x - 5y = 8 \qquad ②$$

To eliminate x, multiply ① by 2 and ② by 3.

$$6x + 8y = -22 \qquad ③$$
$$6x - 15y = 24 \qquad ④$$

(a) Subtract ④ from ③ to obtain an equation with y as the only variable. Solve for y.

(b) Substitute the value of y in ① to find x.

3   Solve.

(a) $x - y = 1$ (b) $y - 7 = 2(x - 1)$ (c) $y - x + 1 = 0$
     $3x + y = 7$ $x + y = 2$ $x - 2y = 1$

(d) $x + 3y = 4$ (e) $3x - y = 16$ (f) $x = 2y - 2$
     $2x - y = 8$ $2(x + 3) = y + 16$ $9y - 6x = 11$

4   Find the co-ordinates of the point of intersection.

(a) $x = 3y - 10$ (b) $3x - y = 4$ (c) $3x = 9 - 4y$
     $y - 4 = x$ $x = 2y + 3$ $4 + 8y = 5x$

5   Solve each system and verify.

(a) $3x + y - z = -4$ (b) $2x + y - z = 0$ (c) $x + y = 3$
     $2x - y + 3z = 5$ $2x - 3y - z = -2$ $z - y = 1$
     $x + y - 2z = -5$ $5x - y + z = 5$ $x - z = -2$

6   Solve each system.

(a) $4x - 2y + 3z = 5$ (b) $2z - y + 3x = 1$ (c) $x + 2 = 2y + 3z$
     $5x + y - 2z = 4$ $2x - 3y + 4z = 6$ $4x - 2y = z + 3$
     $6x - 3y - z = 2$ $-y - 4x + 5z = 32$ $2x - 3y = 3 + z$

Expressing Answers

In order to make a final statement that answers the original problem, you often need to round your answers to the precision given in the problem. Always indicate the accuracy to which you are expressing your final answer.

The distance is 696.3 km (to 1 decimal place). ⟵ means correct to 1 decimal place

Measurements are approximations and are assumed to be precise to ± 0.5 in the last significant place. A measure of 76.3 cm indicates that the digits 7, 6, and 3 are significant and the measure is between 76.25 cm and 76.34 cm. A significant digit is any digit which has a purpose other than placing the decimal point. Each of the following numbers has 2 significant digits.

$$92 \qquad 92\,000 \qquad 0.000\,92$$

These digits only place the decimal point and are thus not significant.

Numbers written in exponential form show the significant digits.

3.60×10^8 km   This digit is significant.

To round answers you should observe the following procedure.

A 6.83 21   rounds to 6.83 (correct to 2 decimal places)

If the first digit to be discarded is less than 5, then the last digit to be kept is not changed.

B (i) 6.83 68   rounds to 6.84 (correct to 2 decimal places)

If the first digit to be discarded is greater than 5, then increase the last digit to be kept by 1.

(ii) 6.83 506   rounds to 6.84 (correct to 2 decimal places)

If the first digit to be discarded is five followed by at least one digit other than zero, then increase the last digit to be kept by 1.

However, if the digit is 5 followed by zeroes, use

6.83 5   rounds to 6.84   last digit retained is odd, add 1.
6.84 5   rounds to 6.84   last digit retained is even, no change made.

C When you add or subtract numbers which occur as the result of measurements, your final answer is rounded to be as precise as the least precise number used in the computation.

The perimeter, in metres, is shown by the calculation $1.253 + 1.91 + 3.681 = 6.834$ ⟵ Round the answer to 2 decimal places, namely 6.83 m.

D When you multiply or divide numbers that occur as a result of measurements your final answer is rounded so that it contains as many significant digits as occurs in the number occurring with the fewest significant digits.

The area, in square metres, is shown by the calculation $136.923 \times 42.38 = 5802.7967$ ⟵ Round the answer to 4 significant digits: 5802 m^2 or (5.802×10^3) m^2.

Review and provide examples to illustrate the meaning of each.

1 Correct to 1 decimal place.
2 Rounded to the nearest tenth
3 Correct to the nearest hundredth
4 Correct to 2 significant digits
5 Rounded to the nearest cent
6 Rounded to the nearest metre
7 Correct to 2 decimal places
8 Correct to 4 significant digits

Calculator Skills

A scientific calculator is invaluable when you have to do many tedious calculations. Calculators have different features and you must refer to the booklet that is provided with the calculator to specifically learn the operation of the various functions on your calculator.

When finding answers, always ask yourself "Is my answer reasonable?" Remember, before you do any calculations with your calculator push $\boxed{CE/C}$.

1 (a) Refer to your manual. Review any information about these features.

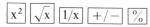

$$\boxed{x^2} \quad \boxed{\sqrt{x}} \quad \boxed{1/x} \quad \boxed{+/-} \quad \boxed{\%}$$

(b) Calculate. (i) $2\sqrt{12} + \dfrac{1}{6}$ (ii) $(3.5)^2 - 5\sqrt{6}$ (iii) $6\pi - \dfrac{1}{2\pi}$

2 It is often convenient to use the memory features on your calculator.

$$\boxed{MS} \qquad\qquad \boxed{MR} \qquad\qquad\qquad \boxed{M+} \qquad\qquad\qquad\qquad \boxed{M-}$$

memory in   memory recall   add to memory   subtract from memory

(a) Refer to your manual. Review the particular feature of these functions.

(b) Follow these steps.   A: Calculate $\dfrac{(3.5)^2 \sqrt{8}}{\pi}$.

B: Use your answer A and calculate $A^2 - 3\sqrt{\pi}$.

C: Add $3\sqrt{5} - \dfrac{1}{\sqrt{2}}$ to your answer in B.

D: Subtract $2\pi^2 - \dfrac{1}{(6.5)^2}$ from your answer in C.

3 Your calculator probably has a constant feature. If you push any of the keys $\boxed{+}\,\boxed{-}\,\boxed{\times}$ or $\boxed{\div}$ twice you establish a constant. The symbol K appears in the display. The constant sum function, $\boxed{+}\,\boxed{+}$, is useful when you want to add the same number each time to different inputs.

		Input		Output				Input		Output
$\boxed{Ce/C}$ 18.6 $\boxed{+}$ $\boxed{+}$		19.2 $\boxed{=}$		37.8		$\boxed{Ce/C}$ 12.2 $\boxed{\times}$ $\boxed{\times}$		6.3 $\boxed{=}$		76.86
		17.2 $\boxed{=}$		35.8				8.3 $\boxed{=}$		101.26
		15.2 $\boxed{=}$		33.8				10.3 $\boxed{=}$		125.66

For each of these, describe what steps you are following. Then calculate.

		Input		Output				Input		Output
(a) $\boxed{CE/C}$ 12.3 $\boxed{+}$ $\boxed{+}$		6.8 $\boxed{=}$		?		(b) $\boxed{CE/C}$ 9.3 $\boxed{-}$ $\boxed{-}$		3.2 $\boxed{=}$		?
		13.2 $\boxed{=}$		?				6.1 $\boxed{=}$		?
		8.9 $\boxed{=}$		?				5.6 $\boxed{=}$		?
(c) $\boxed{CE/C}$ 8.5 $\boxed{\times}$ $\boxed{\times}$ 4.2 $\boxed{=}$				?		(d) $\boxed{CE/C}$ 6.5 $\boxed{\div}$ $\boxed{\div}$ 190.80 $\boxed{=}$				?
		3.6 $\boxed{=}$		?				113.25 $\boxed{=}$		?
		5.3 $\boxed{=}$		?				96.86 $\boxed{=}$		?

1 Functions: Process and Properties

concept of functions, functions and their graphs, sketching graphs, inverse
of a function, reciprocal functions, composition of functions, applications,
strategies, problem solving

Introduction

Often, the study of mathematics involves the learning of the contributions
of many people to the body of knowledge called mathematics. Many of
these contributors, who originated from almost every nation of the world,
had no notion that their new ideas would have such a lasting effect. Truly,
mathematics is an international language.

In studying the mathematics they contributed, you will relive some of
their ideas and thinking processes. You will also develop your own
methods and strategies for solving a problem, and you will notice that
mathematics influences almost every aspect of your modern world.

- in sports

- in architecture

- in communications

- in space technology

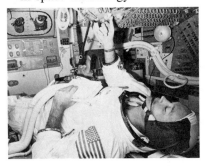

1.1 Skills and Concepts: Functions

The study of relationships occurs in many of the activities of people. For example, in business the relations between sales and time can be shown in different ways. Three ways are shown below.

on a graph

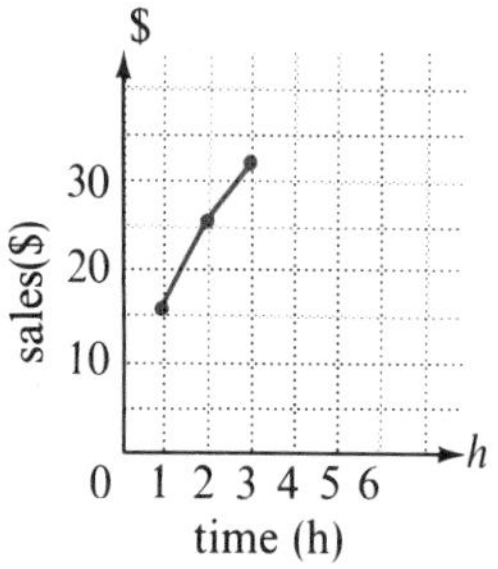

in a chart

hour	sales
1	15
2	25
3	32

as a set of ordered pairs

(1, 15), (2, 25), (3, 32)

In the study of mathematics, the meaning of words is very carefully defined. For example, a *relation, h*, is defined as a set of ordered pairs, as follows.

$$h = \{(x, y)\,|\,x \in \text{domain},\ y \in \text{range}\}$$

The **domain** is the set of all first components, x, of the relation.

The **range** is the set of all second components, y, of the relation.

Example 1   Relations f and g are defined by

$$f = \{(1, 2), (2, 3), (3, 4)\}, \quad \text{and}$$
$$g = \{(1, 2), (2, 1), (2, 3), (3, 0), (3, 4)\}.$$

What is the domain and the range of each relation?

Solution   domain = set of all first components
      domain of $f = \{1, 2, 3\}$
      domain of $g = \{1, 2, 3\}$

range = set of all second components
      range of $f = \{2, 3, 4\}$
      range of $g = \{0, 1, 2, 3, 4\}$

The previous relation f has a special property.

   For each member of the domain there is a unique member of the range.

Thus, the relation f is called a **function**.

For some relations or functions there may be a rule which assigns a value to y for each value of x. This rule may be expressed in several ways, A, B, or C as follows.

A: as an equation

The ordered pairs of the relation, f, in Example 1 are listed in a chart.

first component, x	1	2	3
second component, y	2	3	4

By examining the chart you can see that each value of y is one more than the corresponding value of x. Thus, you can write

$$y = x + 1, \quad x \in \{1, 2, 3\} \longleftarrow \text{domain of the relation, } f$$

B: in mapping notation

The above relation is expressed as

$$f: x \rightarrow x + 1, \quad x \in \{1, 2, 3\}.$$

C: in function notation

For the function f,

write $\quad f(x) = x + 1, \quad x \in \{1, 2, 3\}.$

Thus, $\quad f(2) = 2 + 1$

$\qquad f(2) = 3$

Thus, $f(2) = 3$, and the corresponding ordered pair of the function is $(2, 3)$. That is, if $x = 2$, then $f(2) = 3$.

Symbols are used to express or provide instructions in compact ways. To do mathematics, you must understand clearly the exact meaning of the terms and symbols. The function notation, $f(x) = 2x^2 - 3x + 1$ is a compact form that saves time and words. For example, the instruction

Find the value of the expression $2x^2 - 3x + 1$ for $x = 3$.

can be expressed concisely in the form

Find $f(3)$ if $f(x) = 2x^2 - 3x + 1$.

Example 2 A relation h is defined by $h(x) = 2x^2 - x$. Find each of the following.

(a) $h(-2)$ (b) $h(2m)$ (c) $h(h(x))$

If the domain of a relation is not specified, the variable represents any real number.

Solution (a) Use $h(x) = 2x^2 - x$.   Replace x by -2.

$\qquad h(-2) = 2(-2)^2 - (-2)$

$\qquad\qquad = 2(4) + 2$ Use ( ) to avoid an error in the substitution.

$\qquad\qquad = 8 + 2$

$\qquad\qquad = 10$

(b) Use $h(x) = 2x^2 - x$.   Replace x by $2m$.

$\qquad h(2m) = 2(2m)^2 - (2m)$

$\qquad\qquad = 2(4m^2) - 2m$ Express in simplest terms.

$\qquad\qquad = 8m^2 - 2m$

(c) Use $h(x) = 2x^2 - x$.   Replace x by $h(x)$.

$$
\begin{aligned}
h(h(x)) &= 2(h(x))^2 - (h(x)) \qquad \text{Replace } h(x) \text{ by } 2x^2 - x. \\
&= 2(2x^2 - x)^2 - (2x^2 - x) \\
&= 2(4x^4 - 4x^3 + x^2) - 2x^2 + x \\
&= 8x^4 - 8x^3 + 2x^2 - 2x^2 + x \\
&= 8x^4 - 8x^3 + x
\end{aligned}
$$

1.1 Exercise

A Remember, if the domain of a relation or function is not indicated, then it is understood to be the real numbers.

1 For each relation, a defining equation is given. Express each in function notation.

(a) f, $y = 2x + 1$ (b) g, $y = x^2 - 3$

(c) h, $y = \dfrac{2}{x + 1}$ (d) k, $y = \sqrt{x^2 + 1}$

2 Use function notation to define each of the following.

(a) $f : x \to 3x - 2$ (b) $h : x \to \dfrac{1}{2x - 1}$ (c) $g : x \to \sqrt{9 - x^2}$

3 A relation, f, is given by the set of ordered pairs

$$f = \{(-1, 2), (-2, 2), (1, 3), (2, 4)\}.$$

Find each of the following.

(a) $f(-1)$ (b) $f(1)$ (c) $f(2)$

(d) If $f(n) = 2$, what are the values of n?

4 The domain and range are shown for certain relations. Which are functions?

(a)   (b)   (c)   (d)

5 A relation, f, is given by $f = \{(x, y) \mid y = 2x - 1\}$.

Find each missing value.

(a) $(-4, ?)$ (b) $(0, ?)$ (c) $(?, 3)$

(d) $(1, ?)$ (e) $(?, -5)$ (f) $(-5, ?)$

B   Review the different forms used to express a function or relation in mathematics.

6   A relation g is given by $g(x) = 2x^2 - 3x + 1$. Find each of the following.

   (a) $g(1)$ (b) $g(-1)$ (c) $g(0)$ (d) $g(m)$ (e) $g\left(\dfrac{1}{m}\right)$ (f) $g(3m)$

7   For the relation $g(x) = 3(2x - 1)$, find the following.

   (a) $g(4)$ (b) $g(-4)$ (c) $g(0)$ (d) $g(h)$ (e) $g\left(\dfrac{1}{k}\right)$ (f) $g(2k)$

   -3

8   If $f(x) = x^2 - 2$, find expressions for each of the following.

   (a) $f(-x)$ (b) $f(-x) - 2$ (c) $f(x - 2)$ (d) $-2f(x)$ (e) $f(-2x)$

9   A relation, f, is given by $f(x) = \dfrac{x - 2}{x}$.

   Find expressions for each of the following.

   (a) $\dfrac{2}{f(2)}$ (b) $\dfrac{4}{f(-2)}$ (c) $f\left(\dfrac{1}{x}\right)$ (d) $\dfrac{2}{f(x)}$ (e) $f[f(x)]$ (f) $f[f(-x)]$

10   (a) Find an expression for $f(m - 5)$ if f is defined by $f: x \rightarrow (x + 1)(x + 2)$.

   (b) A function, g, is defined by $g(x) = \dfrac{2}{x - 3}$. Determine $g\left(\dfrac{1}{a + 2}\right)$.

   (c) What expression is represented by $h\left(\dfrac{x + 1}{x}\right)$ if h is given by $h(x) = \dfrac{x + 2}{x - 3}$?

11   A relation, g, is defined by $g: x \rightarrow 2x - 7$. Find m if $g(m) = 5$.

12   Find k if $f(k) = 5$ and f is defined by $f: x \rightarrow x^2 - 4$.

13   For what values of y is $h(y) = -4$, where $h(y) = y^2 - 5y$?

C 14   Two relations are defined as $f(x) = \dfrac{2x + 1}{3}$,   and $g(x) = \dfrac{x - 1}{2}$.

   For what values of m is $f(m) = g(m)$?

Calculator Tip

Remember: Refer to the various Calculator Tips that occur in this text. Make a summary of the various features on your calculator that will reduce the need to do tedious calculations.

1.2 Functions and Graphs

To examine the many patterns and relationships in mathematics, you combine your skills in algebra and your skills in co-ordinate geometry.

To draw the graph of a relation or function, given only the defining rule or equation, your must know the domain.

Example
(a) Draw the graph of the relation given by the defining equation $y = x^2 - 1$, $x \in R$.

(b) Is the relation a function?

Think: to draw the graph of a relation, construct a table of representative values. Plot the points given by the table of values, and then draw a smooth curve through the points.

Solution
(a) *Step 1*
Construct a table of values.

x	y
-3	8
-2	3
-1	0
0	-1
1	0
2	3
3	8

$y = x^2 - 1$
$y = (-1)^2 - 1$
$y = 1 - 1$
$\leftarrow y = 0$

$\leftarrow y = x^2 - 1$
$y = (1)^2 - 1$
$y = 1 - 1$
$y = 0$

Step 2
Plot the points.

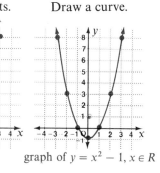

Step 3
Draw a curve.

graph of $y = x^2 - 1$, $x \in R$

(b) For each x there is only one value of y. Thus, the relation defined by $y = x^2 - 1$ is a function.

Once you have the graph of a relation, you can test whether the relation is a function by using a vertical line placed anywhere on the graph.

A: If a vertical line cuts the graph in one point, then f is a function.

B: If a vertical line cuts the graph in more than one point, then f is *not* a function.

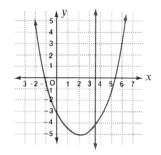

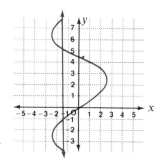

Vertical Line Test

If any vertical line crosses the graph of a relation in one and only one point, then the relation is a function.

1.2 Exercise

A 1   For each of the following graphs,
  • is the relation a function?   • what is the domain?   • what is the range?

(a) 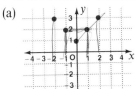   (b) 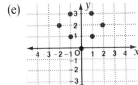   (c)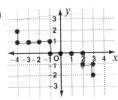

(d) 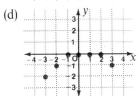   (e) 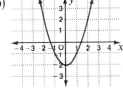   (f)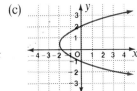

2   A relation f is given by $f = \{(1, 3), (2, 3), (5, -1), (4, -2)\}$.
   (a) Draw its graph.   (b) What is its domain? range?
   (c) Is the relation a function? Why or why not?

3   Draw the graph for each relation below. Indicate whether each relation is a function. Give a reason for your opinion.
   (a) $\{(1, -1), (1, 0), (1, -2), (1, 1)\}$ (b) $\{(-2, 1), (-1, 1), (0, 1), (1, 1), (2, 1)\}$
   (c) $\{(-1, 2), (-1, 3), (2, 4), (2, 5)\}$

4   For each of the following,
   • is the relation a function?
   • what is the domain?   • what is the range?

(a) 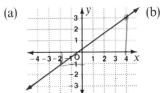   (b) 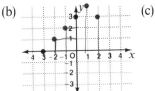   (c)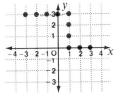

(d) 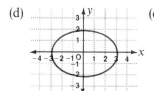 (e) 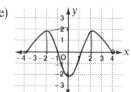 (f)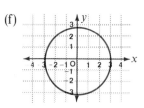

B If the domain is not specified, it is understood to be the real numbers.

5 In the graph, the area, A, of a square
is related to the length of the side, s.

(a) What is the domain? the range?

(b) Find the missing values.

(i) (2, ?) (ii) (3.5, ?) (iii) (?, 9)

(c) Write an equation to relate A and s.

(d) Is the relation a function?

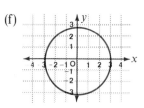

6 The orbit of a "stationary" satellite
is shown in the graph.

(a) What is the domain? the range?

(b) Find the missing values.

(i) (6, ?) (ii) (−6, ?) (iii) (0, ?)

(c) Is the relation a function?

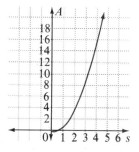

7 The rise in temperature, T, of a substance
with respect to time, t, is shown.

(a) What is the domain? the range?

(b) Find the missing values.

(i) (1.5, ?) (ii) (?, 25) (iii) (3, ?)

(c) Is the relation a function?

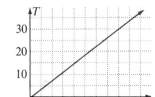

8 Draw the graph of each relation below.

(a) $y = 3x - 1$ (b) $f: x \to 2x + 1$ (c) $g(x) = 2x^2 - 8$

(d) $f: x \to 9 - x^2$ (e) $h(x) = \sqrt{x^2 - 1}$ (f) $y = \frac{1}{2}\sqrt{x}$

Math Tip

The particular skills of graphing have their roots in the work of Rene
Descartes (1596–1650). His contribution to the study of mathematics
was honored by having the Cartesian plane named after him. The
honoring of a mathematician by naming a law or curve or something
mathematical after him or her is very common in mathematics.

1.3 Concepts of Sketching Functions

The process of mathematics includes many steps. For example, people spend a lot of time exploring particular situations and then use the results to make generalizations. In this section, you will follow this process by examining a function, f, given by $y = f(x)$. What is the effect on the graph of the function for each of the following where $a \in R$?

$$y = f(x) + a \qquad y = f(x + a) \qquad y = af(x) \qquad y = f(ax)$$
$$y = -f(x) \qquad y = f(-x) \qquad y = |f(x)|$$
$$y = \frac{1}{f(x)}$$

To determine the effect on the graph of $y = f(x)$, you will need to work through Explorations 1 to 5 that follow. To do so, you need to use the skills illustrated in the following example.

Example   The graph of $y = f(x)$ is shown.

(a) Find the value of
   (i) $f(1)$ (ii) $f(-3)$ (iii) $f(0)$
(b) Draw the graph of $y = f(x + 2)$

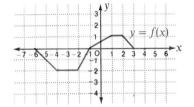

Solution   (a) From the graph,
   (i) If $x = 1$, then $f(1) = 1$.
   (ii) If $x = -3$, then $f(-3) = -2$.
   (iii) If $x = 0$, then $f(0) = \frac{1}{2}$.

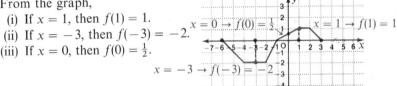

(b) *Step 1:*
   To draw the graph of $y = f(x + 2)$, construct a table of values.

x	-6	-5	-4	-3	-2	-1	0	1	2	3
$x + 2$	-4	-3	-2	-1	0	1	2	3	4	5
$f(x + 2)$	-2	-2	-2	0	$\frac{1}{2}$	1	1	0	0	0

$$x + 2 = -4$$
Then $f(x + 2) = f(-4) = -2$.

Use the value given in the graph for $f(-4)$.

Then draw the graph as follows.

Step 2:

Use the table of values
to draw the graph of
$y = f(x + 2)$.

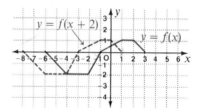

1.3 Exercise

Exploration 1: What is the relationship between $y = f(x)$ and
$y = f(x) + p$, p a constant?

1 The graph defined by $y = f(x)$ is shown.
   On the same set of axes, draw the
   graphs of

   (a) $y = f(x) + 2$.   (b) $y = f(x) + 3$.
   (c) $y = f(x) - 2$.   (d) $y = f(x) - 3$.

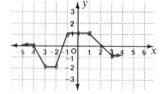

2 Repeat Question 1 for each of the following graphs.
   (a)

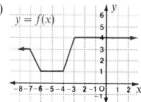

   (b)

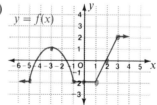

3 Base your observations on your answers to Questions 1 and 2.
   (a) Describe how the graph of $y = f(x)$ is related to the graphs of
       $y = f(x) + q$, if $q > 0$, if $q < 0$.
   (b) What is the effect on the graph of $y = f(x) + q$ if q increases? Decreases?
   (c) Write a mapping to relate the graph of $y = f(x)$ and $y = f(x) + q$.
   (d) Why may the mapping in (c) be described as a translation?

4 Use the graph $y = f(x)$ defined by $f(x) = x + 3$. Sketch the graphs of
   $y = f(x) + 5$ and $y = f(x) - 5$.

5 Use the graph of $y = f(x)$ defined by $f(x) = x^2$ to sketch the graphs of
   $y = f(x) + 3$, $y = f(x) - 2$.

Exploration 2: What is the relationship between $y = f(x)$ and $y = f(x + p)$, p a constant?

6   The graph defined by $y = f(x)$ is shown. On the same set of axes draw the graph of

(a) $y = f(x + 2)$.   (b) $y = f(x + 3)$.
(c) $y = f(x - 2)$.   (d) $y = f(x - 3)$.

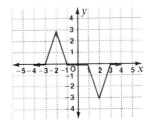

7   Repeat Question 6 for the graph of $y = f(x)$ shown in the diagram.

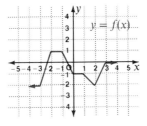

8   Use your results in Questions 6 and 7.

(a) Describe how the graph of $y = f(x)$ is related to the graph of $y = f(x + p)$ if $p > 0$? If $p < 0$?

(b) Describe the effect on the graph of $y = f(x + p)$ if p decreases and if p increases.

(c) Write a mapping to relate the graph of $y = f(x)$ and $y = f(x + p)$.

(d) Why may the mapping in (c) be described as a translation?

9   Use the graph of $y = f(x)$, defined by $f(x) = -2x$, to sketch the graphs of $y = f(x + 2)$ and $y = f(x - 3)$.

10 Use the graphs of $y = f(x)$, defined by $f(x) = -\dfrac{1}{2}x^2$, to sketch the graphs of $y = f(x + 2)$ and $y = f(x - 3)$.

Exploration 3: What is the relationship between $y = f(x)$ and $y = af(x)$, where a is a constant?

11 Use the graph defined by $y = f(x)$ as shown. On the same set of axes, draw the following graphs.

(a) $y = 2f(x)$ (b) $y = 3f(x)$

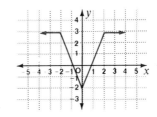

12 Repeat Question 11 for the graph of $y = f(x)$ as shown.

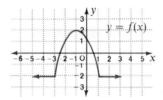

13 Base your answers on the observations in Questions 11 and 12.

(a) Describe how the graphs of $y = f(x)$ and $y = af(x)$, $a > 0$ are related.

(b) What is the effect on the graph of $y = af(x)$ if $a > 0$ and a increases? a decreases?

(c) Write a mapping to relate the graphs of $y = f(x)$ and $y = af(x)$.

(d) An **invariant point** of a mapping is a point not affected by the mapping. Are there any invariant points of the above mapping?

14 Use the graph of $y = f(x)$, defined by $f(x) = x^2$, to sketch the graphs of $y = 2f(x)$ and $y = 3f(x)$.

Exploration 4: What is the relationship between $y = f(x)$ and $y = f(bx)$, where b is a constant?

15 Use the graph defined by $y = f(x)$, as shown. On the same set of axes, draw the graphs of

(a) $y = f(2x)$ (b) $y = f(3x)$

(c) $y = f\left(\dfrac{1}{2}x\right)$ (d) $y = f\left(\dfrac{1}{3}x\right)$

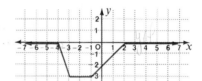

16 Repeat Question 15 for $y = f(x)$, shown in the diagram.

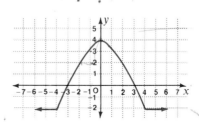

17 Base your answers on your observations in Questions 15 and 16.

(a) Describe how the graphs of $y = f(x)$ and $y = f(bx)$, $b > 0$, are related.

(b) What is the effect on the graph of $y = f(bx)$ if $b > 0$ and b increases? b decreases?

(c) Write a mapping to relate the graphs of $y = f(x)$ and $y = f(bx)$.

(d) Are there any invariant points of the mapping?

18 Use the graph of $y = f(x)$, defined by $f(x) = x^2$, to sketch the graphs of $y = f(2x)$ and $y = f(\frac{1}{2}x)$.

Exploration 5: What is the relationship between $y = f(x)$ and $y = f(-x)$ or $y = -f(x)$?

19　The graph defined by $y = f(x)$ is shown. On the same set of axes draw the graphs of
(a) $y = -f(x)$.　　(b) $y = f(-x)$.

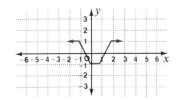

20　Repeat Question 19 for $y = f(x)$, shown in the diagram.

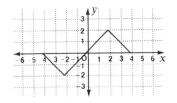

21　Base your answers on your observations in Questions 19 and 20.
(a) What is the relationship of $y = f(x)$ to $y = -f(x)$, and to $y = f(-x)$?
(b) Write a mapping to relate the graph of $y = f(x)$ to $y = -f(x)$.
(c) Write a mapping to relate the graph of $y = f(x)$ to $y = f(-x)$.
(d) Are there any invariant points of the mapping?

22　Use the graph of $y = f(x)$, defined by $f(x) = x^2$, to sketch the graphs of $y = -f(x)$ and $y = f(-x)$.

Math Tip

Once you learn a new topic in mathematics you should

A: make a summary in your own words, of the vocabulary, concepts, and skills of the topic
B: illustrate everything in A with an example of your own.

Now summarize the results of Explorations 1 to 5 as suggested in A and B above. Include the results of

- Given $y = f(x)$, examine $y = f(x) + p$, and $y = f(x + p)$, p constant
- Given $y = f(x)$, examine $y = af(x)$, $y = f(bx)$, a, b, constants.
- Given $y = f(x)$. What is its relationship to $y = f(-x)$? to $y = -f(x)$?

To learn and remember mathematics, look for similarities and differences in the concepts. For example, for the concepts of sketching functions, ask yourself these questions to help you remember the concepts: How are they the same? How are they different?

Applying Strategies: Sketching

On the previous pages you have explored the effect of each of the following on the graph of $y = f(x)$, $a \in R$.

(i) $y = f(x) + a$ (ii) $y = f(x + a)$ (iii) $y = af(x)$
(iv) $y = f(ax)$ (v) $y = -f(x)$ (vi) $y = f(-x)$

In your subsequent work, you will also find the result for $y = \dfrac{1}{f(x)}$ and $y = |f(x)|$.

You can use a combination of the above to sketch the graph of a function given in the general form $y = af(x + b) + c$ where a, b, $c \in R$.

Example   Sketch the graph of the function defined by $y = -\dfrac{1}{2}(x + 2)^2 + 3$.

Solution

Step A Sketch the graph of the basic curve given by $y = x^2$.

Step B Then apply (iii) to sketch the curve given by $y = \dfrac{1}{2}x^2$.

Step C Then apply (iv) to sketch the curve given by $y = -\dfrac{1}{2}x^2$.

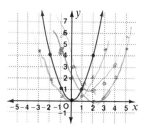 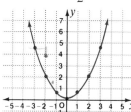 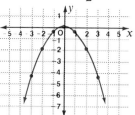

Step D Then apply (ii) to sketch the curve given by $y = -\dfrac{1}{2}(x + 2)^2$.

Step E Then apply (i) to sketch the required curve given by

$$y = -\dfrac{1}{2}(x + 2)^2 + 3$$

 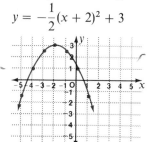

23   Sketch the graph of each of the following on the same co-ordinate axes.
(a) $y = x^2$ (b) $y = 2x^2$ (c) $y = -2x^2$
(d) $y = -2(x + 3)^2$ (e) $y = -2(x + 3)^2 + 5$

24   Sketch the graph of each of the following.
(a) $y = 2(x - 3)^2$ (b) $y = -2(x - 1)^2$ (c) $y = 2(x - 1)^2$
(d) $y = \dfrac{3}{2}(x - 3)^2$ (e) $y = 3(x + 2)^2 - 5$ (f) $y = -2(x - 3)^2 + 6$

1.4 Inverse of a Function

For the domain D and range E, a relation, or function, f, is shown.

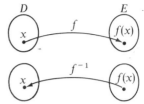

If there is a relation that maps $f(x)$ onto x then that relation is called the **inverse** of f, denoted by f^{-1}, or simply called f inverse.

Note that f^{-1} does not mean $\dfrac{1}{f}$.

If $(x, y) \in f$ then $(y, x) \in f^{-1}$. Thus to find the inverse of f, you interchange the components of each ordered pair (x, y) of f.

Example   A function f is given by $f = \{(-2, 2), (-1, 3), (0, 3)\}$.

(a) Sketch the graph of f and f^{-1}.   (b) Is f^{-1} a function?

Solution   (a) $f^{-1} = \{(2, -2), (3, -1), (3, 0)\}$
From the graph, you can see that the graph of f^{-1} is the reflection of the graph of f in the line $y = x$.

(b) f^{-1} is not a function since $(3, 0)$ and $(3, -1) \in f^{-1}$.

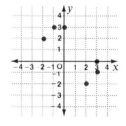

This is a compact way of saying that two values of y occur for one particular value of x. Thus, f^{-1} is not a function.

Based on the results of the above example, and other graphs, you are able to make the following observations.

> - The graph of f^{-1} is the reflection of the graph of f in the line $y = x$.
> - If f is a function, f^{-1} may or may not be a function.

1.4 Exercise

A   Record the meaning of f^{-1}. Use an example to illustrate its meaning.

1   The function f is shown.
(a) Draw a graph of f^{-1}.
(b) Find the values of $f^{-1}(-1), f^{-1}(0), f^{-1}(1)$.
(c) Is f^{-1} a function?

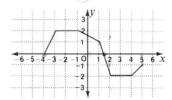

2   A function is given as ordered pairs. Write the domain and range of the inverse of f.
   (a) $(-2, 3), (0, 2), (2, 1), (4, 0)$ (b) $(-3, 3), (-2, 1), (0, 0), (2, 1), (3, 3)$
   (c) For which of the above is f^{-1} a function?

3   A function is given by $f: x \rightarrow x^2$. The domain of f is $D = \{-2, -1, 0, 1, 2\}$.
   (a) List the ordered pairs of f.   (b) Write the ordered pairs of f^{-1}.
   (c) Draw the graph of f and f^{-1}.   (d) Is f^{-1} a function?

4   A relation, R, is given by the vertices of a quadrilateral A$(-2, 1)$, B$(1, 2)$, C$(4, 0)$, D$(1, -1)$.
   (a) Draw a graph of the inverse relation.
   (b) Describe the graph of R^{-1} in relation to the graph of R.

B 5   (a) If $f(x) = 2x + 1$, then complete the ordered pairs.
        $(-1, ?)$ $(0, ?)$ $(?, 0)$
   (b) Write the corresponding ordered pairs of f^{-1} for those in (a). Use these ordered pairs to draw the graph of f^{-1}.
   (c) Write the equation of the graph of f^{-1}.

6   A function is defined by $f(x) = x^2 - 2$.
   (a) Draw a mapping diagram to show f and f^{-1}.
   (b) Draw the graphs of f and f^{-1}.
   (c) Is f^{-1} a function?

   $y = 2x + 1$
   $x = 2y + 1$

7   For each relation, f, defined by an equation, draw its graph, as well as the graph of its inverse on the same axes.
   (a) $y = 2x - 3$ (b) $y = x^2$ (c) $x = y^2$ (d) $x^2 + y^2 = 16$
   (e) Which inverse relations above are functions?

8   A function, f, is given by the equation $f(x) = 2x - 3$.
   (a) Draw the graph of f and its inverse f^{-1}.
   (b) Find the defining equation of f^{-1}.

9   A function, h, is given by $y = x^2$.
   (a) Draw the graph of h and its inverse h^{-1}.
   (b) Find the defining equation of h^{-1}.

10   Find the defining equation of the inverse of each of the following.
   (a) $3x - 2y = 6$ (b) $y = x^2 - 6$

C 11   Prove the following statement. The inverse of a linear function is a function.

1.5 The Reciprocal Function, $\dfrac{1}{f}$

The reciprocal function of a function, f, is defined as $\dfrac{1}{f}$. Thus, if the defining equation of a function is $y = 2x - 3$, then the defining equation of the reciprocal function is $y = \dfrac{1}{2x - 3}$. To draw the graph of the reciprocal function of $y = x$, you organize a table of values. The defining equation of the reciprocal function of $y = x$ is $y = \dfrac{1}{x}$.

x	-3	-2	-1	0	1	2	3
$y = x$	-3	-2	-1	0	1	2	3
$y = \dfrac{1}{x}$	$-\frac{1}{3}$	$-\frac{1}{2}$	-1	undefined	1	$\frac{1}{2}$	$\frac{1}{3}$

To help you sketch the graph you can use the following observations:
Notice that for $f(x) = x$

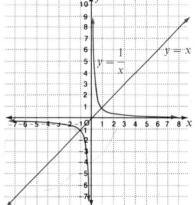

- as $f(x)$ increases, $\dfrac{1}{f(x)}$ decreases.

- as $f(x)$ decreases, $\dfrac{1}{f(x)}$ increases.

- for $f(x) > 0$, $\dfrac{1}{f(x)} > 0$, and

 for $f(x) < 0$, $\dfrac{1}{f(x)} < 0$.

- for $f(x) = 0$, $\dfrac{1}{f(x)}$ is undefined.

- as x increases, $\dfrac{1}{f(x)}$ comes closer to, but never reaches, zero.

- as x decreases, $\dfrac{1}{f(x)}$ comes closer to, but never reaches, zero.

Notice that for very large positive and negative values of x, the graph of the function defined by $y = \dfrac{1}{x}$ comes very close to $y = 0$ (x-axis) but never touches the x-axis. Also the graph comes close to, but never touches, the line $x = 0$ (y-axis). Lines such as $x = 0$ and $y = 0$ are called **asymptotes** of the function $y = \dfrac{1}{x}$.

Example The graph defined by $y = f(x)$ is shown.

(a) Draw the graph of $y = \dfrac{1}{f(x)}$, $-4 \leq x \leq 6$, $x \in R$.

(b) What are the asymptotes of the function $y = \dfrac{1}{f(x)}$?

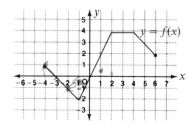

Solution To draw the graph of the function defined by $y = \dfrac{1}{f(x)}$ shown, organize a table of values. Use the values to help you sketch the curve.

x	-4	-3	-2	-1	$-\frac{1}{2}$	0	1	2	4	5	6
$f(x)$	1	0	-1	-2	-1	0	2	4	4	3	2
$\dfrac{1}{f(x)}$	1	undefined	-1	$-\frac{1}{2}$	-1	undefined	$\frac{1}{2}$	$\frac{1}{4}$	$\frac{1}{4}$	$\frac{1}{3}$	$\frac{1}{2}$

Think: Use your earlier observations for sketching the graph of a reciprocal function.

From the graph you can see that the asymptotes of $y = \dfrac{1}{f(x)}$ are $x = -3$ and $x = 0$.

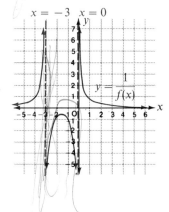

1.5 Exercise

A 1 What is the defining equation of the reciprocal of each of the following?

(a) $y = 2x$　　(b) $y = x^2$　　　　(c) $y = 3x - 2$

(d) $y = \dfrac{1}{2x}$　(e) $y = \dfrac{2}{4x + 1}$　(f) $y = 2x^2 + x - 3$

2 Sketch the graph of the reciprocal function of each of the following.

(a) 　(b) 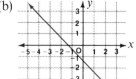　(c)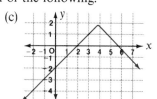

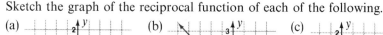

(d)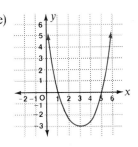

(e)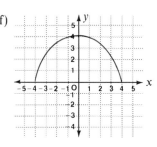

(f)

B 3   (a) The graph defined by $y = f(x)$ is shown.

Draw the graph of $y = \dfrac{1}{f(x)}$.

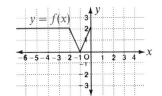

(b) Is $y = \dfrac{1}{f(x)}$ a function?

(c) What are the asymptotes of $y = \dfrac{1}{f(x)}$?

4   (a) The graph defined by $y = f(x)$ is shown.

Draw the graph of $y = \dfrac{1}{f(x)}$.

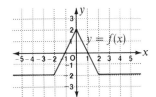

(b) What are the asymptotes of $y = \dfrac{1}{f(x)}$?

5   Base your answers on your observations in Questions 3 and 4.

(a) Describe how the graph of $y = f(x)$ is related to the graph of $y = \dfrac{1}{f(x)}$.

(b) Write a mapping to relate the graph of $y = f(x)$ to $y = \dfrac{1}{f(x)}$.

(c) Are there any invariant points of the mapping?

6   (a) Use the graph of $y = f(x)$, given by $f(x) = 2x - 3$. Draw the graph of
$y = \dfrac{1}{f(x)}$, namely, $y = \dfrac{1}{2x - 3}$.

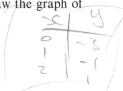

(b) What are the asymptotes of $y = \dfrac{1}{2x - 3}$?

7   (a) If $f(x) = x^2 - 4$, draw the graph of $y = \dfrac{1}{f(x)}$, namely, $y = \dfrac{1}{x^2 - 4}$.

(b) What are the equations of the asymptotes of $y = \dfrac{1}{x^2 - 4}$?

Working with Calculators Efficiently

Whenever you work with a calculator, it is important
- to estimate your answer and ask yourself "Is my answer reasonable?"
- to properly reflect the accuracy of your answer based on the given information. (Refer to *Expressing Answers* on page 17.)
- to refer to the manual provided with your calculator. Often the manual will verify any special features and operation of your calculator.

Sometimes you may need to evaluate an expression for many values of the variable. Using the memory feature may not be the most efficient method of evaluating the expression.

Calculation Problem
Evaluate the expression $4.6x^3 - 3.2x^2 + 6.1x + 10.8$. Use $x = 3.5$.
When using a calculator to do the calculation, you can rewrite the expression in a more convenient form, called **nested form**.
$$4.6x^3 - 3.2x^2 + 6.1x + 10.8 = (4.6x^2 - 3.2x + 6.1)x + 10.8$$
$$= ((4.6x - 3.2)x + 6.1)x + 10.8$$
written in nested form

In this form, you do not need to use the memory feature on your calculator. Begin the calculation in the innermost parentheses. Remember $x = 3.5$.

$$\boxed{\text{CE/C}}\; 4.6 \;\boxed{\times}\; 3.5 \;\boxed{-}\; 3.2 \;\boxed{=}\;\boxed{\times}\; 3.5 \;\boxed{+}\; 6.1 \;\boxed{=}\;\boxed{\times}\; 3.5 \;\boxed{+}\; 10.8 \;\boxed{=}$$

Often when expressions are written in nested form often only the right parentheses are shown,

$$((4.6x - 3.2)x + 6.1)x + 10.8 \text{ is written as } 4.6x - 3.2]x + 6.1]x + 10.8$$

8 Write each expression in nested form.
(a) $x^3 + 2x^2 - 3x + 5$ (b) $2x^4 - 3x^2 + 6x - 3$
(c) $3.1x^3 + 2.5x^2 - 6.5x - 3.6$ (d) $9.8x^4 - 3.6x^3 + 6.5x + 9.3$

9 (a) Calculate $x^3 - 3x^2 + 2x + 12$ when $x = 1, 3, 5, 6.8$.
(b) Calculate $2x^3 + 3x^2 - 4x + 12$ for $x = 1.5, 2.5, 3.5$.
(c) Calculate $2.1x^4 - 3.5x^3 + 8.9x + 36.5$ for $x = 2.1, 3.4, 4.7$.

10 You need to use the $\boxed{+/-}$ key to do these calculations.
(a) Calculate $x^3 - 5x^2 - 2x + 9$ for $x = -3, -4.5, 6.9$.
(b) Calculate $2.6x^3 - 9.8x + 13.6$ for $x = -1.8, 3.2$.

11 You need to use your $\boxed{\dfrac{1}{x}}$ key to do these calculations.
(a) Calculate $1 \div (3x^3 - 2x^2 + 5x + 10)$ for $x = 3, 5.2$.
(b) Calculate $\dfrac{2.5}{x^3} - \dfrac{3.6}{x^2} + \dfrac{4.2}{x} + 12.2$ for $x = 3.2, -4.5$.

1.6 Operations With Functions

Often scientific phenomena can be described mathematically. In an experiment diagrammed as shown, two sounds are produced separately, recorded and displayed visually on an oscilloscope.

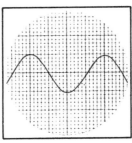

Equation of the Curve
$y = f(x)$

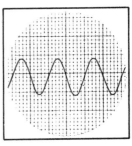

Equation of the Curve
$y = g(x)$

When both sounds are produced at the same time, the oscilloscope displays the pattern shown. The oscilloscope has electronically added the two sounds to obtain the picture shown.

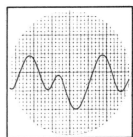

Equation of the Curve
$y = f(x) + g(x)$

You can add or subtract any two functions, f and g (defined on the real numbers), to form a new function

Addition of Functions

The sum $f + g$ is defined by $(f + g)(x) = f(x) + g(x)$.
For the above, $f + g$ is defined for the interval where f and g are defined. That is, any x belonging to the domain of f **and** to the domain of g belongs to the domain $f + g$. In other words, the domain of $f + g$ is the intersection of the domain of f and the domain of g.

function	f	g	$f + g$
domain	S_1	S_2	$S_1 \cap S_2$

In the intervals where $f + g$ is defined, you can find the y co-ordinates of $f + g$ by adding $f(x) + g(x)$ as shown in the following example.

Example 1   Find $f + g$ if $f = \{(2, 3), (3, 5), (4, 7), (-3, -7), (0, -1)\}$
$g = \{(-2, -5), (0, 1), (1, 4), (2, 7)\}$

Solution   The domains intersect for the points with $x = 0$ and $x = 2$.

$$f + g = \{[0, 1 + (-1)], [2, (3 + 7)]\} \qquad f + g = \{(0, 0), (2, 10)\}$$

Notice that when the x values were the same the y values were added.

The technique of **adding ordinates** (y co-ordinates) is a useful technique for obtaining the graph of functions that may not easily be obtained by plotting points.

Example 2 Find the graph of the function, f, defined by $f(x) = x + \dfrac{1}{x}$, $x > 0$.

Solution Let $h(x) = x$ and $g(x) = \dfrac{1}{x}$.

Think of $f(x)$ as consisting of two parts.

Step 1

Construct each graph on the same set of axes.

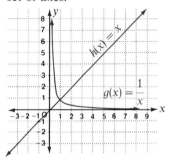

Step 2

For representative values in the domain find the sum $h(x) + g(x)$. Plot the particular points.

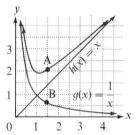

Step 3

You can use the points as a guide to sketch the resulting graph of

$$f(x) = x + \frac{1}{x}$$

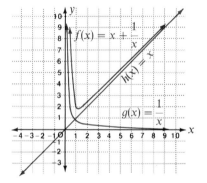

Subtraction of Functions

The difference of two functions, $f - g$, is defined by $(f - g)(x) = f(x) - g(x)$.

function	f	g	$f - g$
domain	S_1	S_2	$S_1 \cap S_2$

In drawing the graph of some functions, like the one in Example 2, the technique of interpreting the defining equation in parts is useful. The following example is done in parts. You can check the graph by directly finding the graph of the given function.

Example 3   Draw the graph of the function f defined by $f(x) = 2x - x^2$.

Solution   *Step 1*

Draw the graphs of $h(x) = 2x$
and $g(x) = x^2$ individually.

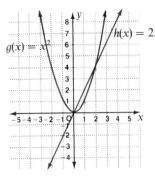

Step 2

For representative values in the
domain, find the difference
$h(x) - g(x)$.

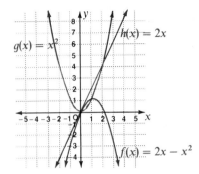

The resulting graph is the graph of $f(x) = 2x - x^2$.

1.6 Exercise

A 1   Graphs of functions f, g are drawn. Make a copy of each graph and
find the graph of $f + g$.

(a)

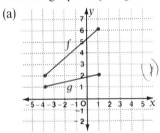

(b)

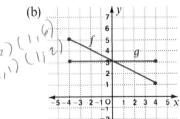

2   Find the graph of $f - g$ for each of the following.

(a)

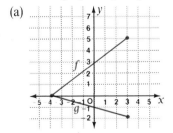

(b)

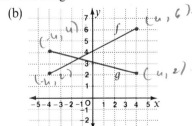

3   Two functions f and g are defined

$$f = \{(-4, 4), (-2, 4), (1, 3), (3, 5), (4, 6)\}$$
$$g = \{(-4, 2), (-2, 1), (0, 2), (1, 2), (2, 2), (4, 4)\}$$

(a) Find $f + g$. What is its domain?
(b) Find $f - g$. What is its domain?
(c) Find $g - f$. What is its domain?

4   The graphs of two functions are shown.

$f(x) = \frac{3}{4}x + 1$

$g(x) = \frac{1}{4}x + 3$

Find each of the following.
(a) $f(0) + g(0)$ (b) $f(4) + g(4)$
(c) $f(0) - g(0)$ (d) $g(0) - f(0)$
(e) $f(4) - g(4)$

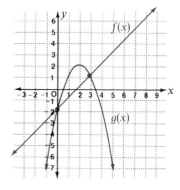

5   The graphs of two functions are shown.

$f(x) = x - 2$

$g(x) = -x^2 + 4x - 2$

Find each of the following.
(a) $f(-2) + g(-2)$ (b) $f(3) + g(3)$
(c) $f(-2) - g(-2)$ (d) $f(3) - g(3)$

6   Two functions are defined by

$$g(x) = 3x - 2, \qquad h(x) = 2x - 3, \quad x \in R.$$

Find an expression for each of the following.
What is the corresponding domain?
(a) $(g + h)(x)$ (b) $(g - h)(x)$ (c) $(h - g)(x)$

7   Two functions are defined as

$$f(x) = 3x - 1, \qquad 0 \le x \le 6$$
$$h(x) = x^2 - 6x, \qquad -1 \le x \le 4$$

Find an expression for each of the following.
What is the corresponding domain?
(a) $(f + h)(x)$ (b) $(f - h)(x)$ (c) $(h - f)(x)$

B 8 For each oscilloscope, draw the graph of $f + g$.

(a)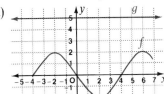

(b)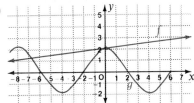

9 Make a copy of the graphs
   (a) Sketch the graph of $f + g$.
       What is its domain?
   (b) Sketch the graph of $f - g$.
       What is its domain?

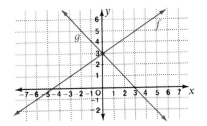

10 Two functions f and g are defined as follows.
$$f(x) = 3x - 2, \qquad g(x) = 2x^2, \qquad \text{domain} = \text{real numbers}$$
   (a) What is the domain of f, g, and $f + g$?
   (b) Draw the graphs of f, g, and $f + g$.

11 Two functions are $g(x) = -3x + 1$, $h(x) = 2x + 3$, $x \in R$.
   (a) Draw the graphs of g, h, and $h + g$.
   (b) Describe the properties of the resulting graph of $h + g$.

12 The functions, h and g, are defined as follows.
$$h(x) = x^2 - 4x, \quad x \in R \qquad g(x) = 3x - 2, \quad x \in R$$
   (a) Draw the graphs of g, h, $h + g$, and $h - g$.
   (b) Describe the properties of the resulting graphs of $h + g$ and $h - g$.

13 For each function the domains are more restrictive.
$$h(x) = \tfrac{1}{2}x^2 - 2x, \quad -2 \le x \le 3$$
$$g(x) = -3x, \quad -3 \le x \le 5$$
   (a) Draw the graphs of h, g, $h + g$, and $h - g$.
   (b) What is the domain of $h + g$ and of $h - g$?

C 14 Functions are given by $f(x) = \dfrac{1}{x^2 - 1}$, $g(x) = x$, $x \in R$.

   (a) What are the domains of f and g?
   (b) Draw the graph of $f + g$. What is its domain?

1.7 Composition of Functions

Previously, two operations were defined as functions.

function,	domain	function,	domain	new function,	domain
f,	D_1	g,	D_2	$f + g$,	$D_1 \cap D_2$
				$f - g$,	$D_1 \cap D_2$

To find the graph of these new functions, the following procedure is followed.

- The graphs for f and g are drawn on the same axes.
- The ordinates for each graph are combined to sketch the graph of $f + g$ and of $f - g$.

By taking a mapping point of view, you can define yet another operation on functions. For example, two functions are defined as follows.

$$f(x) = 2x \quad \text{and} \quad g(x) = 3x^2$$

The mapping diagrams are drawn for the functions.

A new function, denoted by the symbol $f \circ g$ is invented. The function $f \circ g$ maps x as

$$f \circ g: x \to f(g(x))$$

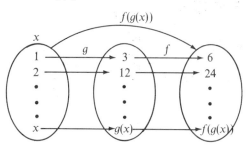

It is named the **composite** of the functions f and g.

Write $f(g(x))$ as $f \circ g(x)$. The operation that combines the function f and g to obtain $f \circ g$ is called the **composition of functions**.

Example 1 f and g are defined by

$f = \{(3, 2), (5, 1), (7, 4), (9, 3), (11, 5)\}$ $g = \{(1, 3), (2, 5), (3, 7), (4, 9), (5, 11)\}$

(a) Find $f(g(3))$.   (b) Find $g(f(9))$.

Solution (a) From (3, 7)   (b) $g(f(9)) = g(3)$
   $f(g(3)) = f(7)$ $= 7$
   $= 4$

Example 2 $f(x) = x^2 - 3 \quad \text{and} \quad g(x) = 2x$

Find each of the following.

(a) $(f \circ g)(1)$ (b) $(g \circ f)(1)$ (c) $(f \circ g)(3k)$

Solution $f(x) = x^2 - 3$ and $g(x) = 2x$

(a) $f \circ g(1) = f(g(1))$ ⟵ $g(1) = 2(1)$
$= f(2)$ ⟵ $= 2$
$f \circ g(1) = (2)^2 - 3$ This means
$= 4 - 3$ find f at 2.
$= 1$

(b) $g \circ f(1) = g(f(1))$ ⟵ $f(1) = (1)^2 - 3$
$g \circ f(1) = g(-2)$ $f(1) = 1 - 3$
$= 2(-2)$ $f(1) = -2$
$= -4$

$f \circ g(x) \neq g \circ f(x)$, which means that the operation of composition of functions is not commutative.

(c) $f \circ g(3k) = f(g(3k))$
$= f(2(3k))$ ⟵ This means use $x = 3k$ for $g(x) = 2x$.
$= f(6k)$ ⟵ This means use $x = 6k$ for $f(x) = x^2 - 3$.
$= (6k)^2 - 3$
$= 36k^2 - 3$

Your previously acquired skills for sketching graphs are applied to find the graph of the composition of functions f and g.

Example 3 Sketch the graph of $f \circ g$ for $f(x) = x^2 - 2x$ and $g(x) = x + 1$.

Solution *Step 1*

Find the defining equation of $f \circ g$ in terms of the variable x.
For any real x,
$f \circ g(x) = f(g(x))$
$= f(x + 1)$
$= (x + 1)^2 - 2(x + 1)$
$= x^2 + 2x + 1 - 2x - 2$
$f \circ g(x) = x^2 - 1$

Step 2

Sketch the graph of the function $f \circ g$ given by $y = x^2 - 1$.
Use your previously acquired skills for sketching graphs.

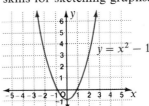

$y = x^2 - 1$

1.7 Exercise

A Record the meaning of $f \circ g$. Illustrate its meaning with an example of your own.

1   f and g are defined by mapping diagrams. Draw a mapping diagram for
(a) $f \circ g$ (b) $g \circ f$

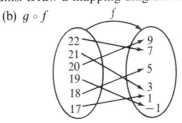

2   Use f and g given by the diagrams.

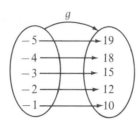 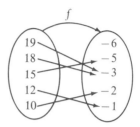

Find the following.
(a) $f(g(-4))$ (b) $f(g(-2))$ (c) $f(g(-1))$
(d) $g(f(10))$ (e) $g(f(19))$ (f) $g(f(18))$

3   f and g are defined as $f = \{(3, 1), (5, 2), (7, 3), (9, 4), (11, 5)\}$
$g = \{(1, 3), (2, 5), (3, 7), (4, 9), (5, 11)\}$

Find
(a) $f(g(1))$ (b) $f(g(4))$ (c) $f(g(5))$ (d) $f(g(3))$

4   Two functions are given by $f(x) = 3x - 2$ and $g(x) = 2x - 3$. Calculate each of the following.
(a) $f(g(1))$ (b) $g(f(-1))$ (c) $f(g(2))$ (d) $g(f(3))$ (e) $f(f(2))$

B   Once you define a new function, you explore its properties by looking for similarities and differences.

5   Functions f and g are defined. For each of the following find expressions for $(f \circ g)(x)$ and $(g \circ f)(x)$.
(a) $f: x \rightarrow 4x,\quad g: x \rightarrow x^2$ (b) $f: x \rightarrow x - 3,\quad g: x \rightarrow 3x^2$
(c) $f = \{(x, y)\,|\,y = x + 5\},\quad g = \{(x, y)\,|\,y = x - 2\}$
(d) $f(x) = 2x - 5,\quad g(x) = (x + 3)^2$ (e) $f(x) = 2,\quad g(x) = 2x - 1$
(f) Based on parts (a) to (e), why is $g \circ f \neq f \circ g$ for all f and g?

6   f and g are defined by each of the following. Find expressions for $f \circ g$ and $g \circ f$.
(a) $f(x) = 3x,\quad g(x) = 2x - 3$ (b) $f(x) = 3x^2,\quad g(x) = 2x - 3$
(c) $f(x) = 3x,\quad g(x) = 2x^2 - 3$ (d) $f(x) = 3x^2,\quad g(x) = 2x^2 - 3$
(e) $f(x) = 3x,\quad g(x) = \dfrac{1}{2x - 3}$ (f) $f(x) = 3x^2,\quad g(x) = \dfrac{1}{2x^2 - 3}$

Examine your results. How are they alike? How are they different?

7   (a) What is a function g that has the property $g \circ f = f$ for all f?
(b) What is a function g that has the property $f \circ g = f$ for all f?

8 Refer to your results in the previous question. The **identity mapping** is given by $i: x \rightarrow x$.
   (a) For a function $f: x \rightarrow 2x + 1$, find expressions for $i \circ f$ and $f \circ i$. What do you notice about the expressions?
   (b) For $g(x) = 3x^2 - 5$, find expressions for $i \circ g$ and $g \circ i$. What do you notice about your expressions?

9 If $f(x) = x + 1$ and $g(x) = x^2 - 4$, $x \in R$, find each of the following.
   (a) $(f \circ g)(x)$ (b) $(g \circ f)(x)$ (c) domain of $f \circ g$
   (d) domain of $g \circ f$ (e) range of $f \circ g$ (f) range of $g \circ f$

10 f and g are defined by $f(x) = 2x$ and $g(x) = 3x^2 - 1$.
   (a) Find expressions for $f \circ g$ and $g \circ f$.
   (b) What are the domain and range of $f \circ g$? $g \circ f$?
   (c) Sketch the graphs of $f \circ g$ and $g \circ f$.

11 Two functions are defined by $f(x) = \sqrt{x - 1}$ and $g(x) = x^2 - 2$.
   (a) Find the defining equation for $f \circ g$ and $g \circ f$.
   (b) Find the domain and range of $f \circ g$ and $g \circ f$.

12 For each of the following • find the defining equation for $f \circ g$ and $g \circ f$.
                                • write the domain and range of $f \circ g$ and $g \circ f$.
   (a) $f(x) = 3x,$ (b) $f(x) = \sqrt{x},$ (c) $f(x) = \sqrt{9 - x^2},$
       $g(x) = \sqrt{9 - x^2}$ $g(x) = 3x + 1$ $g(x) = x^2$

13 (a) If $f(x) = 2x + 3$, find a function g so that $f \circ g = x$.
   (b) If $f(x) = 2x - 2$ and $k(x) = 2x^2 + 4x + 4$, find $g(x)$ so that $f \circ g = k$.

14 (a) If $f: x \rightarrow ax + b$ and $g: x \rightarrow cx + d$, find an expression for $f \circ g$ and $g \circ f$.
   (b) For the following mappings, find an expression for $f \circ g \circ h$.

   $$f: x \rightarrow x + 2, \qquad g: x \rightarrow 3x, \qquad h: x \rightarrow x^2$$

15 If $f: x \rightarrow x - 3$, $g: x \rightarrow 2x$ and $h: x \rightarrow x$, find the expressions for each of the following.
   (a) $f \circ g \circ h$ (b) $f \circ h \circ g$ (c) $g \circ f \circ h$
   (d) $g \circ h \circ f$ (e) $h \circ g \circ f$ (f) $h \circ f \circ g$

C 16 (a) If $f(x) = 3x - 2$, $x = 3t + 2$ and $t = 3k - 2$, find an expression for $y = f(k)$.
   (b) Express y as a function of k if $y = 2x + 5$, $x = \sqrt{3t - 1}$ and $t = 3k - 5$.

Applications: Pendulums and Temperature

You have probably noticed that the time, T, in seconds, taken for one complete swing of a pendulum is less if the length, L, in centimetres, of the pendulum is shortened. The relationship between T and L is given by

$$T = 2\pi \sqrt{\frac{L}{980}}.$$ Thus T is some function of L. $T = f(L)$.

The length, L, in centimetres, of a pendulum is related to the temperature by the following relationship.

$$L = l + 0.0035\ C,\quad \text{where } C \text{ is the temperature, in degrees Celsius}$$
$$l \text{ is the length at } 0°C, \text{ a constant}$$

Thus, L is some function of C.

$$L = g(C).$$

Use a calculator to simplify your calculations.

17   Express period, T, as a function of C.

18   A pendulum is 100 cm in length when the temperature is 0°C. C, L, and T are defined by the above equations.
(a) Calculate the period, T, of the pendulum if the temperature is $0\,°C$.
(b) Calculate the period, T, if the temperature is $20\,°C$.
(c) How has the period of the pendulum been affected by the change in temperature?

19   The pendulum in the previous question is accurate at room temperature $(20\,°C)$.
(a) By how much time is the clock inaccurate, in the summer, per week when the temperature is, on the average, $28\,°C$?
(b) Over a weekend trip, the temperature of the house is lowered to $12\,°C$. How inaccurate is the clock after a 3-d weekend?

Problem Solving

As Jean walked down the escalator (at a uniform rate), she counted 50 steps but Simon counted 75 steps. Since she walked 3 times as fast (uniformly), down the escalator. If the escalator stopped, how may escalator steps would show?

Practice and Problems: A Chapter Review

At the end of each chapter, this section will provide you with additional questions to check your skills and understanding of the topics dealt with in this chapter. An important step for problem-solving is to decide which skills to use. For this reason, these questions are not placed in any special order. When you have finished the review, you might try the *Test for Practice* that follows.

1   A relation f is given by $f: x \to x(3x + 1)$.
   Find the following.

   (a) $f(-1)$ (b) $f(1)$ (c) $f(0)$ (d) $f(k)$ (e) $f\left(\dfrac{1}{k}\right)$ (f) $f(2k)$

2   Two functions are given by $f(x) = x - 1$, $g(x) = 2x + 1$. Calculate.
   (a) $f(1)$ (b) $f^{-1}(1)$ (c) $f[f^{-1}(1)]$ (d) $g(3)$
   (e) $g^{-1}(3)$ (f) $g[g^{-1}(3)]$ (g) $f^{-1}[g^{-1}(-2)]$ (h) $g^{-1}[f^{-1}(-2)]$

3   A relation is given by $9x^2 + 4y^2 = 36$,
   (a) list the properties of the graph, and
   (b) draw a sketch of the graph.

4   Two functions are $f(x) = 2x^2 - 3x$, $g(x) = 3x$, $x \in R$. What is the defining equation and domain for
   (a) $f + g$?   (b) $f - g$?   (c) $g - f$?

5   Find an expression for $f \circ g \circ h$, given that f, g, and h are defined by

$$f: x \to -2x \qquad g: x \to x \qquad h: x \to x^2 - 2$$

6   Find the equation of the inverse for each of the following, $a \neq 0$.

   (a) $f(x) = \dfrac{a}{x}$ (b) $f(x) = \dfrac{1}{x + a}$ (c) $f(x) = \dfrac{x + a}{x}$ (d) $f(x) = \dfrac{x}{x + a}$

7   Sketch the graph of each of the following.
   (a) $y = x^2$ (b) $y = 2x^2$ (c) $y = (x - 2)^2$
   (d) $y = (x - 2)^2 + 3$ (e) $y = (x - 2)^2 - 3$

8   A graph is defined by $f(x) = x^2 - 9$. Sketch the graph given by $y = \dfrac{1}{f(x)}$.

9   (a) If $f: x \to \dfrac{2}{x}$, find the value of $\dfrac{f(x + a) - f(x)}{a}$.

   (b) If $f(x) = 2x + 3$ and $k(x) = 2x^2 + 6x + 9$, find $g(x)$ so that $f \circ g = k$.

Test for Practice

Try this test. Each test will be based on the mathematics you have learned in the chapter. Try this test later in the year as a review. Keep a record of those questions that you were not successful with and review them periodically.

1   For each of the following:
- determine whether or not the relation is a function.
- What is the domain?
- What is the range?

(a) 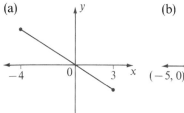   (b) 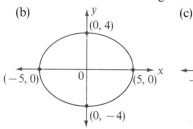   (c)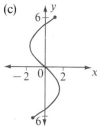

2   A relation is defined by $x^2 + y^2 = 100$.
(a) List the intercepts, the domain, and the range.
(b) Sketch the graph of the relation.
(c) Justify whether the relation is a function.

3   The graph defined by $y = f(x)$ is drawn. What is the effect of each of the following, $a \in R$, on the graph of $y = f(x)$?
(a) $y = f(x) + a$ (b) $y = f(x + a)$ (c) $y = af(x)$
(d) $y = f(ax)$ (e) $y = |f(x)|$ (f) $y = -f(x)$

4   The functions $f(x) = x^2 - 1$ and $g(x) = 2x - 3$ have domain the real numbers. What is the domain of each of the following functions?
(a) $f + g$ (b) $f - g$ (c) $g - f$

5   Given that f and g, are defined by $f(x) = 2x - 3$ and $g(x) = x^2 - 1$. Find an expression for each of the following.
(a) $f[g(x)]$ (b) $g[f(x)]$ (c) $f[f(x)]$ (d) $g[g(x)]$

6   If $f : x \to \dfrac{2x + 1}{3}$ and $g : x \to \dfrac{x - 1}{2}$, determine m so that $f(m) = g(m)$.

7   (a) Find the equation of f^{-1} if $f(x) = 2x^2 + 3$.
(b) What is the domain and range of f^{-1}?

8   Sketch the graph of $y = f(x)$ and $y = \dfrac{1}{f(x)}$ where $f(x) = 3x - 2$.

2 Using Polynomials and Equations

skills for factoring polynomials, common factor, grouping, trinomials, difference of squares, factor theorem, solving quadratic equations, language of mathematics, solving polynomial equations, applications, problem-solving

Introduction

In solving problems, you must know the answers to these two important questions:

- What am I asked to find?
- What am I given?

Whether you are studying mathematics, or solving a problem in physics or chemistry, the answer to these two questions will give a clue to the procedure to be followed to arrive at an answer. In mathematics, or any of your studies, an important goal is to be able to organize your written solution so that you, or others, can understand them at a later time. The *Steps for Solving Problems* be used to organize your solution.

Steps for Solving Problems

Step A: Understand the problem:
- What are you asked to find?
- What are you given?

This is the time to introduce variables for what is unknown.

Record the information on a diagram.

Step B: Decide on a method.
This is the time to write your variables into equations.

Step C: Find the answer.
This is the time to solve the equations.

Step D: Check your answer.
- Is it reasonable?
- Are your calculations correct?

Step E: Write a final statement.
- Have you used the correct units?
- Have you rounded correctly?

As you learn different skills and strategies, and formulate methods of your own, eventually, you will acquire your own techniques for organizing the written record of your solutions. However, throughout your work in this chapter and those that follow, you must develop some method of organizing your work. Your skills in organization which you have developed from the solutions of earlier problems form a useful foundation for developing skills for solving problems you are confronted with for the first time.

2.1 Process of Factoring: Developing Skills

There are very specific steps involved in developing skills in mathematics. An important first step is to understand clearly the meaning of the words used in mathematics.

To **expand** means to write a product of expressions as a sum or difference of terms.

$$m(a + b) = ma + mb$$
$$4x(x - 3) = 4x^2 - 12x$$

To **factor** means to write a sum or difference of terms as a product of expressions. (This is the reverse of expanding.)

$$ma + mb = m(a + b)$$

m is called a common factor of $ma + mb$.

$$4x^2 - 12x = 4x(x - 3)$$

$4x$ is called a common factor of $4x^2 - 12x$

In mathematics you develop a particular skill, apply it, and then extend it to new situations. For example, a first step in learning factoring skills is to develop skills for identifying monomial common factor expressions, as shown in the following example.

Example 1 Factor each of the following.

(a) $6ap - 9aq$ (b) $-5axy - 5bxy + 10cxy$

Solution
(a) $6ap - 9aq = 3a(2p - 3q)$
(b) $-5axy - 5bxy + 10cxy = -5xy(a + b - 2c)$ Remember
$$-a - b = -(a + b)$$

With practice, you will be able to identify, quickly, monomial common factors. Before you do any factoring, *always check for a common factor.* Common factors may also occur as binomials or as polynomials. For example,

$$p(r + s) + 2q(r + s) = (r + s)(p + 2q)$$

common factor

$$(a + c - 2d)(3x) - (a + c - 2d)(2y) = (a + c - 2d)(3x - 2y)$$

common factor

In order to find common factors, your powers of observation must be developed. You must learn to identify particular groupings that will yield a common factor, as shown in the following example.

Example 2 Factor (a) $2ax - bx + 6ay - 3by$ (b) $2tp - rq + rp - 2tq - sp + sq$

Solution (a) $2ax - bx + 6ay - 3by$
$= (2ax - bx) + (6ay - 3by)$
$= x(2a - b) + 3y(2a - b)$
$= (2a - b)(x + 3y)$

(b) $2tp - rq + rp - 2tq - sp + sq$
$= (2tp + rp - sp) - (rq + 2tq - sq)$
$= p(2t + r - s) - q(r + 2t - s)$
$= (2t + r - s)(p - q)$

There may be more than one way of finding a common factor. It depends on how you group the terms. The following illustrates two other methods of grouping which result in the same factors.

(a) Alternative Grouping
$2ax - bx + 6ay - 3by$
$= 2ax + 6ay - bx - 3by$
$= (2ax + 6ay) - (bx + 3by)$
$= 2a(x + 3y) - b(x + 3y)$
$= (x + 3y)(2a - b)$

(b) Alternative Grouping
$2tp - rq + rp - 2tq - sp + sq$
$= (2tp - 2tq) + (rp - rq) - (sp - sq)$
$= 2t(p - q) + r(p - q) - s(p - q)$
$= (p - q)(2t + r - s)$

The same factors are obtained regardless of how the original terms are grouped.

2.1 Exercise

A 1 Find the missing factor in each of the following.
(a) $12x^2y = 6y(\ ?\ )$
(b) $-27ab^2 = -9a(\ ?\ )$
(c) $48xyz^2 = 12(\ ?\ )$
(d) $125pq^3 = 25q^2(\ ?\ )$
(e) $49x^5 = -49x(\ ?\ )$
(f) $35r^3s^2t = -7r^2st(\ ?\ )$
(g) $-36d^4e^3 = 6d^2e^3(\ ?\ )$
(h) $16xyz^4 = -16z^3(\ ?\ )$

2 Find the missing factor.
(a) $5p + 5q = (\ ?\ )(p + q)$
(b) $xy + x^2 = (\ ?\ )(y + x)$
(c) $3xr - 9xs = 3x(\ ?\ )$
(d) $d^2wx + dx = (\ ?\ )(dw + 1)$
(e) $-3ab^2 - 9a^2b + 27a^2b^2 = -3ab(\ ?\ )$
(f) $4m^4n^2 + 16m^3n^5 - 8mn = (\ ?\ )(m^3n + 4m^2n^4 - 2)$

3 Find the greatest common factor of each of the following expressions.
(a) $2p^2 - 6pq$
(b) $9xy^3 - 18x^2y$
(c) $-5w^2v - 25vw^3$
(d) $-7a^3b + 17a^3$
(e) $-36xy - 6z$
(f) $27x^2yz + 3y^5$
(g) $63cd^4 + 9xy + 18c^2x$
(h) $-13m^3n - 65mn^3 - 39mn$
(i) $25g^3h - 75g^2h^2 + 5$
(j) $132s^3t^3 + 44s^3t - 11s^3t^5$

4 (a) Find the factors of $b(k + p) + a(k + p)$.
(b) Find the factors of $k(a + b) + p(a + b)$.
(c) Why are the answers in (a) and (b) the same?

5 (a) Factor the expression by using the common factor indicated. $ax + bx + ay + by$

(b) Factor the expression by using the common factor indicated. $ax + bx + ay + by$

(c) Why are the answers in (a) and (b) the same?

B 6 (a) Decide how to group the terms in order to find the factors of
$ay - by + 2bx - 2ax.$ Then factor the expression.

(b) Check your factors by finding the product.

7 (a) Find the factors of $x^2 + y - x - xy$ in two different ways.
(b) Find the factors of $a^2 + ab - ac - bc$ in two different ways.

8 For each of the following

▶ first decide how to group the terms
▶ then factor.

(a) $15cv + 10cw + 12dv + 8dw$ (b) $-35yz + 14wy + 40xz - 16wx$
(c) $22vx - 6vy + 11wx - 3wy$ (d) $20bp - 12cp - 3cq + 5bq$

9 Factor each of the following.

(a) $ap + aq + bp + bq$ (b) $-2st + rt - rv + 2sv$
(c) $3cm - 12dm + nc - 4dn$ (d) $20gk - 5hk + 8gm - 2hm$

10 Factor.
(a) $16bg - 4gp + 24bh - 6hp$ (b) $2vx^2 - wx^2 - 4vxy + 2wxy$
(c) $a^2b^2 + 2a^2b + 3ab^2 + 6ab$ (d) $-5m^2n^3 - 15mn^3 + 5mn^4 + 15n^4$
(e) $x^2y^2 + 2 + 2x^2 + y^2$ (f) $3ap + aq + 6bp - 3cp + 2bq - cq$
(g) $21vx + 7vy - 28vz - 9wx - 3wy + 12wz$
(h) $x^2 + y - xy - x$ (i) $a^2 - ac + ab - bc$
(j) $3mxy - 6mx - 3nxy + 6nx$ (k) $x^4 + 2x^2y^2 + x^2 + 2y^2$

Math Tip

It is important to understand clearly the vocabulary of mathematics when solving problems.
• Make a list of all the new words you meet in this chapter
• Provide an example to illustrate each word.

2.2 Working Backwards: Factoring Trinomials

Factoring and expanding are related as shown.

$$\xrightarrow{\quad\text{expanding}\quad}$$
$$(x - 2)(x + 5) = x^2 + 3x - 10$$
$$\xleftarrow{\quad\text{factoring}\quad}$$

You can begin with what you know, namely how to expand binomials, and then work backwards. This strategy is often used in mathematics.

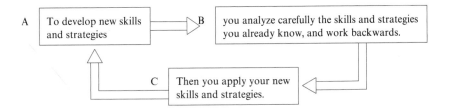

Thus, begin with B since you know how to find the product of $(x - 2)$ and $(x + 5)$.

$$(x - 2)(x + 5) = x^2 + 3x - 10$$

In order to find the factors, work backwards.

$$x^2 + 3x - 10 = (\ ? \ )(\ ? \ )$$

Carefully analyze the steps. Look at the general case.

$$\overset{\displaystyle\overbrace{}^{\text{product of integers}}}{(x + a)(x + b) = x^2 + (a + b)x + ab}$$
$$\underbrace{}_{\text{sum of integers}}$$

Work backwards from what you know by asking yourself:

In order to factor $x^2 + 2x - 3$ what two integers have
- a sum of $+2$ and   • a product of -3?

The integers $+3$ and -1 have a sum of $+2$ and a product of -3. Thus,

$$x^2 + 2x - 3 = (x + 3)(x - 1)$$ You can check your answer by multiplying the binomial factors.

Now you can use this strategy that you have just developed to factor any trinomial. But remember, before you factor, always check to see if there is a common factor of the terms.

Example 1   Factor   (a) $y^2 + 9y + 14$ (b) $m^2n - mn^2 - 6n^3$

Solution   (a) $y^2 + 9y + 14 = (\ ?\ )(\ ?\ )$

Thus, $y^2 + 9y + 14 = (y + 2)(y + 7)$

Think: What two integers have
- a sum of $+9$
- a product of $+14$?

Answer $+7, +2$

(b) $m^2n - mn^2 - 6n^3 = n(m^2 - mn - 6n^2)$

common factor

Thus, $m^2n - mn^2 - 6n^3 = n(m - 3n)(m + 2n)$

Think: What two integers have
- a sum of -1
- a product of -6?

Answer $-3, +2$

You can apply a similar strategy of working backwards in order to develop a method of factoring a trinomial of the form

$$ax^2 + bx + c \quad \text{where } a \neq 1.$$

Again, you can compare the coefficients of the trinomial with the coefficients of the terms of the factors. Thus, analyze an example whose results you know and make useful observations.

Think: $(3)(-3) = -9$

$(2)(1) = 2$

$(-9) + 2 = -7$

$(3x + 2y)(x - 3y) = 3x^2 - 7xy - 6y^2$

$(3)(2)(1)(-3) = -18$ $(3)(-6) = -18$

Note: these products are the same.

After you analyze the example you know, you can ask yourself:

What two integers have
- a product of -18 and
- a sum of -7?

answer: $-9, +2$

Now you can use these two integers to **decompose** the given trinomial into 4 terms, as shown.

$3x^2 - 7xy - 6y^2 = 3x^2 - 9xy + 2xy - 6y^2$
$\qquad = 3x(x - 3y) + 2y(x - 3y)$
$\qquad = (x - 3y)(3x + 2y)$

Once you recognize the binomial common factor of $(x - 3y)$ you now have a familiar problem which you have solved before.

Example 2 Factor (a) $27g^2 - 24gh + 4h^2$ (b) $20a^3 - 6a^2b - 8ab^2$

Solution (a) $27g^2 - 24gh + 4h^2$
$= 27g^2 - 18gh - 6gh + 4h^2$
$= 9g(3g - 2h) - 2h(3g - 2h)$
$= (3g - 2h)(9g - 2h)$

Think: What two integers have
- a sum of -24 and
- a product of $+108$?
Answer $-18, -6$

(b) $20a^3 - 6a^2b - 8ab^2$
$= 2a(10a^2 - 3ab - 4b^2)$
$= 2a[10a^2 - 8ab + 5ab - 4b^2]$
$= 2a[2a(5a - 4b) + b(5a - 4b)]$
$= 2a(5a - 4b)(2a + b)$

Think: What two integers have
- a sum of -3 and
- a product of -40?
Answer $-8, +5$

2.2 Exercise

A 1 What two integers have
  (a) a sum of -1 and a product of -6?
  (b) a sum of -16 and a product of $+63$?
  (c) a sum of $+9$ and a product of $+20$?
  (d) a sum of $+7$ and a product of $+6$?
  (e) a sum of -23 and a product of $+132$?
  (f) a sum of $+10$ and a product of -39?
  (g) a sum of -10 and a product of -39?

2 Find the missing factor.
  (a) $y^2 + 4y + 3 = (y + 3)(\ ?\ )$
  (b) $b^2 - b - 20 = (\ ?\ )(b - 5)$
  (c) $x^2 - 9x + 14 = (\ ?\ )(x - 2)$
  (d) $p^2 + 13p + 42 = (\ ?\ )(p + 6)$
  (e) $z^2 - 16zy + 63y^2 = (z - 7y)(\ ?\ )$
  (f) $m^2 - 5mn + 6n^2 = (m - 2n)(\ ?\ )$

3 Find the missing factor. Check your answer by finding the product.
  (a) $2s^2 + 25st + 63t^2 = (2s + 7t)(\ ?\ )$
  (b) $4q^2 - 21qr + 5r^2 = (\ ?\ )(4q - r)$
  (c) $6p^2 - 25pq - 9q^2 = (3p + q)(\ ?\ )$
  (d) $7a^2 - 39ab - 18b^2 = (a - 6b)(\ ?\ )$
  (e) $10m^2 - 9mn + 2n^2 = (\ ?\ )(2m - n)$
  (f) $11x^2 + 38xy + 15y^2 = (11x + 5y)(\ ?\ )$

4 Find the common factor of each group of terms.
  (a) $3x, 9xy, 12y$
  (b) $2pr, 6p^2, -7pq$
  (c) $m^2n, -m^3n^2, m^4$
  (d) $-7a^4, -21ab, -42a^3b^2$
  (e) $25s^4t^3, 75s^7t^4, 125$
  (f) $-48k^4m^2n, 16k^4mn, -64k^4mn^2$

5   (a) What two integers have a sum of $+10$ and a product of $+21$?

      (b) Use your answer in (a) to factor $v^2 + 10vw + 21w^2$.

6   (a) What two integers have a sum of $+17$ and a product of -84?

      (b) Use your answer in (a) to factor $6a^2 + 17ab - 14b^2$.

B   Remember, check for a common factor first.

7   Factor each of the following. Check your answers by finding the product.

      (a) $p^2 - 6p - 7$ (b) $s^2 + 4s - 12$

      (c) $x^2 - 14x + 33$ (d) $y^2 + 13y + 40$

8   Factor.

      (a) $a^2 + 8a - 9$ (b) $m^2 + 9m - 52$ (c) $k^2 - 9kp + 14p^2$

      (d) $a^2 + 8ab + 15b^2$ (e) $y^2 - 9yz + 18z^2$ (f) $g^2 + gh - 12h^2$

9   Factor each of the following. Check for a common factor.

      (a) $3x^2 + 27x - 66$ (b) $y^3 - 6y^2 - 7y$

      (c) $4p^3 + 32p^2 + 60p$ (d) $mq^2 - 10mq + 16m$

      (e) $r^5 - r^4s - 2r^3s^2$ (f) $7t^2v^2 - 63tv^3 + 140v^4$

10   Factor each of the following.

      (a) $5x^2 - 17x + 6$ (b) $6p^2 + 23p + 7$

      (c) $10 + 19q - 15q^2$ (d) $7a^2 + 29a - 30$

      (e) $10 - 51b + 27b^2$ (f) $30m^2 + 71mn + 11n^2$

      (g) $12g^2 - 19gh - 21h^2$ (h) $14x^2 + 57xy - 27y^2$

      (i) $52s^2 - 57st - 10t^2$ (j) $11p^2 + 40pq + 21q^2$

11   Factor each of the following.

      (a) $4m^3 - 6m^2 - 70m$ (b) $99n^2 + 153n^3 - 90n^4$

      (c) $-35p^2x - 185pqx - 50q^2x$ (d) $45x^5 - 37x^4y + 6x^3y^2$

      (e) $-96a^3b + 152a^2b^2 - 60ab^3$ (f) $2s^4t^2 + 21s^3t^3 - 65s^2t^4$

12   Factor.

      (a) $77 - 18a + a^2$ (b) $10p^2 - 35p - 75$

      (c) $33x^2 + 86xy + 21y^2$ (d) $-12s^2 - 13sr + 35r^2$

      (e) $10m^4n^2 + 27m^3n^2p + 18m^2n^2p^2$ (f) $-75w^3 + 45w^2x + 54wx^2$

      (g) $8v^2 - 14vw + 5w^2$ (h) $-36c^2m^2 - 39cdm^2 + 35d^2m^2$

      (i) $15x^4 - 27x^2 - 132$ (j) $36g^3h^3 - 98g^2h^2 - 98gh$

      (k) $16m^2 + 64mn + 39n^2$ (l) $3p^4q^2r - 36p^2qr^2 + 81r^3$

2.3 Writing Special Factors

When you are solving problems, you must remind yourself, constantly, of the skills and strategies you have already learned. An example of your own, recorded with the summary, will provide an overview for you. For example, you have already developed these methods of factoring.

Common Factors $2pq + 2qr = 2q(p + r)$
$$4m^2n + 16mn^2 - 12m^2n^2 = 4mn(m + 4n - 3mn)$$

Trinomials $x^2 + xy - 2y^2 = (x + 2y)(x - y)$
$$2a^4 + 5a^2b + 3b^2 = (a^2 + b)(2a^2 + 3b)$$

Some expressions, when factored, follow a pattern. You can study this pattern and, from it, develop a method of writing these special factors.

Perfect Square Trinomial

You know that $(a + b)(a + b) = a^2 + 2ab + b^2$
$$(a - b)(a - b) = a^2 - 2ab + b^2.$$
When you write $(x + 3y)^2 = x^2 + 6xy + 9y^2$, notice how the terms of the binomial and of the trinomial are related.

$$(x + 3y)^2 = x^2 + 6xy + 9y^2$$

first term squared   second term squared

twice the product of the terms

A trinomial which can be written as the square of a binomial is called a **perfect square trinomial**.

Example 1   Factor   (a) $4p^2 + 12pq + 9q^2$ (b) $100x^2 - 80xy + 16y^2$

Solution   (a) $4p^2 + 12pq + 9q^2 = (2p + 3q)^2$ Think: $(3q)^2$
Think: $(2p)^2$

Middle term: $2(2p)(3q)$ Checks ✓

(b) $100x^2 - 80xy + 16y^2$
$= 4(25x^2 - 20xy + 4y^2)$ Think: Is there a common factor?
$(2y)^2$ ⎱ middle term
$(5x)^2$ ⎰ $2(2y)(5x)$ Checks ✓

$= 4(5x - 2y)^2$

Difference of Squares

You know that $(a - b)(a + b) = a^2 - b^2$.

An expression of the form $a^2 - b^2$ is called a **difference of squares**. An important skill in factoring a difference of square is being able to recognize the difference of squares pattern.

$$9x^2 - 4y^2 = (3x)^2 - (2y)^2 \quad \text{— Look for a difference.}$$
$$\text{— Look for squares.}$$
$$= (3x - 2y)(3x + 2y)$$

Remember, when factoring any expression,

- first check to see if the expression has a common factor,
- check your answer to see if any of the factors can be factored further.

Example 2   Factor   (a) $49x^4 - 16x^2$ (b) $16x^4 - 81y^4$

Solution   (a) $49x^4 - 16x^2 = x^2(49x^2 - 16)$ common factor
$$= x^2[(7x)^2 - (4)^2]$$
$$= x^2(7x - 4)(7x + 4) \quad \text{Check: Factors cannot be factored further. } \checkmark$$

(b) $16x^4 - 81y^4 = (4x^2)^2 - (9y^2)^2$

Think: Can factors be factored further?
$$= (4x^2 - 9y^2)(4x^2 + 9y^2) \qquad 4x^2 - 9y^2$$
$$= (2x - 3y)(2x + 3y)(4x^2 + 9y^2) \qquad = (2x)^2 - (3y)^2$$
$$= (2x - 3y)(2x + 3y)$$

2.3   Exercise

A   Remember, the first step is to check for a common factor.

1. Each of the following are perfect trinomial squares. What is the missing factor?

   (a) $(m + 2)(\ ? \ )$ (b) $(\ ? \ )(3x + 1)$ (c) $(a - 2b)(\ ? \ )$

   (d) $(5p + 3q)(\ ? \ )$ (e) $(\ ? \ )(x^2 + 2)$ (f) $(3s^2 - 4t^2)(\ ? \ )$

   (g) $(mn - 4)(\ ? \ )$ (h) $(\ ? \ )\left(z - \dfrac{1}{2}\right)$ (i) $(0.75n + 3m)(\ ? \ )$

   (j) $(\ ? \ )(c + d - e)$ (k) $(3g - h + 2k)(\ ? \ )$

2. Each of the following is a difference of squares. Find the missing factor.

   (a) $(b + 4)(\ ? \ )$ (b) $(\ ? \ )(x - 3y)$ (c) $(2g + 3h)(\ ? \ )$ (d) $(11 - 4k)(\ ? \ )$
   (e) $(\ ? \ )(p + q + r)$ (f) $(a - 2b + 3c)(\ ? \ )$ (g) $(x^2 - 7)(\ ? \ )$

   (h) $(c^2d^2 + g)(\ ? \ )$ (i) $\left(3n + \dfrac{1}{3}\right)(\ ? \ )$ (j) $(\ ? \ )\left(m + \dfrac{1}{4}\right)$

3 Complete each of the following.

(a) $x^2 - 25y^2 = x^2 - (\ ?\ )^2$

(b) $4p^2 - 9q^2 = (2p)^2 - (\ ?\ )^2$

(c) $49a^2 - 16b^2 = (\ ?\ )^2 - (4b)^2$

(d) $16m^4 - 36n^2 = (\ ?\ )^2 - (6n)^2$

(e) $x^8 - 64y^4 = (\ ?\ )^2 - (\ ?\ )^2$

(f) $(r + 2s)^2 - 4t^2 = (r + 2s)^2 - (\ ?\ )^2$

(g) $81c^2d^2 - (g + h)^2 = (\ ?\ )^2 - (g + h)^2$

(h) $144x^4y^2z^6 - 100v^2w^2 = (\ ?\ )^2 - (\ ?\ )^2$

4 Each of the following is to be a perfect square trinomial. Find each missing term.

(a) $4a^2 + 4a + (\ ?\ )$

(b) $x^2 - (\ ?\ ) + 4$

(c) $(\ ?\ )^2 + 6pq + q^2$

(d) $25m^2 - (\ ?\ ) + 4n^2$

(e) $c^2 - cd + (\ ?\ )$

(f) $(\ ?\ )^2 + 20x^2y + 4y^2$

B What is the first step you should do when you are asked to factor an expression?

5 Write each of the following as a perfect square trinomial. Check by finding the product.

(a) $a^2 + 8a + 16$

(b) $4b^2 + 12b + 9$

(c) $9y^2 - 12y + 4$

(d) $49p^2 - 56p + 16$

(e) $9a^2 + 12ab + 4b^2$

(f) $\dfrac{x^2}{4} - 5xy + 25y^2$

6 Factor.

(a) $4m^2 - \dfrac{4mn}{3} + \dfrac{n^2}{9}$

(b) $x^2y^2 + 6xy + 9$

(c) $4a^2b^2 + 12abc + 9c^2$

(d) $g^4h^2 - 8g^2h + 16$

(e) $x^2 - 4xyz + 4y^2z^2$

(f) $p^4q^2r^2 + 6p^2qrt + 9t^2$

7 Factor.

(a) $b^2 - 16$ (b) $4x^2 - 1$ (c) $x^2 - 9y^2$ (d) $4g^2 - 9h^2$ (e) $121 - 16k^2$

(f) $m^2 - \dfrac{1}{16}$ (g) $9n^2 - \dfrac{1}{9}$ (h) $x^4 - 49$ (i) $x^2y^2 - 1$

8. Write each of the following polynomials as a difference of squares, then factor. The first one is done for you.

(a) $z^2 + 4z + 4 - y^2 \longrightarrow\ = (z^2 + 4z + 4) - y^2$
$= (z + 2)^2 - y^2$
$= (z + 2 + y)(z + 2 - y)$

(b) $1 + 6x + 9x^2 - 4y^2$

(c) $49a^2 - 1 + 8b - 16b^2$

(d) $p^4 + 4p^2 + 4 - q^4$

(e) $x^2y^2 + \dfrac{1}{2}x - x^2 - \dfrac{1}{16}$

(f) $25 + 36m^2 - 4n^2 + 60m + 4n - 1$

(g) $25g^2 + 49h^2 - 70gh - 12kp - 9k^2 - 4p^2$

9   Factor each of the following. Remember:
- first check for common factors.
- check that the factors in your answer cannot be factored further.

(a) $7x^2 - 28$ (b) $200 - 50y^4$ (c) $y^2 - 49z^2$ (d) $1 - 4p^2q^2$

(e) $3m^2 - 48n^2$ (f) $a^3 - a$ (g) $kg^2h^2 - 4k$ (h) $16p^4 - 25q^2$

(i) $36c^2 - 16d^4$ (j) $m^5 - m^3$ (k) $13b^4 - 52c^4$ (l) $\dfrac{x^4}{8} - 2$

10   Factor.

(a) $-4 + y^8$ (b) $256w^{16} - v^{16}$ (c) $\dfrac{p^2}{4} - \dfrac{q^2}{9}$ (d) $(a + 2)^2 - 9$

(e) $-b^2 + (2c + 1)^2$ (f) $(3g - 2h)^2 - (3g - h)^2$ (g) $0.04m^2 - 6.25n^2$

(h) $3d^4 - 12g^4$ (i) $3(w - z)^2 - \dfrac{1}{3}v^2$ (j) $(a + b)^2 - 16a^2$

(k) $(g - h + 2k)^2 - (g - h - 2k)^2$ (l) $9(x + 2y)^2 - 16(2x - y)^2$

11   Write each of the following in a factored form.

(a) $a^2 + 6a + 9 - b^2$ (b) $x^2 - 2xy + y^2 - 9z^2$

(c) $d^4 - 2d^2e^2 + e^4 - 16g^4$ (d) $4m^2 - 4n^2 - 8np - 4p^2$

(e) $x^4 - y^2 + 2yz - z^2$ (f) $2p^2 + 8qr - 8r^2 - 2q^2$

(g) $9a^2 + 18ab + 9b^2 - 4c^2 - 8cd - 4d^2$

12   Factor each of the following completely. Which ones cannot be factored?

(a) $100 - (x - 3)^2$ (b) $4p^4 - 36p^2 + 9$ (c) $\dfrac{a^2}{64} - \dfrac{b^2}{49}$ (d) $\dfrac{c^4}{16} - \dfrac{d^4}{81}$

(e) $625x^8y^4 - 16z^8$ (f) $6g^2 - 54h^2$ (g) $4x^4 + 24x^2 + 9$

(h) $4x^2 + y^2$ (i) $8x^2 - 50$ (j) $-48 + 27w^2$

(k) $16m^2 - 32mn + 16n^2 - r^2 - 2rs - s^2$

Calculator Tip

Often you can learn certain algebraic techniques to help you more efficiently use a calculator.

Step A: Evaluate $2.1m^3 - 3.2m^2 + 6.5m - 3.8$ for $m = 1.5$.

Did you use the memory feature $\boxed{\text{MS}}$ on your calculator?

Step B: You can rewrite the expression as shown.

$$m(m(2.1m - 3.2) + 6.5) - 3.8$$

The following calculator procedure does not use the memory feature to evaluate the expression for m = 1.5. Can you devise yet a more efficient procedure?

Output

$\boxed{^{CE}\!/_C}$ 2.1 $\boxed{\times}$ 1.5 $\boxed{-}$ 3.2 $\boxed{=}$ $\boxed{\times}$ 1.5 $\boxed{+}$ 6.5 $\boxed{=}$ $\boxed{\times}$ 1.5 $\boxed{-}$ 3.8 $\boxed{=}$

2.4 Dividing by a Binomial

You can use your skills with fractions and your factoring skills to divide a polynomial by a binomial. Compare

Arithmetic

$$\frac{4}{6} = \frac{2 \times 2}{2 \times 3} = \frac{2}{3} \qquad \frac{2}{3} + \frac{1}{2} = \frac{4 + 3}{6} = \frac{7}{6}$$

Algebra

$$\frac{ab}{ac} = \frac{b}{c} \qquad \frac{a}{b} + \frac{c}{d} = \frac{ad + bc}{bd}$$

Example 1 Divide (a) $\dfrac{2x^2 + 7x + 3}{x + 3}$ (b) $\dfrac{4x^2 - 9y^2}{2x - 3y}$

Solution (a) $\dfrac{2x^2 + 7x + 3}{x + 3} = \dfrac{(2x + 1)(x + 3)}{x + 3}$ (b) $\dfrac{4x^2 - 9y^2}{2x - 3y} = \dfrac{(2x - 3y)(2x + 3y)}{2x - 3y}$

$$= \dfrac{(2x + 1)(x + 3)^1}{x + 3_1} \qquad\qquad\qquad = \dfrac{(2x - 3y)(2x + 3y)}{2x - 3y_1}$$

$$= 2x + 1 \qquad\qquad\qquad\qquad\qquad = 2x + 3y$$

Whenever a polynomial cannot be divided evenly by a binomial you can use your skills in arithmetic with long division to divide the polynomial by the binomial.

Example 2 Divide $3x^3 - 4x^2 + 2x - 5 \div (x + 4)$.

Solution

$$\begin{array}{r}
3x^2 - 16x + 66 \\
x + 4 \overline{)\, 3x^3 - 4x^2 + 2x - 5} \\
3x^3 + 12x^2 \\
\hline
-16x^2 + 2x \\
-16x^2 - 64x \\
\hline
66x - 5 \\
66x + 264 \\
\hline
-269
\end{array}$$

Step ①: Divide $3x^3 \div x = 3x^2$.
Step ②: Multiply $3x^2(x + 4)$.
Step ③: Subtract. Write the term $2x$.
Step ④: Divide $-16x^2 \div x = -16x$.
Step ⑤: Multiply $-16x(x + 4)$.
Step ⑥: Subtract. Write the term -5.
Step ⑦: Divide $66x \div x = 66$.
Step ⑧: Multiply $66(x + 4)$.
Step ⑨: Subtract. This is the remainder.

Thus, $(3x^2 - 4x^2 + 2x - 5) \div (x + 4)$ is $3x^2 - 16x + 66$ with remainder -269.

Before dividing a polynomial by a binomial you must check that

- the terms of the polynomial and of the binomial are ordered in descending powers of the variable. Thus, $2x - 5 + x^2$ should be written as $x^2 + 2x - 5$.
- if any terms of the polynomial or binomial are missing then the missing terms must be inserted with coefficient 0. Thus, $x^3 + x - 7$ is written as $x^3 + 0x^2 + x - 7$.

2.4 Exercise

A 1 Arrange each polynomial in descending powers of x.

(a) $-2x + 4x^2 + 9$

(b) $11 + 2x^2 - x - 5x^3$

(c) $x^4 - 2x + 12x^2 + 5 + 7x^3$

(d) $x^2 + 1 - x - \dfrac{1}{x^2} + \dfrac{2}{x}$

2 Fill in any missing terms of each polynomial.

(a) $4p^2 - 5$　　(b) $3x^3 + x - 2$　(c) $y^5 + y + 7$　　(d) $4w^4 + 3w^3 - 75$

B 3 Use your factoring skills to divide.

(a) $\dfrac{q^2 - q - 6}{q + 2}$

(b) $\dfrac{2a^2 - 25a + 63}{a - 9}$

(c) $\dfrac{x^2 + 4xy + 3y^2}{x + y}$

(d) $\dfrac{2ac - ad + 2bc - bd}{a + b}$

(e) $\dfrac{b^2 - c^2}{b + c}$

(f) $\dfrac{18x^2 - 2y^2}{3x + y}$

(g) $\dfrac{2g^2h^2 - 5gh - 3}{gh - 3}$

(h) $\dfrac{m^2n^2 + 2mnp + p^2}{mn + p}$

(i) $\dfrac{6v^2 + 17vw - 14w^2}{3v - 2w}$

(j) $\dfrac{49p^2 + 56pq + 16q^2}{7p + 4q}$

(k) $\dfrac{x^4 - y^4}{x + y}$

(l) $\dfrac{6u^4 + u^2z - z^2}{2u^2 + z}$

4 Find each remainder.

(a) $(x^2 + 2x - 30) \div (x - 4)$　　(b) $(3y^2 - 19y - 8) \div (y - 7)$

(c) $(6p^2 + 17p - 7) \div (3p + 1)$　　(d) $(10m^2 - 69m + 92) \div (5m - 7)$

5 Find each quotient.

(a) $(12n^2 + 17n + 11) \div (3n + 2)$　　(b) $(a^2 - 12a + 21) \div (a - 8)$

(c) $(6d^2 - 2d - 58) \div (2d + 6)$　　(d) $(10k^2 + 23k - 11) \div (5k - 1)$

6 Divide.

(a) $(2x^2y^2 + xy - 10) \div (xy - 2)$　　(b) $(m^4 + 3m^2 - 2m + 5) \div (m + 3)$

(c) $(-8 + 18q^2 - 45q) \div (3q - 8)$　　(d) $(5 + 2s^2 - 11s) \div (s - 5)$

(e) $(1 + a - 3a^2 - 2a^3) \div (1 + 2a)$　　(f) $(d^4 + d^5) \div (1 + d^2)$

7 Divide.

(a) $(x^5 + 1) \div (x + 1)$　　(b) $(z^6 - 1) \div (z - 1)$

(c) $(y^4 + 1) \div (y^2 + 1)$　　(d) $(6q^3 + 18q^2 - 2q - 6) \div (3q^2 - 1)$

(e) $(a^5 - 3a^4 - 2a - 2) \div (a^4 - 1)$　　(f) $(d^6 + 2d^4 - 14d^2 + 5) \div (d^2 + 5)$

(g) $(m^7 + 3m^6 + 6) \div (m^5 + 2)$　　(h) $(w^6 - w^2 + 6) \div (w^2 + 2)$

8 (a) $8x^3 + 10x^2 - px - 5$ is divisible by $2x + 1$. There is no remainder. Find the value of p.

(b) When $x^6 + x^4 - 2x^2 + k$ is divided by $1 + x^2$ the remainder is 5. What is the value of k?

Synthetic Division

The technique of synthetic division is used to reduce the labour of writing the symbols each time. The method is illustrated as follows. Compare the actual division with synthetic division.

Actual Division

$$
\begin{array}{r}
2x^2 + 11x + 59 \\
x - 5 \overline{)\, 2x^3 + 1\,x^2 + 4x - 100} \\
2x^2 - 10\,x^2 \\
\hline
11x^2 + 4x \\
11x^2 - 55x \\
\hline
59x - 100 \\
59x - 295 \\
\hline
195
\end{array}
$$

Synthetic Division

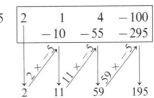

Compare how the corresponding numbers are obtained in each part.

9 The division of a polynomial is done using synthetic division.

$$
\begin{array}{r|rrrr}
1 & 1 & -4 & 1 & 9 \\
  & \downarrow & 1 & -5 & 6 \\
\hline
  & 1 & -5 & 6 & 3
\end{array}
$$

(a) What is the given polynomial?
(b) What is the quotient? remainder?

10 Divide. Use synthetic division.

(a) $(x^3 + 6x^2 + 11x + 6) \div (x + 2)$ (b) $(x^3 - 4x^2 + x + 6) \div (x - 2)$
(c) $(x^3 - 7x + 6) \div (x - 1)$ (d) $(x^3 - 9x - 4x^2 + 36) \div (x - 4)$

Restrictions on the variable
So far, you have assumed that, for division, the divisor is not equal to zero. When you write a rational expression there are certain restrictions on the variables. $\dfrac{a + b}{a}$ $\longleftarrow$ Why is $a \neq 0$? $\dfrac{y^2 + 4y + 4}{y - 3}$ $\longleftarrow$ Why is $y \neq 3$?

The denominator of each rational expression must be non-zero. Thus $a \neq 0$ and $y \neq 3$ for the above expressions.

11 For each expression • indicate the restrictions on the variables.
                         • simplify the rational expression.

(a) $\dfrac{x^2 + 15x + 50}{x + 5}$ (b) $\dfrac{m^2 - 9m - 90}{m - 15}$ (c) $\dfrac{x^2 - 4xy - 32y^2}{x - 8y}$

(d) $\dfrac{t^2 + 7t + 12}{t + 4}$ (e) $\dfrac{a^2 - 5ab - 24b^2}{a + 3b}$ (f) $\dfrac{x^2 - 2xy - 8y^2}{x + 2y}$

(g) $\dfrac{2x^2 - 5x - 3}{x - 3}$ (h) $\dfrac{5k^2 + 22k - 48}{k + 6}$ (i) $\dfrac{6m^2 + mn - 35n^2}{3m - 7n}$

(j) $\dfrac{6x^2 - 107x + 35}{3x - 1}$ (k) $\dfrac{2x^2 + 5xy - 12y^2}{2x - 3y}$ (l) $\dfrac{18x^2 + 9xy - 5y^2}{3x - y}$

2.5 Patterns and Processes: The Factor Theorem

When you study mathematics, certain patterns in your answers may suggest a method of developing other useful mathematical ideas.

For example, you notice that, for trinomials,

$$f(x) = x^2 - 5x + 4, \qquad f(1) = 0$$
$$f(x) = (x - 4)(x - 1), \qquad f(4) = 0.$$

It seems that if $f(1) = 0$, and $f(4) = 0$ then $x - 1$ and $x - 4$ are factors. With the above observation you test other examples.

$$f(x) = x^2 - x - 12 \qquad f(4) = 0$$
$$f(x) = (x - 4)(x + 3) \qquad f(-3) = 0 \qquad \text{True again}$$

The mathematician then asks whether the observation extends to cubic polynomials.

$$f(x) = x^3 - x^2 - 14x + 24$$

Step A: By trial and error $f(2) = 0$.
Step B: Test whether $x - 2$ is a factor.

$$
\begin{array}{r}
x^2 + x - 12 \\
x - 2 \,\overline{)\, x^3 - x^2 - 14x + 24} \\
\underline{x^3 - 2x^2} \\
x^2 - 14x \\
\underline{x^2 - 2x} \\
-12x + 24 \\
\underline{-12x + 24} \\
0
\end{array}
$$

Thus, $x^3 - x^2 - 14x + 24$
$= (x - 2)(x^2 + x - 12)$
$= (x - 2)(x + 4)(x - 3)$

Since the remainder is 0, $\longrightarrow$ then $x - 2$ is a factor.

Conclusion: it appears that the method does extend to cubic polynomials.

Indeed the above observation results in an important tool for factoring polynomials, namely, the factor theorem.

Factor Theorem

$x - a$ is a factor of $f(x)$ if and only if $f(a) = 0$.
That is, • if $f(a) = 0$, then $x - a$ is a factor.
• if $x - a$ is a factor, then $f(a) = 0$.

The above theorem can be used to do the following example.

Example 1 Factor $y^3 - 4y^2 + y + 6$.

Solution Test $y = 1$. $f(y) = y^3 - 4y^2 + y + 6$
$f(1) = (1)^3 - 4(1)^2 + (1) + 6$
$f(1) = 1 - 4 + 1 + 6$
$\quad = 4 \neq 0$

Think: Use the factor theorem. Test values of y.

Thus, $y - 1$ is not a factor.

Test $y = -1$. $f(-1) = (-1)^3 - 4(-1)^2 + (-1) + 6$
$f(-1) = -1 - 4 - 1 + 6 = 0$

Thus, $y + 1$ is a factor.

$y^3 - 4y^2 + y + 6$
$= (y + 1)(y^2 - 5y + 6)$
$= (y + 1)(y - 3)(y - 2)$

Divide to find the other factor.

$$\begin{array}{r} y^2 - 5y + 6 \\ y + 1 \overline{)\ y^3 - 4y^2 + \ y + 6} \\ \underline{y^3 + \ y^2} \\ -5y^2 + \ y \\ \underline{-5y^2 - 5y} \\ 6y + 6 \\ \underline{6y + 6} \\ 0 \end{array}$$

In doing the previous example, you may have wondered how many trials you need to do before you identify the factor. Again by using known examples, you can make observations when the coefficient of y^3 is 1.

$$y^3 - 4y^2 + y + 6 = (y + 1)(y - 3)(y - 2)$$

These numbers produce the product 6.

Thus the numbers you need to test are only all the factors of 6, namely $+1, -1, +2, -2, +3, -3, +6, -6$, at most 8 trials.

The factor theorem states that if $x - a$ is a factor then $f(a) = 0$. You may use this result to complete the next example.

Example 2 Find k if $x^4 + 2kx^3 - k^2x + 2$ is divisible by $x - 1$.

Think of the clue: Since $x - 1$ is a factor, then this suggests the factor theorem.

Solution *Step 1* $x - 1$ is a factor. Thus $f(1) = 0$.

Step 2 Substitute $x = 1$.
$f(x) = x^4 + 2kx^3 - k^2x + 2$,
$f(1) = (1)^4 + 2k(1)^3 - k^2(1) + 2$
$\quad = 3 + 2k - k^2$

Step 3 But $f(1) = 0$. Thus
$k^2 - 2k - 3 = 0$
$(k - 3)(k + 1) = 0$
$k = 3$ or $k = -1$

The mathematician then asks: *What if the polynomial is not divisible by a given factor?* For example, $f(x) = x^4 + 6x^3 - 9x + 2$ is divisible by $x - 1$ since $f(1) = 0$.

But $f(x) = x^4 + 6x^3 - 9x + 4$ is not divisible by $x - 1$ since $f(1) \neq 0$.
Note that $f(1) = (1)^4 + 6(1^3) - 9(1) + 4$
$$f(1) = 2$$

$$\begin{array}{r}
x^3 + 7x^2 + 7x - 2 \\
x - 1 \overline{\smash{)}\ x^4 + 6x^3 - 9x + 4} \\
\underline{x^4 - x^3} \\
7x^3 + 0 \\
\underline{7x^3 - 7x^2} \\
7x^2 - 9x \\
\underline{7x^2 - 7x} \\
-2x + 4 \\
\underline{-2x + 2} \\
2
\end{array}$$

Compare.

With the above result and other examples, the following theorem is suggested and can be proved.

Remainder Theorem

If $f(x)$ is divided by $x - n$, then the remainder is $f(n)$.

If $f(x)$ is divided by $ax - b$, then the remainder is $f\left(\dfrac{b}{a}\right)$.

If $f(m) = 0$, then, m is said to be a **zero** of $f(x)$.
For example, $f(x) = x^3 - 3x^2 + 4x - 4$
$$f(2) = (2)^3 - 3(2)^2 + 4(2) - 4$$
$$= 0$$
Thus, 2 is a zero of $f(x)$.

2.5 Exercise

A Review the meaning of the factor theorem.

1 Show that
(a) $x - 2$ is a factor $x^3 - 2x^2 + 3x - 6$.
(b) $y - 1$ is a factor of $y^3 - 2y + 3y^2 - 2$.
(c) $a - 2$ is a factor of $a^3 + 2a^2 - 16$.

2 Which of the following has $x - 2$ as a factor?
(a) $x^3 - 2x + 5x^2 - 24$ (b) $2x - x^2 + 12 + 3x^3$ (c) $2x^3 - 10 - 3x$

3 Find the factors of each polynomial.
(a) $3x^3 + 4x^2 - 5x - 2$ (b) $4x^3 + 8x^2 + x - 3$
(c) $2x^3 + 13x + 9x^2 + 6$ (d) $2x^4 + 3x^3 - x^2 - 3x - 1$

4 Find the remainder for each of the following.
  (a) $(a^3 + 3a^2 - 9a - 12) \div (a + 4)$ (b) $(4m^3 + 7m^2 - 3m - 20) \div (4m - 5)$

5 Remember: the constant term of a polynomial is related to the following product. $(x + a)(x + b)(x + c) = x^3 + \cdots + abc$

Use this observation to find the factors of each of the following.
  (a) $2x^3 + 15x^2 + 4x - 21$ (b) $6x^3 - 2x^2 - 12x + 4$
  (c) $2x^3 + 7x^2 - 9$ (d) $6x^3 + 5x^2 - 3x - 2$

B 6 (a) Why is $x - 1$ not a factor of $f(x) = 4x^3 + 8x^2 + x - 3$?
  (b) Find the factors of $f(x)$. $4 + 8 - 1 - 3$

7 (a) Use the factor theorem to determine a zero of $f(x) = x^2 - 8x + 7$.
  (b) What is another zero of $f(x)$?

8 Use the factor theorem to determine which of the three values $-2, 1, \frac{1}{2}$, is a zero of $g(x) = 2x^3 + 5x^2 + x - 2$.

9 Find a zero of each of the following.
  (a) $g(x) = x^2 + x - 12$ (b) $h(y) = y^3 + 5y^2 + 7y + 3$
  (c) $f(x) = x^3 + x^2 - 9x - 9$ (d) $g(x) = 2x^3 - 16$

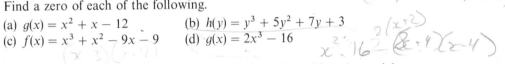

10 (a) The values 2, 3, and -2 are the only zeroes of a cubic polynomial, $g(x)$. Find $g(x)$.
  (b) A polynomial has the property that $f(-1) = 0$, $f(2) = 0$, and $f(-3) = 0$. Find $f(x)$.

11 $f(x) = kx^3 - 8x^2 - x + 3k + 1$ is divisible by $x - 2$.
  (a) Find the value of k.   (b) Find the other factors.

12 (a) Show that $x + a$ is a factor of $x^3 + a^3$.
  (b) Use the result in (a) to write the factors of $x^3 + y^3$ and $x^3 - y^3$.

13 (a) If $x + 1$ is a factor of $x^3 - 2x^2 + 5x + c$, find the value of c.
  (b) If $x - 2$ is a factor of $x^3 - 3x^2 + bx + 6$, find the value of b.

14 Find the value of k so that each polynomial is divisible by the binomial.
  (a) $x^3 + 2x^2 + kx + 6$; $x + 2$ (b) $x^3 + kx^2 - 5x - 7$; $x - 1$

C 15 Show that $x - m$ is a factor of $x^2(1 - m^2) + m^2(x^2 + 1) - (x^2 + m^2)$.

16 When a polynomial $f(x)$ is divided by $2x + 1$, the quotient is $x^2 - 3x - 1$ and the remainder is 4. What is $f(x)$?

Problem-Solving: What If . . .?

In your earlier work you developed skills for factoring a difference of squares. What if you are given a difference of cubes? Can you find a pattern for factoring the expression? You can use the factor theorem to develop factoring rules for cubes.

Difference of cubes

$$f(x) = x^3 - 27$$
$$f(3) = (3)^3 - 27$$
$$= 0$$

Sum of cubes

$$f(x) = x^3 + 27$$
$$f(-3) = (-3)^3 + 27$$
$$= 0$$

▶ Since $f(3) = 0$ then $x - 3$ is a factor of $x^3 - 27$.

▶ Since $f(-3) = 0$ then $x + 3$ is a factor of $x^3 + 27$.

▶ Divide to find the other factor.
$$x^3 - 27 = (x - 3)(x^2 + 3x + 9)$$

▶ Divide to find the other factor.
$$x^3 + 27 = (x + 3)(x^2 - 3x + 9)$$

Cannot be factored further

In general, the factors are written in the following form.

$$(x^3 - y^3) = (x - y)(x^2 + xy + y^2) \qquad (x^3 + y^3) = (x + y)(x^2 - xy + y^2)$$

The above results are known as the factors of a *sum or difference of cubes*. Note that a sum (or difference) of cubes of two numbers is divisible by the sum (or difference) of the numbers themselves.

17 Divide.
(a) $(-x^3 - 8) \div (x + 2)$ (b) $(-y^3 + 27) \div (y - 3)$

18 Show that $x + a$ is a factor of $x^3 + a^3$.

19 (a) Substitute $-y$ for x in $x^3 + y^3$. What result should you expect?
(b) Write the factors of $x^3 + y^3$.

20 Find the factors of each of the following.
(a) $y^3 + 1$ (b) $x^3 + 8$ (c) $y^3 - 1$ (d) $x^3 - 8$
(e) $8x^3 + 1$ (f) $8y^3 - 1$ (g) $x^3 + 64$ (h) $x^3 - a^3$
(i) $x^3 + y^3$ (j) $27x^3 + y^3$ (k) $1 - 27y^3$ (l) $\dfrac{y^3}{8} + 1$

21 Factor fully. Remember, look for common factors first.
(a) $a^3 + 64b^3$ (b) $2x^3 - 2$ (c) $2 - 54x^3$
(d) $y^4 + y$ (e) $x^6 - y^6$ (f) $27a^3 - 125$
(g) $216m^3 - 27$ (h) $1000m^3 - y^3$ (i) $a^3b^3 + \dfrac{1}{8}c^6d^6$

2.6 Solving Quadratic Equations by Factoring

Very often you can apply skills which you have already learned to new situations. For example, you can apply your skills in factoring to solve equations of the following form.

$$x^2 - 6x + 8 = 0 \qquad 2x^2 - 3x + 1 = 0 \qquad x^2 - 9 = 0$$

Each of these equations is of *degree* 2 and is called a quadratic equation.

Since you are already able to solve linear equations, try to transform a quadratic equation into equivalent linear equations.

To solve a quadratic equation you use the principle:

If $ab = 0$ then either $a = 0$ or $b = 0$.

Thus to solve quadratic equations, you need to first apply your factoring skills.

Example 1   Solve $x^2 - 4x + 3 = 0$.

Solution

$$x^2 - 4x + 3 = 0$$
$$(x - 1)(x - 3) = 0$$

Factor the trinomial

$$x - 1 = 0 \quad \text{or} \quad x - 3 = 0$$
$$x = 1 \qquad\qquad x = 3$$

Replace the quadratic equation by two other linear equivalent equations.

You must verify both roots in the original equation.

For $x = 1$	For $x = 3$
$RS = 0$	$RS = 0$
$LS = x^2 - 4x + 3$	$LS = x^2 - 4x + 3$
$\quad = (1)^2 - 4(1) + 3$	$\quad = (3)^2 - 4(3) + 3$
$\quad = 1 - 4 + 3 = 0$	$\quad = 9 - 12 + 3 = 0$
$LS = RS$	$LS = RS$

The roots of the quadratic equation are $\{1, 3\}$.

In order to solve for the variable, using the same principle, the quadratic equation in the following example needs to be rewritten so that one side of the equation is equal to zero.

Example 2   Find the solution set of $(3x - 1)(2x + 3) = -5$.

Solution $(3x - 1)(2x + 3) = -5$ ←——— You cannot use your factoring skills to solve an
$6x^2 + 7x - 3 = -5$ equation given in this form.
$6x^2 + 7x + 2 = 0$
$(3x + 2)(2x + 1) = 0$ ←——— Equate each factor to zero.

$3x + 2 = 0$ or $2x + 1 = 0$
$\quad 3x = -2$ $\quad 2x = -1$
$\quad\quad x = -\dfrac{2}{3}$ $\quad\quad x = -\dfrac{1}{2}$

The solution set is $\left\{ -\dfrac{2}{3}, -\dfrac{1}{2} \right\}$.   Check your answers in
the original equation.

To solve $3x^2 - 3x - 36 = 0$ first remove the common factor of 3.

$\quad\quad x^2 - x - 12 = 0$ Divide by 3.
$\quad\quad (x - 4)(x + 3) = 0$

$x - 4 = 0$ or $x + 3 = 0$
$\quad x = 4$ or $\quad\quad x = -3$ The roots are 4, -3.

In the preceding examples, you were given a quadratic equation and you found the roots. You can reverse the procedure. If you know what the roots of a quadratic equation are you can then write the equation. For example, if the roots are -3 and 2, then you may introduce a variable, say x, and write the following using a reverse procedure.

$$x = -3 \quad \text{or} \quad\quad x = 2$$
$$x + 3 = 0 \quad\quad\quad x - 2 = 0$$

$$\text{Then } (x + 3)(x - 2) = 0$$
$$x^2 + x - 6 = 0$$

A quadratic equation with roots -3 and 2 is $x^2 + x - 6 = 0$.

2.6 Exercise

A 1   What are the roots for each of the following equations?
(a) $x + 5 = 0$ (b) $(x - 2)(x + 5) = 0$ (c) $2y + 1 = 0$
(d) $(3y - 2)(2y + 1) = 0$ (e) $(m - 3)(3m + 2) = 0$ (f) $(2y - 1)(3y - 2) = 0$

2   Solve each of the following.
(a) $(x - 3)(x + 2) = 0$ (b) $(y + 5)(y - 5) = 0$ (c) $(3k - 2)(k + 3) = 0$
(d) $a(a - 5) = 0$ (e) $(m - 3)(2m + 1) = 0$ (f) $3y(y - 5) = 0$

3 Factor. Then find the roots.

(a) $x^2 + 8x + 15 = 0$ (b) $m^2 + 7m - 30 = 0$ (c) $m^2 - 25 = 0$

(d) $5x^2 + 17x + 6 = 0$ (e) $16m^2 - 1 = 0$ (f) $2y^2 + 9y + 4 = 0$

(g) $3y^2 + 10y + 3 = 0$ (h) $6 + 7y - 5y^2 = 0$

4 (a) What might be your first step in solving $y^2 - 7y = -12$?

(b) Solve the equation in (a).

5 Solve each of the following.

(a) $a^2 - 2a = 48$ (b) $3x^2 + 13x = 30$ (c) $3 = 6x^2 - 7x$

(d) $2x^2 - 7x + 6 = 0$ (e) $15 + x = 2x^2$ (f) $m^2 + 7m = 30$

6 (a) To simplify your calculations, what might be your first step in solving
$$x^2 + \frac{5}{2}x = \frac{33}{2}?$$
(b) Solve the equation in (a).

7 (a) Solve $9y^2 - 3y - 6 = 0$ (b) Verify your results in (a).

8 (a) Find the solution set for $x^2 - 4x = 21$.   (b) Verify your results in (a).

B Remember: to solve quadratic equations you use the principle: if $ab = 0$ then $a = 0$ or $b = 0$.

9 (a) What might be your first step in solving $4x^2 - 4x - 48 = 0$?

(b) Solve the equation in (a).

10 Solve. Watch for common factors.

(a) $2y^2 + 12y + 16 = 0$ (b) $3a^2 - 3a - 216 = 0$

(c) $6x^2 - 23x + 21 = 0$ (d) $2x^2 + 10x + 12 = 0$

(e) $15 + x - 2x^2 = 0$ (f) $6x^2 - 22x + 20 = 0$

11 Solve.

(a) $y^2 - y - 6 = 0$ (b) $x^2 - 2x - 15 = 0$

(c) $m^2 + 9m + 20 = 0$ (d) $a^2 - 13a + 42 = 0$

(e) $6x^2 - 28x - 10 = 0$ (f) $y^2 + 4y - 21 = 0$

(g) $5m^2 + 12m + 7 = 0$ (h) $25 - 16y^2 = 0$

12 Solve.

(a) $x^2 - 4x = 21$ (b) $y^2 + 7y = 18$ (c) $15 = 3m^2 + 4m$

(d) $9y = 2y^2 + 7$ (e) $42 = x^2 - x$ (f) $26x = 5x^2 + 24$

13 An important skill is deciding which skill is required to solve equations given in different forms. Find all roots of the following equations.

(a) $5m^2 - 4 = 19m$ (b) $3(x^2 - 5) = -4x$ (c) $x^2 - 16 = 0$

(d) $y^2 + 2y + 1 = 0$ (e) $(2y + 3)(4y + 1) = 18$ (f) $5(2y + 5) + y^2 = 0$

(g) $1 - 7x + 6x^2 = 0$ (h) $x^2 - 10 = \dfrac{x}{3}$ (i) $x^2 + 4x + 4 = 0$

(j) $\dfrac{m + 3}{2} = m^2$ (k) $x(6x + 11) = 10$ (l) $\dfrac{3}{2}(p + 2) = 9p^2$

14 Construct an equation in the form $ax^2 + bx + c = 0$ that has the following roots.

(a) $3, -2$ (b) $-4, -6$ (c) $3, \dfrac{1}{2}$ (d) $0, 4$ (e) $-\dfrac{1}{2}, \dfrac{1}{3}$

15 For each equation, one root is given.

• Find the value of p.   • Find the other root.

(a) $3x^2 + 5x + p = 0,\ -2$ (b) $2x^2 - 9x + p = 0,\ 5$

(c) $4x^2 - px + 6 = 0,\ 6$ (d) $px^2 + 24x - 5 = 0,\ \dfrac{1}{5}$

16 (a) How are these equations alike? different?

$$m^2 - 3m + 2 = 0$$
$$(x - 1)^2 - 3(x - 1) + 2 = 0$$

(b) Solve $(x - 1)^2 - 3(x - 1) + 2 = 0$.

C 17 Solve each of the following.   Hint: Substitute m for the expression in parentheses.

(a) $(x - 1)^2 - 9 = 0$ (b) $(2y + 1)^2 - 16 = 0$

(c) $(a + 3)^2 - 3(a + 3) - 4 = 0$

Problem Solving

A parabola is given by the equation $y = ax^2 + bx + c$.

What relationship must be true among the coefficients of a, b, and c for the graph of the parabola

• to touch the x-axis?
• to intersect the x-axis in two points?

Generalization: Using the Quadratic Formula

Factoring a trinomial, in order to obtain the roots of a quadratic equation, can be time-consuming. For this reason, a formula was developed to obtain directly the roots for given values of a, b, and c in the equation $ax^2 + bx + c = 0$.

Development of formula

Quadratic Formula

The roots of the quadratic equation $ax^2 + bx + c = 0$, $a \neq 0$ are given by the formula

$$x = \frac{-b \pm \sqrt{b^2 - 4ac}}{2a}.$$

For $2x^2 + 5x + 1 = 0$,
$a = 2$, $b = 5$, $c = 1$.

Then $x = \dfrac{-5 \pm \sqrt{25 - 4(2)(1)}}{2(2)}$

Thus, $x = \dfrac{-5 \pm \sqrt{17}}{4}$

$$ax^2 + bx + c = 0$$

$$x^2 + \frac{b}{a}x + \frac{c}{a} = 0$$

$$x^2 + \frac{b}{a}x + \frac{b^2}{4a^2} = \frac{b^2}{4a^2} - \frac{c}{a}$$

$$\left(x + \frac{b}{2a}\right)^2 = \frac{b^2 - 4ac}{4a^2}$$

$$x + \frac{b}{2a} = \pm\sqrt{\frac{b^2 - 4ac}{4a^2}}$$

$$x + \frac{b}{2a} = \pm\frac{\sqrt{b^2 - 4ac}}{2a}$$

$$x = -\frac{b}{2a} \pm \frac{\sqrt{b^2 - 4ac}}{2a}$$

$$x = \frac{-b \pm \sqrt{b^2 - 4ac}}{2a}$$

18. Use the formula to solve each of the following.
    (a) $x^2 + 2x - 35 = 0$ (b) $x^2 + 2x - 15 = 0$
    (c) $3x^2 - 10x + 3 = 0$ (d) $2x^2 + 5x - 3 = 0$
    Check each of the above answers by factoring.

19. Use the formula to find the roots of each quadratic equation.
    (a) $2x^2 - 4x - 1 = 0$ (b) $3x^2 - 6x - 5 = 0$
    (c) $0 = 2x^2 - 5x - 1$ (d) $3x^2 - 3x = 4$

20. Find the zeroes of the function, f, defined by $f : x \rightarrow 5x^2 - x - 3$.

21. Find the roots of each quadratic equation to 1 decimal place.
    (a) $3x^2 - 5x - 1 = 0$ (b) $4x^2 - 6x - 3 = 0$

22. Use the formula to find the roots of each of the following to 2 decimal places.
    (a) $2.1x^2 - 3.2x - 5.2 = 0$ (b) $1.8x^2 - 9.8x + 12.2 = 0$
    (c) $-1.2x^2 - 1.3x + 1.4 = 0$

2.7 Solving Problems: Quadratic Equations

In the study of mathematics, the following process is carried out over and over again.

| A | You learn the vocabulary, skills, and concepts in a particular topic of mathematics. |

$\Longrightarrow$

| B | You apply these skills and concepts to solving problems. |

The skills you have acquired in this chapter give you facility in working with quadratic equations. These new skills are now combined with your earlier skills for solving problems. The steps for solving problems remain the same, and help you organize your solution.

Steps for Solving Problems

Step A Read the problem carefully. Can you answer these two questions?
   I   What information am I asked to find (information I don't know)?
   II What information am I given (information I know)?
   Be sure to understand what it is you are to find, then introduce the variables.

Step B Translate from English to mathematics and write the quadratic equation.

Step C Solve the equation.

Step D Check the answers in the original problem.

Step E Write a final statement as the answer to the problem.

Example 1   Three consecutive integers are squared and the sum of the squares is 245. What are the integers?

Solution   Let the consecutive integers be represented by n, $n + 1$, and $n + 2$.

$$n^2 + (n + 1)^2 + (n + 2)^2 = 245 \longleftarrow \text{Think: These integers are to}$$
$$n^2 + n^2 + 2n + 1 + n^2 + 4n + 4 = 245 \quad \text{be squared and added together}$$
$$3n^2 + 6n + 5 = 245 \quad \text{to yield a sum of 245.}$$
$$3n^2 + 6n - 240 = 0 \longleftarrow \text{Recognize that the equation}$$
$$n^2 + 2n - 80 = 0 \quad \text{can be solved by factoring.}$$
$$(n + 10)(n - 8) = 0$$

$n + 10 = 0$ or $n - 8 = 0$
$\quad n = -10$ $\quad n = 8$

Use $n = -10$.   Use $n = 8$.
The numbers are $-10, -9, -8$.   The numbers are 8, 9, 10

Check for $n = -10$ Check for $n = 8$
$\quad (-8)^2 + (-9)^2 + (-10)^2$ $\quad 8^2 + 9^2 + 10^2$
$= 64 + 81 + 100$ $= 64 + 81 + 100$
$= 245$ checks $= 245$ checks

The numbers are -10, -9, and -8 or 8, 9 and 10.

In the previous example, both roots of the quadratic equation provide a solution.
In the next example an inadmissible root is obtained because time cannot be negative.

Example 2 Two cyclists, Kim and Jacob, are heading toward Edmonton. Kim travels 240 km from the east across the prairies. Jacob travels 180 km from the west through the mountains. Jacob's average speed is 6 km/h less than Kim's. Jacob takes 3 h longer than Kim to make the trip. How long does each travel?

Solution Let t represent the time (in hours) that Kim travels.
$(t + 3)$ h represents the time Jacob travels.

	distance	speed	time taken
Kim	240	$\dfrac{240}{t}$	t
Jacob	180	$\dfrac{180}{t + 3}$	$t + 3$

Use the relationship

$d = s \times t$, or $s = \dfrac{d}{t}$.

Kim's speed $-$ Jacob's speed $= 6$

$$\frac{240}{t} - \frac{180}{t + 3} = 6$$

$240(t + 3) - 180t = 6t(t + 3)$ ⟵ Simplify. Multiply by $t(t + 3)$.
$240t + 720 - 180t = 6t^2 + 18t$ Then $t \neq 0$ and $t + 3 \neq 0$
$-6t^2 + 42t + 720 = 0$
$t^2 - 7t - 120 = 0$
$(t - 15)(t + 8) = 0$

$t - 15 = 0$ or $t + 8 = 0$
$\qquad t = 15$ $\qquad\qquad t = -8$ ⟵

This root is inadmissible.
Time cannot have a negative value.

Kim travels for 15 h and Jacob for 18 h.

To solve problems effectively you must organize the steps of the solution and interpret the clues in the problem. You can organize data in charts or on diagrams depending on the nature of the problem.

2.7 Exercise

A   Express answers correct to one decimal place.

1   The product of two consecutive even whole numbers is 440. What are the numbers?

2   The sum of two numbers is 16, but the sum of their squares is 130. What are the numbers?

3   For each triangle, calculate the length, x.

(a)

(b)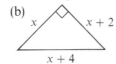

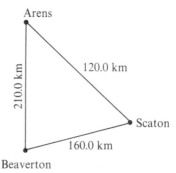

4   To travel from Arens to Beaverton you may use the super highway or connect through Scaton as shown. On the super highway you are able to travel 30.0 km/h faster and take 4 h less time. What are the average speeds for each route?

B   Remember to organize your solutions. Refer to the *Steps for Solving Problems*.

5   The perimeter of a right triangle is 36.0 cm. If the hypotenuse is 15.0 cm, find the length of the other two sides.

6   Find two consecutive whole numbers such that the sum of their squares is 265.

7   The height of the cover of a speaker cabinet exceeds twice its width by 10.0 cm. If the area of the cover is 2100.0 cm², find its dimensions.

8   Four consecutive even integers have the property that the square of the fourth is equal to the square of the sum of the first and third. Find the numbers.

9   The largest telescope in the world is on Mount Semirodriki, U.S.S.R. The diameter of its mirror is 6.0 m. The area of the mirror is equal to the cross sectional area of the metal casing surrounding the mirror. What is the width of the casing?

10 A square lawn is surrounded by a walk 1.5 m wide. If the area of the walk equals the area of the lawn, what are the dimensions of the lawn?

11 A rectangular ice surface is flooded by a machine starting at the centre. After 10% of the surface has been flooded (leaving a uniform strip around the outside) the flooding machine breaks down. If the rink is 25.0 m wide and 40 m long, find the width of the unflooded strip.

12 The *Trend Ahead* magazine prints its photographs in two square sizes. The larger photographs measure 3.0 cm longer than the smaller photographs. Three of the smaller photographs have a combined area 9.0 cm^2 more than the area of one of the larger size. Find the dimensions of each size of photograph.

13 On a 42.0 km go-kart course Alice goes 0.4 km/h faster than Bob, but Alice has trouble with her back wheels so she stops to fix them. This costs Alice 0.5 h and so she arrives 15 min after Bob. Determine the speeds of each and how long it took them to run the course.

14 A trucker drives a distance of 480.0 km. She leaves at the same time as another driver in a jeep. The jeep travels an average speed of 20.0 km/h faster than the truck. The jeep arrives at their common destination 2 h sooner. How fast is each vehicle going?

C 15 José purchased a number of tickets to a rock concert for $180, hoping to sell them for a profit. He kept two for himself and sold each of the others for $9 more. If he made a profit of $36 on the transaction, what was the original price of each ticket?

Computer Tip

The following computer program, written in the language of BASIC, calculates the values of the function $f(g(x))$, $x \in R$, where

$$f(x) = a_1x^2 + b_1x + c_1, \quad \text{and} \quad g(x) = a_2x^2 + b_2x + c_2.$$

```
10 INPUT A1, B1, C1, A2, B2, C2, K
20 LET G = A2 * K↑2 + B2 * K + C2
30 LET F = A1 * G↑2 + B1 * G + C1
40 PRINT "THE VALUE  IS", F
50 END
```

Use the program to obtain values to help you draw the graph of $y = f(g(x))$, where

$$f(x) = x^2 + 2x + 1 \qquad g(x) = x^2 - 4x + 3.$$

Problem-Solving: Using Computers

You often need to do complex operations in subjects such as business, scientific experiments, weather forecasting, etc. You can use a scientific calculator to do computations. However if you have many calculations of a similar nature to do, or if you have a pile of data to sort, you can program a computer to help you do the repetitive and tedious calculations and compilation of data. Although a computer can do many things, it cannot think or plan for itself. You must program it to do things, and the computer, having been programmed, will do exactly as it is told. When you prepare instructions for a computer in some computer language you are *writing a program*. There are different computer languages for different families of computers. The one you will use is called BASIC.

These symbols are used in computer language.

$+$ add	$*$ multiply	$\uparrow$ exponential	$\neq$ not equal to
$-$ subtract	$/$ divide	$>$ greater than	$<$ less than
$>=$ greater than or equal to		$<=$ less than or equal to	

The following program is written in BASIC. Each statement is numbered. The computer will do the steps involved in each statement in order. The statement numbers are not written consecutively in case we may add other steps later. These statements are then coded in computer language on cards, or typed at a terminal which tells the computer what to do. The following program, in BASIC, finds the value of the function $y = ax^2 + bx + c$.

Provide this information for the computer, called the INPUT. →

The computer will stop. All computer programs must have an end. →

```
10 PRINT "FUNCTION VALUES"
20 INPUT A,B,C
30 IF A = 0 THEN 90
40 FOR X = -10 TO 10
50 LET Y = A*X↑2 + B*X + C
60 PRINT "X = ", X
70 PRINT "Y = ", Y
80 NEXT X
90 GOTO 20
100 END
```

The computer will print the information between the quotation marks " ".

The computer provides the answers, called the OUTPUT.

The computer returns to line 20 to allow you to obtain more values. You type A = 0 and the program will end.

Only a brief introduction to computers and the computer language BASIC has been provided above.

1   Obtain more information about computers and the BASIC language from the many books written about them.

2   In subsequent pages you will find some programs written in BASIC. List the mathematics you need to know to write the computer program in the *Computer Tips* you locate.

2.8 Recognizing Equations in Quadratic Form

To solve a problem, you need to know when to apply a particular skill or strategy. For example, quadratic equations can be written in different forms. The skills you have practised for solving quadratic equations can be applied to equations that may not appear to be factorable.

In the following examples, you will make substitutions so your skills for quadratics can be used.

Example 1 Find the solution set for $x^4 - 13x^2 + 36 = 0$.

Solution Use the substitution $m = x^2$.

$$x^4 - 13x^2 + 36 = 0$$
$$(x^2)^2 - 13(x^2) + 36 = 0$$
$$(m)^2 - 13(m) + 36 = 0$$
$$(m - 9)(m - 4) = 0$$

$m - 9 = 0$ or $m - 4 = 0$
$m = 9$ $m = 4$

But $m = x^2$

$x^2 = 9$ $x^2 = 4$
$x = \pm 3$ $x = \pm 2$

In the next example, a substitution simplifies the equation, making it much easier to work with. Remember, you want the equation to look like a quadratic in general form, $ax^2 + bx + c = 0$.

Example 2 Find the solution set for the equation $(x^2 - x)^2 = 26(x^2 - x) - 120$

Solution Write the equation in general form. $(x^2 - x)^2 - 26(x^2 - x) + 120 = 0$.

Let $y = x^2 - x$ $y^2 - 26y + 120 = 0$
and substitute. $(y - 6)(y - 20) = 0$

$y - 6 = 0$ or $y - 20 = 0$
$y = 6$ $y = 20$

But $y = x^2 - x$.

$y = x^2 - x = 6$ or $y = x^2 - x = 20$
$x^2 - x - 6 = 0$ $x^2 - x - 20 = 0$
$(x - 3)(x + 2) = 0$ $(x + 4)(x - 5) = 0$

$x - 3 = 0$ or $x + 2 = 0$ $x + 4 = 0$ or $x - 5 = 0$
$x = 3$ $x = -2$ $x = -4$ $x = 5$

The solutions are $x = -4$, $x = -2$, $x = 3$, or $x = 5$.

In the next example, the equation is written in a simpler form by using the substitution,

$$t = x^2 + x.$$

Once the substitution is made, the equation is again rewritten as a quadratic equation in general form.

Example 3   What are the roots of the equation $x^2 + x + \dfrac{12}{x^2 + x} = 8$?

Solution   Let $t = x^2 + x$ Think: Use the substitution $t = x^2 + x$.
                                          Then the equation is rewritten.

$$x^2 + x + \frac{12}{x^2 + x} = 8$$

$$t + \frac{12}{t} = 8 \quad \longleftarrow \quad \text{Simplify by multiplying by } t, t \neq 0.$$

$$t^2 + 12 = 8t$$
$$t^2 - 8t + 12 = 0$$
$$(t - 2)(t - 6) = 0$$

$$t - 2 = 0 \quad \text{or} \quad t - 6 = 0$$
$$t = 2 \qquad \qquad t = 6$$

But $t = x^2 + x$.

$t = x^2 + x = 2$	$t = x^2 + x = 6$
$x^2 + x - 2 = 0$	$x^2 + x - 6 = 0$
$(x + 2)(x - 1) = 0$	$(x + 3)(x - 2) = 0$
$x + 2 = 0 \quad \text{or} \quad x - 1 = 0$	$x + 3 = 0 \quad \text{or} \quad x - 2 = 0$
$x = -2 \qquad \quad x = 1$	$x = -3 \qquad \quad x = 2$

The roots are $x = -3, x = -2, x = 1$ or $x = 2.$ ⟵ Be sure to check the values in the original equation.

2.8   Exercise

A 1   (a) Use the substitution $m = x^2$ to rewrite the equation $x^4 - 17x^2 + 16 = 0$.
(b) Solve the new equation in part (a).

2   (a) Find the solution set of the equation, $m^4 - 26m^2 + 25 = 0$.
(b) Verify your roots in part (a).

3   (a) Use the substitution $k = x^2 + x$ to write the equation
$(x^2 + x)^2 - 8(x^2 + x) + 12 = 0$ in general form.
(b) Solve the equation in part (a).

4 (a) Use the substitution $k = x^2 + x$. Find the solution set of
$$(x^2 + x) + \frac{24}{x^2 + x} = 14.$$
(b) Verify the roots in part (a).

5 Rewrite each in a general quadratic form and then solve.
(a) $m^4 - 5m^2 + 4 = 0$ (b) $2^{2x} - 12(2^x) + 32 = 0$
(c) $y^4 - 29y^2 + 100 = 0$ (d) $x^4 + 64 = 20x^2$
(e) $5^{2x} + 5 = 6(5^x)$ (f) $5x^2 = x^4 + 6$ (g) $x^2(x^2 - 1) + 9 = 9x^2$
(h) $30(3^x) - 81 = 3^{2x}$ (i) $x^2 + \frac{36}{x^2} = 13$ (j) $x^2 + \frac{4}{x^2} = 5$

B 6 (a) Solve the equation $m^2 - 10m + 16 = 0$.
(b) Use the solution in part (a) to solve $(2^x)^2 - 10(2^x) + 16 = 0$.

7 (a) Find the solution set for $(3^x)^2 - 30(3^x) + 81 = 0$.
(b) Verify your roots in part (a).

8 Solve each of the following. Verify the roots.
(a) $(x^2 + 2x)^2 - 11(x^2 + 2x) + 24 = 0$
(b) $(x^2 - 3x)^2 - 2(x^2 - 3x) - 8 = 0$

9 Find the solution set of each of the following.
(a) $\left(x + \frac{6}{x}\right)^2 - 2\left(x + \frac{6}{x}\right) - 35 = 0$ (b) $\left(x + \frac{4}{x}\right)^2 - 9\left(x + \frac{4}{x}\right) + 20 = 0$

10 Find the roots of each of the following.
(a) $2^{2x} - 6(2^x) + 8 = 0$ (b) $2^{2x} = 10(2^x) - 16$
(c) $3^{2x} + 27 = 12(3^x)$ (d) $4(4^{2x-1} + 1) = 5(4^x)$

11 Solve.
(a) $x^4 - 16x^2 + 60 = 0$ (b) $x^4 - 36x^2 + 35 = 0$
(c) $(x^2 - x)^2 - 18(x^2 - x) + 72 = 0$ (d) $5^{2x} - 30(5^x) + 125 = 0$
(e) $x^2 - 2x + \frac{1}{x^2 - 2x} = 2$ (f) $2^{2x} - 18(2^x) + 32 = 0$
(g) $(x^2 - 2x)^2 - 2(x^2 - 2x) - 3 = 0$ (h) $3(3^{2x}) - 10(3^x) + 3 = 0$

C 12 (a) Solve $\left(x^2 + \frac{1}{x^2}\right) + \left(x + \frac{1}{x}\right) = 10.$
(b) Verify your root in part (a).

Language of Mathematics

To do mathematics you must carefully translate from words to symbols. Each of the following results in an equation that needs to be written in quadratic form.

13 If $(3y^2 - y)^2$ equals the difference between $6(3y^2 - y)$ and 8, find y.

14 If $11(2m^2 + m)$ exceeds $(2m^2 + m)^2$ by 10, find m.

15 The square of $2x^2 + 3x$ exceeds $4(2x^2 + 3x)$ by 5. Find the values of x.

16 For what values of x does $7\left(x + \dfrac{2}{x}\right)$ exceed $\left(x + \dfrac{2}{x}\right)^2$ by 12?

17 For what values of m does the product of 7 and $m^2 + 1$ exceed the square of $m^2 + 1$ by 10?

Computer Tip

Step 1 To find the root of an equation such as $2x^2 - 9.42x + 6.622 = 0$, you can sketch the function given by $y = 2x^2 - 9.42x + 6.622$, and notice that $f(0) > 0$ and $f(2) < 0$. Thus a root x_1 occurs $0 < x_1 < 2$. As well, $f(3) < 0$ and $f(4) > 0$. Thus a root occurs $3 < x_2 < 4$. ⟵———— Remember — Use the nested form to help you calculate the value of polynomials. Refer to the skills on page 38.

Step 2 With the information in step 1, you can use the following computer program to find the values of x between

```
10 PRINT "ROOT OF THE EQUATIONS"
20 FOR X=0 TO 2 STEP .1
30 LET Y=2*X↑2-9.42*X+6.622
40 PRINT X, Y
50 NEXT X
60 END
```

which x_1 and x_2 occurs. Run the program shown.

Step 3 In the computer run in Step 2, you learned from the printout that the root x_1 lies between 0.9 and 0.8. Modify the previous computer program. Use the line

20 FOR X = .8 to .9 STEP .01

Find the value of the root x_1.

A Repeat a similar procedure to find the root x_2 above.

B Find the roots of the following equations to 2 decimal places.
   (a) $x^2 - 1.06x - 18.95 = 0$ (b) $3.68x^2 + 35.32x - 9.87 = 0$
   (c) $0.65x^2 - 11.23x + 48.15 = 0$ (d) $3x^2 + 18.39x + 26.45 = 0$

C Extend the program to find the roots to 2 decimal places of
   (a) $x^3 - 2.4x^2 - 10.6x + 23.4 = 0$
   (b) $7.36x^3 - 20.97x^2 - 24.03x + 15.12 = 0$

2.9 Solving Polynomial Equations

First you learned to factor trinomials. Then you used this skill to solve quadratic equations. Similarly, you learned the Factor Theorem in order to factor polynomials. You then used the Factor Theorem to solve polynomial equations. Skills in mathematics are very often developed for some practical reason. For example, in scientific problems, the roots of polynomials often yield the solution to the problem.

Example 1 Find the roots of the equation $x^3 - 7x + 6 = 0$.

Solution Use $f(x) = x^3 - 7x + 6$.
Test $x = 1$, $f(1) = (1)^3 - 7(1) + 6$
$\qquad\qquad\qquad = 0$
Thus, $(x - 1)$ is a factor of $f(x)$.

$$x^3 - 7x + 6 = 0$$
$$(x - 1)(x^2 + x - 6) = 0 \longleftarrow \text{Divide the polynomial } f(x)$$
$$(x - 1)(x + 3)(x - 2) = 0 \qquad \text{to find the other factors.}$$

$x - 1 = 0$ or $x + 3 = 0$ or $x - 2 = 0$
$\quad x = 1 \qquad\qquad x = -3 \qquad\qquad x = 2$

Thus the roots of the cubic equation are 1, -3, and 2.

Often in the study of scientific problems, the roots of polynomial equations are not integral. One approach that can be used is a skill that you have used many times, namely, estimating roots from a graph. There is always **another strategy** for solving a problem.

> Remember: Use your calculator to help you find appropriate values in order to draw the graph of the equation.

Example 2 Estimate the roots of the polynomial equation $x^3 - 8x^2 + 16.25x - 6.25 = 0$.

Solution Draw the graph of the polynomial $f(x)$.
From the graph, the roots are estimated as 0.5, 2.5, and 5.

A calculator is used to check the roots.
The roots are $f(0.5) = 0$, $f(2.5) = 0$, $f(5) = 0$

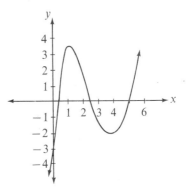

Often, sets of instructions can appear to be different, but are, in fact, the same. This is because each set of instructions requires, basically, the same skills to produce the results. For example, these instructions require the same algebraic skills to be used. However, your final statements in answer to the question, will vary.

- Solve . . .
- Find the solution set . . .
- Find the x-intercept . . .
- Find the root . . .
- Find the zero . . .

2.9 Exercise

A 1   What are the roots of each equation?
(a) $(x - 3)(x - 4) = 0$ (b) $(x + 5)(x - 1)(x - 7) = 0$
(c) $(x + 3)(x - 5)(2x + 5)(3x - 1) = 0$ (d) $(x^2 - 9)(x^2 - 4) = 0$

2   (a) If $f(2) = 0$, find a root of $x^3 - 3x^2 - 4x + 12 = 0$.
(b) Find the other roots of the polynomial in (a).

3   If $x(x - 1)(x - 2) = 0$, then you can write

$$x = 0 \quad \text{or} \quad x - 1 = 0 \quad \text{or} \quad x - 2 = 0.$$

Find the roots of each of the following.
(a) $(m - 1)(m - 2)(m - 3) = 0$ (b) $p(p - 1)(p + 3) = 0$
(c) $x^3 + 5x^2 + 6x = 0$ (d) $2y^3 + 7y^2 - 15y = 0$
(e) $15y = 3y^3 + 4y^2$ (f) $m^3 - 4m^2 = 21m$

4   (a) Solve and verify $x^4 - 10x^2 + 9 = 0$.
(b) Find the solution set for $x^4 - 13x^2 + 36 = 0$.

B 5   (a) Find the factors of $x^3 + 2x^2 - 9x - 18$.
(b) Solve the cubic equation $x^3 + 2x^2 - 9x - 18 = 0$.

6   (a) Show that 3 is a root of the equation $x^3 - 2x^2 - 5x + 6 = 0$.
(b) Find the other roots.

7   Solve and verify $2x^3 - 5x^2 + x + 2 = 0$.

8   (a) What is the first step in solving $x(1 - 4x^2) + 12x^2 = 3$?
(b) Solve the equation in (a).

9 Solve each equation.
  (a) $x^3 - 2x^2 - x + 2 = 0$ (b) $x^3 - x = 2(1 - x^2)$
  (c) $3(x^2 - 4) = 4x - x^3$ (d) $5x^2 - x^3 - 2 - 2x = 0$
  (e) $2(x^3 - 1) = x^2 + 5x$

10 Find the zeroes of each of the following polynomials.
  (a) $x^3 - 7x + 6 = 0$ (b) $x^3 - 5x^2 - x + 5 = 0$

11 What are the x-intercepts of each function?
  (a) $y = x^3 + 2x^2 - 9x - 18$ (b) $y = x^3 - 2x^2 - 5x + 6$

12 Solve and verify.
  (a) $2x^4 - x^3 - 14x^2 - 5x + 6 = 0$
  (b) $x^5 + 3x^4 - 5x^3 - 15x^2 + 4x = -12$

> Remember: Use your calculator skills to help
> you draw the graphs in the following questions.

13 (a) Draw a graph of $y = x^3 - 5x + 2$.
   (b) Use the graph in (a) to find the real roots of $x^3 - 5x + 2 = 0$.

14 What are the real roots of each of the following? Use a graph.
  (a) $x^3 - 4x^2 - 9x + 36 = 0$ (b) $2x^3 - x^2 - 3x + 2 = 0$

15 For each of the following,
  • draw a graph.   • estimate the real zeroes of the function.
  (a) $y = x^3 - 3x + 1$ (b) $y = x^3 - 2x^2 + x - 5$

16 Estimate the real roots of each of the following.
  (a) $x^3 + 2x - 5 = 0$ (b) $x^3 - 3x^2 - 2x - 2 = 0$

C 17 (a) Draw the graph of $y = x^4 - 2x^3 - 13x^2 - 14x + 24$.
   (b) From the graph, what are the real roots of
     $x^4 - 2x^3 - 13x^2 - 14x + 24 = 0$?

Computer Tip

You have learned various strategies for solving equations. On the
next page a particular method is used to solve an equation
using a computer. In various other computer tips, throughout
the book, other suggestions are proved to solve equations using
computer tips. Make a list of these computer tips.

Computer Techniques for Finding Roots

Of course if the problem solver wants a better estimate of the roots of an equation, a graph has limitations. Thus, a computer program is designed to solve the various polynomial equations, and generate their roots.

18 The following computer program, written in the language of BASIC, calculates the roots of the quadratic equation $ax^2 + bx + c = 0$.

To write the BASIC program you need to know your mathematics. Namely, the roots of the equation are given by the formula $x = \dfrac{-b \pm \sqrt{b^2 - 4ac}}{2a}$.

```
10  INPUT A, B, C
20  LET D = B ↑ 2 − 4 * A * C
30  IF D < 0 then 80
40  LET X1 = (−B + SQR(D))/(2 * A)
50  LET X2 = (−B − SQR(D))/(2 * A)
60  PRINT "THE ROOTS ARE", X1, X2
70  GOTO 90
80  PRINT "THE ROOTS ARE NOT REAL"
90  END
```

Use the program to find the roots of the following quadratic equations.

(a) $6x^2 + 17x + 5 = 0$ (b) $-x^2 + 4x = 0$ (c) $\dfrac{1}{2}x^2 - \dfrac{5}{6}x + \dfrac{1}{3} = 0$

(d) $4 - 3x - x^2 = 0$ (e) $2.1x^2 - 3.6x - 1.5 = 0$ (f) $6.2x^2 + 8.3x - 4.5 = 0$

19 The following computer program, written in BASIC, applies the factor theorem to find the integral zeroes of

$$f(x) = ax^3 + bx^2 + cx + d \quad \text{for } -10 \le x \le 10,$$

```
10  INPUT A, B, C, D
20  FOR X = −10 TO 10
30  LET Y = A * X ↑ 3 + B * X ↑ 2 + C * X + D
40  IF Y = 0 THEN 60
50  GOTO 70
60  PRINT "ONE ROOT IS X = ", X
70  NEXT X
80  END
```

Use the program to find the roots of the following polynomial equations.

(a) $x^3 + 4x^2 - 17x - 60 = 0$ (b) $2x^3 + 5x^2 - 4x = 3$

20 How would you modify the BASIC program in the previous question to find the roots of the following? Solve.

(a) $x^4 - 10x^3 + 35x^2 - 50x + 24 = 0$

(b) $24x^5 - 26x^4 - 15x^3 + 25x^2 - 9x + 1 = 0$

Practice and Problems: A Chapter Review

An important step for problem-solving is to decide which skills to use. For this reason, these questions are not placed in any special order. When you have finished the review, you might try the *Test for Practice* that follows.

1 Factor each of the following.
   (a) $rt - rv + 2sv - 2st$ (b) $x^2y^2 + 2x^2 + y^2 + 2$
   (c) $y^2 + 14y + 48$ (d) $7k^2 - 29k - 30$ (e) $4m^2 + 12m + 9$
   (f) $4x^2 - 1$ (g) $4p^4 - 37p^2 + 9$ (h) $32a^4 - 162b^4$

2 Solve.
   (a) $2m^2 + 19m + 35 = 0$ (b) $2y^2 - 10y - 28 = 0$
   (c) $(m + 1)^2 + 11(m + 1) + 28 = 0$

3 (a) Use the formula to find the roots of $2x^2 - 3x - 1 = 0$.
   (b) Check your answers in (a) by completing the square.

4 (a) The zeroes of a polynomial, $f(x)$, are -1, 1, and 2.
       What is the polynomial $f(x)$?
   (b) The zeroes of a polynomial are -3, 1, and 4. Write an expression for the polynomial.

5 (a) Solve $(2^x)^2 - 18(2^x) + 32 = 0$.
   (b) Verify your roots in part (a).

6 When $ax^3 - x^2 - x + b$ is divided by $x - 1$, the remainder is 6. When it is divided by $x + 2$ the remainder is 9. What are a and b?

7 A bag of concrete will make a concrete slab 1.55 m². To build a set of steps, three square slabs are needed with a combined area of 1.55 m². If each slab is 20.0 cm longer than the one above it, what are the dimensions of each slab?

8 After various tests, it is found that the safe stopping distance, d (in metres) for a boat in calm water travelling at v km/h is given by
$$d = 0.002(2v^2 + 10v + 3000).$$
   (a) What is the safe stopping distance of a boat in calm water travelling at 12.0 km/h?
   (b) Use the equation to determine the speed at which the boat is travelling to take 15.0 m to stop safely.

Test for Practice

Try this test. Each *Test for Practice* is based on the mathematics you have learned in this chapter. Try this test later in the year as a review. Keep a record of those questions that you were not successful with, get help in obtaining solutions and review them periodically.

1 Factor each of the following.
   (a) $ax + by + bx + ay$
   (b) $y^2 - 6y - 27$
   (c) $6p^2 - 5p - 6$
   (d) $45a^5 - 37a^4b - 6a^3b$
   (e) $a^2 - 2ab + b^2 - 9c^2$
   (f) $x^4 - y^4$

2 Simplify. (a) $\dfrac{m^2 - 10m + 21}{m - 3}$ (b) $\dfrac{k^4 + k^5}{k + 1}$

3 Divide.
   (a) $(m^5 - 3m^4 - 2m - 2) \div (m^4 - 1)$ (b) $(y^3 - 7y + 6) \div (y - 1)$

4 Without dividing, show that the factors of $6x^3 + 7x^2 - 9x + 2$ are $(2x - 1)$, $(3x - 1)$ and $(x + 2)$.

5 Solve. (a) $2x^2 + 7x - 15 = 0$ (b) $4y^2 + \dfrac{11}{3}y = 5$

6 (a) Find the roots of the equation, $k^4 - 5k^2 + 4 = 0$.
   (b) Verify your roots in part (a).

7 Solve.
   (a) $2^{2x} - 12(2^x) = -32$ (b) $\left(x - \dfrac{1}{x}\right)^2 - 7\left(x - \dfrac{1}{x}\right) + 12 = 0$

8 Solve and verify.
   (a) $x^3 + 5x^2 + 7x + 3 = 0$ (b) $y(1 - 4y^2) + 12y^2 = 3$

9 Three consecutive odd integers have the property that the square of the third is less than the square of the sum of the first two by 351. Find the numbers.

Math Tip

It is important to understand clearly the vocabulary of mathematics when solving problems about quadratic equations. Bring your list of new words up to date. Provide an example to illustrate the meaning of each word. Continue to add to your vocabulary list.

3 Language in Mathematics: Absolute Value and Radicals

absolute value, skills and equations, properties of exponents, vocabulary of radicals, operations with radicals, solving radical equations, strategies, solving problems, applications, problem-solving

Introduction

Symbols, like any other human invention, are introduced to serve specific needs. For example, the invention and acceptance of hieroglyphics, alphabets and other systems of written language came about only after thousands of years of human thought and non-written language. As the need to communicate became more complex and sophisticated, written language became standardized, accepted and convenient.

| 1 | 2 | 3 | 4 | 5 | 6 | 7 | 8 | 9 | 0 |

The numbers you use in your everyday work have been transformed over the years to the form as you see them today. The last numeral to be invented was the numeral 0. This was because it was not important in Ancient Times to record a number for something you didn't have.

Symbols are used in mathematics as a shorthand way of showing ideas. It has taken many centuries to produce the symbols you have accepted as part of your study in mathematics. For example, although the concept of radical numbers was known to the Ancient Greeks, it was not until late in the fifteenth century that the symbol, $\sqrt{}$, as you know it today, appeared, as a convenient method of showing a radical number.

Imagine how difficult it would be to talk about radical numbers without the mathematical symbol $\sqrt{}$.
Imagine how difficult (or impossible) it would be to develop or use any mathematical ideas without mathematical symbols. The inventors of some of the more common symbols are shown on this page

Mathematics is a type of shorthand. It uses symbols to describe and use "big" ideas simply, precisely, and in ways that anyone who knows the language of mathematics can understand.

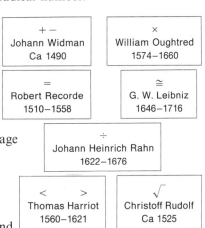

+ −
Johann Widman
Ca 1490

×
William Oughtred
1574–1660

=
Robert Recorde
1510–1558

≅
G. W. Leibniz
1646–1716

÷
Johann Heinrich Rahn
1622–1676

< >
Thomas Harriot
1560–1621

√
Christoff Rudolf
Ca 1525

3.1 Working with Absolute Value

The first successful crossing of the Atlantic Ocean in a hot air balloon depended greatly on the wind currents.

The magnitude, or size of the wind that pushed the balloon shown in the diagram below is the same, but the directions are different.

Represent the speed by 15 km/h.

15 km/h east

Represent the speed by −15 km/h.

15 km/h west

If the direction of the wind is not important, then you can use these absolute value symbols | | to show the size or magnitude of the wind.

$$|+15| = 15 \qquad |-15| = 15$$

This is read as "the absolute value of −15"

The absolute value is carefully defined for any real number x.

If $x > 0$ If $x = 0$ If $x < 0$
$|x| = x$ $|x| = 0$ $|x| = -x$

Example 1   Evaluate $|8 - 3 - 5 - 6 + 2|$.

Solution $|8 - 3 - 5 - 6 + 2| = |-4| = 4$.

The absolute value symbols are used in working with the principal (positive) square root. Since $(-2x)^2 = 4x^2$ and $(2x)^2 = 4x^2$, then $-2x$ and $2x$ are both square roots of $4x^2$. Since the principal (or positive) square root of $4x^2$ must be positive, then you use absolute value symbols to write

$$\sqrt{4x^2} = |2x| \qquad$$ Write this as $|2x| = 2|x|$. Why?

Remember: $\sqrt{4x^2}$, written in this form, is called an entire radical.

Example 2   Simplify $\sqrt{36x^4y^2}$

└──────────── Remember: $\sqrt{}$ is called the radical sign.

Solution $\sqrt{36x^4y^2} = 6|x^2y|$ Why is $|x^2| = x^2$?

$\phantom{\sqrt{36x^4y^2}} = 6x^2|y|$

By working these examples, you can see that

$$|(-2)(-3)| = |(-2)|\,|(-3)|$$
$$|(-6)(7)| = |(-6)|\,|(7)|$$

Numerical examples such as those above suggest that you can write the following general statement.

If $a, b \in R$, then $|ab| = |a|\,|b|$.

To prove something is true in mathematics may be difficult, but often it is much easier to prove something is not true.

For example, choose examples to test whether the following is likely to be true.

If $a, b \in R$, then $|a| + |b| = |a + b|$.

The following example indicates that the above *is likely to be* true.

$$|4| + |8| = |4 + 8|$$
$$|-4| + |-8| = |-4 - 8|$$

However, this example $|-4 + 8| \neq |-4| + |8|$

shows us that the above is *not* true for *all* $a, b \in R$.

3.1 Exercise

A   Review the meaning of absolute value.

1   Find the value of each of the following.
   (a) $|-3|$
   (b) $2|-3|$
   (c) $4|-9|$
   (d) $-2|-8|$
   └ What operation is understood here?

   (e) $-|16|$
   (f) $|-4| + 2|-6|$
   (g) $|-8| - |-14|$
   (h) $-|8| + 3|-3|$
   (i) $-|-6| - |-3| - 3$
   (j) $-|-6| + 2|-2| - |3| - |-6| - 3|-3|$
   (k) $-|-3| + 2|3| - 6|-5| - |-6| + 2|0|$
   (l) $3|-4| - 3|-1| + 5|-3| + 2|4|$

$- 3 + 6 - 6 + 5 - 6 + 0 - 6 - 3 - 3$

2   (a) Simplify $-(3 - 2 - 5 - 6 + 3)$.
    (b) Simplify $-|3 - 2 - 5 - 6 + 3|$.
    (c) Why do your answers in (a) and (b) differ?

3   (a) Simplify $-(-6 - 3 + 8 - 2 + 14)$.
    (b) Simplify $-|-6 - 3 + 8 - 2 + 14|$.
    (c) Why in this case, are your answers in (a) and (b) the same?

4   Simplify.
    (a) $3|-6 + 2| - 3|8 - 2| - 3|-2|$ (b) $-2|8 - 3| + 3|-6 + 2| + |3 - 8|$
    (c) $|4 - 9| + 4|6 - 3| - 8|2 - 5|$ (d) $8|8 - 8| + 3|9 - 9| - 3|6 - 6|$

5   If $x = -3$, $y = 6$, find the value of
    (a) $|2xy|$ (b) $2|xy|$
    (c) What do you notice about your answers in (a) and (b)?

6   If $x = -3$, $y = -6$, find the value of
    (a) $|-3x^2y|$ (b) $x^2|-3y|$
    (c) Why are the answers in (a) and (b) equal?

B   Remember: $\sqrt{32}$ is called an entire radical.
           $4\sqrt{2}$ is called a mixed radical.

7   If $a = -4$, $b = 2$, $c = -5$, find the value of each of the following.
    (a) $|a| - |b|$ (b) $|a - b|$ (c) $|a| + |b|$ (d) $|a + b|$
    (e) $2|a| - b$ (f) $3a - |b|$ (g) $|a + b + c|$ (h) $|a + b| + |c|$
    (i) $|2a + 3b|$ (j) $|2a| + |3b|$ (k) $a|b + c|$ (l) $|ab| + |3c|$
    (m) $|a^2b| - 3|a^2b^2|$ (n) $a^2|b| - 3a^2b^2$
    (o) $2a^2|b| - 4|ab|$ (p) $|-2a^2b - 4ab|$

8   Simplify.
    (a) $\sqrt{36y^2}$ (b) $\sqrt{36y^4}$ (c) $\sqrt{36y^6}$ (d) $\sqrt{25x^2y^2}$
    (e) $\sqrt{25x^4y^3}$ (f) $\sqrt{25x^4y^4}$ (g) $\sqrt{36x^5y^2}$ (h) $\sqrt{100x^2y^6}$
    (i) $\sqrt{81x^5y^4}$ (j) $\sqrt{49x^2y^5}$ (k) $\sqrt{16x^7y^5}$ (l) $\sqrt{144x^3y^5}$

9   Simplify. Express each radical in simplest form.
    (a) $\sqrt{25y^2} + \sqrt{16y^2}$ (b) $\sqrt{25y^4} + \sqrt{16y^4}$ (c) $\sqrt{16x^4 + 9x^4}$
    (d) $\sqrt{16x^2 + 9x^2}$ (e) $\sqrt{25x^4} - \sqrt{9x^4}$ (f) $3\sqrt{100y^2} - 2\sqrt{25y^2}$
    (g) $4\sqrt{64x^3} - 3\sqrt{16x^3}$ (h) $4\sqrt{25x^2y^3} - 3\sqrt{16x^2y^3}$

10   Write each radical as an entire radical.

   (a) $a\sqrt{a}$ (b) $y\sqrt{xy}$ (c) $ab\sqrt{a}$ (d) $a^2b\sqrt{b}$ (e) $xy^3\sqrt{3x}$

11   (a) Simplify $\sqrt{16x^2} + \sqrt{9x^2}$.
   (b) Simplify $\sqrt{16x^2 + 9x^2}$.
   (c) Is $\sqrt{16x^2} + \sqrt{9x^2} = \sqrt{16x^2 + 9x^2}$?

12   (a) Evaluate $|-2||-6|$, $|-2||6|$, $|2||6|$.
   (b) Evaluate $|(-2)(-6)|$, $|(-2)(6)|$, $|(2)(6)|$.
   (c) What do you notice about your answers in (a) and (b)?
   (d) Why are we likely to write $|a||b| = |ab|$ for all $a, b \in R$?

13   (a) Evaluate $|-6| \div |-3|$, $|-6| \div |3|$, $|6| \div |3|$.
   (b) Evaluate $|(-6) \div (-3)|$, $|(-6) \div 3|$, $|6 \div 3|$.
   (c) What do you notice about your answers in (a) and (b)?
   (d) Why are we likely to write $\left|\dfrac{a}{b}\right| = \dfrac{|a|}{|b|}$ for all $a, b \in R$?

14   (a) Evaluate $|5 + 4|$, $|5 - 4|$, $|-5 - 4|$.
   (b) Evaluate $|5| + |4|$, $|5| + |-4|$, $|-5| + |-4|$.
   (c) Based on (a) and (b), is it likely to be true that for all $a, b \in R$
      $|a + b| = |a| + |b|$? Why?

C 15   Use different values for a and b. Test which of the following are likely to
   be true.

   (a) $|a + b| = a + b$ (b) $|a - b| = a - b$ (c) $|a + b| > 2|a|$
   (d) $|a - b| > 2|b|$ (e) $|a| + |b| \geq |a + b|$ (f) $|a| - |b| = |a - b|$

Math Tip

Remember: The absolute value of a real number, r, is defined as

	Example
$|r| = r$ if $r > 0$	$|5| = 5$
$|r| = 0$ if $r = 0$	$|0| = 0$
$|r| = -r$ if $r < 0$	$|-5| = -(-5) = 5$

Read this as:
The absolute value of -5 is 5.

3.2 Applying Process: Absolute Value Equations

To solve equations using absolute value you must understand the meaning of the symbol $|x|$.

If $x > 0$ then $|x| = x$

If $x = 0$ then $|x| = x$

If $x < 0$ then $|x| = -x$ Since x is negative then $-x$ is positive.

To solve $|x| = 8$ note that $|8| = 8$ and $|-8| = 8$. Then the solution set for $\{x \mid |x| = 8, x \in R\} = \{8, -8\}$.

In developing mathematics or applying skills, the following strategy is used. Steps A, B, and C are applied to develop a strategy for solving equations involving absolute value.

Step A		Step B		Step C
Examine a straight-forward example. Apply your understanding of the skill to develop the solution.	$\Rightarrow$	Look for a pattern to apply to other examples. Develop a strategy.	$\Rightarrow$	Apply your strategy to more advanced problems. Ask yourself: Do I understand what I am doing?

To solve equations involving absolute value, you can apply the definition of $|x|$ and the principle shown in the following two cases.

Either   **Case 1**   for all $x \geq 0$ or   **Case 2**   for all $x < 0$

If the absolute value equation involves an expression such as $|x + 3|$, then you apply these two cases.

Either   **Case 1**   for all $x + 3 \geq 0$ or   **Case 2**   for all $x + 3 < 0$

Solve $|x + 3| = 2$.

Case 1

For all $x + 3 \geq 0$,
$$|x + 3| = x + 3.$$
Thus solve $x + 3 = 2$
$$x = -1$$
Satisfies *Case 1*.

Case 2

For all $x + 3 < 0$,
$$|x + 3| = -(x + 3).$$
Thus solve $-(x + 3) = 2$
$$-x - 3 = 2$$
$$x = -5$$
Satisfies *Case 2*.

The solution set is $\{-1, -5\}$.

You will be able to identify a strategy as follows after you have completed other solutions such as the one above.

Strategy: To solve $|x + 3| = 2$ you need to solve
$$x + 3 = 2 \text{ or } -(x + 3) = 2.$$

This strategy is applied to more advanced equations involving absolute value.

Example   Solve and verify $2|x - 3| + |x - 3| = 10 - 2|x - 3|$, $x \in R$.

Solution

$$2|x - 3| + |x - 3| = 10 - 2|x - 3|$$
$$2|x - 3| + |x - 3| + 2|x - 3| = 10$$ First step: simplify the equation.
$$5|x - 3| = 10$$
$$|x - 3| = 2$$

Thus,
$$x - 3 = 2 \quad \text{or} \quad -(x - 3) = 2$$
$$x = 5 \qquad\qquad -x + 3 = 2$$
$$x = 1$$

Verification   For $x = 5$, LS $= 2|x - 3| + |x - 3|$ RS $= 10 - 2|x - 3|$
$$= 2|5 - 3| + |5 - 3| \qquad\qquad = 10 - 2|5 - 3|$$
$$= 4 + 2 = 6 \quad \text{Verifies} \checkmark \qquad = 10 - 4 = 6$$

For $x = 1$, LS $= 2|x - 3| + |x - 3|$ RS $= 10 - 2|x - 3|$
$$= 2|1 - 3| + |1 - 3| \qquad\qquad = 10 - 2|1 - 3|$$
$$= 4 + 2 = 6 \quad \text{Verifies} \checkmark \qquad = 10 - 4 = 6$$

3.2   Exercise

A 1   Solve. All variables represent real numbers.

(a) $|x| = 6$ (b) $|y| = 3$ (c) $2|b| = 8$ (d) $-\frac{1}{2}|w| = -10$

(e) $|2x| = 4$ (f) $|-3x| = 9$ (g) $-2|4x| = -24$ (h) $-\frac{3}{4}|-8y| = 18$

(i) $|x| + 5 = 7$ (j) $|x| + 3 = 8$ (k) $|a| - 6 = 3$

(l) $2|x| - 1 = 7$ (m) $2|a| - 3 = 7$ (n) $3|y| + 4 = 13$

2   Solve and verify.

(a) $|x + 4| = 3$ (b) $|y - 3| = 6$ (c) $|2m - 1| = 9$

3   (a) What is your first step in solving $4|x + 3| = 12$?
(b) Solve and verify the equation in (a).

4   (a) What is your first step in solving $|y| + 3 = 5|y| - 2$?
(b) Solve and verify the equation in (a).

5 (a) What is your first step in solving $2|y - 1| + 7 = 3|y - 1| - 5$?
   (b) Solve and verify the equation in (a).

B Remember: To solve equations involving absolute value, you must understand the strategy you are using.

6 Solve and verify.
   (a) $|x + 6| = 8$ (b) $|3 - x| = 7$ (c) $|x - 5| = 1$
   (d) $|1 + 2x| = 6$ (e) $|2 - 3x| = 11$ (f) $|3x + 1| = 10$

7 Solve.
   (a) $|3 - 2x| = 5$ (b) $|2x - 1| = -3$ (c) $|2x + 3| = 0$
   (d) $-|x + 6| = 10$ (e) $-|2x - 5| = 3$ (f) $-|4x + 2| = -3$

8 (a) Solve and verify $4|x + 2| = 3$ (b) Solve and verify $2|b - 3| + 1 = 4$

9 Find the solution set for each of the following.
   (a) $\{x \mid |3x + 4| = 13, x \in \mathbf{R}\}$ (b) $\{b \mid |2b - 1| = 7, b \in \mathbf{R}\}$
   (c) $\{y \mid |1 - 3y| = 10, y \in \mathbf{R}\}$ (d) $\{z \mid -|5z - 1| = -16, z \in \mathbf{R}\}$
   (e) $\{k \mid 3|3k + 1| = 21, k \in \mathbf{R}\}$ (f) $\{a \mid -\frac{1}{2}|5a - 2| = -4, a \in \mathbf{R}\}$

10 Solve.
   (a) $|x - 4| = 5x$ (b) $|2x - 5| = -3x$ (c) $\frac{1}{3}|x + 3| + 5x = 0$

   (d) $|x + 4| = -2x$ (e) $\left|\dfrac{x - 6}{3}\right| + 4x = 0$ (f) $|2x - 3| = 4x - 2$

11 Solve.   Did you check your answers?
   (a) $3|x| = |x| + 4$ (b) $5|x| - 3 = |x| + 2$ (c) $8|y| - 15 = 5|y| + 9$
   (d) $3|y| - 8 = 5|y| - 26$ (e) $6|k| - 14 = 4|k| + 6$ (f) $3|x| - 7 = 2|x| + 5$

12 (a) Solve $3|y| - 3 = 2 - |y|$.   (b) Solve $3|k + 1| - 3 = 2 - |k + 1|$.
   (c) How are the solutions in (a) and (b) alike? How are they different?

13 Solve.
   (a) $6|p + 1| - 14 = 4|p + 1| + 6$ (b) $3|k - 3| - 7 = 2|k - 3| + 5$
   (c) $3|p - 2| + 2|p - 2| - 5 = 2|p - 2| + 7 + |p - 2|$.

14 Solve and verify.
   (a) $|2x + 1| = |x + 5|$ (b) $|3x - 2| = |2x + 1|$ (c) $|2x - 5| = |x - 1|$

C 15 Solve and verify $|2(x + 1) + 3(x - 5)| = 8, x \in R$.

3.3 Extending Strategies: Solving Inequations

To solve inequations involving absolute value, you can apply, again, the definition of absolute value of x and use the principle

Either | Case 1 for all $x \geq 0$ | or | Case 2 for all $x < 0$

Find the solution set given by $|x - 6| \leq 3$.

> To develop a strategy for solving the inequation, you examine the two cases again.

Case 1

For all $x - 6 \geq 0$ ①,
 then $|x - 6| = x - 6$.
Thus solve $x - 6 \leq 3$
$$x \leq 9 \quad ②$$
In this case the solution
must satisfy ① *and* ②.

$$x - 6 \geq 0 \quad \text{and} \quad x \leq 9$$
or $\quad x \geq 6 \quad \text{and} \quad x \leq 9$
or $\quad\quad 6 \leq x \leq 9$

Case 2

For all $x - 6 < 0$ ③,
 then $|x - 6| = -(x - 6)$.
Thus solve $-(x - 6) \leq 3$
$$-x + 6 \leq 3$$
$$-x \leq -3$$
$$x \geq 3 \quad ④$$

In this case, the solution
must satisfy ③ *and* ④.

$$x - 6 < 0 \quad \text{and} \quad x \geq 3$$
or $\quad x < 6 \quad \text{and} \quad x \geq 3$
or $\quad\quad 3 \leq x < 6$

From the two cases you obtain the solution for $x \in R$, which satisfies

$$\underbrace{3 \leq x < 6 \quad \text{and} \quad 6 \leq x \leq 9}_{3 \leq x \leq 9}$$

Thus, $\{x \mid |x - 6| \leq 3, x \in R\} = \{x \mid 3 \leq x \leq 9, x \in R\}$.

By completing solutions such as the one above, you will be able to identify a strategy.

Strategy: To solve $|x - 6| \leq 3 \implies$ you solve the
equivalent inequations
$$-3 \leq x - 6 \leq 3$$
or $-3 \leq x - 6 \quad \text{and} \quad x - 6 \leq 3$
or $\quad 3 \leq x \quad\quad \text{and} \quad\quad x \leq 9$
or $\quad\quad\quad 3 \leq x \leq 9$

The properties of inequations such as $|m| \geq 2$ will suggest a strategy for solving more advanced inequations.

$$\text{To solve } |m| \geq 2$$

Case 1 $m \geq 0$ Then $|m| = m.$ or   *Case 2* $m < 0$ Then $|m| = -m.$

$m \geq 0$ and $|m| \geq 2$ $m < 0$ and $|m| \geq 2$

$\underbrace{m \geq 0 \quad \text{and} \quad m \geq 2}$ $m < 0$ and $-m \geq 2 \; m < 0.$

       equivalent to $\underbrace{m < 0 \quad \text{and} \quad m \leq -2}$

         $m \geq 2$ equivalent to

                                         $m \leq -2$

Strategy:

Thus, to solve $|m| \geq 2 \implies$ you solve the equivalent inequations

$$m \geq 2 \quad \text{or} \quad m \leq -2.$$

Example   Solve $\{x \,|\, |6 - x| \geq 2, x \in R\}$. Draw the graph of the solution set.

Solution

$|6 - x| \geq 2 \implies (6 - x) \geq 2$ or $-(6 - x) \geq 2$

                   $6 - 2 \geq x$ or $-6 + x \geq 2$

                       $4 \geq x$ or $x \geq 2 + 6$

                       $x \leq 4$ or $x \geq 8$

Thus, $\{x \,|\, |6 - x| \geq 2, x \in R\}$ Graph:

$= \{x \,|\, x \leq 4 \text{ or } x \geq 8\}.$

                    2   3   4   5   6   7   8   9   10

You can write inequations in a compact form by using the absolute value symbols.

$|x| \leq 2$ means $-2 \leq x$ and $x \leq 2$ $|x| \geq 2$ means $x \geq 2$ or $x \leq -2.$

             $-2 \;\; -1 \;\;\; 0 \;\;\; 1 \;\;\; 2 \;\;\; 3$ $-3 \;\; -2 \;\; -1 \;\;\; 0 \;\;\; 1 \;\;\; 2 \;\;\; 3 \;\;\; 4$

3.3   Exercise

A   All variables represent real numbers.

1   Use the absolute value to write each of the following in a compact form.

    (a) $-3 \leq x \leq 3$ (b) $-6 \leq x \leq 6$ (c) $x \geq 2$ or $x \leq -2$

    (d) $-4 \leq x$ and $x \leq 4$ (e) $-5 \leq x$ and $x \leq 5$ (f) $-3 \geq x$ or $x \geq 3$

    (g) $3 \leq x$ or $x \leq -3$ (h) $3 \geq x$ and $x \geq -3$

2 (a) Draw a graph to show $|x| < 2$, $x \in R$.

     (b) Draw the graph to show $|x| \leq 2$, $x \in R$.

3 Draw a graph for each of the following, $y \in R$.

     (a) $|y| < 2$ (b) $|y| \geq 3$ (c) $|y| > 1$

     (d) $|y| \leq 3$ (e) $2 < |y|$ (f) $3 > |y|$

4 Solve.

     (a) $|y| \geq 6$ (b) $|y| < 3$ (c) $|y| > 4$ (d) $|y| \leq 2$

     (e) $|2m| \leq 4$ (f) $|3k| > 6$ (g) $2|p| > 5$ (h) $\frac{1}{2}|x| \geq 3$

5 Solve. Check your answers.

     (a) $|y| + 5 \leq 7$ (b) $|y| - 3 \geq 8$ (c) $2|y| - 3 \leq 5$

     (d) $3 - 2|y| > 1$ (e) $2|y| + |y| < 6$ (f) $3|y| - |y| \leq 6$

B Remember: To solve inequations involving absolute value, you must *understand* the strategy you are using.

6 Solve and draw the graph.

     (a) $|x - 2| \leq 4$ (b) $|x - 2| \geq 4$ (c) $|3 - x| < 6$ (d) $|3 - x| > 6$

7 (a) What is your first step in solving $6|x - 2| < 12$?

     (b) Solve the inequation in (a).

8 (a) What is your first step in solving $3|y - 2| + |y - 2| < 8$?

     (b) Solve the inequation in (a).

9 Solve.

     (a) $|3 - x| < 8$ (b) $|x + 2| - 3 > 0$ (c) $|2x + 1| \leq 4$

     (d) $|2x - 1| \geq 3$ (e) $|4x + 5| < 13$ (f) $|2x - 3| \geq 0$

10 Solve.

     (a) $|x + 3| + 3 \leq 9$ (b) $|2x + 8| + 1 \geq 7$

     (c) $3|x - 2| - 1 < 10 + |x - 2|$ (d) $23 + |x + 1| > 4|x + 1| + 5$

C 11 Solve and verify.

     (a) $|3y + 2| \geq -3y$ (b) $|2m| < m + 3$ (c) $|2x - 1| \geq x$

     (d) $|3k - 5| \leq k$ (e) $|1 - 3y| < y + 5$ (f) $|3a + 1| > 3a + 5$

3.4 Absolute Value Function

For any real number, a, the absolute value was defined as

If $a > 0$, $|a| = a$ If $a = 0$, $|a| = 0$ If $a < 0$, $|a| = -a$

To discover how $y = f(x)$ and $y = |f(x)|$ are related, you compare the graphs of the functions defined by $y = f(x)$ and $y = |f(x)|$. In particular use a table of values for $y = x$, $x \in R$, and $y = |x|$, $x \in R$ and draw the graphs.

x	$y = x$	$y = |x|$
-3	-3	3
-2	-2	2
-1	-1	1
0	0	0
1	1	1
2	2	2
3	3	3

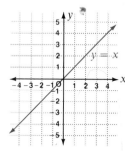

Graph 1

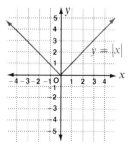

Graph 2

Graph 2 is the graph of the absolute value function, defined as follows.

Absolute Value Function: $y = |x| \begin{cases} y = x & \text{for } x \geq 0 \\ y = -x & \text{for } x < 0 \end{cases}$

Notice from graphs 1 and 2 that points of $y = x$ with positive or zero y co-ordinates are also points of $y = |x|$. But the points of $y = x$ with negative y co-ordinates belong to the graph of $y = |x|$ when they are reflected through the x-axis. This principle is used to draw the graphs of $y = |f(x)|$ as shown in the following example.

Example   The graph defined by $y = f(x)$ is shown. Draw the graph of $y = |f(x)|$.

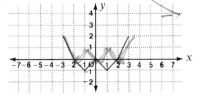

Solution   *Step 1*   All points of the graph $y = f(x)$ with $y \geq 0$ are on the graph of $y = |f(x)|$ as shown.

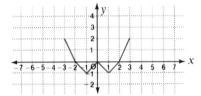

Step 2   All points of the graph of $y = f(x)$ with $y < 0$ are reflected in the x-axis.

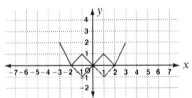

3.4 Exercise

A 1 (a) The graph defined by $y = f(x)$ is shown. Draw the graph of $y = |f(x)|$.

(b) Write a mapping to relate the graph of $y = f(x)$ to $y = |f(x)|$.

(c) Are there any invariant points of the mapping?

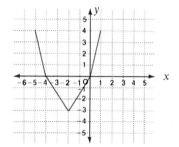

2 For each function defined by $y = f(x)$, draw the graph of the function defined by $y = |f(x)|$.

(a)

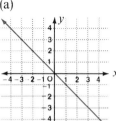

(b)

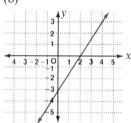

(c)

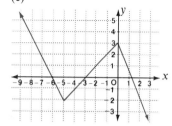

(d)

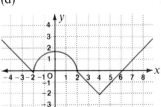

(e)

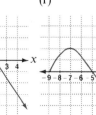

(f)

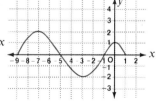

3 For each mapping of $f(x)$ in the previous question, list the co-ordinates of any invariant points.

B 4 (a) Use the graph of $y = f(x)$ given by $f(x) = 3x + 2$. Draw the graph of $y = |f(x)|$.

(b) If $f(x) = 2x^2 - 4$, draw the graph of $y = |f(x)|$.

5 Use the graph of $y = |x|$. Sketch each of the following graphs using your skills with transformations.

(a) $y = |x| + 2$ (b) $y = |x| - 2$ (c) $y = |x + 2|$ (d) $y = |x - 2|$

(e) $y = 2|x|$ (f) $y = \frac{1}{2}|x|$ (g) $y = -|x|$ (h) $y = |-x|$

6 Draw each graph.

(a) $y = |x + 1|$ (b) $y = \left|\dfrac{x}{2}\right|$ (c) $y = |2x - 1|$

(d) $y = 3|x| - 1$ (e) $y = |3x| - 1$ (f) $y = \dfrac{1}{|x|}$

7 Draw each graph.

(a) $y = -2|x + 1| - 1$ (b) $y = 3|x| + 2$ (c) $y = |3x + 2| - 4$

C 8 (a) Sketch the graph of $|x| + |y| = 1$.

(b) What geometric figure have you drawn?

9 (a) Predict what you think the graph of $|x| - |y| = 1$ will show?

(b) Check your prediction in (a). Draw the graph.

Problem-Solving

A calculator is a useful tool to have when working with iterative methods.

Often in the study of mathematics an iterative method repeats the same step over and over until the answer obtained is sufficiently accurate. Newton's method for calculating square roots is an illustration of an iterative method.

Step 1: Make an estimate E_1, of the value of $\sqrt{N}$.

Step 2: Calculate $N \div E_1$, obtain the value of E_2.

Step 3: Calculate the average A_1, of E_1 and E_2. That is, $A_1 = \dfrac{E_1 - E_2}{2}$.

Step 4: (a) If $A_1 = E_1$ for the accuracy desired, then use A_1 as the required square root of N, namely $A_1 = \sqrt{N}$.

(b) If the desired accuracy is not obtained in (a), then use A_1 for E_1 in *Step 1* and repeat all steps again.

Illustrate the significance of each of the above steps by finding the square root of 53 and plotting the values of E_1 and A_1 on an appropriate number line.

3.5 Properties of Exponents: Inventory

You have already studied these properties of exponents.

▶ You also discovered a meaning for a^0 as suggested by examples such as

$$\frac{a^n}{a^n} = a^{n-n} = a^0$$

But $\dfrac{a^n}{a^n} = 1$.

Thus $a^0 = 1$.

▶ As well, you discovered a meaning for a^{-n} suggested by

$a^n \times a^{-n} = 1$

Thus,

$$\frac{a^n \times a^{-n}}{a^n} = \frac{1}{a^n}$$

$$a^{-n} = \frac{1}{a^n}$$

Since
$a^n \times a^{-n}$
$= a^{n-n}$
$= a^0$
$= 1$

> **Laws of Exponents with $m, n \in I$**
>
> *product of powers*
> $\quad a^m \times a^n = a^{m+n}$
> *quotient of powers*
> $\quad a^m \div a^n = a^{m-n}$, for $a \neq 0$
> *power of a power*
> $\quad (a^m)^n = a^{mn}$
> *power of a product*
> $\quad (ab)^m = a^m b^m$
> *power of a quotient*
> $\quad \left(\dfrac{a}{b}\right)^m = \dfrac{a^m}{b^m}$, for $b \neq 0$

> **Definitions of Exponents with $m \in I$**
>
> *zero exponents*
> $\quad a^0 = 1$, for $a \neq 0$
> *integral exponents*
> $\quad a^{-m} = \dfrac{1}{a^m}$, or $a^m = \dfrac{1}{a^{-m}}$, $a \neq 0$

A meaning for rational exponents is related to the definition of radicals as follows.

A Radical

n, index of the radical $\nearrow \sqrt[n]{a} \nwarrow$ the radical sign

a, the radicand

$\sqrt{a} = b \quad$ if $b^2 = a$, $a \geq 0$
$\uparrow$

$\quad b$ is called the principal square root of a.
$\quad \sqrt{a}$ is called a second order radical.

$\sqrt[n]{a} = b \quad$ if $b^n = a$ where, if n is even, then $a \geq 0$.
$\uparrow$

$\quad b$ is called the principal nth root of a.
$\quad \sqrt[n]{a}$ is an nth order radical.

Rational exponents are defined as follows.

> If $n \in N$, then $b^{\frac{1}{n}} = \sqrt[n]{b}$.
> If n is even, then $b \geq 0$,

> If $n \in N$, $m \in I$, then $b^{\frac{m}{n}} = \sqrt[n]{b^m}$ or $(\sqrt[n]{b})^m$.
> If n is even, then $b \geq 0$.

Example Evaluate (a) $\left(\dfrac{1}{\sqrt{3}}\right)^4$ (b) $(32)^{-\frac{3}{5}}$ (c) $(0.25)^{-1.5}$

Solution (a) $\left(\dfrac{1}{\sqrt{3}}\right)^4 = \dfrac{1}{(\sqrt{3})^4}$ (b) $32^{-\frac{3}{5}} = \dfrac{1}{32^{\frac{3}{5}}}$ (c) $(0.25)^{-1.5} = \left(\dfrac{25}{100}\right)^{-1.5}$

$= \dfrac{1}{\sqrt{3^4}}$ $= \dfrac{1}{(\sqrt[5]{32})^3}$ $= \dfrac{1}{\left(\dfrac{25}{100}\right)^{\frac{3}{2}}}$

$= \dfrac{1}{\sqrt{81}}$ $= \dfrac{1}{2^3}$ $= \left(\dfrac{100}{25}\right)^{\frac{3}{2}}$

$= \dfrac{1}{9}$ $= \dfrac{1}{8}$ $= (\sqrt{4})^3$

$= 2^3$

$= 8$

Remember that when n, the index of the radical, is even, then $a \geq 0$. Thus, you must remember any restrictions which may have to be placed on the variable. For example, for $\sqrt{x+1}$ the radicand must be positive or equal to zero. Thus $x + 1 \geq 0$ or $x \geq -1$.

3.5 Exercise

A 1 Evaluate each of the following.

(a) 3^2 (b) $\left(\dfrac{1}{4}\right)^3$ (c) 5^4 (d) $(0.2)^3$

(e) $(-0.2)^3$ (f) -0.2^3 (g) $(-5)^{-2}$ (h) 4^{-3}

(i) $(0.21)^0$ (j) $\left(\dfrac{2}{3}\right)^{-2}$ (k) $(1.5)^{-4}$ (l) $(0.001)^0$

2 Write each of the following in radical form.

(a) $7^{\frac{1}{2}}$ (b) $x^{\frac{1}{5}}$ (c) $a^{\frac{1}{4}}$ (d) $8^{\frac{2}{5}}$ (e) $\left(\dfrac{1}{4}\right)^{\frac{1}{2}}$

(f) $b^{\frac{3}{7}}$ (g) $5x^{\frac{2}{3}}$ (h) $x^{-\frac{1}{3}}$ (i) $a^{-\frac{2}{5}}$ (j) $(0.001)^{\frac{1}{3}}$

3 Write each of the following in exponent form.

(a) $\sqrt{6}$ (b) $\sqrt{10}$ (c) $\sqrt[3]{7}$ (d) $\sqrt{\dfrac{1}{4}}$ (e) $\sqrt[3]{3^2}$

(f) $(\sqrt[7]{12})^3$ (g) $(\sqrt[3]{7})^4$ (h) $(\sqrt[7]{8})^9$ (i) $\sqrt[3]{-0.027}$ (j) $\left(\dfrac{1}{\sqrt{7}}\right)^2$

4   Express the following with exponents.

(a) $\sqrt[5]{x}$ (b) $\sqrt{x^3}$ (c) $\dfrac{1}{\sqrt[3]{x}}$ (d) $\dfrac{1}{\sqrt{x^3}}$ (e) $\sqrt[6]{a^5}$

(f) $\sqrt[7]{p^8}$ (g) $\dfrac{1}{\sqrt[4]{a^3}}$ (h) $\dfrac{1}{\sqrt[5]{a^4}}$ (i) $\left(\dfrac{1}{\sqrt[3]{a}}\right)^{-2}$ (j) $\left(\dfrac{1}{\sqrt{b^8}}\right)^{-0.25}$

B   Review the definitions and properties of exponents.

5   Evaluate.

(a) $27^{\frac{1}{3}}$ (b) $8^{\frac{2}{3}}$ (c) $81^{\frac{3}{4}}$ (d) $0.008^{\frac{1}{3}}$ (e) $4^{0.5}$

(f) $32^{\frac{3}{5}}$ (g) $64^{\frac{5}{6}}$ (h) $0.0001^{0.25}$ (i) $81^{0.75}$ (j) $(16^2)^{\frac{1}{4}}$

6   Evaluate.

(a) $32^{\frac{3}{5}}$ (b) $(27^2)^{-\frac{1}{3}}$ (c) $64^{-\frac{5}{6}}$ (d) $81^{-0.75}$ (e) $10\,000^{\frac{3}{4}}$

(f) $0^{1.78}$ (g) $625^{-\frac{1}{4}}$ (h) $\left(\dfrac{25}{4}\right)^{\frac{5}{2}}$ (i) $81^{-0.25}$ (j) $(0.04)^{-\frac{1}{2}}$

7   Simplify.

(a) $(y^{\frac{1}{2}})^3 \div (4y^4)^{\frac{1}{2}}$ (b) $\sqrt[4]{\dfrac{1}{y}}(y^{-\frac{3}{4}})$ (c) $\left(\dfrac{\sqrt[4]{y^2}}{\sqrt{y}}\right)^{10}$

(d) $\left(\dfrac{x^3}{16}\right)^{\frac{1}{4}}\left(\dfrac{81^{\frac{3}{4}}}{x}\right)$ (e) $\dfrac{(x^2y^4)^{\frac{1}{2}}(x^4y^2)^{\frac{1}{2}}}{(x^{\frac{1}{2}}y^{\frac{1}{2}})^4}$ (f) $\sqrt[4]{\sqrt{81x^8}}$

8   Simplify each of the following.

(a) $3^{\frac{1}{2}} \times 3^{\frac{1}{4}}$ (b) $5^{\frac{1}{9}} \times 5^{\frac{1}{3}}$ (c) $(a^{\frac{1}{4}} \times b^{\frac{1}{2}})^2$ (d) $(a^{\frac{1}{3}}b^{\frac{1}{2}})^6$

(e) $(x^5y^{10}z^{15})^{\frac{1}{5}}$ (f) $(y^{\frac{1}{4}} + 2y^{\frac{3}{4}})y^{\frac{3}{4}}$ (g) $(27x^9y^6)^{\frac{1}{3}}$ (h) $(16x^{12}y^8)^{0.25}$

9   Simplify.

(a) $49^{\frac{1}{2}} + 16^{\frac{3}{4}}$ (b) $128^{-\frac{2}{7}} \div 16^{-0.25}$

(c) $16^{\frac{3}{4}} + 16^{\frac{3}{4}} - 81^{-\frac{3}{4}}$ (d) $(0.16)^{\frac{1}{2}} + 16^{0.75} + 5 \times 27^{\frac{2}{3}}$

10   Write in simplest terms.

(a) $(32x^{10}y^{15}z^{20})^{\frac{2}{5}}$ (b) $\sqrt[3]{\dfrac{\sqrt{x}\sqrt{x^3y}}{x^{\frac{4}{3}}}}$ (c) $(\sqrt[3]{x^{3n+1}})(\sqrt[3]{x^{-1}})$

C 11   Find any restrictions on the variable x, if the following radicals are to represent real numbers.

(a) $\sqrt{x+3}$ (b) $\sqrt{x-4}$ (c) $\sqrt{2x+5}$ (d) $\sqrt{\dfrac{1}{2}x+2}$

(e) $\sqrt{\dfrac{3}{4}x-1}$ (f) $\sqrt{x^2-9}$ (g) $\sqrt{3x-2a}$ (h) $\sqrt{ax+3b}$

Applications: Exponents and Stopping Distances

When a vehicle is driven, the conditions of the weather affect how far a vehicle travels in coming to a complete stop when the brakes are applied.

On a dry surface   • the speed of a car with a trailer is 20 km/h.

                          • distance covered in coming to a stop is 12.3 m.

On a wet surface   • the speed of a car with a trailer is 20 km/h.

                          • distance covered in coming to a stop is 16.8 m.

The actual mathematics involved in calculating stopping distances is complex. However, the following questions illustrate how stopping distances vary significantly depending on the road surface. Drive carefully!

Express your answers to 1 decimal place.

12   The distance, in metres, covered by a tandem truck in coming to a stop on a dry paved road is given by $0.02v^2$, where v is the speed of the truck in kilometres per hour.

(a) How far will the truck travel if the brakes are applied at a speed of 30 km/h?

(b) By how much does the required stopping distance increase if the speed is doubled?

13   (a) A car pulling a trailer on a gravel road is able to stop completely in a distance given by $0.015v^2$, in metres. How far will the car travel before coming to a full stop if the brakes are applied at a speed of 40 km/h?

(b) The same vehicle on a paved road requires a distance given by $0.01v^2$, in metres. By how much does the distance decrease for stopping on a paved road at the same speed?

14   (a) A large oil tanker approaches the dock in calm water at 15 km/h. If the braking distance covered in coming to a complete stop is given by $0.095v^2$, in metres, how far from the dock must the tanker begin to stop for a successful docking?

(b) If the speed of the tanker increases by 5 km/h by how much does its stopping distance increase?

15   On a wet slippery road, the stopping distances, in metres, are shown for 2 vehicles.

Motorcycle: Stopping distance, $0.016v^2$	*Car:* Stopping distance $0.026v^2$	v is the speed of the vehicle in kilometres per hour (km/h).

(a) The motorcycle, travelling at 55 km/h, is 20 m behind a car whose speed is 50 km/h. If the drivers apply their brakes simultaneously, will they collide?

(b) If the car and the motorcycle travel at the same speed, and they both apply their brakes simultaneously, will the motorcyclist stop safely?

3.6 Vocabulary of Radicals

Radicals occur in many scientific formulas.

- For a simple pendulum of length, L, in centimetres, the period of swing, T, in seconds, is calculated from

$$T = 2\pi \sqrt{\frac{L}{g}}$$

where g is the acceleration due to gravity.

- In deep water, the velocity, V, in kilometres per hour, of a water wave with length, L, in metres, is given by

$$V = \sqrt{\frac{gL}{2\pi}}$$

where g is the acceleration due to gravity.

In order to work with radicals, you need to understand the meaning of various terms used with radicals.

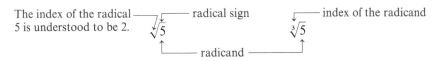

The index of the radical 5 is understood to be 2. ⟶ radical sign $\sqrt{5}$ ⟵ radicand

index of the radicand ⟶ $\sqrt[3]{5}$ ⟵ radicand

$\sqrt{5}$ is said to be a radical of order 2. $\sqrt[3]{5}$ is said to be a radical of order 3, and so on.

Like Radicals $\sqrt{5},\ -4\sqrt{5},\ -\frac{1}{2}\sqrt{5}$

Same order, like radicands

Unlike Radicals $\sqrt{5},\ 2\sqrt{3},\ \sqrt[3]{5},\ \sqrt{5}$

different radicands
different order

Entire Radicals $\sqrt{8},\ \sqrt{12},\ \sqrt{20}$

Mixed Radicals $2\sqrt{2},\ 2\sqrt{3},\ 2\sqrt{5}$

A radical is in **simplest form** if it fulfills the following conditions:

A   For a radical of nth order, the radicand contains no factor that is the nth power of an integer. For example, $\sqrt{3}$ and $\sqrt[3]{6}$ are in the simplest form, but $\sqrt{27}$ and $\sqrt[4]{48}$ are not.

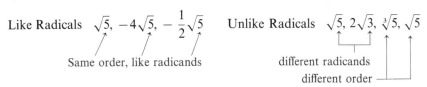

$\sqrt{27}$ ⟵ radical order 2
$= \sqrt{9 \times 3}$
$= \sqrt{3^2 \times 3}$ ⟵ a factor of power 2 occurs
$= 3\sqrt{3}$ ⟵ in simplest form

$\sqrt[4]{48}$ ⟵ radical of order 4
$= \sqrt[4]{16 \times 3}$
$= \sqrt[4]{2^4 \times 3}$ ⟵ a factor of power 4 occurs
$= 2\sqrt[4]{3}$ ⟵ in simplest form

B   The radicand contains no fractions. Thus, $\sqrt{\dfrac{3}{5}}$ and $\sqrt[3]{\dfrac{7}{4}}$ are not in simplest form. You can write radicals of order n, such as the two above, in the simplest form, by writing an equivalent fraction for the radicand whose denominator is the nth power of an integer.

$\sqrt{\dfrac{3}{5}}$ ⟵ radical of order 2

$= \sqrt{\dfrac{3}{5} \times \dfrac{5}{5}}$

$= \sqrt{\dfrac{15}{5^2}}$ ⟵ denominator is expressed to the power 2

$= \dfrac{\sqrt{15}}{5}$ ⟵ in simplest form

$\sqrt[3]{\dfrac{7}{4}}$ ⟵ radical of order 3

$= \sqrt[3]{\dfrac{7}{4} \times \dfrac{4^2}{4^2}}$

$= \sqrt[3]{\dfrac{112}{4^3}}$ ⟵ denominator is expressed to the power 3.

$= \dfrac{\sqrt[3]{112}}{4}$ ⟵ in simplest form

C   The radicand contains no factors which have negative exponents. Thus, $\sqrt{2a}$ and $\sqrt[5]{x^2 y}$ are in simplest form, but $\sqrt[3]{a^{-2}}$ and $\sqrt[4]{x^2 y^{-3}}$ are not. They are expressed in simplest form as follows.

$\sqrt[3]{a^{-2}} = \sqrt[3]{\dfrac{1}{a^2}}$

$= \sqrt[3]{\dfrac{1}{a^2} \times \dfrac{a}{a}}$

$= \sqrt[3]{\dfrac{a}{a^3}}$

$= \dfrac{\sqrt[3]{a}}{a}$

$\sqrt[4]{x^2 y^{-3}} = \sqrt[4]{\dfrac{x^2}{y^3}}$

$= \sqrt[4]{\dfrac{x^2}{y^3} \times \dfrac{y}{y}}$

$= \sqrt[4]{\dfrac{x^2 y}{y^4}}$

$= \dfrac{\sqrt[4]{x^2 y}}{y}$

D   The index of the radical must be as small as possible.
   $\sqrt[4]{2}$ and $\sqrt[3]{y^2}$ are in simplest form, but $\sqrt[4]{2^2}$ and $\sqrt[4]{36 x^6}$ are not.

$\sqrt[4]{2^2} = \sqrt{\sqrt{2^2}}$

$= \sqrt{2}.$

Thus $\sqrt[4]{2^2} = \sqrt{2}$ in simplest form.

$\sqrt[4]{36 x^6} = \sqrt{\sqrt{36 x^6}}$

$= \sqrt{6|x^3|}$

$= |x|\sqrt{6|x|}.$

Thus, $\sqrt[4]{36 x^6} = |x|\sqrt{6|x|}$ in simplest form.

Example Write in simplest form (a) $\sqrt[5]{32x^{10}}$ (b) $\dfrac{1}{2\sqrt{3}}$ (c) $\sqrt[3]{x^{-5}y}$ (d) $\sqrt[4]{x^6y^8}$

Solution

(a) $\sqrt[5]{32x^{10}}$
$= \sqrt[5]{2^5x^{10}}$
$= 2x^2$

(b) $\dfrac{1}{2\sqrt{3}}$
$= \dfrac{1}{2\sqrt{3}} \times \dfrac{\sqrt{3}}{\sqrt{3}}$
$= \dfrac{\sqrt{3}}{6}$

(c) $\sqrt[3]{x^{-5}y}$
$= \sqrt[3]{\dfrac{y}{x^5}}$
$= \sqrt[3]{\dfrac{y}{x^5} \times \dfrac{x}{x}}$
$= \dfrac{1}{x^2}\sqrt[3]{xy}$

(d) $\sqrt[4]{x^6y^8}$
$= |x|y^2\sqrt[4]{x^2}$
$= |x|y^2\sqrt{\sqrt{x^2}}$
$= |x|y^2\sqrt{|x|}$

In order to compare second order radicals numerically, you must write them as entire radicals. For example, $\sqrt{112} > \sqrt{92}$ since $112 > 92$. To see if $4\sqrt{3}$ is greater than $3\sqrt{4}$ you write the radicals as entire radicals. Thus, $4\sqrt{3} = \sqrt{48}$ and $3\sqrt{4} = \sqrt{36}$. Since $48 > 36$, then, $4\sqrt{3} > 3\sqrt{4}$.

3.6 Exercise

A Review the meanings of the vocabulary of radicals: radicand, index, order, like, unlike, entire and mixed.

1 Write each of the following as a mixed radical in simplest form.

(a) $\sqrt{32}$ (b) $-3\sqrt{32}$ (c) $\dfrac{1}{2}\sqrt{32}$ (d) $-\dfrac{1}{5}\sqrt{125}$

(e) $-4\sqrt{27}$ (f) $2\sqrt{75}$ (g) $\sqrt[3]{16}$ (h) $-\sqrt[3]{192}$

(i) $\sqrt[4]{48}$ (j) $-\sqrt{45}$ (k) $\sqrt[5]{96}$ (l) $-3\sqrt{8}$

2 Write each of the following as an entire radical.
(a) $3\sqrt{2}$ (b) $-4\sqrt{3}$ (c) $5\sqrt{27}$ (d) $6\sqrt{8}$
(e) $-2\sqrt{27}$ (f) $2\sqrt[3]{3}$ (g) $-3\sqrt[3]{2}$ (h) $2\sqrt[4]{27}$

B Review the meaning of simplest form of a radical.

3 Write in simplest form.
(a) $\sqrt{8}$ (b) $\sqrt[4]{32}$ (c) $\sqrt[3]{-125z^5}$ (d) $\sqrt[5]{64y^3}$
(e) $\sqrt[6]{128x^7}$ (f) $\sqrt[4]{81^3}$ (g) $\sqrt[6]{64y^{13}}$ (h) $\sqrt[3]{-72x}$
(i) $\sqrt[3]{27a^9}$ (j) $\sqrt[4]{16p^8q^5}$

4   Write in simplest form.

(a) $\sqrt{\dfrac{1}{3}}$ (b) $\sqrt{\dfrac{4}{3}}$ (c) $\sqrt[4]{\dfrac{16}{9}}$ (d) $\sqrt[3]{\dfrac{8}{125}}$ (e) $\sqrt{\dfrac{a^2}{b^3}}$

(f) $\sqrt[3]{-\dfrac{a^4}{3b^2}}$ (g) $\sqrt[4]{\dfrac{pq^4}{8r^8}}$ (h) $\sqrt[5]{\dfrac{-32w^3x^5}{y^7z^5}}$ (i) $\dfrac{1}{\sqrt[3]{4}}$ (j) $\sqrt[3]{\dfrac{x^3y^2}{\sqrt{7z^5}}}$

5   Write in simplest form.

(a) $\sqrt{3^{-2}}$ (b) $\sqrt[3]{-125a^{-6}}$ (c) $\sqrt{ab^{-3}}$

(d) $\sqrt[3]{p^{-2}q^6}$ (e) $\sqrt[3]{-125m^{-4}n^5}$ (f) $\sqrt[3]{-27c^{-1}d^3e^{-6}}$

(g) $\sqrt[5]{\dfrac{-32x^{-6}}{y^{-10}}}$ (h) $\sqrt[6]{27w^{-12}v^{-6}}$ (i) $\sqrt[4]{\dfrac{a^{-4}b^3}{81}}$ (j) $\dfrac{\sqrt[3]{2x^{-2}y^{-3}}}{\sqrt{xy}}$

6   Write in simplest form.

(a) $\sqrt[4]{128}$ (b) $\sqrt[4]{4b^2}$ (c) $\sqrt{8a^3b^6}$ (d) $\sqrt[4]{a^8b^{10}}$

(e) $\sqrt[6]{16a^4}$ (f) $\sqrt[4]{\dfrac{64a^{10}}{b^{14}}}$ (g) $\sqrt[6]{8v^3w^9}$ (h) $\sqrt[3]{\sqrt{a^9b^{18}}}$

7   Simplify.

(a) $\sqrt[3]{16a^2} \times \sqrt{4a}$ (b) $\sqrt[4]{w^3} \times \sqrt{w^3}$ (c) $\dfrac{\sqrt{28p^7q}}{\sqrt{7pq^5}}$

(d) $\sqrt[4]{8ab^4} \times \sqrt[4]{2a^3}$ (e) $\dfrac{\sqrt[3]{-3a^7b^9}}{\sqrt[3]{81a}}$ (f) $\sqrt[4]{a^{-4}-b^{-4}}$

8   Which of the following radicals are equivalent to $\sqrt[3]{128}$?
    A: $8\sqrt[3]{2}$ B: $2\sqrt[3]{16}$ C: $4\sqrt[3]{2}$ D: $4\sqrt[3]{8}$

9   (a) Write $5\sqrt{2}$ as an entire radical.   (b) Write $4\sqrt{3}$ as an entire radical.
    (c) Use your answers in (a) and (b) to show why $5\sqrt{2} > 4\sqrt{3}$.

10 (a) Write   A: $6\sqrt{2}$ B: $5\sqrt{3}$ as entire radicals.
    (b) Which is greater, A or B?

C 11 (a) Show that $4\sqrt{3} < 5\sqrt{2}$.   (b) Show that $4\sqrt{5} > 5\sqrt{3}$.

Math Tip

In the fourteenth century, the century of the Black Death, very little
significant mathematics was developed. However, the greatest
mathematician of the time, Nicole Oresme, used for the first time
fractional exponents as we know them today.

3.7 Radicals: Addition and Subtraction

You can look for similarities and differences in your solutions to help you understand various skills and concepts. For example, you have learned the meaning of "like" in various situations:

like fractions   like terms   like monomials

The meaning of "like" is extended to the radicals and to the operations with like radicals.

To add or subtract like radicals, you add or subtract the coefficients of each radical as shown in the examples.

You can apply the distributive property to simplify radical expressions. Compare:

$$2(3a - 2b) = 6a - 4b \qquad 2(3\sqrt{2} - 4\sqrt{3}) = 6\sqrt{2} - 8\sqrt{3}$$

Example   Simplify.

(a) $4\sqrt{3} - 2\sqrt{5} + 6\sqrt{3} + 5\sqrt{5}$ (b) $-2(2\sqrt{12} - \sqrt{18}) - 5(3\sqrt{32} - \sqrt{27})$

Solution   (a) $4\sqrt{3} - 2\sqrt{5} + 6\sqrt{3} + 5\sqrt{5}$

$= \underbrace{4\sqrt{3} + 6\sqrt{3}} - \underbrace{2\sqrt{5} + 5\sqrt{5}}$

Think: All the terms are expressed in simplest form. Thus, you can add or subtract numerical coefficients.

—— like radicals

$= 10\sqrt{3} + 3\sqrt{5}$ ⟵ —————— unlike radicals

(b) $\quad -2(2\sqrt{12} - \sqrt{18}) - 5(3\sqrt{32} - \sqrt{27})$ ⟵   Think: The terms are not expressed in simplest form.

$= -4\sqrt{12} + 2\sqrt{18} - 15\sqrt{32} + 5\sqrt{27}$

$= -4(2\sqrt{3}) + 2(3\sqrt{2}) - 15(4\sqrt{2}) + 5(3\sqrt{3})$ ⟵ Written in simplest form

$\sqrt{27} = 3\sqrt{3}$

$= -8\sqrt{3} + 6\sqrt{2} - 60\sqrt{2} + 15\sqrt{3}$

$= -8\sqrt{3} + 15\sqrt{3} + 6\sqrt{2} - 60\sqrt{2}$

$= 7\sqrt{3} - 54\sqrt{2}$

3.7 Exercise

A   Remember: add or subtract like radicals.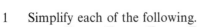

1   Simplify each of the following.

(a) $3\sqrt{2} - 4\sqrt{2} + 5\sqrt{2} - 3\sqrt{2}$ (b) $5\sqrt{2} - 3\sqrt{3} - 6\sqrt{2} + 5\sqrt{3}$

(c) $3\sqrt{5} - 4\sqrt{3} - 3\sqrt{5} + 6\sqrt{3}$ (d) $4\sqrt{3} - 2\sqrt{7} + 3\sqrt{7} - 3\sqrt{3}$

2   Simplify each of the following.
   (a) $2(3 - 3\sqrt{3})$ (b) $3(3\sqrt{2} - 5)$ (c) $-2(2\sqrt{3} - 4\sqrt{2})$
   (d) $5(3\sqrt{5} + 5\sqrt{6})$ (e) $-\frac{1}{2}(2\sqrt{3} + 4\sqrt[3]{2})$ (f) $3(2\sqrt[3]{2} - 3\sqrt{2})$

3   (a) Write each term as a mixed radical in simplest form.
$$6\sqrt{8} + 2\sqrt{18} - \sqrt{72}$$
   (b) Simplify the radical expression in (a).

B   Remember to express your answers in simplest terms.

4   Simplify each of the following.
   (a) $2\sqrt{12} - 5\sqrt{27} + 3\sqrt{48}$ (b) $\sqrt{20} - 3\sqrt{245} - 2\sqrt{20}$
   (c) $-3\sqrt{8} - 2\sqrt{18} + 5\sqrt{72}$ (d) $2\sqrt[3]{16} + 3\sqrt[3]{54} - 2\sqrt[3]{128}$
   (e) $2\sqrt{8} - 3\sqrt{98} - 2\sqrt{200}$ (f) $-3\sqrt{50} - \sqrt{32} + 5\sqrt{200}$

5   Simplify each of the following.
   (a) $6\sqrt{8} - 2\sqrt{27} - 3\sqrt{18} + 2\sqrt{3}$ (b) $-2\sqrt{72} + 3\sqrt{28} - 2\sqrt{112} - 6\sqrt{98}$
   (c) $3\sqrt[3]{81} + \frac{1}{2}\sqrt[3]{128} - 3\sqrt[3]{192} + 4\sqrt[3]{54}$

6   Simplify.
   (a) $-3\sqrt{12} + 5\sqrt{27} - 6\sqrt{48} + 2\sqrt{75}$
   (b) $12\sqrt{8} - 2\sqrt{27} - 2\sqrt{18} + 6\sqrt{3}$
   (c) $2\sqrt{20} - 3\sqrt{245} - 2\sqrt{20} + \sqrt{125}$
   (d) $-2\sqrt[3]{40} - 3\sqrt[3]{135} + 5\sqrt[3]{320} + 8\sqrt[3]{5}$

7   (a) Simplify $3(2\sqrt{8} - 3\sqrt{125})$.   (b) Simplify $3(4\sqrt{2} - 15\sqrt{5})$.
   (c) Why are your answers in (a) and (b) the same?

8   Simplify.
   (a) $-2(5\sqrt{2} + 5) + 5(6 - 3\sqrt{2})$
   (b) $3(3\sqrt{2} - 3\sqrt{3}) - 2(4\sqrt{2} - 2\sqrt{3})$
   (c) $3(3\sqrt[3]{40} - \sqrt[3]{135}) + 4(\sqrt[3]{320} - \sqrt[3]{40})$
   (d) $\frac{3}{2}\sqrt{8} - \frac{4}{3}\sqrt{27} + 6\sqrt{50}$ (e) $\frac{2}{3}\sqrt[3]{81} - \frac{1}{2}\sqrt[3]{24} + \frac{2\sqrt[3]{135}}{3} - \frac{3\sqrt[3]{40}}{2}$

   Remember
$$\frac{3\sqrt{8}}{2} = \frac{3}{2}\sqrt{8}.$$

9   Express the value of each expression in simplest form if
$a = 3\sqrt{2} - 4\sqrt{3}, b = 2\sqrt{12} - 3\sqrt{8}$.

(a) $a + b$ (b) $2a + b$ (c) $3b - 2a$ (d) $\dfrac{1}{2}(2a - 3b)$

10   Which radical expression has the greater value?

A: $4\sqrt{12} - 3\sqrt{27} + 5\sqrt{48} - 3\sqrt{75}$
B: $2\sqrt{27} - 3\sqrt{48} + \sqrt{108} - 2\sqrt{192}$

11   Which radical expression has the lesser value?

A: $2\sqrt{12} - \sqrt{12} + 2\sqrt{27} - 2\sqrt{75}$
B: $2\sqrt{18} - \sqrt{8} - 2\sqrt{8} - \sqrt{18}$

12   Which figure has the greater perimeter?

(a)
$\sqrt{125} - \sqrt{75}$
$\sqrt{108} - 2\sqrt{20}$

(b)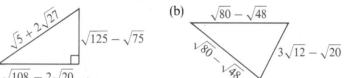
$\sqrt{80} - \sqrt{48}$
$\sqrt{80} - \sqrt{48}$
$3\sqrt{12} - \sqrt{20}$

C 13   Show that $4\sqrt{7} - 3\sqrt{5} > 4\sqrt{5} - 2\sqrt{7}$.

Problem Solving

To check whether a result is true or not we may often use a numerical example. Use a numerical example to check which of the following seem true and which seem false.

A $\sqrt{x^2} + \sqrt{y^2} \overset{?}{=} |x| + |y|$ B $\sqrt{x^2} - \sqrt{y^2} \overset{?}{=} |x| - |y|$
C $\sqrt{x^2} \times \sqrt{y^2} = |xy|$ D $\sqrt{x^2} \div \sqrt{y^2} = |x| \div |y|$

Math Tip

Numbers were invented to aid the solution of equations. To solve $x^2 = 5$, radical numbers were invented, $x = \pm\sqrt{5}$. However, radical numbers were used for many years before the symbol $\sqrt{\ }$ was invented. The Latin word for root is *radix* and from it the word *radical* was derived. The symbol $\sqrt{\ }$ was used for the first time in 1525.

Radical and Literal Expressions

If $a \neq b$, then $\sqrt{a}$ and $\sqrt{b}$ are *unlike radicals*. For $\sqrt{b}$, the values of b are restricted so that $b \geq 0$. For $\sqrt[3]{b}$, b may be positive or negative. To simplify expressions, write mixed radicals as shown.

$$\sqrt{9a} - \sqrt{4b} + \sqrt{36a} + \sqrt{25b} = 3\sqrt{a} - 2\sqrt{b} + 6\sqrt{a} + 5\sqrt{b}$$
$$= 9\sqrt{a} + 3\sqrt{b}$$

If the variable occurs as a radicand then you may need to simplify radicals.

$$\sqrt{36y^2} = 6|y| \qquad \text{Remember: } \sqrt{36y^2} \neq 6y$$

14 Simplify each of the following.
(a) $3a\sqrt{8} - 2b\sqrt{32} + 3b\sqrt{50} - 2a\sqrt{72}$
(b) $-m\sqrt{27} + 3n\sqrt{12} - 2m\sqrt{75} - 3n\sqrt{48}$
(c) $y\sqrt{20} + 3x\sqrt{80} + 2y\sqrt{45} - 3x\sqrt{125}$

15 (a) Evaluate each expression if $p = 4$, $q = -3$.

A: $\sqrt{36pq^2}$ B: $6|q|\sqrt{p}$ C: $6q\sqrt{p}$

(b) Why are the expressions A and B equivalent?
(c) Why are the expressions A and C not equivalent?

16 Express each of the following as mixed radicals in simplest form.
(a) $\sqrt{8x^2}$　　(b) $\sqrt{98x^3}$　　(c) $\sqrt{36a^2b^3}$　　(d) $\sqrt{27a^4b^3}$
(e) $\sqrt{80m^4n^3}$　　(f) $\sqrt{64x^2yz^4}$　　(g) $\sqrt[4]{81x^4y^2}$　　(h) $\sqrt[3]{24p^3q^5}$

17 Simplify.
(a) $3\sqrt{x^2} - 2\sqrt{x^3} - 5\sqrt{x^4}$　　(b) $4\sqrt{b^3} - 3\sqrt{b^5} - 2\sqrt{b^3}$
(c) $3\sqrt{x^2y} - 2\sqrt{x^4y^3}$　　(d) $3\sqrt[3]{x^4y^7} - 6\sqrt[3]{x^7y^4} + 2\sqrt[3]{x^7y^{10}}$

18 Find the value of $\sqrt{4a} + a\sqrt{a^2b} + \sqrt{b^2a} + b\sqrt{9a}$ if $a = 3$, $b = 2$.

19 Simplify each of the following if $a \geq 0$ and $b \geq 0$.
(a) $\sqrt{a^2b} - 2\sqrt{ab^2} - \sqrt{9b}$
(b) $\sqrt{ab} - \sqrt{ab^3} - \sqrt{9a^3b^3} - \sqrt{a^3b}$
(c) $\sqrt{4b} - 2\sqrt{25a^2b} - 3\sqrt{a^2b^3} + 7\sqrt{b}$

20 Simplify each of the following.
(a) $7\sqrt{b^3} + \sqrt{4a^2b} - \sqrt{4b}$
(b) $8\sqrt{49b} - 7\sqrt{9b^3} + \sqrt{4a} + \sqrt{a^3}$
(c) $7\sqrt{a^3b} + 9\sqrt{9a^3b^3} - \sqrt{a^3b^3} + 2\sqrt{4ab}$

3.8 Generalizations: Multiplying Radicals

To show that a statement is not true you need only find one counter example. For example,

Is $\sqrt{a + b} \overset{?}{=} \sqrt{a} + \sqrt{b}$? Test $a = 9$ and $b = 16$.

$$\begin{aligned} \text{LS} &= \sqrt{a + b} & \text{RS} &= \sqrt{a} + \sqrt{b} \\ &= \sqrt{9 + 16} & &= \sqrt{9} + \sqrt{16} \\ &= \sqrt{25} & &= 3 + 4 \\ &= 5 & &= 7 \end{aligned}$$

$$\text{LS} \neq \text{RS}$$

Thus, $\sqrt{a + b} \neq \sqrt{a} + \sqrt{b}$.

However, to show that a general mathematical statement is true, you need to provide a proof. For example, this numerical example suggests a possible general statement.

Is $\sqrt{4} \times \sqrt{9} = \sqrt{36}$?

$$\begin{aligned} \text{LS} &= \sqrt{4} \times \sqrt{9} & \text{RS} &= \sqrt{36} \\ &= 2 \times 3 & &= 6 \\ &= 6 \end{aligned}$$

Since LS = RS, then $\sqrt{4} \times \sqrt{9} = \sqrt{36}$.

The numerical example above suggests that the following statement might be true.

$$\sqrt{m} \times \sqrt{n} = \sqrt{mn}, \quad m \geq 0, n \geq 0$$

To prove the statement for all m and n, use the definition of $\sqrt{m}$. It follows that $\sqrt{m} \times \sqrt{m} = m$.

$$\begin{aligned} \text{Let } x &= \sqrt{m} \times \sqrt{n}. \qquad \text{Thus } x \geq 0. \text{ Why?} \\ \text{Then } x^2 &= (\sqrt{m} \times \sqrt{n})^2 \\ x^2 &= (\sqrt{m})^2(\sqrt{n})^2 \\ x^2 &= mn \end{aligned}$$

Thus $x = \pm\sqrt{mn}$

Since $x \geq 0$, then $x = +\sqrt{mn}$, or $\sqrt{mn}$,

But $\qquad\qquad x = \sqrt{m} \times \sqrt{n}.$

Thus $\boxed{\sqrt{m} \times \sqrt{n} = \sqrt{mn}, m \geq 0, n \geq 0}$

For n^{th} order radicals, you may write

$$\boxed{\begin{array}{l} \sqrt[n]{a} \times \sqrt[n]{b} = \sqrt[n]{ab} \qquad n \in N, a, b \in R, \\ \text{If } n \text{ is even then } a \geq 0, b \geq 0 \end{array}}$$

Example 1   Simplify $(3\sqrt{2})(-2\sqrt{3}) - (5\sqrt{2})(-3\sqrt{3})$.

Solution $(3\sqrt{2})(-2\sqrt{3}) - (5\sqrt{2})(-3\sqrt{3}) = -6\sqrt{6} + 15\sqrt{6}$
$$= 9\sqrt{6}$$

To multiply monomials you may first wish to write the radicals in simplest form. Compare these two procedures.

$$(3\sqrt{8})(2\sqrt{12}) = (6\sqrt{2})(4\sqrt{3}) \qquad (3\sqrt{8})(2\sqrt{12}) = 6\sqrt{96}$$
$$= 24\sqrt{6}. \qquad\qquad\qquad\qquad = 6\sqrt{16 \times 6}$$
$$\qquad\qquad\qquad\qquad\qquad\qquad = 24\sqrt{6}$$

This step often simplifies subsequent calculations.

The skills you learned for multiplying binomial expressions extend to the multiplication of radicals in binomial form.   Remember: $(\sqrt{5})(\sqrt{5}) = 5$

or $(\sqrt{5})^2 = 5$

Practise completing the steps mentally when simplifying binomials involving radicals.

Think: With practice you may do these steps mentally.

Think:

$$(2\sqrt{5} - 3)(3\sqrt{5} + 2) = 30 + 4\sqrt{5} - 9\sqrt{5} - 6$$
$$= 30 - 5\sqrt{5} - 6$$
$$= 24 - 5\sqrt{5}$$

a: $(2\sqrt{5})(3\sqrt{5}) = 30$
b: $(-3)(2) = -6$
c: $(-3)(3\sqrt{5}) = -9\sqrt{5}$
d: $(2\sqrt{5})(2) = 4\sqrt{5}$

The need to substitute occurs frequently. The following example provides practice.

Example 2   If $m = 2\sqrt{3} - 3\sqrt{2}$ and $n = 2\sqrt{2} + 3\sqrt{3}$ find each of the following in simplest form.

(a) mn (b) $(m - n)(m + n)$ (c) $m^2 + mn$

Solution   (a) $mn = (2\sqrt{3} - 3\sqrt{2})(2\sqrt{2} + 3\sqrt{3})$
$$= 4\sqrt{6} + 18 - 12 - 9\sqrt{6}$$
$$= 6 - 5\sqrt{6}$$

Think:
$(2\sqrt{3})(2\sqrt{2}) + (2\sqrt{3})(3\sqrt{3})$
$\quad - (3\sqrt{2})(2\sqrt{2}) - (3\sqrt{2})(3\sqrt{3})$

(b) First, simplify $(m - n)$ and $(m + n)$.
$$m - n = 2\sqrt{3} - 3\sqrt{2} - (2\sqrt{2} + 3\sqrt{3}) \qquad m + n = 2\sqrt{3} - 3\sqrt{2} + 2\sqrt{2} + 3\sqrt{3}$$
$$= 2\sqrt{3} - 3\sqrt{2} - 2\sqrt{2} - 3\sqrt{3} \qquad\qquad = 5\sqrt{3} - \sqrt{2}$$
$$= -\sqrt{3} - 5\sqrt{2}$$

Thus, $(m - n)(m + n) = (-\sqrt{3} - 5\sqrt{2})(5\sqrt{3} - \sqrt{2})$
$$= -15 + \sqrt{6} - 25\sqrt{6} + 10$$
$$= -5 - 24\sqrt{6}$$

(c) $m^2 + mn = m(m + n)$

Think: Use your factoring skills to simplify calculations.

$$= (2\sqrt{3} - 3\sqrt{2})(5\sqrt{3} - \sqrt{2})$$
$$= 30 - 2\sqrt{6} - 15\sqrt{6} + 6$$
$$= 36 - 17\sqrt{6}$$

3.8 Exercise

A Where possible, multiply radicals mentally.

1 (a) Find each product.   A: $(3\sqrt{2})(-2\sqrt{3})$ B: $(3\sqrt{6})(2\sqrt{3})$
      Which product may be written in simpler form?

   (b) Find each product.   A: $(6\sqrt{2})(-12\sqrt{3})$ B: $(3\sqrt{8})(-3\sqrt{48})$
      Why are the answers the same? Write each answer in simplest form.

   (c) To find the product $(-3\sqrt{75})(-2\sqrt{48})$, what first step may be taken?
      Find the product.

2 Write each product in simplest form.
   (a) $(2\sqrt{3})(3\sqrt{2})$ (b) $(4\sqrt{6})(-2\sqrt{5})$ (c) $(3\sqrt{5})(5\sqrt{3})$

   (d) $(2\sqrt{3})\left(-\dfrac{1}{2}\sqrt{5}\right)$ (e) $(-3\sqrt{3})(-5\sqrt{5})$ (f) $(-\sqrt{10})(-2\sqrt{2})$

   (g) $(3\sqrt{2})(5\sqrt{15})$ (h) $\left(-\dfrac{3}{4}\sqrt{8}\right)(2\sqrt{2})$ (i) $\left(\dfrac{2}{5}\sqrt{10}\right)(-3\sqrt{15})$

3 Simplify. Write each product in simplest form.
   (a) $(-3\sqrt{3})(4\sqrt{6})(3\sqrt{7})$ (b) $(2\sqrt{8})(4\sqrt{3})(-3\sqrt{2})$

   (c) $(3\sqrt{48})\left(\dfrac{1}{2}\sqrt{50}\right)(-3\sqrt{2})$ (d) $(-3\sqrt{20})\left(\dfrac{1}{2}\sqrt{48}\right)(5\sqrt{32})$

4 Simplify each of the following.
   (a) $3\sqrt{5}(2\sqrt{2} - 3\sqrt{3})$ (b) $-3\sqrt{3}(3\sqrt{6} - 3\sqrt{2})$ (c) $-4\sqrt{3}(2\sqrt{6} - 2\sqrt{5})$
   (d) $3\sqrt{3}(2\sqrt{2} - 4)$ (e) $3\sqrt{3}(3\sqrt{8} - 2\sqrt{18})$ (f) $-2\sqrt{2}(3\sqrt{12} - 5\sqrt{27})$
   (g) $\dfrac{1}{2}\sqrt{3}(2\sqrt{48} - 3\sqrt{32})$ (h) $\dfrac{3}{2}\sqrt{2}(2\sqrt{18} - 3\sqrt{48})$

5 (a) Find the product $(a + 2b)(3a - b)$.
   (b) Use the results in (a) to help you find $(\sqrt{2} + 2\sqrt{3})(3\sqrt{2} - \sqrt{3})$.
   (c) Use the statement that $(a + b)^2 = a^2 + 2ab + b^2$ to find the square of
      A: $(\sqrt{2} + 3\sqrt{3})^2$ B: $(3\sqrt{6} - 2\sqrt{3})^2$

B 6   Simplify each product.

(a) $(\sqrt{3} + \sqrt{2})(\sqrt{3} - 2\sqrt{2})$ (b) $(2\sqrt{2} - \sqrt{3})(3\sqrt{2} - 3\sqrt{3})$

(c) $(2\sqrt{5} - 2\sqrt{2})(3\sqrt{5} - 3\sqrt{2})$ (d) $(3\sqrt{3} - 2\sqrt{5})(5\sqrt{5} - 3\sqrt{3})$

(e) $(\sqrt{50} - \sqrt{75})(\sqrt{32} - \sqrt{48})$ (f) $(3\sqrt{18} - 3\sqrt{27})(2\sqrt{8} - 2\sqrt{12})$

7   The measures of the rectangles are given. Find a radical expression for their areas.

(a)   (b)   (c)

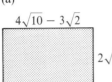
$4\sqrt{10} - 3\sqrt{2}$

$2\sqrt{5}$

$\sqrt{3} + \sqrt{2}$

$4\sqrt{3} - \sqrt{2}$

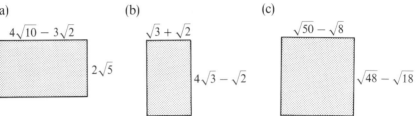
$\sqrt{50} - \sqrt{8}$

$\sqrt{48} - \sqrt{18}$

8   Find the square of each binomial.

(a) $(\sqrt{5} + \sqrt{3})$ (b) $(2\sqrt{3} + 2\sqrt{2})$

(c) $(3\sqrt{2} - 2\sqrt{5})$ (d) $(5\sqrt{7} - 3\sqrt{10})$

(e) $(2\sqrt{3} - \sqrt{6})$ (f) $(5\sqrt{3} + 5)$

9   Simplify each of the following.

(a) $(2\sqrt{5} - \sqrt{3})(\sqrt{5} + 2\sqrt{3}) - (3\sqrt{5} - 2\sqrt{3})^2$

(b) $(3\sqrt{2} - 8)^2 + (5\sqrt{2} + 4)(-2\sqrt{2} - 5)$

(c) $2(2\sqrt{5} - 3\sqrt{3})(3\sqrt{5} + \sqrt{3}) - 4(\sqrt{5} + \sqrt{3})^2$

(d) $4(3\sqrt{6} - 2\sqrt{2})^2 - 8(5\sqrt{6} - 2\sqrt{2})^2$

10   (a) What first step may be taken to simplify

$$(3\sqrt{8} - 2\sqrt{2})^2 - (5\sqrt{8} - 3\sqrt{2})^2?$$

(b) Simplify the expression in (a).

11   Simplify each of the following.   Look for ways to simplify your calculation.

(a) $(\sqrt{32} - \sqrt{50})(\sqrt{18} - \sqrt{2}) - 2(\sqrt{72} - \sqrt{98})^2$

(b) $3(\sqrt{12} - \sqrt{27})^2 + (\sqrt{48} - \sqrt{75})(-3\sqrt{48} - 2\sqrt{75})$

(c) $3(\sqrt{125} - 2\sqrt{48})^2 - 2(3\sqrt{20} - 2\sqrt{27})^2$

12   Simplify.

(a) $(2\sqrt{32} - 4\sqrt{48} + 2\sqrt{18})(2\sqrt{27} - 2\sqrt{50} + 5\sqrt{75})$

(b) $(3\sqrt{48} - 2\sqrt{27} - \sqrt{12})(\sqrt{50} - 3\sqrt{32} - 5\sqrt{8})$.

13 If $a = 3\sqrt{2}$, $b = -2\sqrt{3}$, simplify each of the following.

(a) ab (b) $-3ab$ (c) $\dfrac{1}{2}ab$ (d) $a^2 + b^2$

(e) $a^2 - b^2$ (f) $a^2 - 2b^2$ (g) $a^2 + 2ab$ (h) $b^2 - 2ab$

14 If $m = 3\sqrt{5} - 2$, $n = 2\sqrt{5} + 5$, find the value of
(a) $2mn$ (b) $-3mn$ (c) $m^2 - n^2$

15 (a) Use a numerical example that suggests the generalization in (b).

(b) For all a, $b \in R$ prove that $\sqrt[3]{a}\,\sqrt[3]{b} = \sqrt[3]{ab}$.

(c) For all real numbers p, q prove that

$$\sqrt[n]{p}\,\sqrt[n]{q} = \sqrt[n]{pq} \qquad n, \text{ a natural number.}$$

Computer Tip

The square root function is useful for solving problems about the horizon distance. For example,

The distance d, in metres, to the horizon from a building with height h, in metres, above sea level is given by the following formula.

$$d = \sqrt{2rh} \qquad r \text{ is the radius of the earth, in metres}$$

▶ How far is the horizon from a building, 95.9 m high? The radius of the earth is 12 759.0 km. (You will need a telescope to see the actual horizon.)

▶ Write a computer program to find the distance d.

Problem Solving

Sometimes, at first glance, it appears that a problem does not have enough information to solve it. Often this is not so. Solve the following problem.

On a circular ride, the seats shown are 2.0 m apart. If the person in Seat A goes twice as fast as the person in Seat B, how much further does the person in Seat A go in one loop of the ride?

Problem-Solving: Show That . . .

Very often in mathematics you are asked to show that a fact is true. Often, there is more than one strategy for doing so. For example:

Show that $2\sqrt{2} + 3$ is a root of the quadratic equation $x^2 - 6x + 1 = 0$.

Strategy A

You could use your skills and find the roots of this quadratic equation and show that one of the roots is $2\sqrt{2} + 3$. This strategy would meet the requirements, but, would also involve the additional work of finding a second unasked-for root.

Strategy B

You could show directly that $2\sqrt{2} + 3$ satisfies the equation and is thus a root. For example:

Verify the root in both sides of the equation.
$$LS = x^2 - 6x + 1 \qquad\qquad RS = 0$$
$$= (2\sqrt{2} + 3)^2 - 6(2\sqrt{2} + 3) + 1$$
$$= (8 + 12\sqrt{2} + 9) - 12\sqrt{2} - 18 + 1$$
$$= 12\sqrt{2} - 12\sqrt{2} + 17 - 17 = 0$$
$$LS = RS$$
Thus, $2\sqrt{2} + 3$ is a root of the equation $x^2 - 6x + 1 = 0$.

16 (a) Show that $\sqrt{3} - 1$ is a root of the quadratic equation $x^2 + 2x - 2 = 0$.
    (b) Show that $7 + \sqrt{5}$ is a root of the equation $x^2 - 14x + 44 = 0$.

17 (a) Show that $\sqrt{7} - \sqrt{5}$ is a root of the equation $x^2 - 2\sqrt{7}x + 2 = 0$.
    (b) Test whether $\sqrt{7} + \sqrt{5}$ is a root of the above equation.

18 (a) Show that $6 - \sqrt{3}$ is root of $x^2 - 12x + 33 = 0$.
    (b) Test whether $6 + \sqrt{3}$ is also a root of the above equation.

19 (a) Show that $7 - \sqrt{3}$ is not a root of $x^2 - 14x + 45 = 0$.
    (b) Show that $3 - \sqrt{3}$ is not a root of $x^2 - 3x - 2 = 0$.

20 Show that $\sqrt{6} + \sqrt{3}$ is a root of the equation
$$2x^2 - \sqrt{3}x + 3 = x^2 + \sqrt{3}x + 6.$$

21 Show that $\sqrt{3} - \sqrt{2}$ is a root of the equation
$$x(2x + 5\sqrt{2}) - 4 = x^2 + 3(\sqrt{2}x - 1).$$

3.9 Conjugate Radical Binomials

Patterns in your earlier work provided strategies for doing mathematics such as $(x + y)(x - y) = x^2 - y^2$.

Predict the product.

$(a\sqrt{b} + c\sqrt{d})(a\sqrt{b} - c\sqrt{d}).$

To predict the product, analyze the terms of the binomial radicals.

Find the product.

$$(a\sqrt{b} + c\sqrt{d})(a\sqrt{b} - c\sqrt{d}) = (a\sqrt{b})^2 - (c\sqrt{d})^2$$
$$= a^2 b - c^2 d$$

The result $a^2 b - c^2 d$ is a rational expression. When two radical expressions are multiplied and result in a rational expression, then the radical expressions are called **conjugates.**

opposite signs

$(a\sqrt{b} + c\sqrt{d})$ and $(a\sqrt{b} - c\sqrt{d})$ are called **conjugates.**

same terms

same terms

3.9 Exercise

B 1 Find each product.

(a) $(\sqrt{3} - \sqrt{5})(\sqrt{3} + \sqrt{5})$

(b) $(2\sqrt{3} + \sqrt{2})(2\sqrt{3} - \sqrt{2})$

(c) $(3\sqrt{5} - \sqrt{6})(3\sqrt{5} + \sqrt{6})$

(d) $(2\sqrt{3} + 4\sqrt{2})(2\sqrt{3} - 4\sqrt{2})$

(e) What do you notice about your answers in (a) to (d)?

2 Find the product of each binomial and its corresponding conjugate.

(a) $\sqrt{5} - \sqrt{2}$ (b) $3\sqrt{5} + 2\sqrt{2}$ (c) $5\sqrt{7} + 2\sqrt{10}$ (d) $3\sqrt{2} - 2\sqrt{5}$

(e) $2\sqrt{3} - 8$ (f) $9 + 2\sqrt{5}$ (g) $3\sqrt{5} - 2\sqrt{10}$ (h) $3\sqrt{6} + \sqrt{8}$

(i) $\sqrt{18} - \sqrt{27}$ (j) $\sqrt{50} - 2\sqrt{80}$ (k) $3\sqrt{12} - 2\sqrt{32}$ (l) $4\sqrt{98} - 3\sqrt{45}$

3 Find each product.

(Look for conjugates to simplify your work.)

(a) $(\sqrt{3} - \sqrt{2})(3\sqrt{3} - \sqrt{2})(\sqrt{3} + \sqrt{2})$

(b) $(2\sqrt{5} - 3\sqrt{2})(2\sqrt{5} + 3\sqrt{2})(3\sqrt{5} - \sqrt{2})$

(c) $(4\sqrt{6} - 3\sqrt{2})(4\sqrt{6} + 3\sqrt{2}) - (2\sqrt{6} - \sqrt{8})^2$

(d) $3(2\sqrt{27} - 3)^2 - (2\sqrt{3} - 8)(\sqrt{12} + 8)$

(e) $(5\sqrt{6} - 2\sqrt{3})(3\sqrt{2} - \sqrt{3})(3\sqrt{2} + \sqrt{3})$

(f) $(3\sqrt{2} + 5\sqrt{3})(3\sqrt{2} - 5\sqrt{3}) - 4(\sqrt{18} - \sqrt{27})^2$

3.10 Dividing Radicals

Just as with the multiplication of radicals, numerical examples suggest a rule for dividing radicals.

Is $\dfrac{\sqrt{36}}{\sqrt{4}} \overset{?}{=} \sqrt{\dfrac{36}{4}}$

$\text{LS} = \dfrac{\sqrt{36}}{\sqrt{4}}$

$= \dfrac{6}{2}$

$= 3$

$\text{RS} = \sqrt{\dfrac{36}{4}}$

$= \sqrt{9}$

$= 3$

Since LS = RS then $\dfrac{\sqrt{36}}{\sqrt{4}} = \sqrt{\dfrac{36}{4}}$.

The numerical example suggests the rule.

$\dfrac{\sqrt{a}}{\sqrt{b}} \overset{?}{=} \sqrt{\dfrac{a}{b}}$ $a \geqq 0, b > 0$ which needs to be proved.

To prove the result, let $x = \dfrac{\sqrt{a}}{\sqrt{b}}$ ①

Since $a \geqq 0$ $b > 0$ then $x \geqq 0$. Then $x^2 = \left(\dfrac{\sqrt{a}}{\sqrt{b}}\right)^2$

$= \dfrac{(\sqrt{a})^2}{(\sqrt{b})^2}$

$= \dfrac{a}{b}$

Thus $x = \sqrt{\dfrac{a}{b}}$ ②   since $x \geqq 0$.

You obtain the result you wanted to prove from ① and ②.

$$\dfrac{\sqrt{a}}{\sqrt{b}} = \sqrt{\dfrac{a}{b}}, \quad a, b \in R, a \geqq 0, b > 0.$$

Example 1 Simplify $\dfrac{2\sqrt{10} + 3\sqrt{30}}{\sqrt{5}} + \dfrac{3\sqrt{6} - 4\sqrt{18}}{\sqrt{3}}$.

Solution $\dfrac{2\sqrt{10} + 3\sqrt{30}}{\sqrt{5}} + \dfrac{3\sqrt{6} - 4\sqrt{18}}{\sqrt{3}} = \dfrac{2\sqrt{10}}{\sqrt{5}} + \dfrac{3\sqrt{30}}{\sqrt{5}} + \dfrac{3\sqrt{6}}{\sqrt{3}} - \dfrac{4\sqrt{18}}{\sqrt{3}}$

$= 2\sqrt{2} + 3\sqrt{6} + 3\sqrt{2} - 4\sqrt{6}$

$= 5\sqrt{2} - \sqrt{6}$

In a similar way, you can prove the following for nth order radicals

$$\frac{\sqrt[n]{a}}{\sqrt[n]{b}} = \sqrt[n]{\frac{a}{b}}, \quad a, b \in R, n \in N \quad \text{If } n \text{ is even then } a \geq 0, b > 0.$$

Example 2   Simplify $\dfrac{2\sqrt[3]{15} - 3\sqrt[3]{10}}{\sqrt[3]{5}}$.

Solution $\dfrac{2\sqrt[3]{15} - 3\sqrt[3]{10}}{\sqrt[3]{5}} = \dfrac{2\sqrt[3]{15}}{\sqrt[3]{5}} - \dfrac{3\sqrt[3]{10}}{\sqrt[3]{5}}$

$$= 2\sqrt[3]{\frac{15}{5}} - 3\sqrt[3]{\frac{10}{5}}$$

$$= 2\sqrt[3]{3} - 3\sqrt[3]{2}$$

3.10 Exercise

A 1   Which of the following radical expressions are *not* in simplest form?

(a) $\dfrac{4\sqrt{5}}{3}$ (b) $\dfrac{\sqrt{33}}{\sqrt{11}}$ (c) $\dfrac{8\sqrt{2}}{16}$ (d) $3\sqrt{33}$ (e) $\dfrac{6\sqrt{3}}{2}$ (f) $\dfrac{-\sqrt{48}}{\sqrt{27}}$

(g) $3\sqrt{27}$ (h) $\sqrt{\dfrac{2}{3}}$ (i) $\dfrac{-3\sqrt{27}}{6}$ (j) $\dfrac{\sqrt[3]{16}}{2}$ (k) $\dfrac{\sqrt[3]{9}}{3}$ (l) $\sqrt[4]{\dfrac{1}{8}}$

2   (a) Simplify   A: $\dfrac{\sqrt{120}}{2}$ B: $\dfrac{\sqrt{120}}{\sqrt{2}}$

(b) Why do your answers differ?

3   Which two expressions do not have the same answer when written in simplest form?

(a) $\dfrac{3}{\sqrt{10}}$ (b) $\dfrac{\sqrt{9}}{\sqrt{10}}$ (c) $\dfrac{9}{\sqrt{30}}$ (d) $\dfrac{\sqrt{18}}{\sqrt{20}}$

(e) $\dfrac{6}{\sqrt{40}}$ (f) $\dfrac{3\sqrt{3}}{\sqrt{30}}$ (g) $\dfrac{2\sqrt{3}}{\sqrt{30}}$ (h) $\dfrac{\sqrt{90}}{10}$

B 4   Simplify.

(a) $\dfrac{6\sqrt{10}}{\sqrt{5}}$ (b) $\dfrac{2\sqrt{3}}{\sqrt{5}}$ (c) $\sqrt{\dfrac{2}{3}}$ (d) $8\sqrt{\dfrac{3}{4}}$ (e) $\dfrac{-6}{\sqrt{2}}$ (f) $-\dfrac{2}{\sqrt{12}}$

(g) $\dfrac{-2\sqrt{5}}{3\sqrt{2}}$ (h) $\dfrac{-3}{\sqrt{18}}$ (i) $\dfrac{30}{\sqrt{5}}$ (j) $-\dfrac{2\sqrt{6}}{\sqrt{2}}$ (k) $-\dfrac{2}{3}\sqrt{\dfrac{3}{5}}$ (l) $-\dfrac{1}{2\sqrt{5}}$

5 Simplify.

(a) $\dfrac{8}{\sqrt[3]{2}}$ (b) $\dfrac{12\sqrt[3]{6}}{\sqrt[3]{2}}$ (c) $\sqrt[3]{\dfrac{7}{2}}$ (d) $\dfrac{40}{\sqrt[4]{4}}$ (e) $\sqrt[4]{\dfrac{32}{4}}$ (f) $\sqrt[5]{\dfrac{1}{32}}$

6 Simplify. You may need to divide or multiply.

(a) $1 \div \sqrt{24}$

(b) $8 \times \sqrt{8}$

(c) $\dfrac{2\sqrt{84}}{-\sqrt{12}}$

(d) $(2\sqrt{75}) \div \sqrt{15}$

(e) $3\sqrt{3} \times \sqrt{27}$

(f) $\left(\dfrac{\sqrt{72}}{2\sqrt{8}}\right)(\sqrt{2})$

(g) $\dfrac{5\sqrt{25}}{\sqrt{75}} \div \sqrt{5}$

(h) $(2\sqrt{3})^2 \div \sqrt{3}$

(i) $\dfrac{2\sqrt{3}}{-4} \div \sqrt{27}$

(j) $\left(\dfrac{8\sqrt{5}}{\sqrt{2}}\right) \div \left(\dfrac{6\sqrt{5}}{-\sqrt{10}}\right)$

(k) $\dfrac{25\sqrt[3]{48}}{5\sqrt[3]{6}}$

(l) $\dfrac{\sqrt[4]{125}}{\sqrt[4]{80}} \div (\sqrt[4]{5})$

7 Write each of the following in simplest form.

(a) $\dfrac{-12\sqrt{22}}{4\sqrt{11}}$

(b) $3\sqrt{28} \div (-4\sqrt{7})$

(c) $\dfrac{-5\sqrt{3}}{\sqrt{6}}$

(d) $\dfrac{-4}{2\sqrt{8}}$

(e) $\left(\dfrac{\sqrt{8}}{\sqrt{27}}\right)\left(-\dfrac{\sqrt{25}}{\sqrt{5}}\right)$

(f) $\left(\dfrac{3\sqrt{2}}{2\sqrt{4}}\right) \div 3\sqrt{2}$

(g) $\dfrac{6\sqrt{7}}{3\sqrt{6}} \div \dfrac{4\sqrt{21}}{2\sqrt{3}}$

(h) $\left(\dfrac{-8\sqrt{50}}{2\sqrt{2}}\right)\left(\dfrac{\sqrt{8}}{4}\right)$

(i) $5\sqrt[3]{100} \div 25\sqrt[3]{5}$

(j) $(-3\sqrt[3]{20})(\sqrt[3]{50})$

8 Simplify. Express each answer with a rational denominator.

(a) $\dfrac{-4\sqrt{6} - 2\sqrt{18}}{\sqrt{3}}$

(b) $\dfrac{\sqrt{20} - 3\sqrt{10}}{4\sqrt{5}}$

(c) $\dfrac{36 - \sqrt{6}}{\sqrt{6}}$

(d) $\dfrac{3\sqrt{2} - 3\sqrt{3}}{2\sqrt{2}}$

(e) $\dfrac{2\sqrt{5} - 3\sqrt{2}}{-3\sqrt{2}}$

(f) $\dfrac{5\sqrt{5} - 2\sqrt{2}}{8\sqrt{2}}$

Problem Solving

There are many different functions. The **greatest integer function**, f, has the property that it maps any real number, x, onto the greatest integer which does not exceed x. In symbols, you write $f: x \to [x]$, $x \in R$. Use the definition. Draw the graph of the greatest integer function. What are some of its properties?

3.11 Radical Expressions: Rationalizing Denominators

As you learned earlier, a radical is not in simplest form if the radicand contains a fraction. For example, $\sqrt{\dfrac{2}{3}}$ in simplest form is

$$\sqrt{\frac{2}{3}} = \frac{\sqrt{2}}{\sqrt{3}} = \frac{\sqrt{2}}{\sqrt{3}} \times \frac{\sqrt{3}}{\sqrt{3}} = \frac{\sqrt{2} \times \sqrt{3}}{\sqrt{3} \times \sqrt{3}} = \frac{\sqrt{6}}{3}$$

The process of expressing the denominator as a rational expression and thus eliminating the radical denominator is called **rationalizing the denominator.**

When the denominator of a radical fraction is a binomial expression, you rationalize the denominator by multiplying the denominator by its conjugate.

Thus, $\dfrac{1}{\sqrt{2}+1}$ in simplest form is $\dfrac{1}{\sqrt{2}+1} = \dfrac{1}{\sqrt{2}+1} \times \dfrac{\sqrt{2}-1}{\sqrt{2}-1}$

$$= \frac{\sqrt{2}-1}{2-1}$$

$$= \sqrt{2}-1$$

Thus, $\dfrac{1}{\sqrt{2}+1} = \sqrt{2}-1$ in simplest form.

Example Simplify $\dfrac{3\sqrt{3}-2\sqrt{2}}{\sqrt{3}-\sqrt{2}}$.

Think: To obtain a denominator that is a rational number use a conjugate radical expression as shown.

Solution

$$\frac{3\sqrt{3}-2\sqrt{2}}{\sqrt{3}-\sqrt{2}} = \frac{(3\sqrt{3}-2\sqrt{2})(\sqrt{3}+\sqrt{2})}{(\sqrt{3}-\sqrt{2})(\sqrt{3}+\sqrt{2})}$$

$$= \frac{5+\sqrt{6}}{1}$$

$$= 5+\sqrt{6}$$

3.11 Exercise

A Express your answers in simplest form.

1 Express each radical fraction in simplest form by rationalizing the denominator.

(a) $\dfrac{1}{\sqrt{5}}$ (b) $\sqrt{\dfrac{2}{5}}$ (c) $\dfrac{3}{\sqrt{2}}$ (d) $\dfrac{4\sqrt{3}}{\sqrt{2}}$ (e) $\dfrac{5\sqrt{2}}{3\sqrt{3}}$ (f) $6\sqrt{\dfrac{5}{2}}$

2   Rationalize the denominator of each expression. Write your answer in simplest form.

(a) $\dfrac{\sqrt{3} + \sqrt{5}}{\sqrt{2}}$ (b) $\dfrac{2\sqrt{3} - 3\sqrt{2}}{\sqrt{2}}$ (c) $\dfrac{4\sqrt{3} + 3\sqrt{2}}{2\sqrt{3}}$ (d) $\dfrac{3\sqrt{5} - \sqrt{2}}{2\sqrt{2}}$

B   Express your answers in radical form.

3   Simplify. Express each answer in simplest form.

(a) $\dfrac{3}{\sqrt{5} - \sqrt{2}}$ (b) $\dfrac{2\sqrt{5}}{2\sqrt{5} + 3\sqrt{2}}$ (c) $\dfrac{\sqrt{3} - \sqrt{2}}{\sqrt{3} + \sqrt{2}}$

(d) $\dfrac{2\sqrt{5} - 8}{2\sqrt{5} + 3}$ (e) $\dfrac{2\sqrt{3} - \sqrt{2}}{5\sqrt{2} + \sqrt{3}}$ (f) $\dfrac{3\sqrt{3} - 2\sqrt{2}}{3\sqrt{3} + 2\sqrt{2}}$

4   (a) Rationalize the denominator of $\dfrac{8\sqrt{2}}{\sqrt{20} - \sqrt{18}}$.

(b) Rationalize the denominator of $\dfrac{8\sqrt{2}}{2\sqrt{5} - 3\sqrt{2}}$.

(c) Why are your answers in (a) and (b) the same?

5   Express each of the following in simplest form.

(a) $\dfrac{2\sqrt{2}}{2\sqrt{3} - \sqrt{8}}$ (b) $\dfrac{8}{2\sqrt{75} - 3\sqrt{50}}$ (c) $\dfrac{2\sqrt{6}}{2\sqrt{27} - \sqrt{8}}$

(d) $\dfrac{2\sqrt{2}}{\sqrt{16} - \sqrt{12}}$ (e) $\dfrac{3}{2\sqrt{80} - \sqrt{45}}$ (f) $\dfrac{3\sqrt{2} + 2\sqrt{3}}{\sqrt{12} - \sqrt{8}}$

(g) $\dfrac{3\sqrt{5}}{4\sqrt{3} - 5\sqrt{2}}$ (h) $\dfrac{2\sqrt{3} - \sqrt{2}}{\sqrt{12} + \sqrt{8}}$ (i) $\dfrac{\sqrt{18} + \sqrt{12}}{\sqrt{18} - 3\sqrt{12}}$

6   If $m = 2\sqrt{3} + \sqrt{5}$ and $n = 2\sqrt{3} - \sqrt{5}$, express each of the following expressions in simplest form.

(a) $\dfrac{mn + m^2}{m}$ (b) $\dfrac{m^2 - n^2}{m + n}$

7   If $m = \sqrt{2} - 3\sqrt{3}$ and $n = 2\sqrt{2} + \sqrt{3}$, write each of the following in simplest form.

(a) $\dfrac{m^2 + 3mn}{m}$ (b) $\dfrac{n^2 - 2mn}{n^2}$ (c) $\dfrac{m^2 + 2mn + n^2}{m + n}$ (d) $\dfrac{m^2 - n^2}{mn + m^2}$

8   Express the area of each triangle in simplest form.

(a)

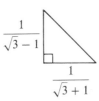

(b)

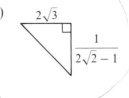

9   Express the area of each rectangle in simplest form.

(a) $3\sqrt{5}$

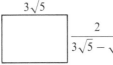

$\dfrac{2}{3\sqrt{5}-\sqrt{2}}$

(b) $3\sqrt{5}-2\sqrt{2}$

$\dfrac{1}{3\sqrt{5}+\sqrt{2}}$

10   Express the hypotenuse of each triangle in simplest form.

(a)

(b)

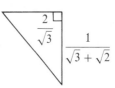

C 11   Without using decimal equivalents, prove that

$2 + \dfrac{1}{\sqrt{2}}$ is greater than $3 - \dfrac{1}{\sqrt{3}}$.

Calculator Tip

Use a calculator to evaluate each expression to 1 decimal place.

A $\dfrac{1}{2\sqrt{3}-\sqrt{2}}$

B $\dfrac{2\sqrt{3}+\sqrt{2}}{10}$

What do you notice about your answers?

Problem Solving

The floor of a gym is 12 m × 30 m and is 12 m high. A spider sits 1 m above the floor on the centre line of the east wall. A sleeping bug is 1 m from the ceiling on the west wall. What is the shortest distance from the spider to the bug? (Remember spiders don't fly.)

3.12 Solving Radical Equations

In each of the following equations, the variables occur in the radicand.

$$\sqrt{y} = 4 \qquad \sqrt{3m + 1} = 7 \qquad \sqrt{x + 8} - \sqrt{x} = 2$$

Equations such as those above are called **radical equations.**

The steps you have learned earlier for solving equations are applied to solving radical equations.

Example 1   Solve $y - 2 = \sqrt{y - 1} + 1$.   Think: Isolate the term with the radical.

Solution
$$y - 2 = \sqrt{y - 1} + 1$$
$$y - 3 = \sqrt{y - 1}$$
$$y^2 - 6y + 9 = y - 1 \qquad \text{Square both sides of the equation.}$$
$$y^2 - 7y + 10 = 0 \qquad \text{Use your earlier skills in algebra to factor.}$$
$$(y - 5)(y - 2) = 0$$
$$y - 5 = 0 \quad \text{or} \quad y - 2 = 0$$
$$y = 5 \qquad\qquad y = 2$$

Verify

$y = 5$ LS $= y - 2$ RS $= \sqrt{y - 1} + 1$ | $y = 2$ LS $= y - 2$ RS $= \sqrt{y - 1} + 1$
$\quad = 5 - 2 \qquad\quad = \sqrt{5 - 1} + 1$ | $\quad = 2 - 2 \qquad\quad = \sqrt{2 - 1} + 1$
$\quad = 3 \qquad\qquad\quad = \sqrt{4} + 1 = 3$ | $\quad = 0 \qquad\qquad\quad = 1 + 1 = 2$

LS = RS | LS ≠ RS
5 is a root. | 2 is not a root.

In solving the radical equation above, the process of squaring the radical in order to eliminate the radical sign was introduced. However, in doing so, values of the variable which are not roots of the equation have now occurred, since 2 is not a root of the original radical equation.

To see why this has occurred, consider the following.

Step 1   The equation $x - 1 = 4$ has the root 5.
Step 2   Square both sides of the above equation.
$$(x - 1)^2 = 4^2$$
$$x^2 - 2x + 1 = 16$$
$$x^2 - 2x - 15 = 0$$
$$(x + 3)(x - 5) = 0$$
$$x + 3 = 0 \qquad \text{or} \quad x - 5 = 0$$
$$x = -3 \quad \text{or} \qquad\quad x = 5$$

In Step 2, another value of the variable, -3, has now occurred, which is not a root of the original equation. The previous example illustrates that the method of squaring both sides of the equation does not always result in an equivalent equation. Squaring may introduce an "extra" value of the variable, which may not be a root. Such values are called **extraneous roots.**

Thus, you *must* verify all values of the variable obtained when solving a radical equation.

To solve certain radical equations, such as those in Example 2, the process of squaring is done twice to simplify the equation.

Example 2   Solve $\sqrt{2m+1} + \sqrt{m} = 5$.

Solution

$\sqrt{2m+1} + \sqrt{m} = 5$ Think: To simplify the solution, rewrite
$\sqrt{2m+1} = 5 - \sqrt{m}$ the equation as shown.

$2m + 1 = 25 - 10\sqrt{m} + m$ Square both sides of the equation.

$10\sqrt{m} = 24 - m$ Isolate the term containing the radical.

$100m = 576 - 48m + m^2$ Square both sides of the equation again.

$m^2 - 148m + 576 = 0$
$(m - 144)(m - 4) = 0$
$m = 144$ or $m = 4$ ⟶   Check: You must verify both answers
The root of the equation is 4.   in the original equation. Which value.
                                  4 or 144, is extraneous in this case?

3.12   Exercise

A   Questions 1 to 4 examine skills needed to solve radical equations.

1   Simplify each of the following.
   (a) $(\sqrt{2x-1})^2$ (b) $(3\sqrt{x}-1)^2$ (c) $(8 - 2\sqrt{m})^2$
   (d) $(2 + \sqrt{x-5})^2$ (e) $(3\sqrt{m-2}-2)^2$ (f) $(8 - 3\sqrt{y-1})^2$

2   If $y = 3$, find the value of each expression.
   (a) $\sqrt{3y} + 3$ (b) $\sqrt{3y-5}$ (c) $\sqrt{15-2y}$
   (d) $3\sqrt{3y} - 2$ (e) $\sqrt{2y-5} + \sqrt{3y}$ (f) $5\sqrt{4-y} - \sqrt{3y-5}$

3   Calculate the value of each radical expression for the given value.

   (a) $\sqrt{2x+1}$, $x = 144$ (b) $2 + \sqrt{4y-1}$, $y = \dfrac{1}{2}$

   (c) $\sqrt{4m+6} - 2m$, $m = -\dfrac{1}{2}$ (d) $\sqrt{2k+1} - \sqrt{k}$, $k = 144$

   (e) $\sqrt{2-m} + \sqrt{11+m}$, $m = -2$ (f) $\sqrt{2p+3} + \sqrt{3-p}$, $p = 3$

4   Two values are given for the variable in each equation. Use verification to indicate whether both, one, or none of them is a root.

(a) $3 + \sqrt{3x + 1} = 10$;   16, 5   (b) $\sqrt{y + 2} = y - 1$;   2, 7

(c) $\sqrt{5x - 6} = x$;   2, 3   (d) $2 + \sqrt{3m - 5} = \sqrt{4m + 1}$;   2, 42

B   Remember: You *must* verify all the values obtained for the variable when solving a radical equation.

5   Solve each equation.   Remember to verify.

(a) $\sqrt{x} = 6$ (b) $\sqrt{m} = 3$ (c) $\sqrt{k} = 2$

(d) $\sqrt{x + 1} = 8$ (e) $\sqrt{y - 1} = 3$ (f) $3\sqrt{y + 1} = 9$

6   Solve.

(a) $\sqrt{y - 1} = 4$ (b) $\sqrt{y - 1} = 4$ (c) $\sqrt{2x - 1} = 3$

(d) $\sqrt{2x - 1} = 3$ (e) $8 - \sqrt{2m} = 0$ (f) $3\sqrt{x} - 1 = 6$

(g) $3\sqrt{y - 2} = 2\sqrt{y + 4}$ (h) $6(\sqrt{y} - 2) = 3(\sqrt{y} + 3)$

7   Two values 6, 14 for the variable are given for $7 - \sqrt{2y + 4} = \sqrt{2y - 3}$.

Which value is a root?

8   Solve. Remember to verify.

(a) $\sqrt{26y - 16} = 3y$ (b) $\frac{1}{2}x = \sqrt{2x - 4}$

(c) $\sqrt{5p + 4} = 5 - 2p$ (d) $2t + 4 = 1 + \sqrt{19t + 6}$

9   Solve.

(a) $\sqrt{3x + 1} = \sqrt{5x + 1}$ (b) $\sqrt{2y + 1} + \sqrt{y} = 5$

(c) $\sqrt{p - 9} + \sqrt{p + 1} = 1$ (d) $\sqrt{2y + 5} - \sqrt{y - 2} = 3$

10   Solve each equation. How many roots does each equation have?

(a) $\sqrt{3 - x} + \sqrt{2x + 3} = 3$ (b) $\sqrt{2y + 1} - \sqrt{y} = 5$

(c) $\sqrt{y - 5} = 6 - \sqrt{y}$

11   Solve.

(a) $\dfrac{\sqrt{2x + 3}}{\sqrt{5x + 9}} = 1$ (b) $1 + \dfrac{\sqrt{y + 4}}{\sqrt{y - 3}} = \dfrac{7}{\sqrt{y - 3}}$

C 12   Solve each of the following equations.

(a) $\sqrt[3]{2x + 1} = 3$ (b) $\sqrt[3]{1 - 3x} - 4 = 0$ (c) $\sqrt[3]{x^2 - 19} = 5$

3.13 Nature of Mathematics: Solving Problems

Throughout your study of mathematics, the process shown in steps A, B, and C often occurs.

| Step A Learn skills and concepts for a topic or branch of mathematics. | Step B Acquire and learn various strategies for solving problems. | Step C Apply these skills and strategies for solving problems and explore applications. |

In order to do an effective job, you need to plan your solution. Refer to the *Steps for Solving Problems* to help you organize your solution.

Steps for Solving Problems

Step A Read the problem carefully. Can you answer these two questions?

    I What information am I asked to find (information you don't know)?

    II What information am I given (information you know)?

    Be sure that you understand what it is you are to find. Plan your work, introduce variables, draw a diagram, record the given information on the diagram, etc.

Step B Translate from English to mathematics. Write any equations.

Step C Do the work. Solve the equations.

Step D Check the answers in the original problem, equation, etc.

Step E Write a final statement as the answer to the problem.

Example

Two rectangles have dimensions in metres expressed as radical expressions.

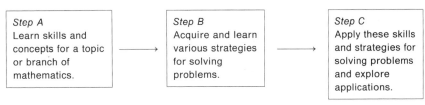

If the difference in the lengths of their diagonals is 2 m, find the dimensions of the rectangles.

Solution

The length of diagonal AC, in metres, is given by

$$AC^2 = (\sqrt{2x + 1})^2 + (\sqrt{x})^2$$
$$= 2x + 1 + x = 3x + 1$$
$$AC = \sqrt{3x + 1}$$

The length of diagonal QS, in metres, is given by

$$QS^2 = (\sqrt{x})^2 + (1)^2 = x + 1$$
$$QS = \sqrt{x + 1}$$

See next page. ⟶

$$\sqrt{3x + 1} - \sqrt{x + 1} = 2 \longleftarrow$$

Think: Translate the given information to obtain the appropriate equation, namely, the difference in the length of the diagonals is 2 m.

$$\sqrt{3x + 1} = 2 + \sqrt{x + 1}$$
$$3x + 1 = 4 + 4\sqrt{x + 1} + x + 1$$
$$2x - 4 = 4\sqrt{x + 1}$$
$$x - 2 = 2\sqrt{x + 1}$$

$$x^2 - 4x + 4 = 4(x + 1)$$
$$x^2 - 4x + 4 = 4x + 4$$
$$x^2 - 8x = 0$$
$$x(x - 8) = 0$$
$$x = 0 \quad \text{or} \quad x - 8 = 0$$
$$x = 8$$

Check the values in the original problem. The value 0 is extraneous. The root is 8.

Verify your results in the original problem.

For rectangle ABCD

$$BD^2 = (\sqrt{17})^2 + (\sqrt{8})^2$$
$$= 17 + 8 = 25$$
$$BD = 5 \longleftarrow \text{Length of diagonal is 5 m.}$$

For rectangle PQRS

$$QS^2 = (\sqrt{8})^2 + (1)^2$$
$$= 8 + 1 = 9$$
$$QS = 3 \longleftarrow \text{The length of the diagonal is 3 m.}$$

$$BD - QS = 5 \text{ m} - 3 \text{ m}$$
$$= 2 \text{ m} \quad \text{Checks} \checkmark$$

Thus, the dimensions of ABCD are $2\sqrt{2}$ m $\times$ $\sqrt{17}$ m and of PQRS are 1 m $\times$ $2\sqrt{2}$ m.

3.13 Exercise

A Refer to the *Steps for Solving Problems* to help you organize your solution.

1 Two rectangles have sides as shown. The diagonal of rectangle A is 2 units longer than the diagonal of rectangle B. Find the length of each rectangle.

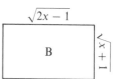

(a) Find a radical expression for the diagonal of rectangle A.

(b) Find a radical expression for the diagonal of rectangle B.

(c) Use the information to write an appropriate radical equation to solve the problem.

(d) Solve the equation. Check the values.

(e) What is the answer to the problem?

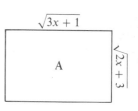

2   Read the following problem.

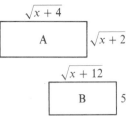

The difference in the lengths of the diagonals in the rectangles shown is 4 units. Find the area of each rectangle, if the diagonal of rectangle A is the greater.

(a) What is your first step in solving the problem?
(b) Write an appropriate radical equation to solve the problem.
(c) Write a complete solution to the problem.

B   For each problem: • decide on the skills needed to solve the problem.
        • write the appropriate radical equation.   • solve the problem.

3   If the difference in length of the hypotenuses of the triangles shown is 3, find the length of each hypotenuse, AB > DF.

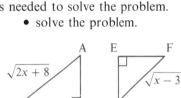

4   The difference in lengths of the hypotenuses shown is 2. Find the length of each hypotenuse, if ST > YW.

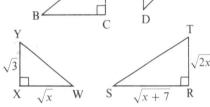

5   The lengths of the hypotenuses of the triangles shown differ by 3. Find the area of each triangle, if UT > MR.

6   Find a number that has the property that the sum of the square root of the number and the square root of 3 less than the number is 3.

7   The square root of, five times a number increased by 34, diminished by the square root of, five times the number increased by six, gives a result of 2. What is the number?

8   The sum of the square root of three times a number increased by 1 and the square root of 4 less than the number is 5. Find the number.

C 9   Carpenters use templates to construct designs on furniture. One such template is in the shape of a rhombus. If the area of the rhombus is 9 cm², calculate the lengths of the diagonals if their dimensions in centimetres are $\sqrt{x} + 2$ and $\sqrt{x} - 1$.   The area of a rhombus is given by

$$\text{Area} = \frac{1}{2} D_1 \times D_2$$

where D_1 and D_2 are the lengths of the diagonals.

3.14 Problem-Solving: Look For Clues

Often the skills and strategies acquired to solve a particular problem can be used in different ways to solve a problem not met before. For example, you are given

$$\sqrt{y + 12} + \sqrt{y} = 6 \qquad \text{①}$$

and are asked to find the value of

$$\sqrt{y + 12} - \sqrt{y} \qquad \text{②}$$

From equation ①, you could directly solve for y and then use this value of y to substitute in ② and eventually find the value of the expression ②.

However, before directly involving yourself with many calculations, ask these questions:

How is   related to

| A: What am I given? | $\Longrightarrow$ | B: What am I asked to find | ? |

In other words, based on the information given in the problem, ask yourself:

How is   related to

| A: $\sqrt{y + 12} + \sqrt{y} = 6$ | $\Longrightarrow$ | B: $\sqrt{y + 12} - \sqrt{y}$ | ? |

In exploring the relationship between A and B, you notice that A and B are conjugates! Thus, you can use this information to simplify your work. Use the equation.

Since $\qquad \sqrt{y + 12} + \sqrt{y} = 6$

then multiply both sides by $\sqrt{y + 12} - \sqrt{y}$

$$\overset{\text{Assume}}{\sqrt{y + 12} - \sqrt{y}} \neq 0$$

$$(\sqrt{y + 12} - \sqrt{y})(\sqrt{y + 12} + \sqrt{y}) = 6(\sqrt{y + 12} - \sqrt{y})$$

$$(y + 12 - y) = 6(\sqrt{y + 12} - \sqrt{y})$$

$$12 = 6(\sqrt{y + 12} - \sqrt{y})$$

Thus. $\qquad \sqrt{y + 12} - \sqrt{y} = 2.$

The strategy for solving a problem of first relating the information you are given to the information you are asked to find, will help you avoid many tedious calculations.

3.14 Exercise

A 1 (a) Simplify $(\sqrt{2x-7} - \sqrt{2x})(\sqrt{2x-7} + \sqrt{2x})$.

   (b) Use your results in (a) to find the value of
   $\sqrt{2x-7} + \sqrt{2x}$ if $\sqrt{2x-7} - \sqrt{2x} = 1$.

2 (a) Simplify $(\sqrt{3x+1} + \sqrt{3x})(\sqrt{3x+1} - \sqrt{3x})$.

   (b) Show that when $\sqrt{3x+1} + \sqrt{3x} = 1$, then $\sqrt{3x+1} - \sqrt{3x} = 1$.

B 3 (a) Solve the equation $\sqrt{x+4} + \sqrt{x} = 8$.

   (b) Use your results in (a) to calculate $\sqrt{x+4} - \sqrt{x}$.

   (c) What other method can be used to find the value of $\sqrt{x+4} - \sqrt{x}$ *without* solving the equation in (a)?

4 (a) If the value of the expression $\sqrt{7y+8} + \sqrt{7y}$ is 8, then find the value of
   $\sqrt{7y+8} - \sqrt{7y}$.

   (b) Find the value of $\sqrt{x+9} + \sqrt{x}$ if you know $\sqrt{x+9} - \sqrt{x} = 3$.

   (c) What is the value of $\sqrt{x-9} - \sqrt{x}$ if you know $\sqrt{x-9} + \sqrt{x} = 9$?

5 (a) If $\sqrt{2x-3} + \sqrt{2x} = 3$, then find the value of $\sqrt{2x-3} - \sqrt{2x}$.

   (b) If $\sqrt{9-x} - \sqrt{4-x} = 1$, then what is the value of $\sqrt{9-x} + \sqrt{4-x}$?

6 (a) If the value of $\sqrt{4x+3} + \sqrt{4x}$ is 3, what is the value of $\sqrt{4x+3} - \sqrt{4x}$?

   (b) Determine the value of $\sqrt{3x-5} - \sqrt{3x}$ if $\sqrt{3x-5} + \sqrt{3x} = 5$.

C 7 Use the given information in A, B, C, D following

   A: $\sqrt{5x+11} + \sqrt{5x} = 7$ C: $\sqrt{x+8} + \sqrt{x} = 4$

   B: $\sqrt{3x+6} - \sqrt{3x} = \dfrac{2}{3}$ D: $\sqrt{x-3} + \sqrt{x} = -1$

   to calculate the value of each of the following expressions.

   (a) $\sqrt{3x} + \sqrt{3x+6}$ (b) $\sqrt{x} - \sqrt{x+8}$

   (c) $-(\sqrt{x} - \sqrt{x-3})$ (d) $\sqrt{5x+11} - \sqrt{5x}$

Math Tip

It is important to understand clearly the vocabulary of mathematics when solving problems. Bring your list of new words up to date. Provide an example to illustrate the meaning of each word. Continue to add to your vocabulary list.

Practice and Problems: A Chapter Review

An important step for problem-solving is to decide which skills to use. For this reason, these questions are not placed in any special order. When you have finished the review, you might try the *Test for Practice* that follows.

1 Simplify each of the following.

(a) $\dfrac{3^{\frac{7}{12}}}{3^{\frac{1}{3}} \times 3^{\frac{1}{4}}}$

(b) $\dfrac{64^{\frac{5}{6}} \times 16^{\frac{3}{4}}}{27^{\frac{2}{3}}}$

(c) $\dfrac{x^{\frac{1}{2}} \times x^{\frac{2}{3}}}{x^{\frac{1}{4}}}$

(d) $\dfrac{x^{\frac{5}{6}} \times x^{\frac{2}{3}}}{x^{\frac{1}{2}}}$

(e) $(-3\sqrt[3]{4})(2\sqrt[3]{4})$

(f) $(-5\sqrt[3]{3})(2\sqrt[3]{9})$

2 Solve. Draw a graph of the solution set.

(a) $2 - |x| < -4$

(b) $\dfrac{2}{3}|x| \geq 6$

(c) $\dfrac{1}{2}|x| - 2 \leq 6$

(d) $3 - |2x| < 0$

3 Simplify.

(a) $\sqrt{12} + 5\sqrt{27} - \sqrt{48}$

(b) $3\sqrt{8} + \sqrt{18} - 5\sqrt{72}$

(c) $13\sqrt{8} + \sqrt{27} - 2\sqrt{18} + \sqrt{3}$

(d) $6\sqrt{27} - \sqrt{396} - 8\sqrt{108} - 7\sqrt{99}$

4 Simplify.

(a) $3a\sqrt{32} - 4a\sqrt{27} + 3a\sqrt{12} - 4a\sqrt{50}$

(b) $3m\sqrt{63} - 2m\sqrt{45} + m\sqrt{20} - 2m\sqrt{28}$

5 Find a radical expression for the shaded area.

$3\sqrt{50} - 2\sqrt{32} + \sqrt{12}$

$\sqrt{12} - \sqrt{8}$

$2\sqrt{18} - \sqrt{8}$

$\sqrt{12} - \sqrt{8}$

6 Solve and verify $3|k| + 2|k| - 5 = 2|k| + 7 + |k|$, $k \in R$.

7 Solve

(a) $4 - \sqrt{x - 2} = x$

(b) $\sqrt{4 - 6x} - 1 = \sqrt{-5x - 1}$

8 Show that $3\sqrt{5} + 1$ is a root of $x^2 - 2x - 44 = 0$.

9 A number b has the property that the square root of 3 more than the number is equal to 5 diminished by the square root of 2 less than the number. Calculate b.

Test For Practice

Try this test. Each test will be based on the mathematics you have learned in the chapter. Try this test later in the year as a review. Keep a record of those questions that you were not successful with and review them periodically.

1 (a) If $a = 2$, $b = -3$, $c = -2$, find the value of $3|a| - 2|b| + 6|c|$.
   (b) Solve and verify $2|x| + |x| = 10 - 2|x|$, $x \in R$

2 Simplify.
   (a) $\left(\dfrac{81}{144}\right)^{-0.5}$ (b) $\dfrac{64^{-\frac{5}{6}}}{81^{-\frac{3}{4}}}$ (c) $\left(\dfrac{27}{216}\right)^{\frac{1}{3}}$ (d) $\dfrac{(0.09)^{-\frac{1}{2}}}{(0.04)^{-\frac{1}{2}}}$

3 Simplify.
   (a) $4b\sqrt{12} - 2b\sqrt{27} + b\sqrt{75}$ (b) $m\sqrt{20} - 4m\sqrt{80} + 2m\sqrt{45}$
   (c) $3(2\sqrt{2} - 1)(3\sqrt{2} + 2) - 3(\sqrt{2} - 1)$
   (d) $(2\sqrt{5} - \sqrt{2})^2 - 3(\sqrt{5} - \sqrt{2})(\sqrt{5} + \sqrt{2})$

4 Simplify.
   (a) $\dfrac{-3\sqrt{5}}{4\sqrt{2}}$ (b) $(2\sqrt[3]{2})(3\sqrt[3]{4})$ (c) $\sqrt[3]{20} \times \sqrt[3]{50}$

5 Express $\dfrac{2\sqrt{18} + \sqrt{12}}{\sqrt{18} + 3\sqrt{12}}$ in simplest form.

6 (a) Write $\sqrt[3]{27y^6}$ in simplest form.   (b) Show that $8\sqrt{3} < 5\sqrt{2}$.
   (c) Show that $7 - \sqrt{5}$ is a root of $x^2 - 14x + 44 = 0$.

7 Solve
   (a) $\sqrt{x - 2} = x - 4$.   (b) $\sqrt{2x + 4} - \sqrt{x - 2} = 2$.

8 Which figure has the greater perimeter, A or B?
   A

   $2\sqrt{2}$ $\sqrt{2}$

   B

   $\sqrt{3}$

9 A number has the following property. The square root of 3 less than the number, increased by the square root of 4 more than the number is 7. Find the number.

Maintaining Skills

1   Write each of the following as an entire radical.
    (a) $2\sqrt{2}$ (b) $-3\sqrt{3}$ (c) $5\sqrt{5}$
    (d) $2\sqrt{27}$ (e) $3\sqrt{8}$ (f) $4\sqrt{27}$

2   Write each of the following as a mixed radical in simplest form.
    (a) $3\sqrt{48}$ (b) $-2\sqrt{27}$ (c) $\dfrac{1}{4}\sqrt{32}$
    (d) $2\sqrt{50}$ (e) $-3\sqrt{150}$ (f) $2\sqrt{98}$

3   Simplify.
    (a) $-4\sqrt{12} + 2\sqrt{27} - 3\sqrt{48} + 6\sqrt{75}$
    (b) $8\sqrt{8} - 4\sqrt{27} - 3\sqrt{18} + 5\sqrt{3}$
    (c) $6\sqrt{20} - 9\sqrt{245} - 2\sqrt{20} + 3\sqrt{125}$
    (d) $3\sqrt{45} - 2\sqrt{63} - 4\sqrt{125} - 3\sqrt{112}$

4   Find each product and simplify.
    (a) $4\sqrt{3}(2\sqrt{8} - 3\sqrt{18})$ (b) $-3\sqrt{2}(4\sqrt{12} - 2\sqrt{27})$
    (c) $-4\sqrt{5}(2\sqrt{20} - 5\sqrt{45})$ (d) $(2\sqrt{3} - \sqrt{2})(3\sqrt{3} - \sqrt{2})$
    (e) $(3\sqrt{6} - 2\sqrt{2})(2\sqrt{6} + 3\sqrt{2}) - (4\sqrt{6} - \sqrt{8})^2$

5   Write the conjugate of each radical expression.
    (a) $2\sqrt{3}$ (b) $\sqrt{3} + \sqrt{2}$ (c) $-2\sqrt{3} - \sqrt{2}$
    (d) $3\sqrt{3} + \sqrt{2}$ (e) $\sqrt{2} - \sqrt{5}$ (f) $-\sqrt{5} + 2\sqrt{2}$

6   Rationalize the denominator of each expression. Write your answer in simplest form.
    (a) $\dfrac{\sqrt{2} + \sqrt{5}}{\sqrt{3}}$ (b) $\dfrac{\sqrt{5} - \sqrt{2}}{\sqrt{5} + \sqrt{2}}$ (c) $\dfrac{2\sqrt{3} - \sqrt{2}}{3\sqrt{2} - \sqrt{3}}$ (d) $\dfrac{2\sqrt{3} - 3\sqrt{2}}{2\sqrt{3} + 3\sqrt{2}}$

7   If $a = -3$ and $b = -2$, find the value of each of the following.
    (a) $3|a| - 4|b|$ (b) $|a| - 2|b|$ (c) $|a - b|$
    (d) $4|a| - 3b$ (e) $2|b - a|$

8   Solve.
    (a) $2|y| - 1 = 7$ (b) $3|k| = |k| + 4$
    (c) $6|y| - 14 = 4|y| + 6$ (d) $3|x| - 8 = 5|x| - 26$
    (e) $4|y + 2| = 3$ (f) $2|y - 3| + 1 = 4$
    (g) $|m - 4| = 5m$ (h) $|2k - 3| = 4k - 2$

4 Trigonometry: Concepts and Skills

definitions of trigonometry, trigonometric values, angles and quadrants, radians, graphs of trigonometric functions, transformations, amplitude, period, phase shift, reciprocal functions, solving problems and strategies, applications, problem-solving

Introduction

With the invention of co-ordinate geometry by René Descartes, the study of angles was transferred to the co-ordinate plane. This development considerably advanced the early Greek studies of angles and sides in triangles. The origins of the word, trigonometry, lie in three Greek words.

trigonometry

tri	**gono**	**metria**
three	**angle**	**measurement**

The study of trigonometry is used to solve many problems. For example
- problems in economics
- problems in engineering
- problems in space
- problems in energy development.

One of the earliest uses of trigonometry was in surveying. Many of the structures of the last century, such as buildings, bridges, tunnels, were constructed using the principles of trigonometry and the solution of triangles. The study of the solution of triangles is dealt with in the next chapter. Modern equipment and trigonometry have helped surveyors refine their art to a science.

Trigonometry plays a part in the study of economic cycles that are periodic. Traders and brokers rely on predictions of future trends in the business cycle when trading stocks. Analysts supply them with this information.

Many of the calculations required in the production of electrical energy require skills in trigonometry. The Pickering Power Station above, is one of the world's largest nuclear power undertakings.

4.1 The Definitions of Trigonometry

The study and applications of trigonometry have a varied and rich history. Many of the early concepts and skills are related to problems based on astronomy and land measurement — all involving triangles. Today, the concepts and skills of trigonometry have a much wider scope, and occur in many fields such as electronics, engineering, and space exploration.

The Cartesian plane is fundamental in the study of trigonometry. Angles are related to the co-ordinate axes and associated with rotation.

The vertex of the angle is placed at the origin and the initial arm placed along the x-axis. $\angle$ AOC is said to be in **standard position**.

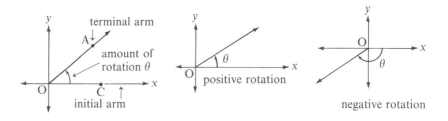

When the acute angle θ was placed in standard position, it was noticed that similar triangles were formed.

$$\triangle OP'Q' \sim \triangle OP''Q'' \sim \triangle OP'''Q'''$$

Thus,

$$\frac{P'Q'}{OP'} = \frac{P''Q''}{OP''} = \frac{P'''Q'''}{OP'''} = \text{constant value}$$

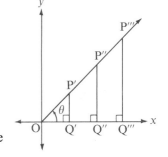

The constant nature of the diagram suggested the very important definitions of trigonometry as follows.

For any point, P, on the terminal arm, and **PQ** perpendicular to the initial arm (x-axis),

$$OQ = x. \qquad PQ = y. \qquad OP = r.$$

$$r^2 = x^2 + y^2$$
$$r = \sqrt{x^2 + y^2}, \quad r > 0.$$

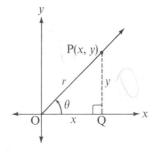

The following definitions then form the basis of trigonometry.

Primary Trigonometric Values

$$\text{sine } \theta = \frac{y}{r} \qquad \text{cosine } \theta = \frac{x}{r} \qquad \text{tangent } \theta = \frac{y}{x}$$

$$\sin \theta = \frac{y}{r} \qquad \cos \theta = \frac{x}{r} \qquad \tan \theta = \frac{y}{x}$$

By writing the reciprocals of the above, other trigonometric values are defined.

Reciprocal Trigonometric Values

$$\text{cosecant } \theta = \frac{r}{y} \qquad \text{secant } \theta = \frac{r}{x} \qquad \text{cotangent } \theta = \frac{x}{y}$$

$$\csc \theta = \frac{r}{y} \qquad \sec \theta = \frac{r}{x} \qquad \cot \theta = \frac{x}{y}$$

To calculate the trigonometric values you need find only a point on the terminal arm.

Example 1 The point (3, 4) is on the terminal arm of angle θ as shown. Calculate the trigonometric values.

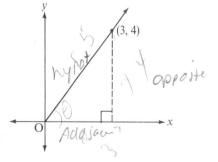

Solution From the diagram $r = \sqrt{x^2 + y^2}$
$$= \sqrt{(3)^2 + (4)^2}$$
$$= 5.$$

Use $x = 3$, $y = 4$, $r = 5$.

$$\sin \theta = \frac{y}{r} = \frac{4}{5} \quad \bigg| \quad \cos \theta = \frac{x}{r} = \frac{3}{5} \quad \bigg| \quad \tan \theta = \frac{y}{x} = \frac{4}{3}$$

Trigonometric values are often expressed in fractional form.

$$\csc \theta = \frac{r}{y} = \frac{5}{4} \quad \bigg| \quad \sec \theta = \frac{r}{x} = \frac{5}{3} \quad \bigg| \quad \cot \theta = \frac{x}{y} = \frac{3}{4}$$

In the next example, the terminal arm of the angle occurs in the second quadrant.

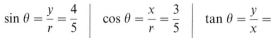

Second quadrant   First quadrant

Third quadrant   Fourth quadrant

Example 2 P(-8, 15) is a point on the terminal arm of angle α. Calculate its trigonometric values.

Solution From the diagram,

$r = \sqrt{x^2 + y^2}$

$r = \sqrt{(-8)^2 + (15)^2}$

$r = 17$

Use $x = -8$, $y = 15$, $r = 17$

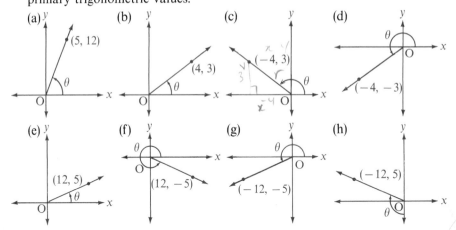

Sketch a diagram with the given information.

$$\sin \alpha = \frac{y}{r} = \frac{15}{17} \qquad \cos \alpha = \frac{x}{r} = \frac{-8}{17} \qquad \tan \alpha = \frac{y}{x} = \frac{15}{-8}$$

$$\csc \alpha = \frac{r}{y} = \frac{17}{15} \qquad \sec \alpha = \frac{r}{x} = \frac{17}{-8} \qquad \cot \alpha = \frac{x}{y} = \frac{-8}{15}$$

4.1 Exercise

A You may express answers for trigonometric values in fractional or radical form.

1 For each angle, a point on the terminal arm is shown. Calculate the primary trigonometric values.

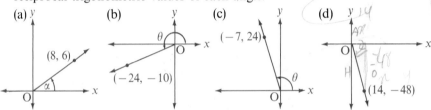

2 A point on the terminal arm for each angle is shown. Calculate the reciprocal trigonometric values of each angle.

3   The point P(2, 7) lies on the terminal arm of θ. For θ, find the
    (a) sine   (b) cosine   (c) tangent

4   The point Q(-9, 11) lies on the terminal arm of α. For α, calculate its
    (a) cosecant   (b) secant   (c) cotangent

B 5   (a) Angle θ is a second quadrant angle, and $\cos \theta = -\dfrac{3}{4}$. Sketch a diagram.

   (b) Find the other primary trigonometric values of θ.

6   (a) Angle θ is in the third quadrant and $\tan \theta = \dfrac{3}{4}$. Find the reciprocal
       trigonometric values.

   (b) Angle α is in the fourth quadrant and $\cos \alpha = \dfrac{8}{15}$. Find the other
       trigonometric values.

7   (a) θ is a first quadrant angle. If $\cos \theta = \dfrac{1}{\sqrt{5}}$, find $\sin \theta$ and $\sec \theta$.

   (b) β is a second quadrant angle. If $\tan \beta = \dfrac{7}{-\sqrt{65}}$, find $\cos \beta$ and $\csc \beta$.

8   Draw a sketch of each angle in standard position. Calculate the other
    trigonometric values of each angle.
    (a) $\sec \theta = -\dfrac{13}{12}$,   (b) $\sin \alpha = \dfrac{7}{25}$,   (c) $\csc \theta = -\dfrac{17}{8}$,
        θ in third quadrant α in first quadrant θ in fourth quadrant

9   P(x, y) is a point on the terminal arm of α. OP $= r$.
    (a) Find the possible values of k in each of the following.
    (b) Find the primary trigonometric values for each value of k.
    (i) P(3, k), $r = 5$ (ii) P(k, 8), $r = 10$ (iii) P(3, k), $r = \sqrt{13}$

10   As in algebra, $\sin^2 \alpha$ means $(\sin \alpha)^2$. α is a third quadrant angle and
    $\tan \alpha = \dfrac{5}{12}$. Find a value for $\sin^2 \alpha + \cos^2 \alpha$.

11   β is a second quadrant angle and $\csc \beta = \dfrac{17}{15}$. Find a value for
    $2 \sin \beta + 3 \cos \beta$.

C 12   For any angle θ, show why
    (a) $\sin \theta = \dfrac{1}{\csc \theta}$ (b) $\cos \theta = \dfrac{1}{\sec \theta}$ (c) $\tan \theta = \dfrac{1}{\cot \theta}$
    (d) $\csc \theta = \dfrac{1}{\sin \theta}$ (e) $\sec \theta = \dfrac{1}{\cos \theta}$ (f) $\cot \theta = \dfrac{1}{\tan \theta}$

4.2 Angles and Quadrants

In the study of mathematics, people invent conventions so that the study of trigonometry is consistent. For example, the co-ordinate axes are used to divide the plane into four quadrants. An angle is shown in each quadrant. Clockwise angles are, by convention, negative and counter clockwise angles are positive.

First Quadrant Angle

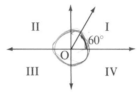

Second Quadrant Angle

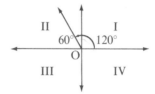

Third Quadrant Angle

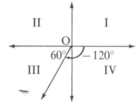

Fourth Quadrant Angle

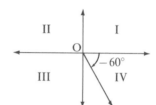

Earlier, you saw that, if you know one of the trigonometric values and the quadrant that the angle is in, you can calculate the remaining trigonometric values.

Example 1 A positive angle θ is in the third quadrant and $\cos \theta = -\frac{8}{17}$. Calculate the primary trigonometric values.

Solution From the diagram $r = 17$, $x = -8$.

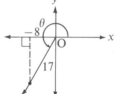

Since $r = \sqrt{x^2 + y^2}$ then
$$17 = \sqrt{(-8)^2 + y^2}$$
$$289 = 64 + y^2$$
$$225 = y^2$$
or $y = \pm 15$

Since P is a point in the third quadrant, $y = -15$.

Use $r = 17$, $x = -8$, $y = -15$.

$$\sin \theta = \frac{y}{r} \qquad \cos \theta = \frac{x}{r} \qquad \tan \theta = \frac{y}{x} = \frac{-15}{-8}$$

$$= \frac{-15}{17}, \qquad = \frac{-8}{17}, \qquad = \frac{15}{8}$$

However, if you do not know what quadrant an angle is in, you can sketch a diagram to show the possibilities, as shown in Example 2.

Example 2 Find $\cos \theta$ if θ is positive and $\sin \theta = -\dfrac{3}{5}$.

Solution If θ is a positive angle and $\sin \theta = -\dfrac{3}{5}$, then θ may be an angle in the third or fourth quadrant.

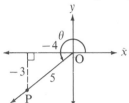

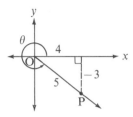

Third Quadrant
$x = -4, y = -3, r = 5$

$$\cos \theta = -\dfrac{4}{5}$$

Fourth Quadrant
$x = 4, y = -3, r = 5$

$$\cos \theta = \dfrac{4}{5}$$

4.2 Exercise

A Throughout the exercise you may leave answers in fractional or radical form.

1 Draw a sketch of each angle in standard position.

(a) $120°$ (b) $-60°$ (c) $135°$ (d) $-225°$ (e) $210°$

2 For each angle θ, a point on the terminal arm is given. Calculate the primary trigonometric ratios for θ.

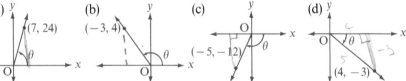

(a) $(7, 24)$ (b) $(-3, 4)$ (c) $(-5, -12)$ (d) $(4, -3)$

3 For each angle, a point on the terminal arm is shown. Calculate the reciprocal trigonometric ratios.

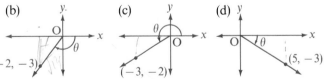

(a) $(-3, 5)$ (b) $(-2, -3)$ (c) $(-3, -2)$ (d) $(5, -3)$

4 α is an angle in the third quadrant and $\cos \alpha = \dfrac{-\sqrt{3}}{2}$.

(a) Write the co-ordinates of a point on the terminal arm.
(b) Find $\sin \alpha$ and $\tan \alpha$.

5 θ is an angle in the second quadrant and $\csc \theta = \dfrac{17}{15}$.

(a) Write the co-ordinates of a point on the terminal arm.
(b) Find $\cos \theta$, $\sec \theta$, and $\cot \theta$.

B 6 Given that $\cos \theta = -\dfrac{7}{25}$,

(a) In which possible quadrants can the terminal arm be placed?
(b) Draw a diagram to show each case in (a).
(c) Calculate the trigonometric values of $\sin \theta$.

7 β is an angle in standard position and $\sin \beta = \dfrac{4}{5}$.

(a) In which quadrants is it possible for the terminal arm to lie?
(b) Draw a diagram to show each case in (a).
(c) Calculate values for $\cos \beta$ and $\tan \beta$.

8 You know that $\cot \alpha = -\dfrac{24}{7}$.

(a) In which quadrants is it possible for the terminal arm to lie?
(b) Draw a diagram to show each case in (a).
(c) Calculate values for $\sin \alpha$ and $\cos \alpha$.

9 Examine the possibilities of each of the following.

(a) If $\sin \theta = \dfrac{-8}{17}$, find two values of $\cos \theta$.

(b) Given that $\cot \alpha = -\dfrac{12}{5}$. Find two values of $\sin \alpha$.

(c) For $\sec \beta = -\dfrac{25}{7}$, find $\tan \beta$.

(d) θ is in standard position. If $\cos \theta = \dfrac{-\sqrt{3}}{2}$, find $\cot \theta$.

10 If $\cos \theta = -\dfrac{7}{25}$, then calculate values for

(a) $\sin \theta$ (b) $(\sin \theta)(\cos \theta)$ (c) $(\cot \theta)(\tan \theta)$

C 11 If $\sin \theta = -\dfrac{3}{5}$, prove that $\sin^2 \theta + \cos^2 \theta = 1$

12 (a) $\cot \theta = -\dfrac{15}{8}$. Prove $\dfrac{\sin \theta}{\cos \theta} = \tan \theta$.   (b) $\cos \theta = -\dfrac{\sqrt{3}}{2}$. Prove $\dfrac{\sin \theta}{\cos \theta} = \tan \theta$.

(c) What probable conclusion seems true based on your results in (a) and (b)?

Exploring Signs of Sin θ, Cos θ, and Tan θ

13 θ is an angle in the second quadrant. What is the sign (positive or negative) of each of the following? Give reasons for your answer.
(a) sin θ (b) cos θ (c) tan θ (d) csc θ (e) sec θ (f) cot θ

14 α is an angle in the third quadrant. What is the sign of each of the following? Give reasons for your answers.
(a) sin α (b) tan α (c) sec α (d) cos α (e) cot α (f) csc α

15 β is an angle in the fourth quadrant. What is the sign of each of the following? Give reasons for your answers.
(a) cot β (b) sec β (c) cos β (d) csc β (e) tan β (f) sin β

16 In which quadrant does the terminal arm of θ lie if
(a) sin θ and cos θ are both positive?
(b) tan θ is positive and sin θ is negative?
(c) cos θ is negative and sin θ is positive?
(d) csc θ and tan θ are both negative?
(e) sec θ and tan θ are both negative?
(f) cot θ and sin θ are both positive?

17 In which possible quadrant(s) does the terminal arm of θ lie if
(a) sin θ is positive? (b) cos θ is negative? (c) tan θ is positive?
(d) csc θ is negative? (e) cot θ is positive? (f) sec θ is positive?

18 For each quadrant, what conclusion can be made about the sign (positive or negative) for each of the following?
(a) sin θ (b) cos θ (c) tan θ
(d) csc θ (e) sec θ (f) cot θ

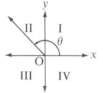

19 Write a rule for the signs of sin θ, cos θ, and tan θ based on your results in the previous questions.

Math Tip

To calculate the trigonometric values, use the observation about the sings of the trigonometric values.

II	I
sin θ is positive	All are positive
tan θ is positive	cos θ is positive
III	IV

CAST RULE

S | A
T | C

4.3 Coterminal Angles

Three positive angles in standard position are shown, each having a terminal arm that passes through $(-3, 4)$.

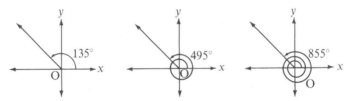

The angles are said to be coterminal since they have the same terminal arm. Associated with coterminal angles is the related principal angle, 135°, which is referred to as the smallest positive angle. For coterminal angles, the related principal angle θ has the limits given by $0° \leq \theta \leq 360°$.

Three negative coterminal angles are shown.

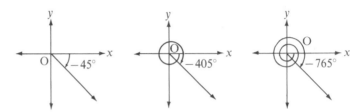

The principal angle θ, related to these coterminal angles is shown in the diagram. $0° \leq \theta \leq 360°$

Thus, the principal angle for the coterminal angles $-45°$, $-405°$ and $-765°$ is 315°.

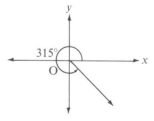

Example For the angle 765°

(a) draw a sketch in standard position.   (b) write the principal angle.

Solution (a)

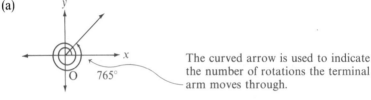

The curved arrow is used to indicate the number of rotations the terminal arm moves through.

(b) Since $765° = 2(360°) + 45°$ then the principal angle is 45°.
   $(0° \leq 45° \leq 360°)$

Based on Example 1, any angle, coterminal with 45°, may be written in general as general as $45° + k(360°)$ where k is an integer.

Thus, coterminal angles of 45° are given by

$$
\begin{array}{cccc}
k = 1 & k = -1 & k = 2 & k = -2 \\
405° & -315° & 765° & -675°
\end{array}
$$

Since coterminal angles have the same terminal arm, then the corresponding trigonometric values are equal. Thus, $\cos 45° = \cos 405° = \cos(-315°)$ and so on since 45°, 405° and $-315°$ are coterminal angles.

4.3 Exercise

Remember: Negative angles are shown as clockwise angles.

A 1 What is the value of θ in each diagram? The reference triangle is marked.
(a) $\angle POM = 60°$ (b) $\angle POM = 45°$ (c) $\angle POM = 30°$ (d) $\angle POM = 60°$

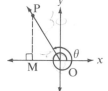 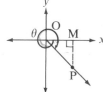 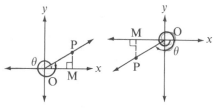

2   Refer to the diagram.
(a) Write two angles that are coterminal with θ.
(b) What is the principal angle?

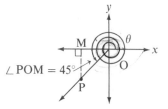

$\angle POM = 45°$

3   Refer to the diagram.
(a) Write two angles that are coterminal with β.
(b) What is the principal angle?

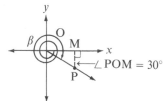

$\angle POM = 30°$

4   Write 2 negative angles that are coterminal with each angle.
(a) 45°   (b) 120°   (c) 380°

5   Write 2 positive angles that are coterminal with each angle.
(a) $-45°$ (b) $-120°$ (c) $-420°$

B   The principal angle θ is shown by $0° \leqq \theta \leqq 360°$.

6   (a) Draw 790° in standard position.
   (b) Write a negative angle, coterminal with 790°.
   (c) What is the smallest positive angle coterminal with 790°?
   (d) What is the principal angle?

7   (a) Draw −510° in standard position.   (b) What is the principal angle?

8   For each angle,
   • sketch the angle in standard position.
   • mark the principal angle.

   (a) 520°   (b) −225°   (c) −390°   (d) −420°   (e) 780°   (f) −840°

9   For each angle, write the value of the principal angle, θ, $0° \leqq \theta \leqq 360°$.
   (a) 390°   (b) −315°   (c) 415°   (d) −105°   (e) −120°
   (f) 780°   (g) 1090°   (h) −340°   (i) 940°   (j) −460°

10   Which of the following do not have the same value as sin 45°?
   (a) sin (−315°)   (b) sin 405°   (c) sin (−680°)   (d) sin 705°

11   Which of the following have the same value as cos (−30°)?
   (a) cos (−390°)   (b) cos 790°   (c) cos 330°   (d) cos (−750°)

12   Which values are equal?
   (a) sin 75°   (b) sin (−285°)≤   (c) sin 795°   (d) sin (−680°)

13   Which values are equal?
   (a) sin 760°   (b) sin (−50°)   (c) sin (−410°)   (d) sin 310°

Calculator Tip

A scientific calculator is an invaluable tool for solving problems in trigonometry. Refer to the manual provided with your calculator for information about ⌈sin⌉⌈cos⌉⌈tan⌉.   The calculator is placed in the degree mode. Refer to your manual.

	Calculator Procedure	
Find the value of	*Input*	*Output*
sin 135°	C 135 SIN	?
sin (−45°)	C 45 +/− SIN	?
tan 75°	C 75 TAN	?
cos (−120°)	C 120 +/− COS	?

4.4 Trigonometric Values: 30°, 45°, and 60°

To calculate a particular trigonometric value of 45°, you can do the construction. The terminal arm passes through the point P(1, 1).

From the diagram, $\triangle ABC$ is isosceles.
Thus $\angle B = \angle A = 45°$.

In $\triangle ABC$, $AB^2 = AC^2 + BC^2$
$$AB^2 = 1^2 + 1^2$$
$$AB = \sqrt{2}$$

Thus, from the diagram,

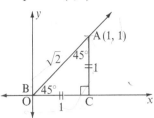

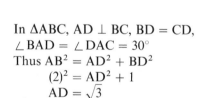

$$\sin 45° = \frac{1}{\sqrt{2}} \quad \cos 45° = \frac{1}{\sqrt{2}} \quad \tan 45° = 1$$

$$\csc 45° = \sqrt{2} \quad \sec 45° = \sqrt{2} \quad \cot 45° = 1$$

Trigonometric values are frequently expressed in fractional or radical form.

To calculate the trigonometric values of 30° and 60° an equilateral triangle of sides 2 units is drawn as shown.

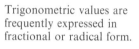

In $\triangle ABC$, $AD \perp BC$, $BD = CD$,
$\angle BAD = \angle DAC = 30°$
Thus $AB^2 = AD^2 + BD^2$
$$(2)^2 = AD^2 + 1$$
$$AD = \sqrt{3}$$

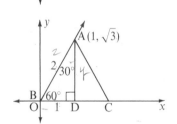

From $\triangle ABD$, you can calculate the primary trigonometric values.

$$\sin 60° = \frac{\sqrt{3}}{2} \quad \cos 60° = \frac{1}{2} \quad \tan 60° = \frac{\sqrt{3}}{1} = \sqrt{3}$$

$$\csc 60° = \frac{2}{\sqrt{3}} \quad \sec 60° = 2 \quad \cot 60° = \frac{1}{\sqrt{3}}$$

Similarly, the diagram is used to calculate the trigonometric values of 30°.

$$\sin 30° = \frac{1}{2} \quad \cos 30° = \frac{\sqrt{3}}{2} \quad \tan 30° = \frac{1}{\sqrt{3}}$$

$$\csc 30° = 2 \quad \sec 30° = \frac{2}{\sqrt{3}} \quad \cot 30° = \sqrt{3}$$

In studying mathematics, you will frequently need to use the trigonometric values of 45°, 30° and 60°. Thus, it is useful to learn these diagrams so that you can mentally obtain the trigonometric values of 45°, 30° and 60°, as shown in the following example.

 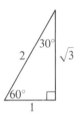

Example 1 Calculate the value of sin 45° cos 45° + cos 30° tan 60°.

Solution

$$\sin 45° \cos 45° + \cos 30° \tan 60°$$

Refer to the diagrams to obtain the values.

$$= \left(\frac{1}{\sqrt{2}}\right)\left(\frac{1}{\sqrt{2}}\right) + \left(\frac{\sqrt{3}}{2}\right)(\sqrt{3})$$

$$= \frac{1}{2} + \frac{3}{2}$$

$$= 2$$

Associated with each rotation angle is a reference triangle, △OPM, as shown in the diagrams.

 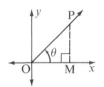 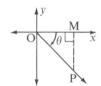 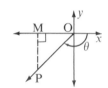

△OPM is used as a reference triangle to aid the calculation of values. In the following example, to show 225° in standard position, use the reference triangle to locate the position of the terminal arm.

Example 2 Find the value of sin 225°.

Solution

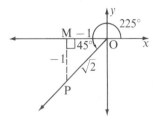

Make a sketch of 225° in standard position. To calculate trigonometric values choose any convenient point on the terminal arm.

Note that in the reference triangle.
∠POM = 45°.

The co-ordinates of P are (−1, −1).

From the diagram

$$x = -1, \ y = -1, \ r = \sqrt{2}. \qquad \sin \theta = \frac{y}{r} \qquad \sin 225° = \frac{-1}{\sqrt{2}}$$

4.4 Exercise

A Throughout this exercise it is helpful to learn the trigonometric values of 30°, 45° and 60°. Express your answers in fractional or radical form.

1 Copy and complete the chart. Use the chart as a reference chart in your later work.

	30°	45°	60°
sin	?	?	?
cos	?	?	?
tan	?	?	?

2 Calculate each of the following.
   (a) sin 45°   (b) cos 60°   (c) cos 30°   (d) sin 60°
   (e) sec 45°   (f) cot 30°   (g) tan 30°   (h) csc 60°
   (i) cot 45°   (j) sec 30°   (k) csc 45°   (l) cot 60°

3 Find the trigonometric values of θ.
   (a) 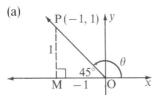   (b)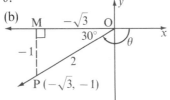

4 A point is shown on the terminal arm of each angle. Write the primary trigonometric values.
   (a) 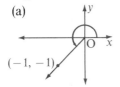 (b)   (c) 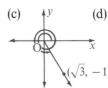 (d)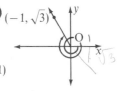

B Express your answers in fractional or radical form.

5 (a) Draw a sketch of 300° in standard position.
   (b) Calculate the primary trigonometric values for 300°.

6 (a) Draw a sketch of −225° in standard position.
   (b) Calculate the reciprocal trigonometric values of −225°.

7 Calculate each of the following.
   (a) sin (−60°)   (b) sin 300°
   (c) What do you notice about your answers in (a) and (b)? Give a reason for your answer.

8 Calculate each of the following.
  (a) cos 225°   (b) sin 120°   (c) tan (−150°)
  (d) tan 135°   (e) sec (−30°)   (f) cot 330°
  (g) sec (−45°)   (h) tan (−225°)   (i) csc 240°

9 Calculate each of the following.
  (a) tan 390°   (b) cos (−480°)   (c) tan 510°
  (d) csc (−495°)   (e) tan 585°   (f) cos 780°

10 Calculate each of the following.
  (a) cos 45° sin 225° + cos 330°   (b) csc 315° sin (−120°) cot 225°
  (c) tan² 225° − sin 60° cos 30°

11 θ is an angle in the second quadrant and $\cos \theta = -\dfrac{1}{2}$.

  (a) Find tan θ.   (b) What is the value of θ, $0° \leq \theta \leq 360°$?

12 (a) If $\cos \theta = -\dfrac{1}{2}$ and $0° \leq \theta \leq 360°$, find sin θ.

  (b) What are the possible values of θ?

13 For each of the following $0° \leq \theta \leq 360°$. Find possible values of θ.

  (a) $\cos \theta = \dfrac{\sqrt{3}}{2}$ (b) $\sin \theta = -\dfrac{1}{2}$ (c) $\cot \theta = 1$

  (d) $\csc \theta = -\sqrt{2}$ (e) $\sec \theta = -2$ (f) $\sin \theta = -\dfrac{1}{\sqrt{2}}$

C 14 The terminal arm of α is in the fourth quadrant. If $\cot \alpha = -\sqrt{3}$, then calculate $\sin \alpha \cot \alpha - \cos^2 \alpha$.

Calculator Tip

You can use a calculator to solve for the missing angles in questions such as those above.

The calculator is in the degree mode.

$0 \leq \theta < 360°$ Find θ.	Calculator Procedure Input	Output	
$\sin \theta = \dfrac{1}{2}$	C 0.5 INV SIN	?	You now need to interpret the output.
$\cos \theta = \dfrac{1}{2}$	C 0.5 +/− INV COS	?	

Use this skill in your later work to solve equations involving trigonometry.

4.5 Finding Trigonometric Values

In the early stages of trigonometry, geometric figures were used to obtain the tables of trigonometric values, which were obtained, tediously, by calculation. Today, these values are obtained, to a high degree of accuracy, with calculators and computers.

From these geometric figures, you can obtain a few exact values, such as the trigonometric values of 30°, 45° and 60° (and related angles).

On a calculator or a computer, not only the trigonometric values of 30°, 45°, and 60° can be obtained to a high degree of accuracy, but the trigonometric values of all angles can be obtained.

$$\sin 45° = \frac{1}{\sqrt{2}} \qquad \cos 60° = \frac{1}{2}$$

$$\tan 30° = \frac{1}{\sqrt{3}}$$

$\sin 45° = 0.707\ 106\ 8$ (7 decimals)
$\cos 60° = 0.5$ (exact)
$\tan 30° = 0.577\ 350\ 3$ (7 decimals)
$\sin 36° = 0.587\ 785\ 3$ (7 decimals)
$\tan 79° = 5.144\ 554\ 1$ (7 decimals)

A computer is used to generate a table of trigonometric values for $0° \leq \theta \leq 90°$. These values, whether obtained on a calculator, or from the tables, will play an important role in solving equations in trigonometry and problems in applied trigonometry.

Part of the table of trigonometric values, found at the back of the book, is shown as follows:

degrees	sin	cos	tan	csc	sec	cot
45	0.7071	0.7071	1.0000	1.4142	1.4142	1.0000
46	0.7193	0.6947	1.0355	1.3902	1.4396	0.9657
47	0.7314	0.6820	1.0724	1.3673	1.4663	0.9325
48	0.7431	0.6691	1.1106	1.3456	1.4945	0.9004
49	0.7547	0.6561	1.1504	1.3250	1.5243	0.8693
50	0.7660	0.6428	1.1918	1.3054	1.5557	0.8391
51	0.7771	0.6293	1.2349	1.2868	1.5890	0.8098
52	0.7880	0.6157	1.2799	1.2690	1.6243	0.7813
53	0.7986	0.6018	1.3270	1.2521	1.6616	0.7536
54	0.8090	0.5878	1.3764	1.2361	1.7013	0.7265

Values are expressed to 4 decimal places.

From the table

A   When $\theta = 48°$, $\sin 48° = 0.7431$
B   When $\theta = 52°$, $\tan 52° = 1.2799$
C   If $\cos \theta = 0.6157$ then $\theta = 52°$
D   If $\sec \theta = 1.4396$ then $\theta = 46°$

However if you wish to find the values for θ for the trigonometric values which do not occur precisely in the table, you can still use the tables to find the value of θ as shown in the following example. The process used is referred to as interpolation.

Example (a) If $\sin \theta = 0.7859$, find θ. (b) If $\cos \alpha = 0.6311$, find α.

Solution From the above tables,

(a) $\sin 51° = 0.7771$ difference
   $\sin \theta = 0.7859$ 0.0088
   $\sin 52° = 0.7880$ difference
   0.0021

Thus $\theta \doteq 52°$. since $\sin \theta$ is nearer $\sin 52°$

(b) $\cos 50° = 0.6428$ difference
   $\cos \alpha = 0.6311$ 0.0088
   $\cos 51° = 0.6293$ difference
   0.0018

Thus $\alpha \doteq 51°$.

Perform these steps with a calculator.

	Input				Display	
To find θ.	C	0.7859	INV	SIN	=	51.803994	$\theta = 52°$ (nearest degree)
To find α.	C	0.6311	INV	COS	=	50.868675	$\alpha = 51°$ (nearest degree)

You can also use the tables to find the trigonometric values for angles greater than 90°. For example, in the diagram if $\theta = 123°$, then

$$\cos \angle POM = \cos 57°$$
$$= 0.5446$$

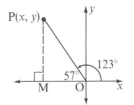

In the second quadrant, $\cos 123°$ is negative. Thus $\cos 123° = -0.5446$. Perform these steps with a calculator.

Input   Display
C 123° COS -0.544639 Thus, $\cos 123° = -0.5446$ (4 decimal places).

4.5 Exercise

A Express answers to 4 decimal places.

1 What is the value of each of the following?
   (a) $\sin 47°$ (b) $\tan 48°$ (c) $\csc 45°$ (d) $\cos 50°$ (e) $\sec 47°$ (f) $\cot 45°$
   (g) $\csc 54°$ (h) $\sin 48°$ (i) $\cos 52°$ (j) $\sin 53°$ (k) $\cot 52°$ (l) $\csc 51°$

2 As θ increases in value from 45° to 54° does the value of each of the following increase (I) or decrease (D)?
   (a) $\sin \theta$ (b) $\cos \theta$ (c) $\tan \theta$ (d) $\csc \theta$ (e) $\sec \theta$ (f) $\cot \theta$

3   Find θ for each of the following.
   (a) $\sin \theta = 0.7547$ (b) $\tan \theta = 1.1504$ (c) $\cot \theta = 0.8391$
   (d) $\cos \theta = 0.6691$ (e) $\sec \theta = 1.6243$ (f) $\sin \theta = 0.8090$

4   Find θ for each of the following.
   (a) $\sin \theta = 0.7443$ (b) $\sec \theta = 1.4718$ (c) $\csc \theta = 1.3478$
   (d) $\cot \theta = 0.9391$ (e) $\tan \theta = 1.2662$ (f) $\cos \theta = 0.6320$

5   Find the value of each of the following.
   (a) $\cos 23°$ (b) $\sin 46°$ (c) $\sec 79°$ (d) $\sec 11°$
   (e) $\csc 63°$ (f) $\cot 72°$ (g) $\tan 81°$ (h) $\sin 55°$

6   Find the value of
   (a) $\sin 127°$ (b) $\cos 272°$ (c) $\cos (-45°)$ (d) $\tan 229°$
   (e) $\cos 135°$ (f) $\sec (-57°)$ (g) $\csc 160°$ (h) $\cot (-172°)$

B 7   Find the value of θ for each of the following.
   (a) 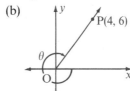   (b)

P(3, 5)   P(4, 6)

8   Use the information in each diagram to calculate all the primary
   trigonometric values of θ.
   (a)   (b)   (c)   (d)

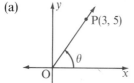 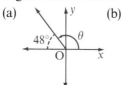 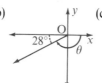

9   For each diagram, find the trigonometric value of θ shown.
   (a)   (b)   (c)

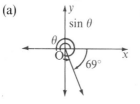 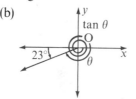 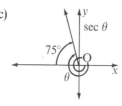

10   $P(x, y)$ is a point on the terminal arm of θ. Find θ to the nearest degree.
   (a) $P(2, 1)$ (b) $P(3, 2)$ (c) $P(4, 3)$ (d) $P(2, 7)$

11 θ is a first quadrant angle. Find each value of θ.

(a) $\sin \theta = \dfrac{1}{2}$ (b) $\cos \theta = \dfrac{2}{3}$ (c) $\tan \theta = \dfrac{3}{2}$ (d) $\cot \theta = \dfrac{4}{3}$

12 Find the value θ for each of the following.

(a) $\sin \theta = 0.1736$ (b) $\sec \theta = 1.0515$ (c) $\tan \theta = 0.7265$
(d) $\csc \theta = 1.0223$ (e) $\cot \theta = 0.4663$ (f) $\cot \theta = 2.1445$
(g) $\sin \theta = 0.3439$ (h) $\sec \theta = 1.1143$ (i) $\cos \theta = 0.2856$

13 For each θ, the quadrant is given. Find the value of θ, $0° \leq \theta \leq 360°$.

(a) $\cos \theta = -0.9063$ II (b) $\tan \theta = 0.5543$ III
(c) $\csc \theta = -1.1547$ III (d) $\sin \theta = -0.9962$ IV
(e) $\sec \theta = 1.7434$ I (f) $\cos \theta = -0.8572$ III

14 Suppose $\sin \theta = -0.9063$. Since $\sin \theta$ is negative then θ is either a third or fourth quadrant angle.

(a) Sketch the angle θ in standard position for each quadrant.
(b) Find θ, $0° \leq \theta \leq 360°$.

15 Draw the diagrams and find the 2 values of θ, $0° \leq \theta \leq 360°$.

(a) $\cos \theta = -0.3420$ (b) $\tan \theta = -0.4663$

16 Find two possible values of θ for each of the following $0° \leq \theta \leq 360°$.

(a) $\sin \theta = -0.8910$ (b) $\sec \theta = 1.0785$ (c) $\tan \theta = -4.3315$
(d) $\cos \theta = -0.4384$ (e) $\tan \theta = 0.2761$ (f) $\csc \theta = 1.1829$

17 θ is a first quadrant angle. As θ increases describe the change in values of

(a) $\sin \theta$ (b) $\cos \theta$ (c) $\tan \theta$ (d) $\csc \theta$ (e) $\sec \theta$ (f) $\cot \theta$

18 Without referring to the tables of trigonometric values, which of the following are false, for $0° \leq \theta \leq 90°$? Give reasons for your answers.

(a) $\cos \theta = 3.2151$ (b) $\tan \theta = 3.2151$ (c) $\cot \theta = 3.2151$
(d) $\sec \theta = 3.2151$ (e) $\sin \theta = 3.2151$ (f) $\csc \theta = 3.2151$

19 Show that

(a) $2 \sin 32° \neq \sin 64°$ (b) $\sin 20° + \sin 40° \neq \sin 60°$ (c) $\dfrac{\tan 75°}{3} \neq \tan 25°$

C 20 If $0° < \theta < 90°$, then find θ if

(a) $\sin 2\theta = 0.3420$ (b) $\cos (\theta + 10°) = 0.1045$
(c) $\tan (90° - 2\theta) = 1.7321$

4.6 Radian Measure

Systems of measurement which already exist, very often are not adequate for the topic in mathematics that you are studying at the time. As a result a new system of measurement may have to be invented. In earlier work, degrees were used as a system of measurement for angles. Since angles in trigonometry suggest circles, another unit of measurement was invented and a relationship between the two systems was developed.

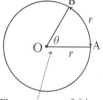

A **radian** is the measure of the angle subtended at the centre of the circle by an arc equal in length to the radius of the circle.

$$\text{measure of an angle in radians} = \frac{\text{length of arc subtending the angle}}{\text{length of radius}}$$

The measure of θ is defined to be 1 radian.

For a circle with radius r, the circumference has the measure $2\pi r$. The radian measure of each angle θ is calculated.

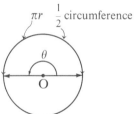

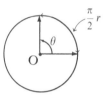

Measure of θ is given by

$$\frac{\text{length of arc}}{\text{length of radius}} = \frac{\pi r}{r} = \pi$$

Measure of θ is given by

$$\frac{\text{length of arc}}{\text{length of radius}} = \frac{\frac{\pi}{2} r}{r} = \frac{\pi}{2}$$

The measure of angles in degrees is related to the measure of angles in radians, as shown.

In degrees
$\angle AOB = 180°$

In radians
$\angle AOB = \pi$ rad

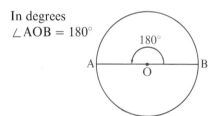

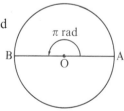

From the diagrams, the relationship is established. Thus

$$\boxed{180° = \pi \text{ rad.}}$$

Throughout your work, the measure of an angle, in radians, is written as follows: 2 rad or 2

π rad or π ← The symbol for radians is often not written, but understood to be radians.

Example 1   (a) Express 30° in terms of radians.   (b) Express $\dfrac{\pi}{3}$ rad in terms of degrees.

Solution   (a) $180° = \pi$ rad

$$1° = \left(\dfrac{\pi}{180}\right) \text{rad}$$

$$30° = 30\left(\dfrac{\pi}{180}\right) \text{rad}$$

$$30° = \dfrac{\pi}{6} \text{rad}$$

(b) π rad $= 180°$

$$\dfrac{\pi}{3} \text{rad} = \dfrac{180°}{3}$$

$$\dfrac{\pi}{3} \text{rad} = 60°$$

You can write radian measures as degree measures.

Example 2   Use $\pi \doteq 3.14$.

(a) Express 2.3 rad in terms of degrees, to the nearest degree.

(b) Express 48° in terms of radians correct to two decimal places.

Solution   (a) π rad $= 180°$

$$1 \text{ rad} = \left(\dfrac{180°}{\pi}\right)^{\!\circ}$$

$$2.3 \text{ rad} = 2.3\left(\dfrac{180}{\pi}\right)^{\!\circ}$$

$$= 132° \text{ (nearest degree)}$$

(b) $180° = \pi$ rad

$$1° = \left(\dfrac{\pi}{180}\right) \text{rad}$$

$$48° = 48\left(\dfrac{\pi}{180}\right) \text{rad}$$

$$= 0.84 \text{ rad (to 2 decimal places)}$$

Throughout your work, write π rad as π.

4.6   Exercise

Throughout the exercise, use $\pi \doteq 3.14$. Remember, π as a measure of an angle is understood to be π radians.

A 1   Write each radian measure as a degree measure

(a) $\dfrac{\pi}{3}$ (b) $\dfrac{\pi}{2}$ (c) $\dfrac{\pi}{4}$ (d) $\dfrac{\pi}{6}$ (e) $\dfrac{3}{4}\pi$

(f) $-\dfrac{\pi}{2}$ (g) $-\dfrac{2}{3}\pi$ (h) $\dfrac{5\pi}{6}$ (i) $\dfrac{4}{3}\pi$ (j) $-\dfrac{3}{4}\pi$

(k) $-\dfrac{5}{3}\pi$ (l) $\dfrac{3\pi}{2}$ (m) 2π (n) -4π (o) 3π

2　Write each degree measure as a radian measure.
(a) 180°　(b) 360°　(c) 90°　(d) 45°　(e) −60°　(f) −150°
(g) 30°　(h) 240°　(i) −330°　(j) 270°　(k) −90°　(l) 120°

3　Write the value of θ in radian measure.

(a) 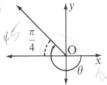　　(b) 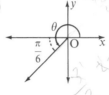　　(c)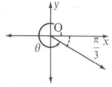

(d) 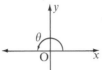　　(e) 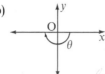　　(f)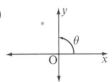

4　In which quadrant does the terminal arm of each angle lie?
(a) $\frac{3}{4}\pi$　(b) $-\frac{2}{3}\pi$　(c) $\frac{7}{6}\pi$　(d) $-\frac{7}{4}\pi$　(e) $-\frac{\pi}{3}$　(f) $\frac{7}{4}\pi$

5　Sketch each angle, given in radian measure, in standard position.
(a) π　(b) $\frac{1}{2}\pi$　(c) $-\frac{1}{4}\pi$　(d) $\frac{5}{4}\pi$　(e) $\frac{2}{3}\pi$　(f) $-\frac{5}{6}\pi$

6　Convert each of the following to radian measure.
(a) $\frac{1}{2}$ revolution　(b) $\frac{2}{3}$ revolution　(c) 2 revolutions

7　Each radian measure is given to 1 decimal place. Express to the nearest degree.
(a) 1.1　(b) 0.9　(c) −0.3　(d) 1.5　(e) −0.8　(f) 0.4

Check the features of your calculator. Can you convert radians to degrees and vice versa?

8　Express the measure of angle θ in radians.

(a) 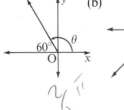　(b) 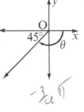　(c) 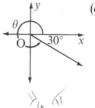　(d)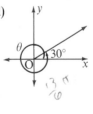

B Check your answers with a calculator.

9   Sector angles are drawn in a unit circle. Find the measure of the arc of the circle that subtends an angle measuring.

(a) 90°   (b) 30°   (c) 1 rad   (d) $\frac{\pi}{2}$ rad   (e) 2.6 rad

10 A point is given on the terminal arm of each angle. Calculate the measure of the principal angle in radians to 1 decimal place.

(a) (2, 3)   (b) (−3, 1)   (c) (−2, −5)   (d) (3, −5)

11 The radian measures of angles are shown. Write the measure of the coterminal angle θ, $-2\pi \leq \theta \leq 2\pi$

(a) $\frac{\pi}{3}$ (b) $-\frac{\pi}{4}$ (c) $\frac{\pi}{6}$ (d) $-\pi$ (e) $\frac{3}{4}\pi$ (f) $-\frac{3}{2}\pi$

12 Write each degree measure as a radian measure expressed to 4 decimal places.

(a) 1°   (b) 8°   (c) 25°   (d) 65°   (e) 170°   (f) −235°

13 Two angles α and $-\frac{3}{4}\pi$ are in standard position and have a terminal arm in common. Find the measure of α if $-2\pi \leq \alpha \leq 2\pi$.

C 14 Calculate each of the following.   *radiant*.

(a) cos 2   (b) tan 1   (c) sec 3   (d) cot (−1)   (e) sin (−2)

Computer Tip

The trigonometric values given in tables have been calculated by using the following formulas.

This means
3 × 2 × 1.

$$\sin x = x - \frac{x^3}{3!} + \frac{x^5}{5!} - \frac{x^7}{7!} + \cdots$$

$$\cos x = 1 - \frac{x^2}{2!} + \frac{x^4}{4!} - \frac{x^6}{6!} + \cdots$$

A computer is then designed to tabulate the values of sin x and cos x. Write a program in the language of **BASIC** to include the first five terms of sin x and cos x which will tabulate values.

Applications: Wheels and Orbits

As a satellite orbits the earth it goes through an increasing number of revolutions as time goes on.

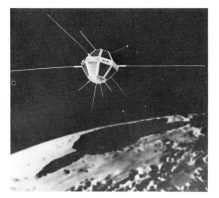

The **angular velocity** of an object is given as the amount of rotation around a central point per unit of time. The amount of rotation may be expressed in terms of degrees or radians. For example,

A fan makes 60 rotations in 5 s. What is its angular velocity?

$$1 \text{ rotation} = 360° = 2\pi \text{ rad} \qquad 60 \text{ rotations} = 120\pi \text{ rad}$$

$$\text{Angular Velocity} = \frac{\text{amount of rotation}}{\text{unit time}}$$

$$= \frac{120\pi \longleftarrow \text{radians}}{5 \longleftarrow \text{seconds}}$$

$$= 24\pi \text{ radians per second (rad/s)}.$$

15   A wheel revolves at 120 rad/min.

(a) What is the angular velocity in radians per second?

(b) A point is 22 cm from the point of rotation. How far does it travel in 3 s?

16   A propeller is rotating at 1800 rad/min. Write this angular velocity in radians per second.

17   A car travels at 80 km/h. Determine the angular velocity of a tire of radius 34 cm.

18   A person on a ferris wheel makes one complete revolution in 22 s. Calculate the angular velocity of the person in radians per second.

19   A satellite makes one complete revolution of the earth in 96 min. If the satellite is situated 300 km above the equator, find the satellite's speed in kilometres per second. Assume the earth's equatorial radius is 6400 km.

20   (a) The radius of a car wheel is 30 cm. If the car is travelling at 50 km/h, how many revolutions will the tire make in 1 s?

(b) Calculate the angular velocity of the tire.

4.7 Basic Graphs of Trigonometric Functions

In the study of mathematics, you often use the same skill and apply the skill in a new situation. For example, for each θ, $\sin \theta$ has a unique value, as you have seen in calculating trigonometric values. You can define a function given by,

$$f:\theta \to \sin \theta, \qquad f(\theta) = \sin \theta \quad \text{or} \quad y = \sin \theta$$

You have already used graphing skills. These can be applied again, to obtain a picture of the function $f:\theta \to \sin \theta$. A table of values needs to by organized and then the ordered pairs $(\theta, \sin \theta)$ plotted on a graph. The values of $\sin \theta$ are expressed to two decimal places.

θ	0°	30°	60°	90°	120°	150°	180°	210°	240°	270°	300°	330°	360°
$\sin \theta$	0	0.50	0.87	1	0.87	0.50	0	−0.50	−0.87	−1	−0.87	−0.50	0

Plot the ordered pairs.

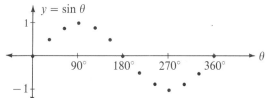

Since there is a corresponding value of $\sin \theta$ for each θ, join the points to form a smooth curve.

A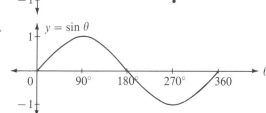

For θ measured in radians, the axes are labelled as shown.

B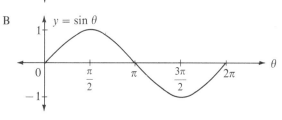

For either graph A or B, note the following characteristics of the graph.

> domain: $\{\theta \mid 0 \leq \theta \leq 2\pi\}$ (as radians)
> $\{\theta \mid 0° \leq \theta \leq 360°\}$ (as degrees)
> range: $\{y \mid -1 \leq y \leq 1\}$, where $y = \sin \theta$
> intercepts: y-intercept $= 0$
> x-intercepts $= 0°, 180°, 360°$
> or $= 0, \pi, 2\pi$

The graphs on the previous page have been drawn with restricted domains even though the curves can be extended indefinitely in both directions, positive or negative. The function is called **periodic** because the values of y repeat at equal intervals of θ.

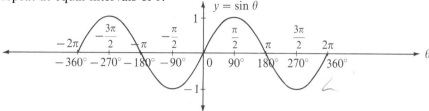

From the graph $f : \theta \to \sin \theta$ note the following.

I   The graph extends indefinitely in either direction as $(\theta, \sin \theta)$ are graphed for all θ.

II The graph is periodic. The period is 2π because for $0° \leq \theta \leq 360°$, or $0 \leq \theta \leq 2\pi$, the function completes one period.

III The y-intercept is 0.

IV The x-intercepts are integral multiples of π or $180°$.
In degrees: $\ldots, -360°, -180°, 0°, 180°, 360°, \ldots$
In radians: $\ldots, -2\pi, -\pi, 0, \pi, 2\pi, \ldots$
In general: the x-intercepts are given by $n\pi$, $n \in I$ or $n(180°)$, $n \in I$.

V   Every value of θ determines a unique value of $\sin \theta$. Thus, $f : \theta \to \sin \theta$ is a function.

Periodic phenomena occur often in many other subject areas. In the following exercise, you will investigate other trigonometric functions, and their properties.

4.7   Exercise

A   Throughout, use $\pi \doteq 3.14$ as needed. Express answers to 1 decimal place or to the nearest degree. Questions 1 to 4 refer to the previously sketched graph of $y = \sin \theta$, $-2\pi \leq \theta \leq 2\pi$ or $-360° \leq \theta \leq 360°$.

1   For what values of θ is
(a) $\sin \theta = 0$?   (b) $\sin \theta = 1$?   (c) $\sin \theta = -1$?

2   What are the values of each of the following?

(a) $\sin \pi$ (b) $\sin \dfrac{\pi}{2}$ (c) $\sin \dfrac{3}{2} \pi$ (d) $\sin 0$

3   Use the graph. Calculate each of the following.

(a) $\sin \dfrac{\pi}{4}$ (b) $\sin \dfrac{3}{4}\pi$ (c) $\sin \left(-\dfrac{\pi}{4}\right)$ (d) $\sin \left(-\dfrac{3}{4}\pi\right)$

4   Use the graph. Find the values of θ in degrees, $-360° \le \theta \le 360°$.

(a) $\sin \theta = \dfrac{1}{2}$ (b) $\sin \theta = -\dfrac{1}{2}$

5   (a) Draw the graph of $f : \theta \to \sin \theta$ for $-4\pi \le \theta \le 4\pi$.
(b) What is the value of the y-intercept?
(c) What are the values of the x-intercepts?
(d) What is the maximum value of $\sin \theta$?
(e) For what values of θ do the maximum values occur?
(f) What is the minimum value of $\sin \theta$?
(g) For what values of θ do the minimum values occur?
(h) Why is the graph of $y = \sin \theta$ considered to be periodic?

B   Throughout the exercise, use trigonometric values to draw graphs. Refer to the tables or use your calculator.

6   Copy and complete each table of values for $(\theta, \cos \theta)$.

(a)

degrees	0°	30°	60°	90°	120°	150°	180°
radians	0	$\frac{\pi}{6}$	$\frac{\pi}{3}$	$\frac{\pi}{2}$	$\frac{2\pi}{3}$	$\frac{5\pi}{6}$	π
$\cos \theta$							

.6   0.50866)

(b)

degrees	180	210	240	270	300	330	360
radians	π	$\frac{7}{6}\pi$	$\frac{4}{3}\pi$	$\frac{3}{2}\pi$	$\frac{5}{3}\pi$	$\frac{11}{6}\pi$	2π
$\cos \theta$							

0   90 180 270

7   Use the table of values in the previous question. Construct a graph of the function $f : \theta \to \cos \theta$, $-2\pi \le \theta \le 2\pi$.

8   For the graph drawn of the function $f : \theta \to \cos \theta$, $-2\pi \le \theta \le 2\pi$, give the following.
(a) the domain   (b) the range
(c) the value of the y-intercept   (d) the values of the x-intercepts
(e) the maximum value of $\cos \theta$ for $-2\pi \le \theta \le 2\pi$
(f) the minimum value of $\cos \theta$ for $-2\pi \le \theta \le 2\pi$
(g) the value(s) of θ when $\cos \theta$ is a maximum value
(h) the value(s) of θ when $\cos \theta$ is a minimum value

9   (a) Study the graph in the previous question. Use it to sketch the graph of the trigonometric function $y = \cos \theta$, $-4\pi \leq \theta \leq 4\pi$.

   (b) Extend the graph in (a) to include the domain $-6\pi \leq \theta \leq 6\pi$.

   (c) Why can the graph given by $y = \cos \theta$ be referred to as periodic?

10 (a) What is meant by this statement?

   The trigonometric function $f:\theta \rightarrow \cos \theta$ is periodic.

   (b) What is the period of the cosine function?

11 (a) Construct a table of values for $(\theta, \tan \theta)$, $-2\pi \leq \theta \leq 2\pi$. (Refer to the plan given in Question 6 as a model for drawing the graph.)

   (b) Use the results in (a) to draw the graph given by $f:\theta \rightarrow \tan \theta$.

12 Refer to the graph in the previous question for $f:\theta \rightarrow \tan \theta$, $-2\pi \leq \theta \leq 2\pi$.

   (a) What is the domain?   (b) Give the values of the x-intercepts.

   (c) Give the value of the y-intercept.

   (d) Why may we say that $f:\theta \rightarrow \tan \theta$ represents a function?

13 Use your graph in Question 11 to justify each of the following statements for the function given by the equation $y = \tan \theta$.

   (a) The tangent function has no maximum or minimum value.

   (b) The function is periodic. The period is π.

14 The graphs of the primary trigonometric functions are shown. In what ways are the graphs alike? different?

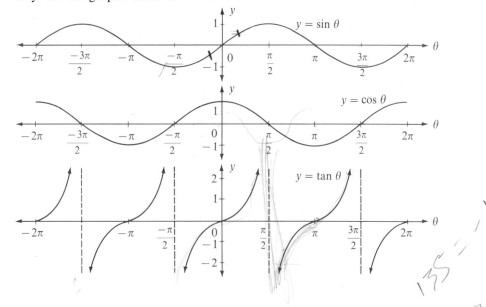

Applications: Tides

The effect of the moon on the tides of the earth is periodic. The cycle of the tides repeats every 24 h.

The Fundy tides rise more than 8m around the "Flower Pot Rocks" at Hopewell Cape, New Brunswick

A graph of the tidal motion is shown for a cycle of any 24-h period.

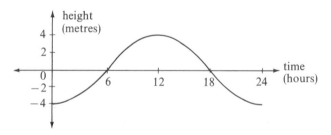

15 What will be the height of the tide at each of the following times?
  (a) 6 h   (b) 12 h   (c) 18 h   (d) 3 h   (e) 9 h   (f) 21 h

16 At what times on the graph will each of the following heights of the tide occur?
  (a) 4 m   (b) −4 m   (c) 0 m   (d) −2 m   (e) 3.5 m   (f) 2 m

17 (a) Extend the above graph to cover a 48-h period.
  (b) At what times on the graph will maximum tide occur?
  (c) At what times on the graph will minimum tide occur?

18 At a certain town, the tide has the same profile as that shown above. By measurement, it is noted that maximum tide occurs at 11:00.
  (a) Draw a graph to show the heights of the tide for a 24-h period.
  (b) Use the graph in (a). At what times does low tide occur?

4.8 Exploring Properties: Trigonometric Graphs

Various principles reoccur in your study of mathematics, such as that shown in the diagram.

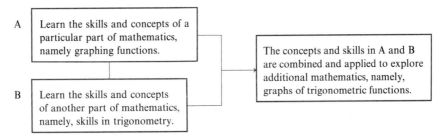

Thus, your skills with graphing functions, and the newly acquired skills with trigonometry are applied to the exploration of the properties of trigonometric functions.

How are the graphs of I and II related when the defining equations are related as shown?

Equation I	Equation II
$y = \sin \theta$	$y = 2 \sin \theta$
$y = \cos \theta$	$y = \cos 2\theta$
$y = \tan \theta$	$y = \tan \left(\theta + \dfrac{\pi}{3} \right)$

Once you learn to graph $y = \sin \theta$, $y = \cos \theta$, and $y = \tan \theta$ you can use the graphs you produce to sketch the graphs of $y = 2 \sin \theta$, $y = \cos (2\theta)$ and $y = \tan \left(\theta + \dfrac{\pi}{3} \right)$.

4.8 Exercise

The exercise that follows explores the various properties of graphs related to the graphs of $y = \sin \theta$, $y = \cos \theta$, and $y = \tan \theta$.

Explore 1 1 (a) Draw the graph of these functions on the same set of axes.
    (i) $y = \sin \theta$ (ii) $y = 2 \sin \theta$ (iii) $y = 3 \sin \theta$
    (b) How are the graphs alike?   (c) How are the graphs different?

2 Use your graphs from Question 1.
    (a) For $y = 2 \sin \theta$, what is the maximum and minimum value of y?
    (b) For $y = 3 \sin \theta$, what is the maximum and minimum value of y?
    (c) What is the period of the graph of   (i) $y = 2 \sin \theta$?   (ii) $y = 3 \sin \theta$?

3 Based on the previous results, predict the maximum value of y for each.
    (a) $y = 4 \sin \theta$ (b) $y = 6 \sin \theta$

Explore 2 4 (a) Draw the graph of these functions on the same set of axes.

   (i) $y = \cos \theta$ (ii) $y = 2 \cos \theta$ (iii) $y = 3 \cos \theta$

   (b) How are the graphs alike?   (c) How are the graphs different?

5 (a) For $y = 2 \cos \theta$, what is the maximum and minimum value of y?

   (b) For $y = 3 \cos \theta$, what is the maximum and minimum value of y?

   (c) What is the period of the graph of (i) $y = 2 \cos \theta$?   (ii) $y = 3 \cos \theta$?

6 Based on the previous results, predict the maximum value of y for each.

   (a) $y = 5 \cos \theta$ (b) $y = 8 \cos \theta$

Explore 3 7 (a) Draw the graph of these functions on the same set of axes.

   (i) $y = \sin \theta$ (ii) $y = \sin (2\theta)$ (iii) $y = \sin \left(\dfrac{1}{2}\theta\right)$

   (b) How are the graphs alike?   (c) How are the graphs different?

8 Use the graphs in Question 7.

   (a) What are the maximum values?   (b) What are the minimum values?

   (c) What is the period of each graph?

9 (a) Based on your answer in Question 8(c), predict the period of $y = \sin (3\theta)$.

   (b) Draw the graph of $y = \sin (3\theta)$.

   (c) Based on your answer in part (b), what is the period of $y = \sin (3\theta)$?

10 Based on the previous results, predict the periods of each.

   (a) $y = \sin (4\theta)$ (b) $y = \sin \left(\dfrac{1}{3}\theta\right)$

Explore 4 11 (a) On the same set of axes, draw the graph.

   (i) $y = \cos \theta$ (ii) $y = \cos (2\theta)$ (iii) $y = \cos \left(\dfrac{1}{2}\theta\right)$.

   (b) How are the graphs alike?   (c) How are the graphs different?

12 Use the graphs in Question 11.

   (a) What are the maximum values?   (b) What are the minimum values?

   (c) What are the periods of each graph?

13 (a) Based on your answer in Question 12(c), predict the period of
       $y = \cos (3\theta)$.

   (b) Draw the graph of $y = \cos (3\theta)$.

   (c) Based on your answer in part (b), what is the period of $y = \cos (3\theta)$?

14   Based on the previous results, predict the periods of

(a) $y = \cos(4\theta)$ (b) $y = \cos\left(\dfrac{1}{3}\theta\right)$

Explore 5   15   (a) On the same set of axes, draw the graphs of

(i) $y = \tan\theta$ (ii) $y = \tan(2\theta)$ (iii) $y = \tan\left(\dfrac{1}{2}\theta\right)$

(b) How are the graphs alike?   (c) How are the graphs different?
(d) What is the period of each graph?

16   (a) Based on the Question 15(d) answer predict the period of $y = \tan(3\theta)$
(b) Draw the graph of $y = \tan(3\theta)$.
(c) Based on your answer in part (b), what is the period of $y = \tan(3\theta)$?

17   Based on the above results, predict the periods of

(a) $y = \tan(4\theta)$ (b) $y = \tan\left(\dfrac{1}{3}\theta\right)$

Explore 6   18   (a) On the same set of axes, draw the graphs defined by

(i) $y = \sin\theta$ (ii) $y = \sin\left(\theta + \dfrac{\pi}{3}\right)$ (iii) $y = \sin\left(\theta - \dfrac{\pi}{3}\right)$

(b) How are the graphs alike?   (c) How are the graphs different?

19   (a) On the same set of axes, draw the graphs defined by

(i) $y = \cos\theta$ (ii) $y = \cos\left(\theta + \dfrac{\pi}{4}\right)$ (iii) $y = \cos\left(\theta - \dfrac{\pi}{4}\right)$

(b) How are the graphs alike?   (c) How are the graphs different?

20   (a) On the same set of axes, draw the graphs defined by

(i) $y = \tan\theta$ (ii) $y = \tan\left(\theta + \dfrac{\pi}{6}\right)$ (iii) $y = \tan\left(\theta - \dfrac{\pi}{6}\right)$

(b) How are the graphs alike?   (c) How are the graphs different?

Math Tip

In the previous work in this section, you explored various principles
and properties associated with trigonometric graphs. Make a
summary of them. This summary will be helpful in the subsequent
sections, when you apply the principles and properties to sketch the
graphs of trigonometric functions.

4.9 Amplitude, Period, and Phase Shift

To work with periodic graphs such as the graphs of trigonometric functions, you need to learn the vocabulary of amplitude, period, and phase shift. In the previous section, you learned about certain properties. The following diagrams summarize the properties of the sine function, and illustrate the meanings of the vocabulary. Similar comments apply for the cosine function.

Amplitude

The amplitude of the graph of $y = \sin \theta$ is 1.

The amplitude of the graph of $y = 2 \sin \theta$ is 2.

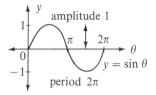

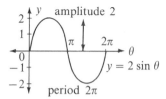

The amplitude of $y = a \sin \theta$ or of $y = a \cos \theta$, is $|a|$.

Period

The values of the trigonometric functions repeat periodically. Namely, at certain values of the domain, the same values of the function reoccur.

The period of the function defined by $y = \sin \theta$ is 2π.

The period of the function defined by $y = \sin 2\theta$ is π.

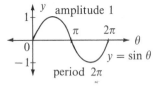

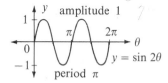

The **period** of a trigonometric function is the interval in the domain that completes one cycle. Namely, the period of a function is the smallest interval for which the values of the function repeat. In general, the period of $y = \sin k\theta$ or of $y = \cos k\theta$ is $\dfrac{2\pi}{k}$, $k > 0$. For $y = \tan k\theta$, the period is $\dfrac{\pi}{k}$, $k > 0$.

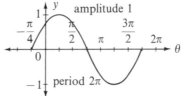

Phase Shift

The phase shift of a graph is the amount by which the graph leads or lags the graph in standard position. The diagrams illustrate a phase shift. One cycle of each graph is shown.

The graph of $y = \sin \theta$ is shown in standard position.

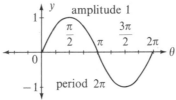

amplitude 1

period 2π

The cycle starts at 0, and ends at 2π.

The graph of $y = \sin\left(\theta + \dfrac{\pi}{4}\right)$ has undergone a phase shift.

amplitude 1

period 2π

The cycle starts at $\dfrac{-\pi}{4}$ and ends at $\dfrac{7\pi}{4}$.

The graph of $y = \sin \theta$ has undergone a phase shift of $-\dfrac{\pi}{4}$ to yield the graph of $y = \sin\left(\theta + \dfrac{\pi}{4}\right)$. In general, the phase shift of $y = \sin(\theta + k)$ is $-k$. The value $-k$ represents a lag of k units, or, a shift of k units to the left, of the graph of $y = \sin \theta$. The phase shift of $y = \cos(\theta - k)$ is k. This value, k, represents a lead of k units, or, a shift of k units to the right, of the graph of $y = \cos \theta$.

To find the phase shift of the graph defined by $y = 3 \sin\left(2\theta - \dfrac{\pi}{4}\right)$, the defining equation needs to be written as $y = 3 \sin 2\left(\theta - \dfrac{\pi}{8}\right)$

The phase shift is $\dfrac{\pi}{8}$.

4.9   Exercise

A   Remember: To work with mathematics, you need to know the meanings of words: amplitude, period, phase shift.

1   What is the amplitude for each of the following?

   (a) $y = \sin(2x)$ (b) $y = 4 \cos\left(\dfrac{1}{2}\theta\right)$ (c) $y = \dfrac{1}{2}\cos(3\theta)$ (d) $y = 2 \sin(\pi x)$

2   What is the period for each of the following?

   (a) $y = 3 \sin(2\theta)$ (b) $y = 2 \cos(3x)$ (c) $y = \dfrac{1}{2}\tan(2\theta)$

   (d) $y = \dfrac{3}{2}\cos\left(\dfrac{1}{2}x\right)$ (e) $y = 5 \tan\left(\dfrac{1}{2}\theta\right)$ (f) $y = 4 \cos(\pi x)$

3   What is the phase shift for each of the following?

(a) $y = 3 \cos \left(x + \dfrac{\pi}{4} \right)$ (b) $y = 2 \sin \left(\theta - \dfrac{\pi}{3} \right)$

(c) $y = 3 \tan \left(x + \dfrac{\pi}{2} \right)$ (d) $y = \dfrac{1}{2} \sin \left(\dfrac{\pi}{3} + \theta \right)$

(e) $y = 2 \sin 2 \left(x - \dfrac{\pi}{3} \right)$ (f) $y = \dfrac{3}{2} \cos \left(2\theta - \dfrac{\pi}{4} \right)$

(g) $y = 3 \tan (3x - \pi)$ (h) $y = 5 \cos \left(\dfrac{1}{2}x - \pi \right)$

B 4   A graph of $y = 2 \sin \theta$ is drawn. Find the missing co-ordinates, $0 \le \theta \le 2\pi$.

(a) $(\pi, ?)$ (b) $(?, 2)$ (c) $(?, 0)$ (d) $\left(\dfrac{3}{4}\pi, ? \right)$

5   Find the missing values for the ordered pairs if $y = \cos 2\theta$, $0 \le \theta \le 2\pi$.

(a) $(?, 1)$ (b) $\left(\dfrac{\pi}{2}, ? \right)$ (c) $(\pi, ?)$ (d) $(?, -1)$

6   Complete the co-ordinates for the graph given by $y = \sin \left(\theta + \dfrac{\pi}{4} \right)$, $0 \le \theta \le 2\pi$.

(a) $\left(\dfrac{\pi}{4}, ? \right)$ (b) $(?, 0)$ (c) $(0, ?)$ (d) $(?, -1)$

7   Find the missing values for each graph, $0 \le \theta \le 2\pi$.

(a) $y = \sin \dfrac{1}{2}\theta$ $(\pi, ?)$ $(?, 0)$

(b) $y = \tan 2\theta$ $(?, 1)$ $\left(\dfrac{\pi}{2}, ? \right)$ $(\pi, ?)$

(c) $y = 3 \cos \theta$ $(0, ?)$ $(?, -3)$ $\left(\dfrac{\pi}{2}, ? \right)$

(d) $y = \cos \left(\theta - \dfrac{\pi}{4} \right)$ $\left(\dfrac{\pi}{2}, ? \right)$ $(?, 1)$ $\left(\dfrac{\pi}{4}, ? \right)$

8   Sketch the graph of each of the following on the same set of co-ordinate axes.

(a) $y = \cos \theta$ (b) $y = 2 \cos \theta$ (c) $y = \dfrac{1}{2} \cos \theta$

9   Sketch the graph of each of the following on the same set of co-ordinate axes.

(a) $y = \sin x$ (b) $y = \sin \left(\dfrac{1}{2}x \right)$ (c) $y = \sin (4x)$

10 Sketch the graph of the following on the same set of co-ordinate axes.

(a) $y = \sin x$ (b) $y = \sin\left(x - \dfrac{\pi}{4}\right)$ (c) $y = \sin\left(x + \dfrac{\pi}{4}\right)$

11 Sketch the graph of the following on the same set of co-ordinate axes.

(a) $y = \cos x$ (b) $y = \cos\left(x + \dfrac{\pi}{3}\right)$ (c) $y = \cos\left(x - \dfrac{\pi}{3}\right)$

12 Each of the following graphs was obtained from the graph of $y = \sin \vartheta$. What is the defining equation of each graph?

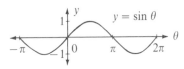

(a)   (b)   (c)

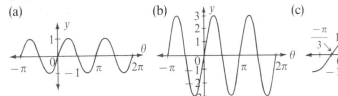

13 Each of the following graphs was obtained from the graph of $y = \cos \theta$. What is the defining equation of each graph?

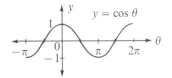

(a)   (b)   (c)

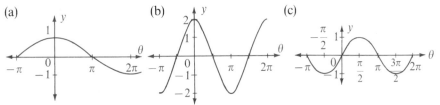

Problem Solving

In the study of mathematics, certain constant numbers, such as π, occur. Another one, occurring in advanced mathematics, is the number represented by e. To obtain an approximation to the value of e, use the formula and your calculator.

$$e = 1 + \frac{1}{1} + \frac{1}{1 \times 2} + \frac{1}{1 \times 2 \times 3} + \frac{1}{1 \times 2 \times 3 \times 4} + \cdots .$$

How many terms do you need to obtain the value of $e \doteq 2.718\,281$ (to 6 decimal places)?

4.10 Sketching Graphs: Trigonometric Functions

You have learned various skills for graphing functions, as well as the basic graphs of trigonometric functions. In this section, you will combine your various skills to draw the graphs of any trigonometric function.

Example 1   Sketch the graph defined by $y = 3 \sin (2\theta)$ for $-2\pi \leq \theta \leq 2\pi$.

Solution   Think of these steps.   Sketch the graph.

Step 1

The graph of $y = \sin \theta$ is used as the reference curve. Graph one cycle only.

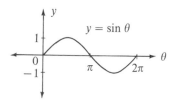

Step 2

For the graph of $y = \sin (2\theta)$, perform a *horizontal stretch* on the graph of $y = \sin \theta$. For the variable, 2θ, the *stretch* is by a factor of $\dfrac{1}{2}$.

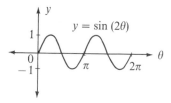

Step 3

For the graph $y = 3 \sin (2\theta)$, perform a *vertical stretch* on the graph of $y = \sin (2\theta)$. For the multiplier, 3, the *stretch* is by a factor of 3.

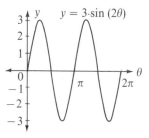

Since the graph of $y = \sin \theta$ is periodic, the above cycle for $y = 3 \sin (2\theta)$ is repeated to obtain the graph of $y = 3 \sin (2\theta)$ for $-2\pi \leq \theta \leq 2\pi$.

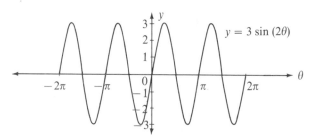

In the next example, the significance of phase shift is examined.

Example 2   Sketch the graph of $y = 2 \cos\left(\theta + \dfrac{\pi}{2}\right)$, for $-2\pi \leq \theta \leq 2\pi$.

Solution   Think of these steps.   Sketch the graph.

Step 1

Graph one cycle of $y = \cos \theta$.

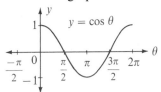

Step 2

For the graph of $y = \cos\left(\theta + \dfrac{\pi}{2}\right)$ perform

a *horizontal translation* on the graph of
$y = \cos \theta$. For the variable expression

$\theta + \dfrac{\pi}{2}$, the *translation* is $\dfrac{-\pi}{2}$. The cycle

begins at $\dfrac{-\pi}{2}$ and ends at $\dfrac{3\pi}{2}$ (The

distance from $-\dfrac{\pi}{2}$ to $\dfrac{3\pi}{2}$ is 2π, the period

of the curve.)

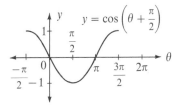

Step 3

For the graph of $y = 2 \cos\left(\theta + \dfrac{\pi}{2}\right)$,

perform a *vertical stretch* on

$y = \cos\left(\theta + \dfrac{\pi}{2}\right)$. The stretch is by a

factor of 2.

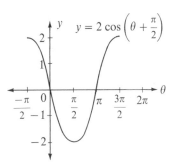

The one cycle is used to extend the graph
to cover the interval $-2\pi \leq \theta \leq 2\pi$.

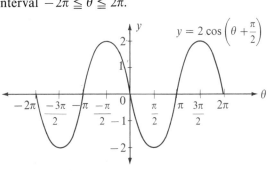

4.10 Exercise

A To draw the graphs, determine the amplitude, period, and phase shift.

1 For the graph given by $y = 3 \sin (2x)$, find the missing co-ordinates for the $0 \le x \le 2\pi$.

(a) $(0, ?)$ (b) $\left(\dfrac{\pi}{4}, ?\right)$ (c) $(\pi, ?)$ (d) $\left(\dfrac{\pi}{2}, ?\right)$ (e) $\left(\dfrac{5}{4}\pi, ?\right)$

2 For the graph given by $y = \sin \left(\theta + \dfrac{\pi}{6}\right)$, find the missing values for the interval $-2\pi \le \theta \le 2\pi$.

(a) $\left(-\dfrac{\pi}{6}, ?\right)$ (b) $\left(\dfrac{5\pi}{6}, ?\right)$ (c) $\left(\dfrac{\pi}{3}, ?\right)$ (d) $\left(\dfrac{4}{3}\pi, ?\right)$ (e) $\left(-\dfrac{7\pi}{6}, ?\right)$

3 For each of the following:

▶ What is the amplitude, period, and phase shift?
▶ Use $y = \sin \theta$ as a basis and draw the graph.

(a) $y = \sin (2\theta)$ (b) $y = 2 \sin \theta$ (c) $y = 2 \sin (3\theta)$

(d) $y = 3 \sin (2\theta)$ (e) $y = \sin \left(\theta - \dfrac{\pi}{3}\right)$ (f) $y = \sin \left(\theta + \dfrac{\pi}{4}\right)$

4 For each of the following:

▶ What is the amplitude, period, and phase shift?
▶ Use $y = \cos \theta$ as a basis and draw the graph.

(a) $y = \cos (3\theta)$ (b) $y = 3 \cos \theta$

(c) $y = 2 \cos (3\theta)$ (d) $y = \cos \left(\theta + \dfrac{\pi}{3}\right)$

5 Use $y = \tan \theta$ as a basis and draw the graph of each of the following.
(a) $y = \tan (2\theta)$ (b) $y = \tan (3\theta)$

(c) $y = \tan \left(\theta + \dfrac{\pi}{3}\right)$ (d) $y = \tan \left(\theta - \dfrac{\pi}{2}\right)$

6 Sketch the graphs of the following for the interval $0 \le x \le 2\pi$.
(a) $y = 4 \sin (3x)$ (b) $y = 5 \cos (2x)$ (c) $y = 6 \sin (4x)$
(d) Choose a point on each graph in parts (a) to (c). Check whether the co-ordinates of the point satisfy the given equation.

B   The following questions combine skills for graphing.

7   Sketch the following for the interval $-2\pi \leq x \leq 2\pi$.

(a) $y = 2 \sin \left(x + \dfrac{\pi}{4} \right)$ (b) $y = 3 \cos \left(x - \dfrac{\pi}{3} \right)$ (c) $y = 2 \cos \left(x + \dfrac{\pi}{6} \right)$

(d) Select a point on the graph of each of the above. Do the co-ordinates of the point satisfy the given equation?

8   Sketch each of the following graphs on the same set of co-ordinate axes.

(a) $y = \sin (x) + 3$ (b) $y = 2 \cos x$ (c) $y = 2 \cos (x) - 3$

9   (a) Sketch the graphs of   (i) $y = \cos (2x)$ (ii) $y = 3 \cos (2x)$

(b) Use the above graphs to sketch the graph of $y = 3 \cos 2 \left(x + \dfrac{\pi}{2} \right)$.

10   (a) Sketch the graphs of   (i) $y = 3 \sin x$ (ii) $y = 3 \sin \left(\dfrac{1}{2}x \right)$

(b) Use the above graphs to sketch the graph of $y = 3 \sin \left(\dfrac{1}{2}x + \dfrac{\pi}{4} \right)$.

11   (a) Sketch the graphs of   A: $y = 5 \sin \left(3x - \dfrac{\pi}{2} \right)$ B: $y = 5 \cos \left(3x + \dfrac{\pi}{2} \right)$

(b) How are the graphs of A and B alike? How do they differ?

12   Sketch the graphs of the following for the interval $0 \leq x \leq 2\pi$.

(a) $y = 3 \sin (2x + \pi)$ (b) $y = 2 \cos (3x - \pi)$

(c) $y = 4 \sin \left(2x + \dfrac{\pi}{2} \right)$ (d) $y = 3 \cos \left(2x - \dfrac{\pi}{4} \right)$

13   Sketch each graph. Use a point to check the graph.

(a) $y = 3 \sin \left(x + \dfrac{\pi}{3} \right)$ (b) $y = 2 \sin \left(2x + \dfrac{\pi}{3} \right)$ (c) $y = 2 \cos \left(x - \dfrac{\pi}{2} \right)$

(d) $y = 2 \cos \left(3x - \dfrac{\pi}{2} \right)$ (e) $f(x) = 3 \sin (2x + \pi)$ (f) $f(x) = 2 \cos (3x - \pi)$

14   Sketch (a) $y = 2 \cos (\theta + \pi) - 3$ (b) $y = \dfrac{1}{2} \sin \left(2\theta - \dfrac{\pi}{2} \right) - 3$

C 15   For the functions given by the defining equation, $x \in R$, $y = a \sin (bx + c)$, where $a > 0$, what is the

(a) range?   (b) period?   (c) amplitude?
(d) phase shift?   (e) minimum value?   (f) maximum value?

Applications: Periodic Phenomena

The rotation of the moon about the earth causes the rise and fall of ocean levels and thus creates tides. Ocean tides are examples of periodic phenomena and can be represented by graphs of trigonometric functions. The following graph describes the rise and fall of an ocean tide at a port.

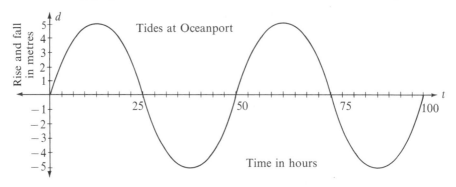

Questions 16 and 17 refer to the periodic tide graph.

16 (a) What is the period of the graph?
    (b) What is the maximum value of d?
    (c) For what values of t is the maximum value of d obtained?
    (d) What is the minimum value of d?
    (e) For what values of t is the minimum value of d obtained?

17 (a) For what value(s) of t is   (i) $d = 4$?   (ii) $d = -6$?
    (b) What are the x-intercepts of the graph?
    (c) What is the y-intercept of the graph?

18 For Seaforth, high tide occurs 3 h before the high tide for Oceanport.
    (a) Draw a graph of the tides for Seaforth.
    (b) At what values of t does maximum tide occur?
    (c) At what values of t does minimum tide occur?

19 For the pendulum, the angle θ as shown is given by the equation
$$\theta = \frac{1}{5} \sin \left(\frac{1}{2} \pi t \right)$$
where t is in seconds and θ is in radians.

    (a) Draw the graph of the function
        given by the above equation.
    (b) For what values of t is $\theta = 0$?
    (c) What is the value of the complete angle
        through which the pendulum swings?

4.11 Graphs of Reciprocal Functions

The reciprocal trigonometric values are defined as

$$\csc \theta = \frac{1}{\sin \theta} \qquad \sec \theta = \frac{1}{\cos \theta} \qquad \cot \theta = \frac{1}{\tan \theta}$$

and give rise to the reciprocal functions

$$f: \theta \rightarrow \csc \theta \qquad g: \theta \rightarrow \sec \theta \qquad h: \theta \rightarrow \cot \theta$$
$$y = \csc \theta \qquad\quad y = \sec \theta \qquad\quad y = \cot \theta$$

You can base the sketching of the graph of $y = \csc \theta$ on the graph of $y = \sin \theta$. Observations A to G are used to construct one cycle of the graph of $y = \csc \theta$.

Sketch one cycle of $y = \sin \theta$.

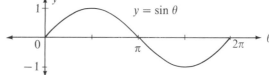

Observations

A   For $\theta = \frac{\pi}{2}$, $\sin \theta = 1$. Thus, $\csc \theta = \frac{1}{\sin \theta} = 1$.

B   For $\theta = \frac{3\pi}{2}$, $\sin \theta = -1$. Thus, $\csc \theta = \frac{1}{\sin \theta} = -1$.

C   For $0 < \theta < \frac{\pi}{2}$, $\sin \theta$ is positive and increases in value. Thus,

   $\csc \theta = \frac{1}{\sin \theta}$ is positive and decreases in value.

D   For $\frac{\pi}{2} < \theta < \pi$, $\sin \theta$ is positive and decreases in value. Thus,

   $\csc \theta = \frac{1}{\sin \theta}$ is positive and increases in value.

E   For $\theta = \pi$, $\sin \theta = 0$. Thus, $\csc \theta = \frac{1}{\sin \theta}$ is undefined.

F   For $\pi < \theta < \frac{3\pi}{2}$, $\sin \theta$ is negative and decreases in value. Thus,

   $\csc \theta = \frac{1}{\sin \theta}$ is negative and increases in value.

G   For $\frac{3\pi}{2} < \theta < 2\pi$, $\sin \theta$ is negative and increases in value. Thus,

   $\csc \theta = \frac{1}{\sin \theta}$ is negative and decreases in value.

Use the observations A to G to sketch the graph of $y = \csc \theta$.

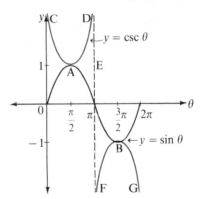

The graph $y = \csc \theta$ is periodic. The one cycle is then used to sketch the graph for $-4\pi \leqq \theta \leqq 4\pi$.

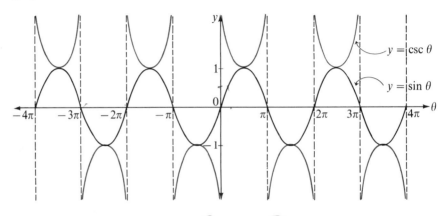

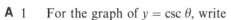

4.11 Exercise

A 1 For the graph of $y = \csc \theta$, write
(a) the domain.   (b) the range.
(c) the values of θ for which $y = 1$.   (d) the values of θ for which $y = -1$.

2 (a) What is the amplitude of $y = \sin \theta$?
(b) What is the amplitude of $y = \csc \theta$, if any?

B 3 (a) Sketch the graph of $y = \cos \theta$.
(b) Use the graph in part (a) to aid your sketch of $y = \sec \theta$.
(c) List the domain and the range of $y = \sec \theta$.

4   (a) Sketch the graph of $y = \tan \theta$.
    (b) Use the graph in part (a) to aid your sketch of $y = \cot \theta$.
    (c) List the domain and the range of $y = \cot \theta$.

5   (a) Sketch the graph of $y = \csc(2\theta)$. (Use the graph of $y = \sin(2\theta)$ to aid you.)
    (b) Use the co-ordinates of a point on the curve to check the graph.
    (c) Write the domain and range of $y = \csc(2\theta)$.

6   (a) Sketch the graph of $y = \csc\left(\dfrac{\theta}{2}\right)$. (Use the graph of $y = \sin\left(\dfrac{\theta}{2}\right)$ to aid you.)
    (b) Write the domain and the range of $y = \csc\left(\dfrac{\theta}{2}\right)$.

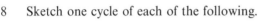

7   (a) Sketch the graph of $y = \sec(2\theta)$.
        (Use the graph of $y = \cos(2\theta)$ to aid you).
    (b) List the domain and the range of $y = \sec(2\theta)$.

8   Sketch one cycle of each of the following.
    (a) $y = \sec(\theta) + 1$ (b) $y = \csc(\theta) - 2$
    (c) $y = \sec\left(\theta + \dfrac{\pi}{3}\right)$ (d) $y = \cot(\theta - \pi)$

C 9   Sketch one cycle of each of the following.
    (a) $y = -\sec\theta$ (b) $y = -\csc\theta$ (c) $y = |\sec\theta|$
    (d) $y = |\csc\theta|$ (e) $y = |\cot\theta|$ (f) $y = \left|\sec\left(\theta + \dfrac{\pi}{3}\right)\right|$
    (g) $y = |\cot(\theta + \pi)|$ (h) $y = \left|\csc\left(\theta - \dfrac{\pi}{3}\right)\right|$

Math Tip

It is important to clearly understand the vocabulary of mathematics when solving problems. *You cannot solve problems if you don't know what the clues are.*
- Make a list of all the words you have learned in this chapter.
- Provide a simple example to illustrate the meaning of each word.

4.12 Problem-Solving: Working Backwards

In the previous sections, you were given the equation of a trigonometric function. By analyzing the equation, you were able to determine the amplitude, period and phase shift. Once this information is known, a sketch of the trigonometric function is completed.

In order to work backwards, you need to identify the amplitude, period and phase shift. With this information, you can work backwards to write the defining equation of the trigonometric function.

1 Write the equation for the sine function that has the following properties.
    (a) amplitude 3, period π, and phase shift 0
    (b) amplitude 2, period 3π, and phase shift 0

    (c) amplitude $\dfrac{1}{2}$, period 2π, and phase shift $\dfrac{\pi}{2}$

    (d) amplitude 4, period 2π, and phase shift $\dfrac{\pi}{3}$

    (e) What is the range for each of the above sine functions?

2 One cycle of each graph is shown.
    • Determine the amplitude, period, and phase shift of each graph.
    • Then construct the defining equation of each trigonometric function.

(a)
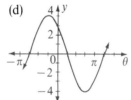

(b)
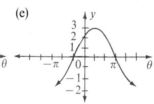

(c)
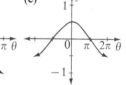

(d)
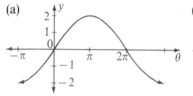

(e)
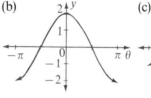

(f)
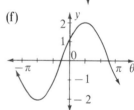

One cycle of each graph is shown.

3 Analyze each graph. Construct a defining equation of each graph.

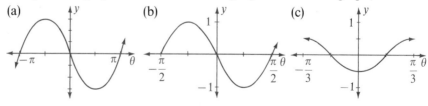

Practice and Problems: A Chapter Review

An important step for problem-solving is to decide which skills to use. For this reason, these questions are not placed in any special order. When you have finished the review, you might try the *Test for Practice* that follows.

1 θ is an angle such that $\cot \theta = -\dfrac{24}{7}$ and $90° \leq \theta \leq 180°$.

   (a) Draw a diagram to show θ.
   (b) From the diagram, find $\sin \theta$, $\tan \theta$ and $\cos \theta$.

2 (a) Find the value of each of the following.
      (i) $\cos 132°$ (ii) $\tan(-137°)$
   (b) Find θ, if $0 \leq \theta \leq 2\pi$, and $\cos \theta = 0.9612$.

3 Calculate each of the following.
   (a) $\sin 45° \cos 45° + \tan^2 225°$ (b) $\sin 30° \cos 60° + \sin 60° \cos 30°$
   (c) $2 \sin 60° \cos 60°$ (d) $\sin^2 60° + \cos^2 60°$

4 θ and β are angles in standard position. θ has its terminal arm in the first quadrant; β has its terminal arm in the second quadrant. If $\cos \theta = \dfrac{3}{5}$ and $\cos \beta = -\dfrac{3}{5}$, what is the value of $\cos \beta + \cos \theta + \sin \beta + 2 \sin \theta + \tan \beta$?

5 Calculate the angular velocity of Saturn if it circles the sun once every 29.5 years. Express your answer in degrees per day to 2 decimal places.

6 (a) Sketch each graph on the same axes.

      A: $y = \sin 2\left(\theta - \dfrac{\pi}{3}\right)$ B: $y = \sin\left(2\theta + \dfrac{\pi}{3}\right)$

   (b) How are the graphs alike? How are they different?

7 Solve for θ and α if $0° < \theta + \alpha < 90°$.
   (a) $\sin(\theta + \alpha) = 0.9397$ (b) $\tan(\theta + \alpha) = 3.7321$
      $\cos(\theta - \alpha) = 0.9848$ $\tan(\theta - \alpha) = 0.2679$

8 The current, I, in amperes, of an electrical circuit, is given by the formula
      $I = 4.5 \sin(120\pi t)$,   where t is the time in seconds.

   (a) Draw a graph defined by the above equation.
   (b) What is the period?
   (c) At what values of t is the current a maximum? A minimum?

Test for Practice

Try this test. Each *Test for Practice* is based on the mathematics you have learned in the preceding chapter. Try this test later in the year as a review. Keep a record of those questions that you were not successful with, get help in obtaining solutions and review them periodically.

1 Given $\cos \theta = \dfrac{24}{25}$, $0 \leqq \theta \leqq \dfrac{\pi}{2}$.

   (a) Draw a diagram to show the terminal arm.
   (b) Find $\sin \theta$, $\tan \theta$, and $\cot \theta$.

2 Calculate each of the following. Express your answer in radical form.
   (a) $\tan(-60°)$ (b) $\sin 225°$

3 Write each of the following in radian measure.
   (a) $150°$ (b) $-310°$ (c) $750°$

4 If $\cos \theta = -\dfrac{5}{13}$, and θ is a second quadrant angle, then calculate

   (a) $\sin \theta \cos \theta$ (b) $2 \cot \theta$ (c) $\sin^2 \theta$ (d) $\cot \theta \tan \theta$

5 If β is an angle in the third quadrant and $\cos \beta = -\dfrac{1}{\sqrt{2}}$, then calculate

   $\sin \beta \cos \beta + \tan^2 \beta$.

6 For each θ, the quadrant is given. Find the value of θ, $0° \leqq \theta \leqq 360°$.
   (a) $\tan \theta = -3.7161$, II   (b) $\cos \theta = -0.9636$, III
   (c) $\sec \theta = 1.0954$, IV

7 Sketch a graph of the following.
   (a) $y = 2 \sin \theta$ (b) $y = \sin 3\theta$

8 (a) What is the amplitude, period and phase shift of each graph?

   A: $y = 3 \cos\left(x - \dfrac{\pi}{4}\right)$ B: $y = 2 \sin\left(2\theta + \dfrac{\pi}{3}\right)$

   (b) Sketch the graph of each trigonometric function in (a).

9 The sketch of a curve is shown. Find its defining equation.

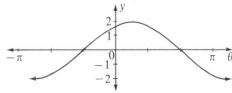

5 Applying Trigonometry: Solving Problems

working with identities, solving trigonometric equations, solving triangles, inaccessible distances, Law of Sines, Law of Cosines, ambiguous case, problems in 3 dimensions, solving problems and strategies, applications, problem-solving

Introduction

The following process is important in the study of mathematics.
Concepts and skills with trigonometry are also used to solve problems involving triangles. Triangles are one of the essential components of construction.

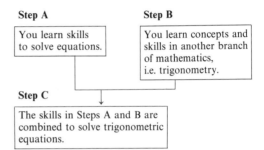

Step A

You learn skills to solve equations.

Step B

You learn concepts and skills in another branch of mathematics, i.e. trigonometry.

Step C

The skills in Steps A and B are combined to solve trigonometric equations.

Triangles are used to maintain rigidity.

Triangles are used in roof supports because of the special role they play in producing the strength needed to hold massive objects.

- The triangles used in the construction of antennae provide a maximum of strength with a minimum of material.
- The systems of antennae set up send and receive messages. Sometimes such messages are bounced off satellites.

The use of the concepts and skills of trigonometry are required to calculate the precision of satellite communications.

5.1 Applying Identities in Trigonometry

The same vocabulary occurs in different branches of mathematics. In algebra, an identity is true for all values of the variable. For example, this identity occurs for all $x \in R$, $(x + 1)^2 = x^2 + 2x + 1$.
In trigonometry, identities occur that are useful in solving problems.

An angle, θ, is shown in standard position. $P(x, y)$ is any point on the terminal arm. From the basic definition of the trigonometric values, you may derive the fundamental trigonometric identities.

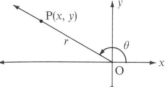

$$\sin \theta = \frac{y}{r} \qquad \csc \theta = \frac{r}{y} \qquad \text{Similarly,}$$

$$\sin \theta = \frac{1}{\dfrac{r}{y}} \qquad \csc \theta = \frac{1}{\dfrac{y}{r}} \qquad\qquad \cos \theta = \frac{1}{\sec \theta} \qquad \sec \theta = \frac{1}{\cos \theta}$$

$$\sin \theta = \frac{1}{\csc \theta} \qquad \csc \theta = \frac{1}{\sin \theta} \qquad\quad \tan \theta = \frac{1}{\cot \theta} \qquad \cot \theta = \frac{1}{\tan \theta}$$

Skills with algebra can be used to develop other identities in trigonometry. For example, for all θ

$$\frac{\sin \theta}{\cos \theta} = \frac{\dfrac{y}{r}}{\dfrac{x}{r}} = \frac{y}{x} = \tan \theta. \qquad\qquad \text{Thus } \frac{\sin \theta}{\cos \theta} = \tan \theta.$$

Pythagorean Identities

From the diagram you can write the identity

$$y^2 + x^2 = r^2 \quad \text{(Pythagorean Theorem)}$$

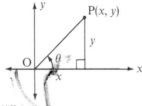

Since $r \neq 0$, then you can write the identity in different forms.

Divide by r^2	Divide by x^2	Divide by y^2
$y^2 + x^2 = r^2$	$y^2 + x^2 = r^2$	$y^2 + x^2 = r^2$
$\dfrac{y^2}{r^2} + \dfrac{x^2}{r^2} = \dfrac{r^2}{r^2}$	$\dfrac{y^2}{x^2} + \dfrac{x^2}{x^2} = \dfrac{r^2}{x^2}$	$\dfrac{y^2}{y^2} + \dfrac{x^2}{y^2} = \dfrac{r^2}{y^2}$
	$\tan^2 \theta + 1 = \sec^2 \theta$	$1 + \cot^2 \theta = \csc^2 \theta$

or $\left(\dfrac{y}{r}\right)^2 + \left(\dfrac{x}{r}\right)^2 = 1$

$(\sin \theta)^2 + (\cos \theta)^2 = 1$
or $\sin^2 \theta + \cos^2 \theta = 1$

Pythagorean Identities

$$\text{For all } \theta \qquad \sin^2 \theta + \cos^2 \theta = 1$$
$$\tan^2 \theta + 1 = \sec^2 \theta$$
$$1 + \cot^2 \theta = \csc^2 \theta$$

Once you have proved the above fundamental trigonometric identities you can use them to deduce other results and prove identities. To show that the following identity is true, simplify the expression and show that LS = RS.

Example 1 For all α, prove that $\dfrac{1}{\cot \alpha} = \sin \alpha \sec \alpha$. Since the RS has sin α, try to express LS and RS so that they involve sin α.

Solution $LS = \dfrac{1}{\cot \alpha} \qquad RS = \sin \alpha \sec \alpha \qquad$ Thus LS = RS.

$= \tan \alpha \qquad\quad = \sin \alpha \dfrac{1}{\cos \alpha} \qquad$ For all α, $\dfrac{1}{\cot \alpha} = \sin \alpha \sec \alpha$.

$= \dfrac{\sin \alpha}{\cos \alpha} \qquad\quad = \dfrac{\sin \alpha}{\cos \alpha} \qquad\qquad$ Make a final concluding statement.

To prove some identities such as the following observe that one side has only sin θ and cos θ. Thus, begin with the other side and attempt to express it in terms of sin θ and cos θ.

Example 2 For all θ, prove that $\tan \theta + \dfrac{1}{\tan \theta} = \dfrac{1}{\sin \theta \cos \theta}$.

Solution $LS = \tan \theta + \dfrac{1}{\tan \theta} \qquad\qquad\qquad\qquad RS = \dfrac{1}{\sin \theta \cos \theta}$

$= \dfrac{\sin \theta}{\cos \theta} + \dfrac{1}{\dfrac{\sin \theta}{\cos \theta}} \longleftarrow$ Use $\qquad\qquad = \dfrac{1}{\cos \theta \sin \theta}$

$\tan \theta = \dfrac{\sin \theta}{\cos \theta}$

$= \dfrac{\sin \theta}{\cos \theta} + \dfrac{\cos \theta}{\sin \theta}$

$= \dfrac{(\sin \theta)(\sin \theta) + (\cos \theta)(\cos \theta)}{\cos \theta \sin \theta}$

$= \dfrac{\sin^2 \theta + \cos^2 \theta}{\cos \theta \sin \theta} \longleftarrow$ Use $\qquad\qquad$ Thus LS = RS

$\sin^2 \theta + \cos^2 \theta = 1.$

$= \dfrac{1}{\cos \theta \sin \theta} \qquad\qquad$ For all θ, $\tan \theta + \dfrac{1}{\tan \theta} = \dfrac{1}{\sin \theta \cos \theta}$.

5.1 Exercise

A Questions 1 to 9 examine skills needed to prove identities in trigonometry.

1 Use the definitions $\sin \theta = \dfrac{y}{r}$, $\cos \theta = \dfrac{x}{r}$, and $\tan \theta = \dfrac{y}{x}$. Prove that

   (a) $\sin \theta = \dfrac{1}{\csc \theta}$ (b) $\cot \theta = \dfrac{1}{\tan \theta}$

2 For all θ, prove that

   (a) $\cos \theta = \dfrac{1}{\sec \theta}$ (b) $\sec \theta = \dfrac{1}{\cos \theta}$

   (c) $\tan \theta = \dfrac{1}{\cot \theta}$ (d) $\csc \theta = \dfrac{1}{\sin \theta}$

3 Use the Pythagorean identity $\sin^2 \theta + \cos^2 \theta = 1$ to show that
   (a) $\tan^2 \theta + 1 = \sec^2 \theta$ (b) $1 + \cot^2 \theta = \csc^2 \theta$

4 (a) Prove that $1 + \cot^2 \alpha = \csc^2 \alpha$ for all α.
   (b) Use $\alpha = 120°$ to verify the identity in (a).

5 Use the definitions $\sin \theta = \dfrac{y}{r}$, $\cos \theta = \dfrac{x}{r}$, $\tan \theta = \dfrac{y}{x}$. Prove that

   (a) $\tan \theta \cos \theta = \sin \theta$ (b) $\cot \theta \sec \theta = \csc \theta$ (c) $\dfrac{1 + \cot^2 \theta}{\csc^2 \theta} = 1$

6 To prove some identities, you need to use your factoring skills. Factor each of the following.
   (a) $1 - \cos^2 \theta$ (b) $1 - \sin^2 \theta$ (c) $\sin^2 \theta - \cos^2 \theta$
   (d) $\sin \alpha - \sin^2 \alpha$ (e) $\tan^2 \alpha - \cot^2 \alpha$ (f) $\sec^2 \theta - 1$

7 To prove some identities, you need to express left or right sides as $\sin \theta$ or $\cos \theta$. Express each of the following in terms of $\sin \theta$ or $\cos \theta$ or both.

   (a) $\dfrac{1}{\sec \theta}$ (b) $\sin^2 \theta + \dfrac{1}{\sec^2 \theta}$ (c) $\cos \theta \dfrac{1}{\sec \theta}$

   (d) $\tan \theta \cos \theta$ (e) $1 - \csc^2 \theta$ (f) $\dfrac{1 + \cot^2 \theta}{\cot^2 \theta}$

B Remember: you may need to use various skills in algebra to work with identities.

8 (a) To show that $\cos \alpha \cot \alpha = \dfrac{1}{\sin \alpha} - \sin \alpha$ express the LS in terms of $\sin \alpha$.

   (b) Prove the identity.

9 (a) To show that $\dfrac{\cos^2 \theta}{1 - \sin \theta} = 1 + \sin \theta$ write $\cos^2 \theta$ in terms of $\sin \theta$.

   (b) What are the factors of $1 - \sin^2 \theta$?

   (c) Use (a) and (b) to prove the identity.

10 (a) Prove that for all angles θ, $\dfrac{\tan^2 \theta}{1 + \tan^2 \theta} = \sin^2 \theta$.

   (b) Check the identity in (a) using $\theta = 45°$.

11 (a) Prove that for all α, $\dfrac{\tan^2 \alpha}{\sin^2 \alpha} = 1 + \tan^2 \alpha$.

   (b) Check the identity in (a) using $\alpha = 120°$.

12 Prove each identity.

   (a) $\tan \theta \cos \theta = \sin \theta$ (b) $1 - \cos^2 \theta = \dfrac{\cos \theta \sin \theta}{\cot \theta}$

   (c) $\sec \theta \cos \theta + \sec \theta \sin \theta = 1 + \tan \theta$

13 Prove each of the following identities.

   (a) $\dfrac{\sin^2 \theta}{1 - \cos \theta} = 1 + \cos \theta$ (b) $\dfrac{1 - \tan^2 \theta}{\tan \theta - \tan^2 \theta} = 1 + \dfrac{1}{\tan \theta}$

14 Prove each of the following identities. (Be aware of any restrictions that might be involved in completing your proof.)

   (a) $\cos^2 \theta = (1 - \sin \theta)(1 + \sin \theta)$ (b) $\dfrac{1}{\sec \theta} + \dfrac{\sin \theta}{\cot \theta} = \dfrac{1}{\cos \theta}$

   (c) $\sec \theta + \dfrac{1}{\cot \theta} = \dfrac{1 + \sin \theta}{\cos \theta}$ (d) $\dfrac{\cos \theta}{\csc \theta} - \dfrac{\sin \theta}{\tan \theta} = \dfrac{\sin \theta - 1}{\sec \theta}$

   (e) $\dfrac{\sec^2 \theta}{\sin^2 \theta} = \dfrac{1}{\sin^2 \theta} + \dfrac{1}{\cos^2 \theta}$ (f) $\dfrac{\cos^2 \theta - \sin^2 \theta}{\cos^2 \theta + \sin \theta \cos \theta} = \dfrac{\cot \theta - 1}{\cot \theta}$

   (g) $\sin \theta \tan \theta + \cos \theta - \sec \theta + 1 = \sec^2 \theta \cos^2 \theta$

C 15 (a) Prove for all θ, $\sin^2 \theta - \cos^2 \theta = \dfrac{2 - \sec^2 \theta}{- \sec^2 \theta}$.

   (b) Prove that for all θ, $\dfrac{\csc \theta}{1 + \csc \theta} + \dfrac{\csc \theta}{1 - \csc \theta} = -\dfrac{2 \sin \theta}{\cos^2 \theta}$.

Problem Solving

If $\tan A = \dfrac{2xy}{x^2 - y^2}$ where $x > y > 0$ and $0 < A < \dfrac{\pi}{2}$, find an expression for $\sin A$.

5.2 Problem Solving: Developing Identities

Often in the study of mathematics, a numerical example suggests a generalization. For example, to calculate sin 240°, a diagram is drawn. The reference triangle is used to obtain the trigonometric values.

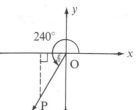

From the reference triangle

$$\sin 60° = \frac{\sqrt{3}}{2}, \cos 60° = \frac{1}{2}, \tan 60° = \sqrt{3}$$

The values are related.

Now calculate the trigonometric values

$$\sin 240° = -\frac{\sqrt{3}}{2}, \cos 240° = -\frac{1}{2}, \tan 240° = \sqrt{3}.$$

Based on these and other numerical examples, general relationships are suggested, as shown in the following example. You can calculate trigonometric values without drawing a diagram if you learn to recognize the relationships developed in the next example.

Example Prove. (a) $\sin(180° + \theta) = -\sin\theta$ (b) $\cos(180° + \theta) = -\cos\theta$
(c) $\tan(180° + \theta) = \tan\theta$

Solution Sketch a diagram to show θ and $180° + \theta$.

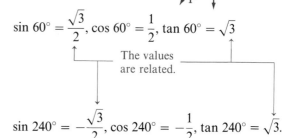

From the diagram, notice

(a) $\sin\theta = \dfrac{b}{\sqrt{a^2 + b^2}}$ ① $\sin(180° + \theta) = \dfrac{-b}{\sqrt{a^2 + b^2}}$

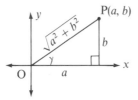

$$= -\dfrac{b}{\sqrt{a^2 + b^2}} \quad ②$$

From ① and ② $\boxed{\sin(180° + \theta) = -\sin\theta.}$

(b) $\cos\theta = \dfrac{a}{\sqrt{a^2 + b^2}}$ ③ $\cos(180° + \theta) = \dfrac{-a}{\sqrt{a^2 + b^2}}$

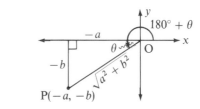

$$= -\dfrac{a}{\sqrt{a^2 + b^2}} \quad ④$$

From ③ and ④ $\boxed{\cos(180° + \theta) = -\cos\theta.}$

(c) $\tan \theta = \dfrac{b}{a}$ ⑤

$\tan (180° + \theta) = \dfrac{-b}{-a}$

$= \dfrac{b}{a}$ ⑥

From ⑤ and ⑥ $\boxed{\tan (180° + \theta) = \tan \theta.}$

Since the above are true for all values of θ they are **identities**. They may be used as follows to calculate trigonometric values.

$$\begin{aligned}
\sin 240° & & \cos 210° & & \tan 225° \\
= \sin (180° + 60°) & & = \cos (180° + 30°) & & = \tan (180° + 45°) \\
= -\sin 60° & & = -\cos 30° & & = \tan 45° \\
= -\dfrac{\sqrt{3}}{2} & & = -\dfrac{\sqrt{3}}{2} & & = 1
\end{aligned}$$

If θ is measured in radians the identities are written as

$$\sin (\pi + \theta) = -\sin \theta \qquad \cos (\pi + \theta) = -\cos \theta \qquad \tan (\pi + \theta) = \tan \theta.$$

5.2 Exercise

A Questions 1 to 5 develop various trigonometric identities based on the method in the example.

1 Sketch a diagram to show θ and $180° + \theta$. Then prove each of the following.
(a) $\csc (180° + \theta) = -\csc \theta.$ (b) $\sec (180° + \theta) = -\sec \theta.$
(c) $\cot (180° + \theta) = \cot \theta.$

2 Prove.
(a) $\sin (\pi - \theta) = \sin \theta$ (b) $\cos (\pi - \theta) = -\cos \theta$ (c) $\tan (\pi - \theta) = -\tan \theta$

3 Use the pattern in Question 2.

▶ Predict a relationship for each of the following.
▶ Then prove the relationships.

(a) $\csc (180° - \theta)$ (b) $\sec (180° - \theta)$ (c) $\cot (180° - \theta)$

4 Prove.
(a) $\sin (-\theta) = -\sin \theta$ (b) $\cos (-\theta) = \cos \theta$ (c) $\tan (-\theta) = -\tan \theta$

5 Use the pattern in Question 4.

▶ Predict a relationship for each of the following.
▶ Then prove the relationships.

(a) $\csc (-\theta)$ (b) $\sec (-\theta)$ (c) $\cot (-\theta)$

B   Make a summary of the various trigonometric identities you have developed. Use them in the following questions.

6   Use your results from the previous questions. Write each of the following in a simpler form.
   (a) $\sin(180° - \theta)$ (b) $\sec(180° + \theta)$ (c) $\tan(-\theta)$
   (d) $\csc(180° + \theta)$ (e) $\sec(180° - \theta)$ (f) $\cot(180° + \theta)$
   (g) $\cos(180° - \theta)$ (h) $\sin(-\theta)$ (i) $\csc(180° - \theta)$ (j) $\sec(-\theta)$

7   Write each of the following in a simpler form.
   (a) $\sin(\pi - \theta)$ (b) $\tan(\pi + \theta)$ (c) $\sec(\pi - \theta)$
   (d) $\csc(-\theta)$ (e) $\cot(\pi + \theta)$ (f) $\cos(\pi - \theta)$
   (g) $\cot(-\theta)$ (h) $\tan(\pi - \theta)$ (i) $\cot(\pi - \theta)$ (j) $\csc(\pi + \theta)$

8   Use the relationships to calculate each of the following.
   (a) $\sin(180° + 30°)$ (b) $\cot(180° - 60°)$ (c) $\sec(180° + 45°)$
   (d) $\csc(180° - 30°)$ (e) $\tan(180° + 60°)$ (f) $\cos(180° - 45°)$

9   Calculate each of the following.
   (a) $\sec\left(\pi - \dfrac{\pi}{6}\right)$ (b) $\cos\left(\pi - \dfrac{\pi}{3}\right)$ (c) $\cot\left(\pi + \dfrac{\pi}{4}\right)$
   (d) $\sin\left(\pi + \dfrac{\pi}{6}\right)$ (e) $\tan\left(\pi + \dfrac{\pi}{3}\right)$ (f) $\csc\left(\pi - \dfrac{\pi}{4}\right)$

10   Use the relationship $\sin(180° + \theta) = -\sin\theta$ to calculate the following.
   (a) $\sin 225°$ $\sin 225° = \sin(180° + 45°)$ (b) $\sin 210°$ (c) $\sin 240°$

11   Use the relationship $\cos(180° - \theta) = -\cos\theta$ to calculate the following.
   (a) $\cos 150°$ $\cos 150° = \cos(180° - 30°)$ (b) $\cos 120°$ (c) $\cos 135°$

12   Use the relationship $\tan(-\theta) = -\tan\theta$ to calculate each of the following.
   (a) $\tan(-60°)$ (b) $\tan(-30°)$ (c) $\tan(-45°)$

13   Calculate each of the following.
   (a) $\sin(-30°)$ (b) $\cot 150°$ (c) $\sec 120°$ (d) $\tan 210°$
   (e) $\cos 135°$ (f) $\csc 120°$ (g) $\sec 240°$ (h) $\cos(-60°)$
   (i) $\tan 225°$ (j) $\csc 150°$ (k) $\cot 135°$ (l) $\sin(-45°)$

14   Calculate each of the following.
   (a) $\csc\dfrac{3}{4}\pi$ (b) $\cot\dfrac{2}{3}\pi$ (c) $\sin\dfrac{5}{4}\pi$ (d) $\cos\left(-\dfrac{\pi}{4}\right)$
   (e) $\tan\dfrac{4}{3}\pi$ (f) $\sec\left(-\dfrac{\pi}{3}\right)$ (g) $\sec\dfrac{7}{6}\pi$ (h) $\cot\dfrac{3}{4}\pi$

5.3 Proving Trigonometric Relationships

The skills you learn with co-ordinate geometry can be applied to deriving new relationships in trigonometry. For example, to derive a relationship involving $\sin\left(\dfrac{\pi}{2}+\theta\right)$, construct a diagram.

From the diagram

$$\sin\left(\frac{\pi}{2}+\theta\right)=\frac{a}{\sqrt{a^2+b^2}}.$$

But $\cos\theta=\dfrac{a}{\sqrt{a^2+b^2}}.$

Thus $\sin\left(\dfrac{\pi}{2}+\theta\right)=+\cos\theta.$

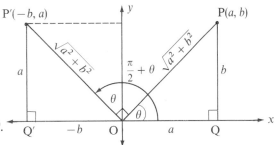

5.3 Exercise

B 1 (a) Use the above diagram to prove that $\sin\left(\dfrac{\pi}{2}+\theta\right)=\cos\theta$.

(b) Use the above diagram to prove that

(i) $\cos\left(\dfrac{\pi}{2}+\theta\right)=-\sin\theta.$ (ii) $\tan\left(\dfrac{\pi}{2}+\theta\right)=-\cot\theta.$

2 Prove.

(a) $\csc\left(\dfrac{\pi}{2}+\theta\right)=\sec\theta$ (b) $\sec\left(\dfrac{\pi}{2}+\theta\right)=-\csc\theta$

3 Prove.

(a) $\sin(270°-\theta)=-\cos\theta$ (b) $\tan(270°-\theta)=\cot\theta$

4 Use the pattern in the previous question.

► Predict a relationship for each of the following.
► Then prove the relationship.

(a) $\csc\left(\dfrac{3}{2}\pi-\theta\right)$ (b) $\sec\left(\dfrac{3}{2}\pi-\theta\right)$ (c) $\cot\left(\dfrac{3}{2}\pi-\theta\right)$

5 Use the pattern in Questions 3 and 4.

► Predict a relationship. ► Then prove the relationship.

(a) $\sin\left(\dfrac{3}{2}\pi+\theta\right)$ (b) $\cos\left(\dfrac{3}{2}\pi+\theta\right)$ (c) $\tan\left(\dfrac{3}{2}\pi+\theta\right)$

(d) $\csc\left(\dfrac{3}{2}\pi+\theta\right)$ (e) $\sec\left(\dfrac{3}{2}\pi+\theta\right)$ (f) $\cot\left(\dfrac{3}{2}\pi+\theta\right)$

5.4 Applying Algebra: Trigonometric Equations

The skills you learn in mathematics are applied over and over. For example, you have earlier used principles of solving equations and applied them to solving problems. Now you will apply these skills to solving trigonometric equations.

Example 1   Solve $\cos \theta = \dfrac{1}{2}, -2\pi \leq \theta \leq 2\pi$.   The equation is solved for this restriction on the domain.

Solution   Use the graph of $y = \cos \theta$.

From the graph,

$$y = \cos \theta = \frac{1}{2}$$

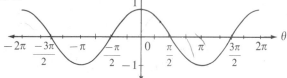

when $\theta = \dfrac{-5\pi}{3}, -\dfrac{\pi}{3}, \dfrac{\pi}{3}$ and $\dfrac{5\pi}{3}$.   Thus the solution is $\left\{ -\dfrac{5\pi}{3}, -\dfrac{\pi}{3}, \dfrac{\pi}{3}, \dfrac{5\pi}{3} \right\}$.

In the example, there was a restriction on the domain, namely $-2\pi \leq \theta \leq 2\pi$. If there is no restriction on θ, you can obtain a general solution of the equation, as follows.

For $\cos \theta = \dfrac{1}{2}, \theta \in R$. Then $\theta = \dfrac{\pi}{3} + 2n\pi, \dfrac{5}{3}\pi + 2n\pi, n \in I$.   This is referred to as the general solution of the above equation.

You need to use your factoring skills to solve the following equation.

Example 2   Solve $\sin^2 \theta - \sin \theta = 0, \quad -\pi \leq \theta \leq \pi$.

Solution $2 \sin^2 \theta - \sin \theta = 0$ Think of $\sin^2 \theta = (\sin \theta)^2$ in order to factor.
$\sin \theta (2 \sin \theta - 1) = 0$

$\sin \theta = 0$ or $2 \sin \theta - 1 = 0$
For $-\pi \leq \theta \leq \pi$, $\sin \theta = \dfrac{1}{2}$
$\theta = -\pi, 0, \pi$.

For $-\pi \leq \theta \leq \pi$,

$\theta = \dfrac{1}{6}\pi, \dfrac{5}{6}\pi$.

You might also need to use trigonometric relationships to solve some equations, as shown in the next example.

Example 3   Solve $1 - 2\cos^2\theta = -\sin\theta, 0 \leq \theta \leq 2\pi$.

Solution

$$1 - 2\cos^2\theta = -\sin\theta$$ — Think: $\cos^2\theta = 1 - \sin^2\theta$
$$1 - 2(1 - \sin^2\theta) = -\sin\theta$$
$$2\sin^2\theta - 1 = -\sin\theta$$
$$2\sin^2\theta + \sin\theta - 1 = 0$$
$$(2\sin\theta - 1)(\sin\theta + 1) = 0$$

$2\sin\theta - 1 = 0$ or $\sin\theta + 1 = 0$

$\sin\theta = \dfrac{1}{2}$ $\sin\theta = -1$

For $0 \leq \theta \leq 2\pi$,   For $0 \leq \theta \leq 2\pi$,

$\theta = \dfrac{\pi}{6}, \dfrac{5}{6}\pi.$ $\theta = \dfrac{3}{2}\pi.$

In the next example, you need to use the properties of $y = \cos x$ to make a decision about your solution.

Example 4   Solve $2\cos^2 x - 7\cos x + 3 = 0$, $0 \leq x \leq 2\pi$.

Solution $2\cos^2 x - 7\cos x + 3 = 0$, $0 \leq x \leq 2\pi$
$$(2\cos x - 1)(\cos x - 3) = 0$$

$2\cos x - 1 = 0$ or $\cos x - 3 = 0$

$\cos x = \dfrac{1}{2}$ $\cos x = 3$ Since the maximum value of $\cos x$ is 1

For $0 \leq x \leq 2\pi$,   there are no values of x so that

$x = \dfrac{\pi}{3}, \dfrac{5\pi}{3}.$ $\cos x = 3.$

5.4   Exercise

A   Throughout, observe the restrictions on the domain of the variable.

1   Use the graph for $0 \leq \theta \leq 2\pi$.
   What is the value
   of θ for each of the
   following?

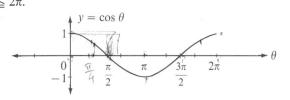

$y = \cos\theta$

(a) $\cos\theta = 1$ (b) $\cos\theta = -1$ (c) $\cos\theta = \dfrac{1}{2}$

2   Use the graph. Solve each of the following for $-2\pi \leq \theta \leq 2\pi$.

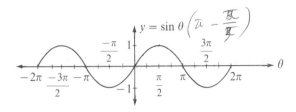

(a) $\sin \theta = 1$ (b) $\sin \theta = -1$ (c) $\sin \theta = -\dfrac{1}{2}$

3   Find the roots of each equation for $0 \leq \theta \leq 2\pi$.
(a) $\sin \theta = 1$ (b) $\sin \theta = -1$ (c) $\csc \theta = 0$

(d) $\sin \theta = \dfrac{1}{2}$ (e) $\csc \theta = \dfrac{1}{2}$ (f) $\sin \theta = \dfrac{1}{\sqrt{2}}$

4   Solve if $-2\pi \leq \theta \leq 2\pi$.
(a) $\cos \theta = \dfrac{-\sqrt{3}}{2}$ (b) $\cos \theta = -1$ (c) $\cos \theta = \dfrac{1}{\sqrt{2}}$

5   Solve each equation for $0° \leq \theta \leq 360°$.
(a) $\sin \theta = \dfrac{1}{2}$ (b) $\cos \theta = -\dfrac{\sqrt{3}}{2}$ (c) $2 \cos \theta = -1$

6   Solve for θ, $0 \leq \theta \leq 2\pi$. Express roots correct to 1 decimal place.
(a) $2 \sin \theta = -1$ (b) $4 \cos \theta = 3$ (c) $\sqrt{2} \cos \theta = 1$

7   (a) Solve $\cos \theta = \dfrac{\sqrt{3}}{2}$ for $0 \leq \theta \leq 2\pi$.

(b) Write the general solution in (a).

B   Where needed, write the solution in the general form.

8   (a) Solve $\sin \theta = -1$, $0 \leq \theta \leq 2\pi$.
(b) Solve $\sin \theta = -1$, $-2\pi \leq \theta \leq 2\pi$.
(c) Why do your answers in (a) and (b) differ?

9   (a) Solve $\tan \theta = 1$, $0 \leq \theta \leq 2\pi$.
(b) Solve $\tan \theta = 1$, $-2\pi \leq \theta \leq 2\pi$.
(c) Why do your answers in (a) and (b) differ?

10 (a) Write $\sin^2 \theta$ in terms of $\cos \theta$.
(b) Use the relationship in (a) to solve $2 \sin^2 \theta - 1 = \cos \theta$, $-\pi \leq \theta \leq \pi$.

11 (a) Write $2 \sin^2 \theta + \sin \theta - 1$ in factored form.

(b) Use the factors in (a) to solve $2 \sin^2 \theta + \sin \theta - 1 = 0$, $-2\pi \leq \theta \leq 2\pi$.

12 Solve for $0 \leq \theta \leq 2\pi$ and $0 \leq A \leq 4\pi$.

(a) $\sin 2\theta = 1$ (b) $\sin 2\theta = -1$ (c) $\sin 2\theta = \dfrac{1}{2}$

(d) $2 \cos 2\theta = 1$ (e) $\cos \dfrac{1}{2} A = 1$ (f) $\sin \dfrac{A}{2} = -1$

13 Solve for $-\pi \leq \theta \leq \pi$.   (a) $\sin^2 \theta = 1$ (b) $\cos^2 \theta = \dfrac{1}{2}$ (c) $\sin^2 \theta = \dfrac{3}{4}$

14 Solve each of the following for $0 \leq \theta \leq 2\pi$.

(a) $\sin \theta \cos \theta = 0$ (b) $\sin \theta (\cos \theta - 1) = 0$

(c) $\sin \theta (\sqrt{3} - 2 \cos \theta) = 0$ (d) $(1 - \sin \theta)(1 - \cos \theta) = 0$

15 (a) Solve $\left(\sin \theta + \dfrac{1}{2} \right) \left(\sin \theta - \dfrac{1}{2} \right) = 0$, $0° \leq \theta \leq 360°$.

(b) Solve $(\cos \theta - 0.5)(\cos \theta + 0.5) = 0$, $-360° \leq \theta \leq 360°$.

(c) Solve $\sin \theta = -\dfrac{1}{2}$, for all θ.

(d) Solve $2 \cos^2 \theta - \cos \theta = 0$ for all θ.

16 Solve. Express your answer to the nearest degree. $0° \leq \theta \leq 360°$.

(a) $\sin \theta = -0.5$ (b) $\sin 2\theta = 0.5$

(c) $2 \sin^2 \theta - \sin \theta - 1 = 0$ (d) $\sin^2 \theta - 0.3 \sin \theta - 0.4 = 0$

17 To solve the following you need to use the different strategies you have practised with algebraic and trigonometric equation solving. Solve for $0° \leq \theta \leq 360°$.

(a) $\cos \theta = \dfrac{\sqrt{3}}{2}$ (b) $\tan 2\theta = -\sqrt{3}$ (c) $2 \sin^2 \theta = 1$ (d) $\sin 2\theta = -\dfrac{1}{2}$

(e) $\cos^2 2\theta = -\cos 2\theta$ (f) $1 - 2 \cos \theta = 0$ (g) $6 \cos^2 \theta + 5 \cos \theta - 4 = 0$

18 Solve for $-\pi \leq \theta \leq \pi$.

(a) $\sin \theta = \dfrac{1}{2}$ (b) $\sin \dfrac{1}{2}\theta = 1$ (c) $\sin^2 \theta = 1$

(d) $\cot^2 \theta = \cot \theta$ (e) $\sin^2 \theta + \sin \theta = 0$ (f) $2 \cos^2 \theta + \cos \theta - 1 = 0$

19 Solve for $-2\pi \leq \theta \leq 2\pi$.

(a) $\csc \theta = -\sqrt{2}$ (b) $\tan^2 \theta = 1$ (c) $\cos^2 \theta = \cos \theta + 2$

(d) $1 - \sin \theta = 2 \sin^2 \theta$ (e) $2 \sin^2 \theta + 5 \sin \theta - 3 = 0$

Applications: Equations of Movement

A ferris wheel is a common part of any fair, exhibition, or midway. The ferris wheel is named after George Ferris. The largest ferris wheel gave 1440 people a ride at the same time!

The ride of a person on a ferris wheel provides an example of periodic phenomena that may be described by trigonometry.

20 A ferris wheel with radius 8 m has a boarding platform, A, 4 m high. A diagram of the wheel is placed on an x-y grid so that the boarding platform is at the origin. The height, d (in metres) of a specific car, C, is given by the equation $d = 8 \sin\left(\dfrac{\pi}{12}t\right)$ where t is expired time in seconds.

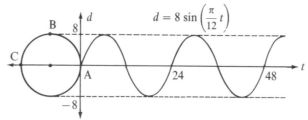

(a) When Shirley boards the ride, she gets on car C. When the ride starts she is directly opposite the boarding platform A. For what values of t is
 (i) Shirley at the boarding platform?
 (ii) Shirley at the top of the ferris wheel?
(b) (i) After how many seconds is Shirley 4 m above *her* starting point for the first turn of the ferris wheel?
 (ii) After how many seconds is Shirley 4 m below *her* starting point on her first turn of the ferris wheel?

21 When Shirley started at C, Betty was at the top of the ride at B. For what values of t is

(a) Betty at the top of the ride?

(b) Betty at the bottom of the ride?

22 The latitude on the earth's surface when the maximum height of a tide is obtained, is given by the roots of the equation, $4\cos^2\theta - 3 = 0$. Find the latitudes.

5.5 Applying Trigonometry: Solving Triangles

The earliest use of trigonometry was to solve problems involving measurement. To calculate the height in the photograph, you need to apply your steps for solving problems, and your skills with trigonometry.

The process of mathematics repeats itself many times.

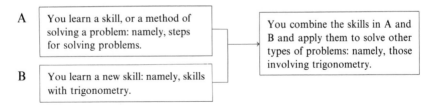

A   You learn a skill, or a method of solving a problem: namely, steps for solving problems.

B   You learn a new skill: namely, skills with trigonometry.

You combine the skills in A and B and apply them to solve other types of problems: namely, those involving trigonometry.

The development of the following skills in trigonometry will enable you to solve practical problems involving measurement.

For acute angles in the first quadrant, convenient words are invented.

Angle in the plane   Angle in a triangle

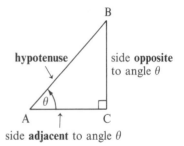

The trigonometric values of angle A in $\triangle ABC$ above are defined as follows.

Primary

$$\sin A = \frac{\text{opposite}}{\text{hypotenuse}}, \qquad \cos A = \frac{\text{adjacent}}{\text{hypotenuse}}, \qquad \tan A = \frac{\text{opposite}}{\text{adjacent}}$$

Reciprocal

$$\csc A = \frac{\text{hypotenuse}}{\text{opposite}}, \qquad \sec A = \frac{\text{hypotenuse}}{\text{adjacent}}, \qquad \cot A = \frac{\text{adjacent}}{\text{opposite}}$$

To **solve** a triangle means to find the measures of the sides and angles whose values are not given. To do these calculations you may need to use the trigonometric values from the tables or from a calculator.

Example (a) Solve for k. Express your answer to 1 decimal place.

(b) Solve for θ. Express your answer to the nearest degree.

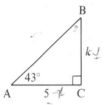

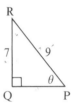

Solution (a) In $\triangle ABC$, $\dfrac{k}{5} = \tan 43°$

$$k = 5 \tan 43°$$
$$k \doteq 5(0.9325)$$
$$k = 4.7 \text{ (to 1 decimal place)}$$

(b) In $\triangle PQR$, $\sin \theta = \dfrac{7}{9}$

$$\theta \doteq 0.7778$$
$$\theta = 51° \text{ (to the nearest degree)}$$

Some problems require that you solve two right triangles before you answer the question. In the diagram, both triangles are solved before w is known.

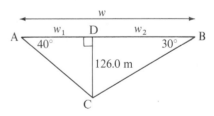

In $\triangle ACD$, $\dfrac{w_1}{126.0} = \cot 40°$. In $\triangle BCD$, $\dfrac{w_2}{126.0} = \cot 30°$.

To find w, use $w_1 + w_2 = w$.

5.5 Exercise

A Express the measure of a side correct to 1 decimal place, and the measure of an angle to the nearest degree.

1 Name the primary or reciprocal trigonometric value of angle P for each of the following.

(a) $\dfrac{25}{7}$ (b) $\dfrac{25}{24}$ (c) $\dfrac{7}{24}$ (d) $\dfrac{7}{25}$ (e) $\dfrac{24}{7}$ (f) $\dfrac{24}{25}$

2 For $\triangle PQR$, in the previous question, what are the trigonometric values for each of the following?
(a) csc Q   (b) cos Q   (c) cot Q

3 Solve for the side indicated in each triangle.

(a)   (b)   (c)   (d)

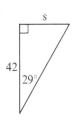
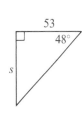

4 Solve for angle θ in each triangle.

(a)   (b)   (c)   (d)

5 For each triangle, find the measure of θ or the side indicated by a variable.

(a)   (b)   (c)   (d)

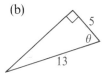

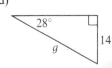

B   The skills you learn with right triangles, will be used to solve problems in the next section. Learn them well.

6   To solve a triangle, it is helpful to record the given information on a sketch of the triangle.

In $\triangle$PQR, find the measure of the missing sides.
(a) $\angle Q = 90°$, $\angle P = 67°$, $q = 17$ (b) $\angle P = 90°$, $\angle Q = 32°$, $q = 3.2$
(c) $\angle R = 90°$, $\angle Q = 58°$, $r = 140$

7   In $\triangle$ABC, find the measure of the missing angles.
(a) $\angle A = 90°$, $a = 17$, $c = 12$ (b) $\angle C = 90°$, $c = 10.2$, $a = 4.7$
(c) $\angle B = 90°$, $c = 72$, $b = 125$

8   Solve each triangle.
(a) $\triangle$PQR: $Q = 90°$, $r = 10$, $p = 7$
(b) $\triangle$ABC: $\angle A = 90°$, $\angle B = 72°$, $b = 22$
(c) $\triangle$ABC: $\angle B = 90°$, $\angle C = 32°$, AC $= 5.2$

9   Find the value of w in each diagram.
(a) 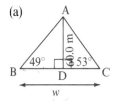   (b) 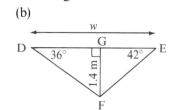   (c)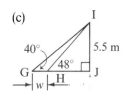

Problem-Solving

You may use trigonometry to develop new formulae for the area of a triangle. Prove that the area, A, of $\triangle$ABC is given by

$A = \dfrac{1}{2} ab \sin C$

$A = \dfrac{1}{2} ac \sin B$

$A = \dfrac{1}{2} bc \sin A$ Use the formula to find the areas of triangles.

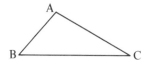

5.6 Solving Problems: Applying Trigonometry

Being able to use mathematics to solve problems is really what studying mathematics is about. To be able to solve a wide variety of problems, it is necessary to have a plan for problem-solving. In other words, you must establish a sequence of steps to organize the solutions of problems. The suggested *Steps for Solving Problems* will help you organize your solution.

Steps for Solving Problems

Step A   Read the problem carefully. Can you answer these two
        questions?
  I   What information are you asked to find? (information
      you don't know)
  II   What information are you given? (information you know)
  Be sure to understand what it is you are to find, *then*
  introduce the variables. Draw a diagram.

Step B   Translate from English to mathematics and write the.
        equations.

Step C   Solve the equations.

Step D   Check the answers in the original problem.

Step E   Write a final statement as the answer to the problem.

You can use your skills in trigonometry and solving problems to calculate distances and angles.

Example   A cable, 200.0 m in length, is secured to the top of an FM transmitting antenna. The cable makes an angle of 37° with the ground. How high is the antenna? (Express your answer to 1 decimal place.)

Solution   Let h, in metres, represent the height of the antenna.

Interpret the given information. Draw a diagram to represent the information accurately.

$$\frac{h}{200.0} = \sin 37°$$

$$h = 200.0 \, (\sin 37°)$$
$$h \doteq 200.0 \, (0.6018)$$
$$h = 120.4 \text{ (to 1 decimal place)}$$

The height of the tower is 120.4 m.

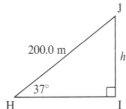

In mathematics, vocabulary is often introduced to provide information more compactly. The following vocabulary occurs often in solving problems about trigonometry.

Angle of Elevation

From a point A an observer can measure an angle of elevation and the distance d, to calculate the height of a hydro tower.

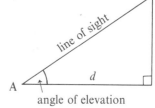

angle of elevation

Angle of Depression

From a tower B, an observer can measure the angle of depression and use the height, d, of the tower to calculate distances.

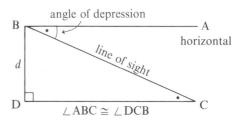

$\angle ABC \cong \angle DCB$

To solve trigonometric problems, you will need to interpret key words or phrases and create a diagram to represent them.

5.6 Exercise

A When necessary, express measures to 1 decimal place, unless otherwise indicated. Express angle measures to the nearest degree.

1. A person is deep-sea fishing. The boat is travelling at a speed such that the fishing line enters the water at a 30° angle. If the rod tip is 2.0 m above the water's surface, how much line must be let out in order that the bait be at a depth of 40.0 m?

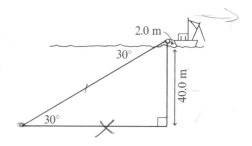

2. At a concert a spotlight is placed at a height of 12.0 m. The spotlight beam shines down at an angle of depression of 35°. How far is the spotlight from the stage?

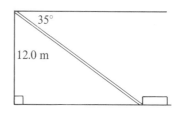

3 To solve problems involving trigonometry, you need to sketch diagrams to record the given information.

(a) Sketch a diagram for the following problem.

A tree casts a shadow of 120.0 m late in the afternoon. If the tree is 25.0 m high, at what angle do the sun's rays strike the ground?

(b) Solve the problem in (a).

B Organize your solutions to the following problems. Refer to the *Steps for Solving Problems*.

4 A television antenna casts a shadow that is 75.0 m long. If the angle of elevation of the sun is 39°, calculate how high above the ground a blackbird perched on the top of the antenna is.

5 From the top of a lighthouse, a hovercraft is sighted at an angle of depression of 47°. If the lighthouse is 42.0 m high, how far is the vessel from the lighthouse?

6 A ladder is in an unsafe position if it makes an angle of less than 15° with a wall. A 10.0 m ladder is placed so that its foot is 3.0 m from the wall. Is the ladder standing safely?

7 The face of a cliff rises vertically to a height of 112.0 m. A sighting is made from a yacht to the top of the cliff. The angle of elevation is read as 14°. How far is the yacht from the base of the cliff? Express your answer to 3 significant digits.

8 An underground parking lot is being constructed 8.00 m below ground level.
(a) If the exit ramp is to rise at a 15° angle, how long will the ramp be?
(b) What horizontal distance is needed for the ramp?

9 A jet is flying at an altitude of 5000 m and a speed of 200 km/h. An observer records the position of the plane when it is immediately overhead and 3 min later. What is the angle of elevation of the plane at the time of the second sighting? Express your answer to the nearest degree.

Calculator Tip

Whenever doing calculations in mathematics you should always think of the following,

▶ Review the procedure you used to do the calculation.
▶ Ask yourself, "Is there a more efficient procedure I could have used?" "Are there any steps that I do not need to include?"

Applications: Famous Heights

To find the heights called for in the following problems you need two measurements: that of an angle and that of a distance from the base of the structure.

10 The tallest totem pole in the world is at Alert Bay, Canada. From a distance 37.0 m from its base, the elevation of the top is 55°. Calculate its height to the nearest tenth of a metre.

11 From a distance 312 m from the base of the Empire State Building, the angle of elevation of the top is 55°. Calculate its height to the nearest metre.

12 The angle of elevation of the top of the Gateway To the West Arch in St. Louis, Missouri is 76°. If the sighting was made 48.0 m from the arch, what is the height of the arch to one decimal place?

13 The Great Pyramid of Cheops is in Giza.
   (a) If the angle of elevation is 23°, measured 348.0 m from its base, calculate the height of the pyramid to one decimal place.
   (b) The actual height of the pyramid is 146.6 m. How much does your answer in (a) differ from the actual height? How would you account for the difference?

14 Two sightseers chose a spot 69 m from the base of the Eiffel Tower. The angle of elevation of the top was 77°.
   (a) Calculate the height of the tower to the nearest metre.
   (b) A television mast was placed on the Eiffel Tower. If the angle of elevation of the top of the mast is about 78°, about how high is the top of the mast?

Problem Solving

Often the information you need to solve a problem is not specifically found in the word problem. Thus, you need to go to another source for more information. Solve the following problem.

The Canadian National (CN) Tower is the world's tallest self-supporting structure. From an island 1.13 km from the base, the angle of elevation is 26°. How much taller is the C.N. Tower than the Empire State Building? Express your answer to the nearest metre.

Where can you get the needed information to solve the problem?

5.7 Problem Solving: Inaccessible Distances

The history of mathematics illustrates how mathematics has been developed in order to solve a particular problem. In the process often new branches of mathematics have emerged. For example, based on the mathematics in the previous section, it would not be easily possible to find the height of the cliff since, to construct a right triangle, you would need to tunnel into the base of the cliff. To find the height of the cliff using more efficient mathematics, you can use two right triangles. The strategy is to choose two points along a line of sight from the base of the cliff.

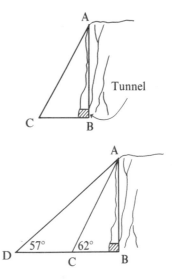

Measurements of the angles at D and C are recorded. If the distance between the points D and C is calculated, the height of the cliff can be determined. Thus, by using the process of mathematics you can solve the problem without tunnelling. A full solution is presented in the example.

Example For the information shown on the diagram, calculate the height of the cliff to 1 decimal place.

Solution Let h, in metres, represent the height of the cliff.
From the diagram

$$\frac{DB}{h} = \cot 57° \qquad \frac{CB}{h} = \cot 62°$$

$$DB = h \cot 57° \qquad CB = h \cot 62°$$

$$DB - CB = h \cot 57° - h \cot 62°$$

$$12.4 = h (\cot 57° - \cot 62°)$$

$$\frac{12.4}{\cot 57° - \cot 62°} = h \qquad h \doteq \frac{12.4}{0.6494 - 0.5317}$$

$$= \frac{12.4}{0.1177}$$

$$= 105.4 \quad \text{to 1 decimal place}$$

Refer to the calculator tip in this section.

The height of the cliff is 105.4 m.

5.7 Exercise

A In Questions 1 to 4 diagrams have been provided. Interpret the information accurately. Express your answers to 1 decimal place when necessary.

1 Use the information in the diagram to calculate the height of the mountain, PS.

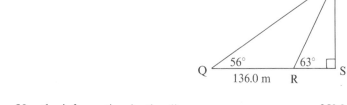

2 Use the information in the diagram to calculate the depth of the valley.

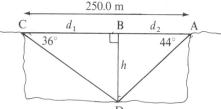

   (a) In the diagram, why is
   $AB + BC = h \cot 44° + h \cot 36°$?

   (b) Use the expression in (a) to calculate the depth, h, in metres.
   $(AB + BC = 250.0 \text{ m.})$

3 An engineer wishes to find the distance across a canyon. She takes a sighting from A and then a sighting from B to a point C on the opposite side of the canyon. The measurements are given on the diagram.

   Find distance d across the canyon.

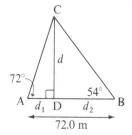

4 The angles of elevation from a point A and a point B to the top of a mountain 780.0 m high are 67° and 54° as shown. Based on the information in the diagram, how long would a tunnel be from A to B?

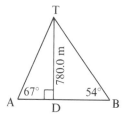

B For the following questions, you need to sketch a diagram and interpret the information accurately. Organize your solution.

5 A forest ranger in a tower 128.0 m high sights two fires in the same line of sight with angles of depression 42° and 61°. How far apart are the fires?

6   From a window 26.0 m above the ground, the angle of elevation of the top of a building is 39°, while the angle of depression to the bottom of the building is 29°. How high is the building?

7   A helicopter, directly above a building, sights a position, A, on the ground at an angle of depression of 38°. The helicopter then rises vertically above the building, a distance of d, in metres, and sights position A, now at an angle of depression of 52°. If point A is 352.0 m from the building, how far has the helicopter risen?

8   The angle of elevation of the top of a building from a point, A, 56.0 m from the building is 58°. A flagpole is on top of the building. The angle of elevation from point A to the top of the flagpole is 62°. What is the length of the flagpole?

9   Two spotlights are placed 10.0 m apart on the same line of sight. The blue spotlight makes an angle of elevation of 45° and hits the bottom of a mirrored ball. The white spotlight makes an angle of elevation of 70° and hits the same area. What is the height of the bottom of the ball?

C 10   For the diagram, prove that $h = \dfrac{d}{\cot \alpha - \cot \theta}$.

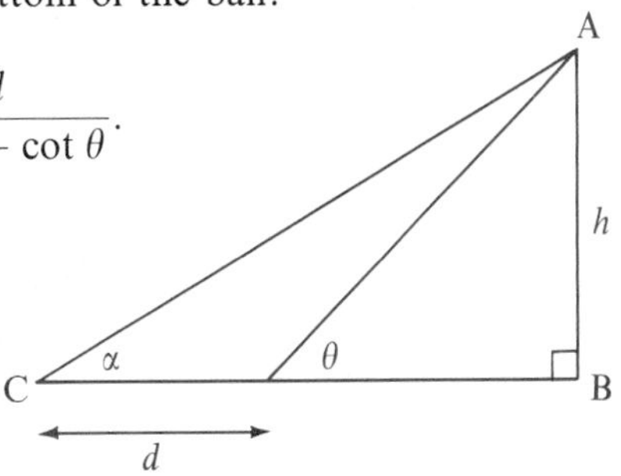

Calculator Tip

The expressions shown below occur doing calculations in trigonometry. As you do your calculations, be sure that the steps are efficient. Can you improve the steps for each of the following? Calculate.

$$A = \frac{12.4}{\cot 57° - \cot 62°}$$

$\boxed{C}$ 62 $\boxed{TAN}$ $\boxed{1/x}$ $\boxed{MS}$ 57 $\boxed{TAN}$ $\boxed{1/x}$ $\boxed{-}$ $\boxed{MR}$ $\boxed{=}$ $\boxed{1/x}$ $\boxed{\times}$ 12.4 $\boxed{=}$ *Output* ?

$$B = \frac{6.5 \sin 20°}{\sin 49°}$$

$\boxed{C}$ 49 $\boxed{SIN}$ $\boxed{MS}$ 20 $\boxed{SIN}$ $\boxed{\times}$ 6.5 $\boxed{\div}$ $\boxed{MR}$ $\boxed{=}$ *Output* ?

5.8 What If . . . ?: The Law of Sines

acute

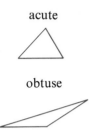

obtuse

In the process of doing mathematics, the question, "What if . . . ?" has often led to the development of new skills and concepts. For example, in the previous sections, right angle triangles were solved. The question, "What if the triangles are not right?" leads to the development of additional powerful skills in doing trigonometry. An oblique triangle does not have a right angle and may either be acute or obtuse.

To solve oblique triangles, the **Law of Sines** for triangles is developed as follows.

To do so, the proof must examine all possibilities, so that two cases are dealt with.

Law of Sines

$$\frac{a}{\sin A} = \frac{b}{\sin B} = \frac{c}{\sin C}$$

Given: $\triangle ABC$ is an oblique triangle.

Required to prove: $\dfrac{a}{\sin A} = \dfrac{b}{\sin B} = \dfrac{c}{\sin C}$.

Proof:

Case 1 $\triangle ABC$ is acute.

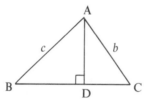

Draw AD ⊥ BC.

In $\triangle ABD$, $\dfrac{AD}{c} = \sin B$

$$AD = c \sin B$$

In $\triangle ADC$, $\dfrac{AD}{d} = \sin C$

$$AD = b \sin C$$

Thus $c \sin B = b \sin C$

or $\dfrac{c}{\sin C} = \dfrac{b}{\sin B}$

Case 2 $\triangle ABC$ is obtuse.

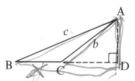

Draw AD ⊥ BC produced.

In $\triangle ABD$, $\dfrac{AD}{c} = \sin B$

$$AD = c \sin B$$

In $\triangle ACD$, $\dfrac{AD}{b} = \sin (180° - C)$

$$= \sin C$$

$$AD = b \sin C$$

But
$\sin (180° - C)$
$= \sin C$

Thus $c \sin B = b \sin C$

or $\dfrac{c}{\sin C} = \dfrac{b}{\sin B}$.

Thus for any $\triangle ABC$, $\dfrac{c}{\sin C} = \dfrac{b}{\sin B}$.

Similarly, by drawing perpendiculars to the other sides, it may be proved that $\dfrac{a}{\sin A} = \dfrac{b}{\sin B} = \dfrac{c}{\sin C}$

The Law of Sines for triangles provides new powerful skills to solve additional problems involving measurements.

You can use the Law of Sines to find the measures of the other parts of the triangle if you are given the following information.

I   two angles and a side opposite one of the given angles

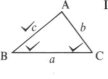

$$\frac{b}{\sin B} = \frac{c}{\sin C}$$

$$\text{or } b = \frac{c \sin B}{\sin C}$$

II   two angles and the contained side

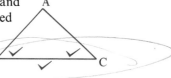

Since $\angle B$ and $\angle C$ are known, then

$$\angle A = 180° - (\angle B + \angle C)$$

$$\frac{b}{\sin B} = \frac{a}{\sin A} \quad \text{or } b = \frac{a \sin B}{\sin A}$$

To *solve an oblique triangle* means to use the given measures of sides and angles to determine the measures of the remaining sides and angles. In the next section, these skills will be applied to the solution of word problems in trigonometry.

Example   In $\triangle ABC$, $\angle A = 105°$, $b = 15.1$, $a = 20.3$.

(a) Find $\angle B$ to the nearest degree.

(b) Find AB to 1 decimal place.

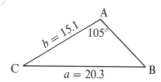

Solution   (a) Use the Law of Sines.

$$\frac{b}{\sin B} = \frac{a}{\sin A}$$

$$\sin B = \frac{b \sin A}{a}$$

$$= \frac{15.1 \sin 105°}{20.3}$$

$$\doteq \frac{15.1(0.9659)}{20.3}$$

$$\doteq 0.7185$$

From the tables or a calculator

$$\angle B \doteq 46° \quad \text{(to the nearest degree)}$$

$\sin 105°$
$= \sin (180° - 75°)$
$= \sin 75°$
$= 0.9659$
↑
from the tables or your calculator

(b) Use the Law of Sines.

$$\frac{c}{\sin C} = \frac{a}{\sin A}$$

$$\text{Thus } c = \frac{a \sin C}{\sin A}$$

$$\angle C = 180° - (\angle A + \angle B)$$
$$= 180° - (105° + 46°)$$
$$= 180° - 151°$$
$$= 29°$$

$$\text{Thus } c = \frac{20.3 \sin 29°}{\sin 105°}$$

$$\doteq \frac{20.3(0.4848)}{(0.9659)}$$

$$= 10.2 \text{ to 1 decimal place.}$$

Thus AB = 10.2 to 1 decimal place.

5.8 Exercise

A Express the measures of distances correct to 1 decimal place and the measures of angles to the nearest degree.

1 For each triangle, find the measure of the side indicated.

(a)

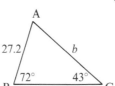

(b)

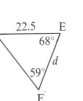

(c)

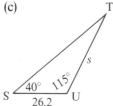

2 For each triangle, find the measure of angle θ.

(a)

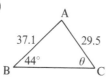

(b)

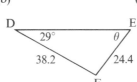

(c)

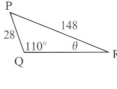

3 For each triangle, find the measure of the parts indicated.

(a)   (b)   (c)   (d)

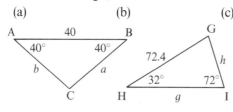

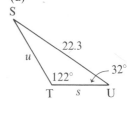

4 Solve △SUN.

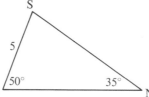

5 For the given triangle, determine the measures of the remaining sides and angles.

B To solve the following triangles, sketch a diagram and record the given information accurately. Read carefully and precisely!

6 In $\triangle$PQR, find the value of q if $\angle$R = 73°, $\angle$Q = 32°, and r = 23.

7 In $\triangle$ABC, a = 22, $\angle$B = 53°, and $\angle$C = 43°. Find b, c, and $\angle$A.

8 In $\triangle$ABC, $\dfrac{a}{b} = \dfrac{1}{1.2}$ and $\angle$A = 40°. Find $\angle$B.

9 The ratio of sides p to q in $\triangle$PQR, is $3:2\sqrt{2}$. If $\angle$Q = 38°, find $\angle$P.

10 Solve each triangle.
(a) $\triangle$MNP: $\angle$M = 107°, m = 22.1, $\angle$N = 17°
(b) $\triangle$ABC: $\angle$A = 32°, $\angle$C = 81°, a = 24.1.
(c) $\triangle$ABC: $\angle$A = 40°, $\angle$C = 15°, a = 35.2.
(d) $\triangle$DEF: $\angle$E = 70°, $\angle$F = 52°, e = 12.7.

11 (a) If $\angle$A = 45° and $\angle$B = 30°, find a value for $\dfrac{a}{b}$ in $\triangle$ABC. Express your answer in radical form.
(b) In $\triangle$PQR, $\angle$Q = 60°, and $\angle$R = 45°. Find a radical expression for the value of $\dfrac{q}{r}$.

12 In $\triangle$PQR, PQ = 45, $\angle$P = 70°, $\angle$R = 49°. Find the length of the altitude from R to PQ.

C 13 Use the Law of Sines to prove that if the measures of two angles of a triangle are equal, then the lengths of the sides opposite these angles are equal.

Computer Tip

To solve an equation such as $x - 2 \sin x - 1 = 0$, you can do these steps.

Step 1 Sketch the graph to find the value of the root x_1 approximately. A $< x_1 <$ B

```
10 PRINT "FIND THE ROOTS"
20 FOR X = A TO B STEP .1
30 LET Y = X − 2∗ SIN X − 1
40 PRINT X, Y
50 NEXT X
60 END
```

Step 2 Use the computer program. Choose the values of A and B from the sketch of the graph.

Step 3 Adjust the step in line 20 to obtain the accuracy you want.
20 For X = C TO D STEP .01

5.9 Solving Problems: Law of Sines

Once you have learned some mathematics and know the particular vocabulary, you can apply the skills and processes to solve problems. When solving a problem, you must clearly understand I and II, as shown.

To solve a problem involving the Law of Sines, the problem must be interpreted correctly. You must understand two important questions.

I What information am I asked to find?
II What information am I given?

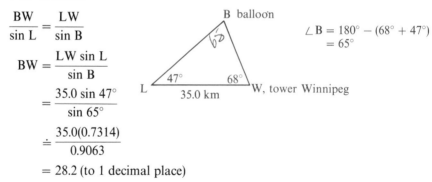

| I What you want to know. |

↕

To use the information given (II) and find what you are asked (I) often requires the use of different mathematical skills and strategies, some of which you already know. The more strategies and skills you know or remember, the better you will be able to solve problems.

↕

| II What you are given. |

Example From an observation tower near Lake Winnipeg, the angle of elevation of a weather balloon is 68°. In the same plane, 35.0 km away, the balloon is sighted from another location with an angle of elevation of 47°. Calculate the distance from the weather balloon to the observation tower to the nearest tenth of a kilometre.

Solution To calculate BW, use the Law of Sines.

$$\frac{BW}{\sin L} = \frac{LW}{\sin B}$$

$$BW = \frac{LW \sin L}{\sin B}$$

$$= \frac{35.0 \sin 47°}{\sin 65°}$$

$$\doteq \frac{35.0(0.7314)}{0.9063}$$

$$= 28.2 \text{ (to 1 decimal place)}$$

Draw a diagram to record the information and help plan the solution of the problem.

$\angle B = 180° - (68° + 47°)$
$= 65°$

Thus, the distance from the weather balloon to the observation tower is 28.2 km.

When solving problems that involve calculations, plan your computations so that you use your calculator efficiently. For example, to find the value of BW in Example 1, use these steps. (Refer to the manual for your calculator.)

Display

$\boxed{C}$ 35 $\boxed{\times}$ 47° $\boxed{SIN}$ $\boxed{=}$ $\boxed{\div}$ 65° $\boxed{SIN}$ $\boxed{=}$ 28.243583

Be sure your calculator is in the degree mode.

The same type of diagram is used in all problems involving the Law of Sines. For example, the diagram used below may represent various situations. In this situation, A and B are lighthouses. The problem is to find the distance from a ship, C, to the shore.

Problem How far is the ship from the coast, AB?

To find CD, you use one more step in the solution of the above problem.

Diagram

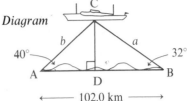

Steps of the Solution **Step 1** To find b, use

$$\frac{b}{\sin B} = \frac{c}{\sin C}.$$

Step 2 Use the value of b, to find CD.

$$\frac{CD}{b} = \sin A$$

5.9 Exercise

A In the following exercise, calculate all angles to the nearest degree and all distances to 1 decimal place.

1 The largest iceberg on record was an Antarctic iceberg with a surface area of over 31 000 km². The approximate shape of the iceberg is shown. Calculate the width of the southern shore.

2 A house under construction is 10.0 m wide. The two sides of the roof are to meet at an 80° angle. To what length should the roof rafters, d, be cut, allowing for 0.5 m overhang?

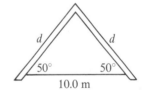

3 A rack for drying fish nets has 5 posts, each 1.7 m long. How many metres of the net can be exposed in one wrap around the rack?

B Translate each of the problems into an appropriate form of the Law of Sines. Organize your solution. Refer to the *Steps For Solving Problems*.

4 An engineer wants to build a bridge over a river from point B to point A. The distance from B to a point C is 520.0 m. A transit is used and ∠B is found to be 78°. ∠C is found to be 54°. How long will the bridge be?

5   A smokestack, AB, is 205.0 m high. From two points, P and Q, on the same side of its base B, the angles of elevation of A are 40° and 36° respectively. Find the distance between P and Q.

6   The largest known crater is in northern Arizona. It is approximately 1500.0 m wide. From one side of the crater a weather plane is spotted with an angle of elevation of 45°. From the opposite side of the crater, the angle of elevation to the plane is 64°. Calculate the distances from the plane to each side of the crater.

7   A 20.0 m tall tree makes an angle of 12° with the vertical. To prevent the tree from falling over any more a 35.0 m rope is attached to the top of the tree and is pegged into the ground some distance away. Find the angle the rope makes with the ground.

8   Three towns (A, B, and C) are located so that B is 25.0 km from A and C is 34.0 km from A. If ∠ABC is 110°, calculate the distance from B to C.

9   The triple-tiered Pont du Gard near Nimes, France is the greatest Roman aqueduct. A balloon flying above the aqueduct measures the angles of depression to each end of the aqueduct to be 54° and 71°. The closest end of the aqueduct is 270.0 m from the balloon. Calculate the length of the aqueduct.

10 The tallest sign on record is the 63.0 m tall sign of the Stardust Hotel, Las Vegas, Nevada. It is supported on one side by two guy wires, one to its top and one to its midsection. Both wires are fastened to the ground at the same point and form an angle of 13° between them. If the angle of depression of the longer wire is 53°, find the length of the shorter wire.

Applications: Astronomy

Trigonometry is often used to solve problems in astronomy. For example, using advanced mathematics, scientists were able to determine in 1910 that Halley's Comet came very close to the earth, so close that the Earth passed through its tail. In fact, the comet came as close as 22.5×10^6 km from the earth. If this seems a long distance, consider the fact that the sun is about 150×10^6 km away from the earth. In astronomy both these distances are very small.

11 The planets Earth and Mercury form a triangle with the sun as shown.

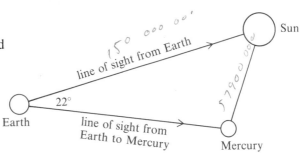

The distance from Mercury to the Sun is 57 900 000 km, while the distance from the Earth to the Sun is 150 000 000 km. The angle between the lines of sight from Earth to the Sun and from Earth to Mercury is 22°. Calculate the distance between Earth and Mercury for this sighting, to the nearest kilometre.

12 The *aphelion* is the maximum distance a planet achieves from the sun. When Mars is at its aphelion distance, a space probe transmits the following information.

▶ The space probe is 203×10^6 km from the sun.
▶ The line of sight from the probe to the sun and the line of sight from the probe to Mars meet at 95°.
▶ The line of sight from Mars to the sun and the line of sight from Mars to the probe meet at 54°.

Calculate the aphelion distance of Mars.

13 From an observation point on the earth, the distance to the moon is approximately 239 000 km, as shown in the diagram.

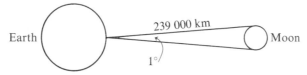

A rough calculation is made and the angle subtended on earth is about 1°. From this data, calculate the approximate diameter of the moon to the nearest 100 km.

5.10 What If . . . ?: The Law of Cosines

What if the given information needed to solve a triangle involves only the sides? In order to use the Law of Sines you need to be given an angle. To overcome this difficulty, additional mathematics is derived so that you have the necessary skills to solve *any* triangle. For example, in the previous section, you used the Law of Sines to solve a triangle when given
- two angles and the contained side (ASA).
- two angles and a side (AAS).

To solve a triangle when given
- two sides and the contained angle (SAS) or
- three sides (SSS)

you can derive another property of triangles known as the Law of Cosines.

Law of Cosines

For any $\triangle ABC$,

$a^2 = b^2 + c^2 - 2bc \cos A$
$b^2 = c^2 + a^2 - 2ac \cos B$
$c^2 = a^2 + b^2 - 2ab \cos C$

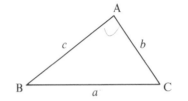

To derive the Law of Cosines, you need to express one side in terms of the other sides and the contained angle.

Case 1:

Thus in $\triangle ABC$,
if both $\angle B$ and $\angle C$ are acute,
construct $AP \perp BC$.

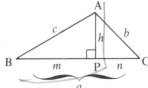

In $\triangle APB$, $c^2 = h^2 + m^2$
$\qquad = h^2 + (a - n)^2 \longleftarrow$ (Since $m = a - n$)
$\qquad = h^2 + a^2 - 2an + n^2$ In $\triangle APC$,
$\qquad = \underbrace{h^2 + n^2} + a^2 - 2an$ $\dfrac{n}{b} = \cos C$
$\qquad$ In $\triangle APC, b^2 = h^2 + n^2$ $n = b \cos C$

Thus, $\qquad c^2 = b^2 + a^2 - 2ab \cos C$.

In a similar manner, you can prove
$$a^2 = b^2 + c^2 - 2bc \cos A$$
$$b^2 = a^2 + c^2 - 2ac \cos B$$

Case 2:

If ∠C is obtuse, draw AP ⊥ BC produced as shown. Develop the proof using the fact that $\cos \angle ACP = \cos (180° - C) = -\cos C$.

In △APB, $c^2 = h^2 + (a + n)^2$
$= h^2 + a^2 + 2an + n^2$
$= \underbrace{h^2 + n^2} + a^2 + 2an \longleftarrow$ In △APC

In △APC, $b^2 = h^2 + n^2$

$\dfrac{n}{b} = \cos (180° - C)$

Thus $c^2 = b^2 + a^2 - 2ab \cos C.$ $n = -b \cos C$

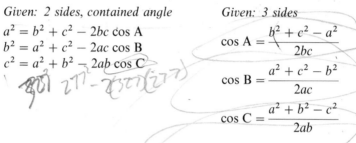

The Law of Cosines is given in various convenient forms, depending on what information is given.

Given: 2 sides, contained angle

$a^2 = b^2 + c^2 - 2bc \cos A$
$b^2 = a^2 + c^2 - 2ac \cos B$
$c^2 = a^2 + b^2 - 2ab \cos C$

Given: 3 sides

$\cos A = \dfrac{b^2 + c^2 - a^2}{2bc}$

$\cos B = \dfrac{a^2 + c^2 - b^2}{2ac}$

$\cos C = \dfrac{a^2 + b^2 - c^2}{2ab}$

To use the Law of Cosines, you need to decide which of the 6 different forms you need to use.

Example 1   Find ∠B if $a = 12$, $b = 18$, and $c = 15$ in △ABC.

Solution

$\cos B = \dfrac{a^2 + c^2 - b^2}{2ac}$

Based on the given information, use the Law of Cosines.

$= \dfrac{12^2 + 15^2 - 17^2}{2(12)(15)}$

$= \dfrac{144 + 225 - 289}{360}$

$= \dfrac{80}{360}$

Draw a diagram to record the given information

$= 0.2222$ (to 4 decimal places)

Thus, ∠B = 77° (to the nearest degree)

from the tables or your calculator

To solve a triangle, you may need to use the Law of Cosines and the Law of Sines as shown in the next example.

Example 2   Solve $\triangle PQR$ if $\angle Q = 74°$, $p = 5.9$ and $r = 3.8$.

Solution   *Step 1:* Use the Law of Cosines.

$q^2 = p^2 + r^2 - 2pr \cos Q$ Record the given information on the diagram.

$q^2 = (5.9)^2 + (3.8)^2 - 2(5.9)(3.8) \cos 74°$

$\doteq 34.81 + 14.44 - 44.84(0.2756)$

$\doteq 34.81 + 14.44 - 12.36$

$= 36.89$

$q \doteq 6.07$

$ = 6.1$ (to 1 decimal place)

 └──── Use this value in your calculations.

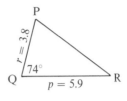

Step 2: Use the Law of Sines.

$$\frac{\sin P}{p} = \frac{\sin Q}{q}$$

$$\sin P = \frac{p \sin Q}{q}$$

$$= \frac{5.9 \sin 74°}{6.1}$$

$$\doteq \frac{5.9(0.9613)}{6.1}$$

$$\doteq 0.9298$$

To do the calculations on a calculator efficiently, use these steps.

 *Display*

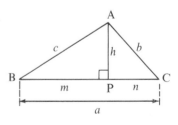

Thus, $\angle P = 68°$ (to the nearest degree), $\angle R = 180° - (68° + 74°) = 38°$.

Thus $p = 5.9$ $q = 6.1$ $r = 3.8$

 $\angle P = 68°$ $\angle Q = 74°$ $\angle R = 38°$

5.10 Exercise

A   Express your answers to 1 decimal place, and angles to the nearest degree.

1   Prove the Law of Cosines in the form given.

$$b^2 = a^2 + c^2 - 2ac \cos B$$

2   In $\triangle PQR$, $\angle R$ is obtuse. Prove that $r^2 = p^2 + q^2 - 2pq \cos R$.

3   (a) Sketch any triangle of your own.
     (b) List the different forms of the Law of Cosines that apply to your triangle.
     (c) List the different forms of the Law of Sines that apply to your triangle.

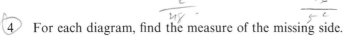

4 For each diagram, find the measure of the missing side.

(a)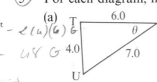

(b) P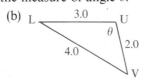

(c)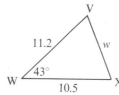

5 For each diagram, find the measure of angle θ.

(a)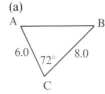

(b)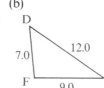

(c)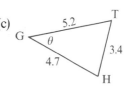

6 Solve each triangle.

(a)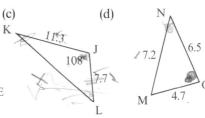

(b)

(c)

(d)

B 7 (a) In $\triangle$PQR, $q = 15.0$, $r = 12.0$, and $\cos P = \dfrac{1}{8}$. Find p.

(b) In $\triangle$ABC, $a = 17.1$, $b = 15.3$, and $c = 8.2$. Find $\cos A$.

8 Solve each triangle.
(a) $\triangle$ABC: $\angle A = 62°$, $b = 62.0$ m, $c = 73.0$ m
(b) $\triangle$DEF: $d = 81.0$ m, $e = 93.0$ m, $f = 102.0$ m
(c) $\triangle$XYZ: $x = 123.0$ m, $y = 108.0$ m, $z = 92.0$ m
(d) $\triangle$PQR: $\angle P = 112°$, $q = 82.0$ m, $r = 63.0$ m

9 In $\triangle$DEF, the sides are given as 6.0, 12.0, and 15.0. Find the measure of the smallest angle for the triangle.

10 Find the perimeter of $\triangle$PQR if $q = 10.1$, $r = 16.3$, and $\angle P = 60°$.

11 A parallelogram has sides 12.0 cm and 18.0 cm and a contained angle of 78°. Find the shorter diagonal.

12 In a triangle, two sides measure 43.0 cm and 28.0 cm and the contained angle is 112°. Find the measure of the smallest angle.

C 13 For the diagram, prove that

$$c^2 = a^2 + b^2 - 2ab \cos C.$$

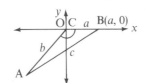

5.11 Solving Problems: Law of Cosines

When you solve word problems, similar principles and processes occur again and again. For example, to solve any problem, you must interpret accurately the problem and translate the word problem into a mathematical model.

In particular, to solve problems in trigonometry, first, you must draw a diagram that shows accurately

I   the information you are asked to find.
II   the information you are given.

Example   An architect is designing a solar heated house which will be 10.0 m wide. The south side of the roof, containing the solar collectors, will rise for 8.0 m at an angle of elevation of 60°. At what angle of elevation will the north side of the roof rise? Express your answer to the nearest degree.

Solution

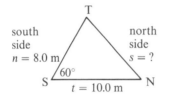

Think: To find the measure of $\angle N$, set up the Law of Sines.

$$\frac{s}{\sin 60°} = \frac{8.0}{\sin N} = \frac{10.0}{\sin T}$$

$s^2 = t^2 + n^2 - 2tn \cos S$
$s^2 = (10.0)^2 + (8.0)^2 - 2(10.0)(8.0)\cos 60°$
    $= 100.0 + 64.0 - (160.0)(0.5)$
    $= 164.0 - 80.0$
    $= 84.0$
$s = 9.17$ (to 2 decimal places)

You do not have enough information to solve for $\angle N$. Use the Law of Cosines first to solve for s.

Use 2 decimal places in your calculations until you complete the solution.

Use the Law of Sines to calculate $\angle N$.

$$\frac{9.17}{\sin 60°} = \frac{8.0}{\sin N}$$

$$\sin N = \frac{8.0 \sin 60°}{9.17}$$

$$\sin N \doteq \frac{8.0(0.8660)}{9.17}$$

Remember: Use your calculator efficiently to do calculations.

$\sin N = 0.7555$ (to 4 decimal places)

Find the measure of $\angle N$ to the nearest degree.

$$\angle N = 49°$$

Thus, the angle of elevation of the north side is 49°.

5.11 Exercise

A Express distances to 1 decimal place, and angles to the nearest degree, when necessary.

1 The posts of a hockey goal are 2.0 m apart. A player attempts to score by shooting the puck along the ice from a point 6.5 m from one post and 8.0 m from the other. Within what angle, θ, must the shot be made?

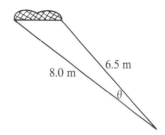

2 The diagram shows information about Lake Volta. Find the length AB across the lake. Express your answer to the nearest metre.

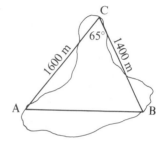

3 Find the size of each angle of the isosceles triangle ABC.

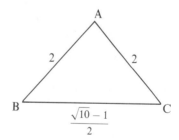

4 A campsite is shown in the diagram at the end of a point in Lake Louise. From the campsite to A is 1.7 km and to B is 1.5 km. If the angle between the sightings from the campsite is 37°, find the distance between A and B.

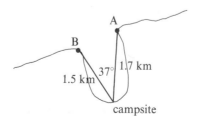

B   Organize your solution. Refer to the *Steps for Solving Problems.*

5   Terry is building an A-frame ski chalet near Blue Mountain. The length of each of two rafters is 9.0 m. If the angle at the apex of the frame is to be 46°, calculate the proposed width of the chalet at the base.

6   The longest car ferry is the 173.0 m Norland operating on the North Sea. What angle would her length subtend when viewed from a point at sea which is 300.0 m from her bow and 220.0 m from her stern?

7   A football player is attempting a field goal. The angle formed by the player's position on the field and a line of sight to each upright is 33°. If the distances to the uprights are 7.5 m and 10.0 m, calculate the width of the uprights.

8   A squash player hits the ball 2.3 m to the side wall. The ball rebounds at an angle of 100° and travels 3.1 m to the front wall. How far is the ball from the player when it hits the front wall? Assume the player does not move after her shot.

9   The largest tanker ever built is the Seawise Giant from Japan. A person on a high crane near one end of the tanker sights a point at the middle. The angle of depression to this point is 23°. If the person is 82.0 m from this point, and the length of the tanker is 765.0 m, find how far the person is from the opposite end of the tanker.

Math Tip

Remember: to solve a word problem requiring an equation
• use a variable to represent what you are to find.
• use a chart, table, etc., to help you organize the given information.
• write an equation.
• check your answer in the *original* problem.
• write a *final* statement to answer what you were asked to find.

Problem Solving

Very often, the information given in a problem will suggest a clue as to how to solve the problem. In the following problem, the expression suggests a particular skill that you have used in this section. Use the clue to solve the problem.

If a, b, and c represent the sides of $\triangle ABC$ and $\dfrac{a^2 + b^2 - c^2}{2ab}$ has a value of zero, prove that the triangle is right-angled.

Applications: Dog Legs, Tees, and Fairways

What do dog legs, tees, and fairways have in common?

Golf, which has been played for about five hundred years, is one of the oldest modern sports. You will learn some of the terms used by the approximately 15 000 000 people who play golf regularly.

On golf courses the term *hole* refers not only to the hole on the green but also to the entire arrangement of tee, fairway and green, with attendant hazards. In the problems below, you will read about various kinds of holes designed to challenge golfers.

10 Various terms are associated with golf. The golf hole shown suggests a "dog leg", and thus the term *dog leg* is used to describe certain parts of a golf course. In the diagram, A is referred to as the dog leg.

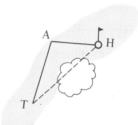

If $\angle T = 32°$, TH = 450.0 m, and AT = 330.0 m, find the length of AH.

11 A golf hole has a dog leg as shown in the diagram. What is the angle at the dog leg ($\angle TAH$), if the distance from the tee, T, to the hole, H, is 362.0 m?

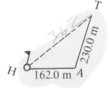

12 A golf hole has a dog leg as shown in the diagram. What is the direct line distance from the tee to the hole?

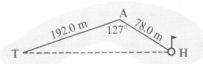

13 A golfer hits a tee shot from T toward a golf hole, H. The shot is sliced 12° (to the right), as shown in the diagram. If the golfer walks 200.0 m to the ball, B, and estimates that the ball is still 200.0 m from the hole, how far is the tee from the hole?

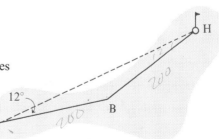

5.12 Solving Triangles: The Ambiguous Case

For each of the following triangles, the given information determines a unique triangle.

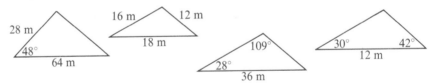

However, if two sides and a non-contained angle are given, then the given information may give no triangle, 1 triangle, or 2 triangles, as summarized in the following.

Given: △ABC, side a, side b, and ∠A. Consider the possibilities.

Case 1: $a > b$

For this case CB > CA. Only one triangle can be constructed.

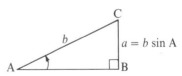

Case 2: $a < b$

(i) If $a < b \sin A$, then no triangle can be drawn.
$$CD = b \sin A$$

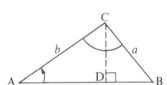

(ii) If $a = b \sin A$, then there is one triangle and one solution.

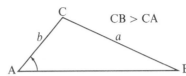

(iii) If $a > b \sin A$, then two triangles occur.

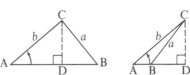

The **ambiguous case** is so named because more than one triangle can be drawn. A diagram is used to summarize the information in a compact way.

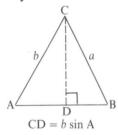

$CD = b \sin A$

The length of $CD = b \sin A$ helps you to decide whether the given information provides the ambiguous case.

Case 1: $a > b$
   1 solution

Case 2: $a < b$
  (i) $a < b \sin A$, no solution
  (ii) $a = b \sin A$, 1 solution
  (iii) $a > b \sin A$, 2 solutions, ambiguous case

To determine whether more than one solution may occur in solving triangles, it is an important step to interpret the given information correctly as shown in the following example.

Example △ABC is given with $\angle A = 41°$, $a = 23$, $b = 28$. Solve △ABC. Express angles to the nearest degree and lengths to 1 decimal place.

Solution

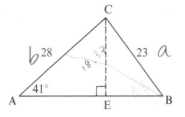

Check Since $a < b$ and the angle is not a contained angle, you need to check for the ambiguous case.
$$CE = b \sin A$$
$$= 28 \sin 41°$$
$$\doteq 28(0.6561)$$
$$= 18.37 \quad \text{to 2 decimal places}$$
Since $a > b \sin A$, then there are two solutions for △ABC.

Now solve the triangle.

$$\frac{28b}{\sin B} = \frac{23a}{\sin 41°A}$$

$$\sin B = \frac{28 \sin 41°}{23}$$

Remember: do the computations efficiently.

$$\doteq \frac{28(0.6561)}{23}$$

$$\sin B = 0.7987 \quad \text{to 4 decimal places}$$

$$\angle B = 53° \quad \text{or} \quad \angle B = 180° - 53°$$
$$\angle B = 127°$$

Draw a diagram for each case.

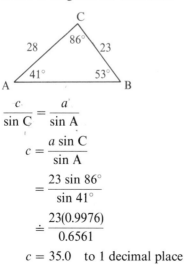

$$\frac{c}{\sin C} = \frac{a}{\sin A}$$

$$c = \frac{a \sin C}{\sin A}$$

$$= \frac{23 \sin 86°}{\sin 41°}$$

$$\doteq \frac{23(0.9976)}{0.6561}$$

$$c = 35.0 \quad \text{to 1 decimal place}$$

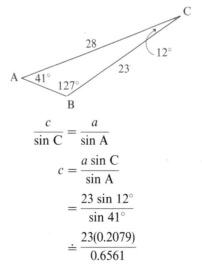

$$\frac{c}{\sin C} = \frac{a}{\sin A}$$

$$c = \frac{a \sin C}{\sin A}$$

$$= \frac{23 \sin 12°}{\sin 41°}$$

$$\doteq \frac{23(0.2079)}{0.6561}$$

$$c = 7.3 \quad \text{to 1 decimal place}$$

5.12 Exercise

A Express the measures of sides correct to one decimal place and angles to the nearest degree.

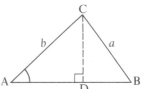

1 For what conditions does △ABC have
(a) no solution?
(b) one solution?
(c) two solutions?

2 For △ABC, $a = 1.2$, $b = 2.7$, and $\angle A = 32°$.
(a) Calculate $b \sin A$.   (b) How many solutions will occur for △ABC?

3 For △ABC, $a = 2.4$, $b = 4.8$, and $\angle A = 30°$.
(a) Calculate $b \sin A$.   (b) How many solutions will occur for △ABC?

4 For △ABC, $a = 6.1$, $b = 8.1$, $\angle A = 32°$.
(a) Calculate $b \sin A$.   (b) How many solutions will occur for △ABC?

5 Each triangle has the given information marked. Determine which triangles have no solution, which have one solution and which have two solutions.

(a)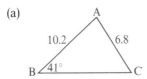

(b)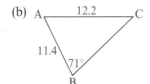

(c)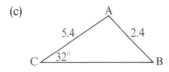

(d)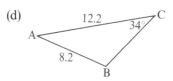

B Be sure to check whether the given information yields the ambiguous case.

6 For each triangle, the data are given. Determine the number of solutions: no solution, one solution, or two solutions. Then solve the triangle.
(a) △ABC: $\angle A = 71°$, $a = 12.2$, $b = 11.4$
(b) △ABC: $\angle A = 55°$, $a = 7.1$, $b = 9.6$
(c) △ABC: $\angle A = 44°$, $a = 9.3$, $b = 12.3$
(d) △DEF: $\angle D = 42°$, $d = 8.5$, $f = 7.3$
(e) △DEF: $\angle E = 38°$, $d = 16.6$, $e = 13.4$
(f) △DEF: $\angle D = 47°$, $d = 8.1$, $f = 12.2$

7   A race enthusiast plans a car racing track as shown.

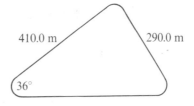

(a) What is the perimeter (length) of the track?

(b) Why is there a problem in constructing this track?

8   A tower, standing on top of a hill that is inclined at an angle of 18°, casts a shadow 45.0 m long down the hill. The angle of elevation of the sun is 47°. Find the height of the tower.

9   Two ships, S and T, 120.0 km apart, pick up a downed spacecraft's homing signal. Ship T estimates that the spacecraft is 70.0 km away and that the angle between the line from T to S and the line from S to the spacecraft is 28°. Find the distance from S to the spacecraft.

10 Two forest fire stations, P and Q are 20.0 km apart. A ranger at station Q sees a fire 15.0 km away. If the angle between the line PQ and the line from P to the fire is 21°, find how far station P is from the fire.

11 Hole 18 of a certain golf course has a tee, T, a hole, H, and a dog-leg at A. If TA is 112.0 m, AH is 78.0 m and the angle at T is 22°, what is the angle at A?

12 A Marathon swimmer starts at Island A, swims 9.2 km to Island B and then 8.6 km to Island C. If $\angle$ BAC = 52°, how far does she have to swim back to Island A?

C 13 (a) Determine the range of values side a can assume so that $\triangle$ABC has two solutions if $\angle$ A = 40° and b = 50.0.

(b) Determine the range of values that side a can assume for $\triangle$ABC to have no solution if $\angle$ A = 56° and b = 125.7.

(c) $\triangle$ABC has exactly one solution. If $\angle$ A = 57° and b = 73.7, what are the values of side a for which this is possible?

Problem Solving

Numbers have intrigued people and in developing strategies to solve problems, often much mathematics has subsequently been created. The number 153 has a strange property. $153 = 1^3 + 5^3 + 3^3$ What other numbers have the same property? To solve the problem you could use trial and error. One of the answers is 407; but you could spend a lifetime checking numbers. Write a computer program in BASIC to check all the numbers to 1000.

5.13 Problems in Three Dimensions

Some problems in trigonometry involve more than one triangle. Thus, it becomes crucial, when solving the problem, that an accurate sketch be used to display the given information. The principles you learned earlier are used again to solve problems: the Law of Sines, the Law of Cosines, and skills with right triangles.

Example
To calculate the height of an inaccessible cliff, data are collected and recorded on the diagram. Calculate the height of the cliff correct to one decimal place.

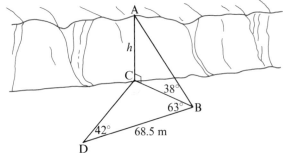

Solution
Let, h, in metres, represent the height of the cliff.

In $\triangle DCB$, $DB = 68.5$ m, $\angle D = 42°$ and $\angle DBC = 63°$

$$\angle DCB = 180° - (42° + 63°)$$
$$\angle DCB = 75°.$$

Use the Law of Sines to determine CB.

$$\frac{CB}{\sin 42°} = \frac{DB}{\sin 75°}$$

$$CB = \frac{DB \sin 42°}{\sin 75°}$$

$$= \frac{68.5 \sin 42°}{\sin 75°}$$

$$\doteq \frac{68.5(0.6691)}{0.9659}$$

$$CB = 47.45 \quad \text{to 2 decimal places}$$

In $\triangle ABC$, $\dfrac{h}{CB} = \tan 38°$ or $\dfrac{h}{47.45} = \tan 38°$

$$h = 47.45 \tan 38°$$
$$\doteq 47.45(0.7813)$$
$$h = 37.1 \quad \text{(to 1 decimal place)}$$

The height of the cliff is 37.1 m.

5.13 Exercise

A Express the measures of lengths correct to 1 decimal place and angles to the nearest degree.

1 Use the diagram. Calculate the length *h*, in metres.

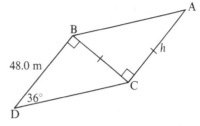

2 Calculate the distance AB shown in the diagram.

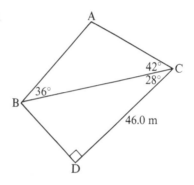

3 How far is A from B?

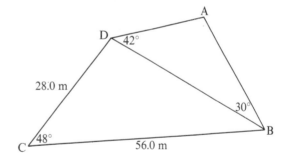

4 An engineer wants to find the height of an inaccessible cliff and takes measurements as shown in the diagram.
△ACB is a horizontal plane.

Find the height of the cliff, DB.

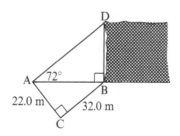

5   A promotion blimp floated above Sheridan Plaza for two days. As a
    project, a group of students was asked to
    determine the altitude of the blimp.
    The data were recorded in their
    diagram shown at the right. What
    was the altitude, *h*, of the blimp?
    △BDC is a horizontal plane.
    △ADB and △ADC are vertical
    planes.

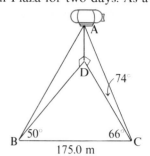

B   Be sure to draw a diagram to record the given information.

6   The crows-nest of the yacht *Mutiny* is 50.0 m above the water level. The
    angle of depression from the crows-nest to a buoy due west of the boat is 40°.
    The angle of depression to another buoy S 70° W of the yacht is 34°. How
    far apart are the buoys?

7   Two roads intersect at 34°. Two cars leave the intersection on different
    roads at speeds of 80 km/h and 100 km/h. After 2 h, a traffic helicopter
    which is above and between the two cars takes readings on them. The angle
    of depression to the slower car is 20° and the distance to it is 100 km. How
    far is the helicopter from the faster car?

8   To estimate the usable lumber in a redwood tree in California, the company
    officials must first estimate the usable height of the tree. A certain tree has
    angles of elevation of 41° and 52° respectively determined from points that
    are 50.0 m apart. If the angle formed at the base of the tree by the positions
    of the two sightings is a right angle, find the height of the tree.

9   Jennifer and Alex were flying a hot air balloon when they decided to
    calculate the straight line distance from Beaverton to Tandy. From a height
    of 340.0 m they recorded the angles of depression of Beaverton and Tandy
    as 2° and 3° respectively. The angle between the lines of sight to the two
    towns was 80°. Find the distance from Beaverton to Tandy to one tenth of
    a kilometre.

10 A Coast Guard helicopter hovers between an island and a damaged sail
    boat. The angle between the lines from the island to the helicopter and
    to the sailboat is 73°. The angle subtended at the helicopter by the lines
    to the island and to the crippled yacht is 40°. A police rescue ship is
    coming toward the sailboat and is now at a point 800.0 m away. From
    this point the angle between the lines to the island and the sailboat is 35°.
    If the angle at the island between the lines of sight to the two boats is 68°,
    find the distance from the helicopter to the sailboat.

5.14 Problem-Solving: Making Decisions

In problem solving, a diagram shows the given information translated from the problem and recorded. Then a decision is made as to which property of triangles applies.

- Right Triangles
- Law of Sines
- Law of Cosines

The same diagram may represent different word problems. For example, each of the following seemingly different problems may be solved using the same diagram.

Problem 1

Two points A and B are located on the same bank of a river, 28.5 m apart. A point C on the other bank is determined by the angle measurements shown in the diagram. Use the data to calculate the width of the river.

Problem 2

The angle of elevation of a jet is 43° and 62° from two locations 28.5 km apart. Calculate the height of the jet.

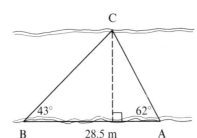
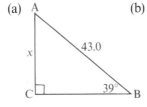

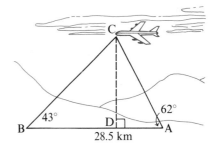
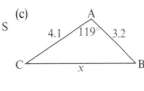

Once you have interpreted the clues given in the word problem and recorded them by constructing a diagram, you can decide which skills to use to solve the problem.

5.14 Exercise

A 1 For each diagram, first decide which skills you need to use to find the missing part and then find the value of x.

(a)

(b)

(c)

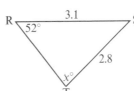

(d)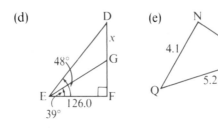

(e)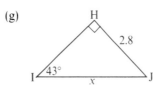

(f)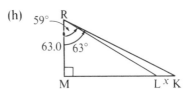

(g)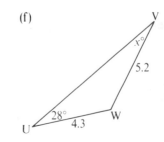

(h)

2 (a) In $\triangle ABC$, $\angle A = 43°$, $\angle B = 90°$, and $b = 42.0$. Find c.
   (b) In $\triangle PQR$, $\sin P = 0.2$, $p = 1.5$, and $q = 6.0$. Find $\sin Q$.
   (c) In $\triangle PQR$, $p = 10.2$, $q = 20.3$, and $r = 12.1$. Find $\cos R$.
   (d) In $\triangle PQR$, $\angle Q = 90°$, $\angle R = 23°$, and $p = 4.1$. Find q.
   (e) In $\triangle MNP$, $\angle M = 78°$, $\angle P = 32°$, and $p = 96.0$. Find m.
   (f) The ratio $s:r$ in $\triangle STR$ is $\sqrt{5}:2$. If $\angle R = 25°$, find $\angle S$.

B Organize the solution to each problem. Refer to the *Steps For Solving Problems.* Remember, to solve a problem you must know the answer to the following questions

   I   What information does the problem ask me to find?
   II What information is given to me in the problem?

3 A roof truss is 9.8 m wide. If the angles formed by the roof beams are 15° and 18°, find the lengths of the roof beams.

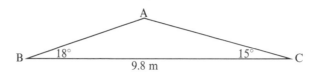

4 A ski resort is installing tows on its new ski runs. The baby hill has a vertical height of 500.0 m and is inclined at 42°. Determine the approximate length of the T-bar cable if it goes to the top of the hill and returns.

5 A surveyor wishes to find the distance, BC, across a river. He selects a position A so that $BA = 86.0$ m, and measures $\angle ABC$ and $\angle BAC$ at 39° and 52° respectively. Find the distance, BC, accurate to the nearest metre.

6   A building is situated on the bank of a river. A point X is located directly opposite the building on the other bank of the river. The angle of depression of point X from the top of the building is 39°. If the building is 22.0 m high, how wide is the river?

7   A new senior citizens home is under construction and a wheel-chair ramp is being built to the front porch. If the angle of inclination of the ramp will be 20° and the height of the porch is 1.5 m, how far back from the porch must the ramp start?

8   When a ship leaves a port, it is given a heading as shown. A trawler leaves Cap d'Espoir, New Brunswick on a heading of S 78° E. Another trawler leaves North Point, Prince Edward Island with a heading of N 15° E. Cap d'Espoir is approximately 90.0 km from North Point. If the trawlers meet at point A, how far is point A from Cap d'Espoir? Cape d'Espoir is due north of North Point.

Heading S 40° W

Heading N 75° E

9   A hovercraft leaves Cape Ommaney, British Columbia at a heading of S 22° W. A tug leaves Cape Knox in the Queen Charlotte Islands and travels on a heading of N 72° W. Cape Ommaney is 215.0 km from Cape Knox. If the vehicles rendezvous at point A, how far is the rendezvous point from Cape Knox? Cape Knox is due south of Cape Ommaney.

10 A tug leaves Ramea Island in Newfoundland and travels 80.0 km into the Gulf of St Lawrence. Another tug leaves Rose Blanche in Newfoundland and travels 70.0 km into the Gulf of St. Lawrence where it meets the first tug. If Ramea Island is 60.0 km due east of Rose-Blanche, what is the heading of the first tug?

Math Tip

It is important to clearly understand the vocabulary of mathematics when solving problems. *You cannot solve problems if you don't know what the clues are.*

- Make a list of all the words you have learned in this chapter.
- Continue to add the remaining new words to your list.
- Provide a simple example to illustrate the meaning of each word.

Practice and Problems: A Chapter Review

1. Use the relationship $\csc(180° + \theta) = -\csc\theta$ to calculate
   (a) $\csc 225°$ (b) $\csc 210°$ (c) $\csc 240°$

2. For all α, prove $\tan\alpha = \dfrac{\sin\alpha + \sin^2\alpha}{\cos\alpha(1 + \sin\alpha)}$.

3. Solve for $0 \le \theta \le 2\pi$ and $0 \le A \le 4\pi$.   (a) $\cos 2\theta = 1$ (b) $\cos 2A = -1$

4. Solve $\cos\theta(1 - 2\sin\theta) = 0$ for $0 \le \theta \le 2\pi$.

5. Solve $\triangle ABC$.

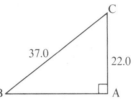

6. In $\triangle MNP$, $\dfrac{p}{n} = \dfrac{4}{3}$. If $\angle N = 35°$, find $\angle P$.

7. Solve each triangle.
   (a) $\triangle ABC$: $a = 16.0$ m, $b = 19.0$ m, $c = 22.0$ m
   (b) $\triangle DEF$: $\angle D = 73°$, $e = 132.0$ m, $f = 112.0$ m

8. At a construction site in downtown Ottawa, two tunnels were excavated starting at the same point. One tunnel was 400.0 m long and the other tunnel was 250.0 m long. Find the distance between the ends of the tunnels if the angle contained between them is 84°.

9. The legs of a step ladder are each 2.0 m long. What is the maximum spread of the legs if the maximum angle at the top is 40°?

10. A surveyor is locating three amusement sections, M, N, and P, around an artificial lake. $\angle MNP$ is measured and recorded as 57°. Length MN is 728.0 m and length MP is 638.0 m. What is the angle at M?

11. In studying the solar system, an astronomical unit (1 A.U.) is given as the distance from the earth to the sun. At one particular position of the planets, Earth, Mars and the Sun form a triangle with the following properties. The angle between the lines of sight from Mars to the Sun and from Mars to Earth is approximately 39°. The angle between the lines of sight from Earth to the Sun and from Earth to Mars is 80°. How many astronomical units is it from Mars to the Sun? Express your answer to two decimal places.

Test for Practice

2   2² = 4

1   For $\triangle ABC$, write the formulas for the
     (a) sine law   (b) cosine law

2   Prove each identity.
     (a) $\cot \theta \cot \theta \times \tan \theta \tan \theta = 1$
     (b) $\tan \theta \sin \theta + \cos \theta = \sec \theta$
     (c) $\sin^4 \alpha - \cos^4 \alpha = \sin^2 \alpha - \cos^2 \alpha$

3   Solve $2 \cos \theta = 1$ for $0 \le \theta \le 2\pi$.

4   Solve $2 \sin^2 \theta + \sin \theta - 1 = 0$ for $0 \le \theta \le 2\pi$ or $0° \le \theta \le 360°$.

5   In $\triangle DEF$, $\angle D = 52°$, $\angle E = 47°$, and $f = 12$. Find the measure of side e.

6   Solve each triangle.
     (a) $\triangle STU$: $\angle T = 90°$, $\angle U = 22°$, $s = 15.1$
     (b) $\triangle DEF$: $\angle E = 90°$, $DE = 17.2$, $DF = 27.3$

7   If two sides of a triangle measure 30.0 cm and 48.0 cm and the contained angle measures 60°, find the third side.

8   From a balloon, situated directly above a point A, the angle of depression to a second point B is 28°. Points A and B are both on level ground. The balloon rises vertically 58.0 m. Now the angle of depression to B is 42°. What is the distance from A to B?

9   Two planes flying at the same altitude are 3000.0 m apart when they spot a raft on the sea below them. The angles of depression to the raft are 47° and 38°. Find the distance from the raft to the closest plane.

10   The bridge over the Royal Gorge of the Arkansas River in Colorado is the highest in the world. A person standing under the bridge finds that the angles of elevation to the ends of the bridge are 62° and 71°. If the distances from the ends of the bridge to the person are 420.0 m and 360.0 m, calculate the length of the bridge.

Looking Back: A Cumulative Review

1 For each relation, make a sketch of the region in which its graph is contained. What is the domain and range?
   (a) $x^2 + y^2 = 36$ (b) $y = x^2 - 1$
   (c) $y^2 = 8 - x$ (d) $y = 2|x| + 1$

2 Solve each of the following.
   (a) $5b^2 - 3b + 1 = 0$
   (b) $(x^2 - 2)^2 - 9(x^2 - 2) + 14 = 0$
   (c) $\sqrt{2x + 1} - \sqrt{x} = 1$
   (d) $x - \sqrt{x - 3} = 5$

3 (a) Draw the graph of the inverse of $y = x^2 - 4$. What is its equation?
   (b) Is the inverse in (a) a function? Give reasons for your answer.
   (c) Sketch the graph of the reciprocal of $f(x) = 18 - 2x^2$.
   (d) Is the reciprocal in (a) a function or not? Give reasons for your answer.

4 Two relations are given by $f(x) = 3x - 2$ and $g(x) = 2x^2 + 3x$.
   (a) Find expressions for $f(g(x))$ and $g(f(x))$.
   (b) Calculate $f(g(-1))$ and $g(f(0))$.

5 Prove.
   (a) $\dfrac{1 - \tan^2 \theta}{\tan \theta - \tan^2 \theta} = 1 + \dfrac{1}{\tan \theta}$ (b) $\sin^4 \theta - \cos^4 \theta = \dfrac{2 - \sec^2 \theta}{-\sec^2 \theta}$

6 Show that the function given by $2f(x) = 3 - \dfrac{2}{x}$ is an increasing function for $x > 0$.

7 A window in a building is 15.0 m above the ground. From this window the angle of elevation to the top of a second building is $32°$ and the angle of depression to the bottom of the second building is $17°$. What is the height of the second building?

8 (a) Show that $x - k$ is a factor of
   $$m^5(x - k) + x^2(m^2 - 1) + k^2(1 - m^2) = 0.$$
   (b) Solve $2x^3 + x^2 - 5x + 2 = 0$.

9 Find the real roots of $x^4 + 4x^3 - 7x^2 - 22x + 24 = 0$.

6 Methods of Geometry: Process and Proof

vocabulary and concepts of geometry, congruence, writing proofs, analyzing deduction, using language, inequalities, parallel lines, indirect proof, concepts of transformations, strategies and proving deductions, problem-solving

Introduction

The two most important questions to ask yourself when you are solving any problem are:

    I What information am I asked to find or to prove?

    II What information am I given?

The diagram at the right will help you to understand the nature of solving a problem.

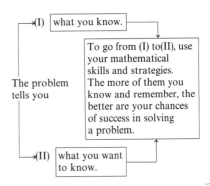

In your study of geometry, you will continue a process that has been used and studied for thousands of years—the use of logical reasoning to prove geometric facts, given in the form of deductions. About two thousand years ago, Euclid, a Greek mathematician, wrote thirteen books called *The Elements*, a compilation of the various facts already known about deductive geometry. This body of facts, taken as a whole, is often referred to today as **Euclidean Geometry**.

The structure of Euclidean Geometry may be represented by the following diagram.

The study of Euclidean Geometry is not unlike your study of mathematics, as shown below.

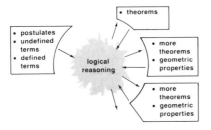

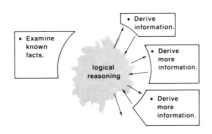

6.1 Geometry: Vocabulary and Concepts

Euclid and his followers began the foundation on which your present work in geometry is built. For example, they organized the ideas of geometry logically into a deductive system, as shown on the right.

In your study of algebra, there are some words, such as *set*, which must be accepted as undefined. There are also words in geometry which must be accepted as undefined.

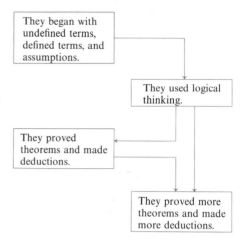

undefined terms   Not all words can be defined. Thus, you must accept some of them as undefined. Some of the undefined terms or words used in geometry are *point, line, plane.*

defined terms   These undefined terms are used to write definitions. For example, the word *circle* is defined as the set of all points that are the same distance from a fixed point. This fixed point is called the **centre**.

A definition may be reversible. For the above you can write:
The set of all points that are the same distance from a fixed point is a circle.

Once you have defined the terms, you can then use the terms to define new ideas, for example, concentric circles with the same centre.

axioms or postulates   Sometimes certain assumptions are made that are accepted as true. These basic assumptions are called **axioms** or **postulates**. For example:

- A line contains at least two points.
- For every two different points there is exactly one straight line that contains them.

In your study of plane geometry, you will learn other axioms and postulates. Statements to be proved are referred to as **deductions**. You will begin to work with deductions in subsequent sections. Some deductions may be used to prove further statements. These proven statements are called **theorems**.

To speak the language of geometry and to solve problems, you need to know the meaning of the words used. Fundamental to your study of geometry is a clear understanding of the vocabulary. You will meet again many of the following in your work in geometry. How many are you familiar with? Which ones are

- defined?   - undefined?

These terms are not in any special order.

point ✓	chord ✓	converse	inscribed angle ✓
line ✓	radius ✓	indirect proof	collinear points ✓
plane ✓	arc ✓	corollary	congruent angles ✓
ray ✓	segment	hypotenuse ·	congruent figures ✓
angle ✓	sector ✓	straight angle ✓	alternate angles ✓
vertex ✓	line segment ✓	reflex angle	interior angles ✓
polygon ✓	arms of angle	adjacent angles ✓	mean proportional
triangle ✓	coplanar points ✓	diameter ✓	complementary angles ✓
pentagon ✓	right bisector ✓	diagonal ✓	vertically opposite angles ✓
hexagon ✓	angle bisector ✓	midpoint ·	congruent line segments ✓
octagon ✓	equidistant ✓	circumference ✓	corresponding angles ✓
altitude ✓	parallel lines ✓	subtended ✓	quadrilateral ✓
median ✓	transversal ✓	exterior angle ✓	regular polygon ✓
theorem ✓	hypothesis	conclusion	postulate
circle ✓			

triangles: acute, equiangular, equilateral, right, obtuse, scalene, isosceles, similar
quadrilaterals: trapezoid, parallelogram, rectangle, rhombus, square

6.1 Exercise

B   In the following questions, you need to record a meaning for each word or term used in geometry. To do geometry you need to know the precise meaning of each word. However, an example of your own to illustrate the meaning will help you learn the words. Follow these steps

Step A   Write a definition for each term.
Step B   Illustrate the term by sketching a diagram.

1   Each of the following describes a certain angle. Write a definition for each.
   (a) acute   (b) obtuse   (c) right
   (d) supplementary (e) straight   (f) adjacent
   (g) complementary (h) vertically opposite

2   Write a definition for each of the following polygons.
   (a) triangle   (b) quadrilateral   (c) hexagon   (d) octagon

3   Write a definition for each type of triangle.
    (a) acute   (b) obtuse   (c) right
    (d) scalene   (e) equiangular   (f) equilateral
    (g) isosceles   (h) right-angled-isosceles

4   Each of the following terms is associated with triangles. Define each term.
    (a) altitude   (b) median
    (c) right bisector of a side   (d) bisector of an angle
    (e) exterior angle   (f) hypotenuse

5   Define each quadrilateral.
    (a) rectangle   (b) square   (c) parallelogram
    (d) rhombus   (e) trapezoid

6   These words are associated with circles. Write a definition for each.
    (a) radius   (b) diameter   (c) circumference
    (d) chord   (e) arc   (f) semi-circle
    (g) sector   (h) segment   (i) tangent
    (j) secant   (k) inscribed angle   (l) central angle
    (m) inscribed angles subtended by the same chord (arc)

7   Define each of the following.
    (a) parallel lines   (b) equidistant points   (c) collinear points
    (d) coplanar points   (e) vertex of a polygon   (f) diagonal of a polygon

8   Certain terms are associated with parallel lines. Write a definition for each.
    (a) transversal   (b) alternate angles   (c) corresponding angles
    (d) interior angles on the same side of the transversal

Math Tip

Up until the 6th century B.C., the known facts of geometry were not organized. One of the best known mathematicians of Greece was Euclid. In *The Elements*, he compiled and systematically arranged his work and that of others. *The Elements* consisted of thirteen books that dealt with not only geometry but also with number theory and algebra. Howard Eves, a historian of mathematics said, "No work except the Bible has been more widely used, edited, or studied and probably no work has exercised a greater influence on scientific thinking."

6.2 Conditions for Congruence

The concept of congruence is basic in your study of geometry.

Congruent Line Segments: If two line segments are congruent, then their lengths are equal. Thus, $\overline{AB} \cong \overline{CD}$ if and only if $\overline{AB} = \overline{CD}$.

↑   ↑

Means $\overline{AB}$ is congruent   Means the length of
to $\overline{CD}$. $\overline{AB}$ and $\overline{CD}$ are equal.

Symbolism in mathematics is important. But often the symbolism may create reading difficulties. For this reason, the following simplified symbolism will be used. Throughout the work in geometry, use AB = CB for $\overline{AB} \cong \overline{CD}$.

Congruent Angles   If two angles are congruent, then their measures are equal. Thus, $\angle ABC \cong \angle DEF$ if and only if $\angle ABC = \angle DEF$.

↑   ↑

Means $\angle ABC$ is   Means the measures of
congruent to $\angle DEF$. $\angle ABC$ and $\angle DEF$ are equal.

Similarly, in the work in geometry, use the simplified symbolism $\angle ABC = \angle DEF$ for $\angle ABC \cong \angle DEF$.

Congruent Triangles   Two triangles are congruent, if and only if all corresponding angles and sides are congruent. Since $\triangle ABC$ and $\triangle DEF$ are congruent, then you can write

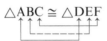

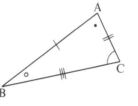 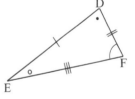

It is important how the congruence relation is written, since it shows which vertices correspond to each other. Thus, the congruence relation shows which parts are related.

Corresponding Sides	Corresponding Angles
AB = DE	$\angle A = \angle D$
BC = EF	$\angle B = \angle E$
CA = FD	$\angle C = \angle F$

Congruence Assumptions   The condition of congruence for triangles requires that all 6 parts be congruent. However, each of the following congruence assumptions may be made since, if certain parts are congruent, then they indicate that other parts are congruent.

side side side congruence assumption (SSS)
If three sides of one triangle are respectively congruent to three sides of another triangle, then the two triangles are congruent.

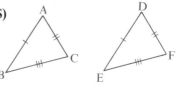

$\triangle ABC \cong \triangle DEF$

side angle side congruence assumption (SAS)
If two sides and the contained angle of one triangle are respectively congruent to two sides and the contained angle of another triangle, then the two triangles are congruent.

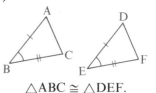

$\triangle ABC \cong \triangle DEF$.

angle side angle congruence assumption (ASA)
If two angles and the contained side of one triangle are respectively congruent to two angles and the contained side of another triangle then the two triangles are congruent.

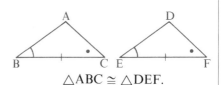

$\triangle ABC \cong \triangle DEF$.

Another form of the above congruence assumption is given by the diagram.
If two angles and a side of one triangle are respectively congruent to two angles and the corresponding side of another triangle, then the two triangles are congruent.

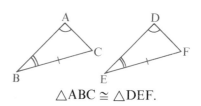

$\triangle ABC \cong \triangle DEF$.

right triangle congruence postulate (HS)
If the hypotenuse and a side of a right-angled triangle are congruent to the corresponding hypotenuse and side of another right-angled triangle, the triangles are congruent.

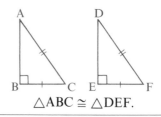

$\triangle ABC \cong \triangle DEF$.

The congruence assumptions are applied in the following Example to show that if certain parts of the triangles are congruent, then the triangles are congruent.

Example   In the diagram, BA bisects
∠CAD and AD = AC.
Show why CB = DB.

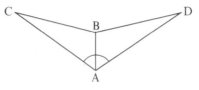

Solution   In △ABC and △ABD

AB = AB (AB is common to both triangles)
∠BAC = ∠BAD (bisected angle)
AC = AD (given information)

Thus, △ABC ≅ △ABD (congruence assumption SAS)

Since the triangles are congruent, then all corresponding parts are congruent. Thus, as required, CB = DB.

In the next section, you will learn additional techniques for writing complete proofs.

6.2 Exercise

A 1   An important skill in proving deductions is recognizing pairs of congruent triangles. For the following triangles

• write the congruence relation for pairs of congruent triangles.
• give reasons for your answer.

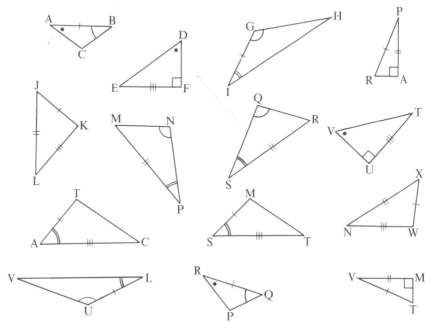

2   To prove triangles congruent, you can use SSS, SAS, ASA or HS. For each
    of the following, what additional information is needed in order to prove
    that the triangles in each pair are congruent? Give reasons for your answers.

(a)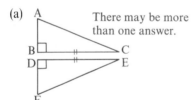
There may be more
than one answer.

(b)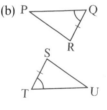

(c)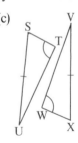

(d)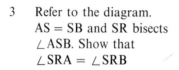

(e)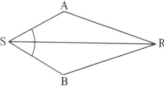

B   For each of the following, sketch the diagram, and record the given
        information on the diagram.

3   Refer to the diagram.
    AS = SB and SR bisects
    ∠ASB. Show that
    ∠SRA = ∠SRB

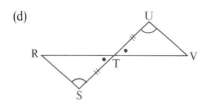

4   In △QAB, QR is the right bisector of AB.
    Show that ∠QAB = ∠QBA.

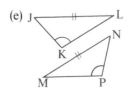

5   In quadrilateral TPAH, ∠THP = ∠APH
    and ∠TPH = ∠AHP. Show that
    AH = TP.

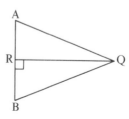

6   From the diagram, RC = RB and
    CD = BA. If ∠RBC = ∠RCB, show that
    △RAD is an isosceles triangle.

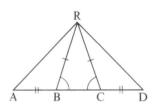

6.3 Writing Proofs

A combination of the congruence assumptions in the previous section and other facts you know can be used to write complete proofs of deductions in geometry. To solve a problem or prove a deduction requires that you understand clearly the answer to two important questions.

  I   What are you asked to find? (What are you asked to prove?)
  II What do you know? (What are you given?)

In Example 1, you are given certain facts, which, along with logical reasoning, will allow you to reach a conclusion or prove the statement. In order that others may understand the reasoning or the process, you must organize your work in a clear and concise way.

Given (or hypothesis): ⟵ This shows the information given in II.

Required to prove ⟵———— This shows the information
  (or Conclusion):   required in I.

Proof: ⟵——————— This shows the process, steps, procedure, by which you
               use the information in II to arrive at the conclusion in I.

Example 1   For quadrilateral
XYZW, if WX = WZ
and XY = ZY, prove
that ∠WXY = ∠WZY.

Solution   Given: Quadrilateral XYZW,

          WZ = WX, XY = ZY ⟵———— This information is
                         given in the original
                         question. Read carefully.

Required to prove: ∠WXY = ∠WZY

                             Give reasons why you may
                          ⌐ deduce each line of the proof.

Proof: In △WXY and △WZY	*Reasons*
WX = WZ	given S
XY = ZY	given S
WY = WY	common side S
Thus, △WXY ≅ △WZY	side-side-side congruence assumption or SSS for short
∠WXY = ∠WZY	congruent triangles

                              ↑
                   Since the triangles are congruent
                   other corresponding parts are congruent.

The process of solving problems is similar in different branches of mathematics. Compare the similarities in the following.

In Algebra

- translate the problem
- construct the equation
- solve the problem
- make a final statement

In Geometry

- translate the deduction
- construct the diagram
- write the proof
- make a final statement

In the information in Example 2, it is important to translate the problem accurately and record it on a sketch of the diagram. In this way, you can plan carefully the solution and apply the needed congruence assumptions to complete the proof.

Example 2   In quadrilateral PQRS, PQ = PS, and QR = SR, prove that QS ⊥ RP.

Solution   Given: Quadrilateral PQRS,
PQ = PS, QR = SR

Required to prove: QS ⊥ RP

Proof: *Step 1*   In △PQR and △PSR

		Reasons	
	PQ = PS	given	S
	QR = SR	given	S
	RP = RP	common side	S
Thus,	△PQR ≅ △PSR	SSS	
	∠QPT = ∠SPT	congruent triangles	

Step 2   In △PQT and △PST

		Reasons	
	PQ = PS	given	S
	∠QPT = ∠SPT	proven	A
	PT = PT	common side	S
Thus,	△PQT ≅ △PST	SAS	
	∠QTP = ∠STP	congruent triangles	
	∠QTP + ∠STP = 180°	QTS straight line	
Thus,	∠QTP = 90°		
and	QS ⊥ RP.		

As seen in the previous example, you may need to recognize different overlapping triangles in a diagram, and use them to prove the deductions.   A diagram may contain hidden triangles.

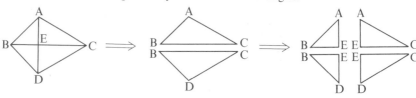

6.3 Exercise

A For each deduction, make a sketch of the diagram. Record the given information on it.

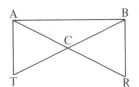

1. If TP = SP and SQ = TR prove that ∠Q = ∠R.

2. For the diagram, RB ⊥ AB TA ⊥ AB and BR = AT. Prove that ∠ATC = ∠BRC.

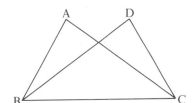

3. △ABC and △DBC are on the same side of the common base BC. If ∠ABC = ∠DCB and ∠DBC = ∠ACB, prove that AB = DC.

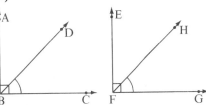

4. Often you will be asked to prove a general statement in geometry. Thus, not only must you construct a diagram, but also you must label the diagram in order to organize the steps of the solution. Use the given diagram to prove the theorem.

Complementary Angle Theorem (CAT)
If two angles are equal then their complements are equal.

Prove: If ∠DBC = ∠HFG then ∠ABD = ∠EFH.

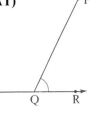

5. Use the diagram to prove the theorem.

Supplementary Angle Theorem (SAT)
If two angles are equal then their supplements are equal.
Prove: If ∠PQR = ∠STV then ∠PQK = ∠STM.

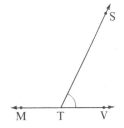

6   Use the diagram to prove the theorem.

Vertically Opposite Angle Theorem (VOAT)
If two lines intersect, the
vertically opposite angles are equal.
Prove: If AB and CD intersect
         at O, then $\angle$ AOC = $\angle$ DOB
         and $\angle$ AOD = $\angle$ COB

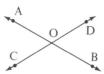

> When proving deductions, use the symbols to refer to these theorems.
>
> • Complementary Angle Theorem   (CAT)
> • Supplementary Angle Theorem   (SAT)
> • Vertically Opposite Angle Theorem   (VOAT)

Once you prove a theorem, you can use it to prove additional deductions.

B   For each of the following, record the given information on a diagram.

7   If A is the midpoint of
CD and BT, and CT = BD,
prove $\angle$ C = $\angle$ D.

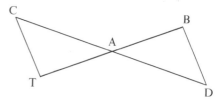

8   Line segments XZ and WY bisect each other at O. Prove XW = ZY.

9   Given $\triangle$CAB. CX bisects $\angle$ C and is perpendicular to AB. CA is
produced to Y and AB is produced to Z. Prove $\angle$ YAX = $\angle$ CBZ.

10   Given $\triangle$XYZ with AY being a bisector of $\angle$ Y and A being a point of XZ.
If XY = ZY, prove XA = ZA.

11   XYZA is a quadrilateral where YA and ZX intersect at C, YA = ZX, and
$\angle$ CYZ = $\angle$ CZY. Prove that YX = ZA.

C 12   The four points BCDE are placed on a circle in the given order with
BC = DE. F is the centre of the circle. Prove that
(a) $\angle$ BFC = $\angle$ EFD   (b) BE = DC

6.4 Process of Geometry: Steps for Solving Deductions

Very often the steps you perform in one branch of mathematics are very similar to the steps you perform in another branch of mathematics. When you are solving problems you need to organize your solution. Similarly, when proving deductions in geometry you need to organize the steps of your proof.

Algebra

Steps for Solving Problems

Step A Read the problem carefully. Can you answer these two questions?
  I What information does the problem ask you to find? (information you don't know)
  II What information are you given?
  Be sure to understand what it is you are to find. Then introduce the variables.

Step B Translate from English to mathematics. Write the equations.

Step C Solve the equations. Organize the flow.

Step D Check your result in the original problem.

Step E Write a final statement as the answer to the problem.

Geometry

Steps for Solving Deductions

Step A Read the problem carefully. Can you answer these two questions?
  I What information does the deduction ask you to prove? (information you don't know)
  II What information are you given? (information you know)
  Be sure you understand what it is you are to find.

Step B Translate from English to mathematics. Draw a diagram. Record the given information on the diagram. Think of a plan.

Step C Solve the deduction. Organize the steps of your proof.

Step D Check your result in the original problem.

Step E Write a final statement as the answer to the problem.

In the process of doing geometry, as shown in the diagram, you will prove many deductions. The results of some deductions are significant and are referred to as theorems. You have already done the theorems which follow. Symbols are used to summarize the title of the theorem, and used as reference in the development of proofs as shown on the following page.

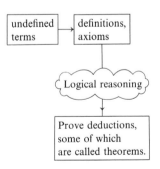

- *Complementary Angle Theorem (CAT)*
  If two angles are equal
  then their complements
  are equal

If ∠DBC = ∠HFG then ∠ABD = ∠EFH.

- *Supplementary Angle Theorem (SAT)*
  If two angles are equal
  then their supplements
  are equal.

If ∠PQR = ∠STV then ∠PQK = ∠STM.

- *Vertically Opposite Angle Theorem (VOAT)*
  If two lines intersect, then
  the vertically opposite
  angles are equal.

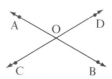

If AB and CD intersect at O, then
∠AOD = ∠COB, ∠AOC = ∠DOB.

- *Angle Sum of a Triangle Theorem (ASTT)*
  The sum of the measures
  of the angles of a triangle
  is 180°.

∠A + ∠B + ∠C = 180°

Often a single fact may be directly deduced from a theorem. For example, based on ASTT you can show that if two angles of a triangle are congruent to two angles of another triangle then the third pair of corresponding angles is also congruent. This is called a **corollary** of the theorem.

Corollaries of ASTT

(a) If two angles of a triangle are congruent to two angles of another triangle, then the remaining angles are congruent.

(b) Each angle of an equilateral triangle is 60°.

6.4 Exercise

A 1 Prove the corollaries of the Angle Sum of a Triangle Theorem (ASTT) as shown previously.

2 Each diagram illustrates an important theorem.
Step A Write the theorem shown in the diagram.
Step B Prove the theorem expressed in Step A.

(a) $a° + b° = x°$ (b) $x° = y° = z° = 60°$ (c) $p° + q° + r° + s° = 360°$

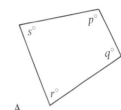

3 The diagram illustrates a theorem.
(a) Write the theorem shown in the diagram.
(b) Prove the theorem in (a).

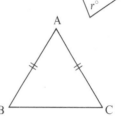

If AB = AC, then ∠B = ∠C.

4 To do deductions in geometry, you must recall the theorems, corollaries and definitions.
(a) Make a list of the theorems and corollaries and their symbols used for reference that you have worked with so far.
(b) Provide a diagram to illustrate each theorem or corollary in (a).

B Remember: give a reason for each result that you record in your deduction.

5 For each triangle, find the value of x.

(a)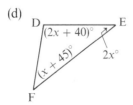

(b)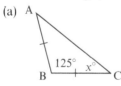

(c)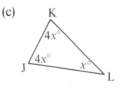

(d)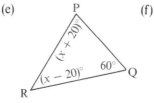

(e)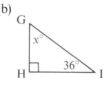

(f)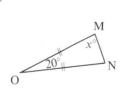

6 Find the values of x and y for each of the following diagrams.

(a)

(b)

(c)

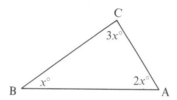

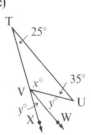

7 For the diagram,
   show that $\angle C$ is
   a right angle.

8 Prove that the sum of the angles of a quadrilateral is 360°.

9 If P is any point in $\triangle DEF$ such that $PD = PE = PF$, prove that
   $\angle EPF = 2\angle EDF$.

10 Prove that in an equilateral triangle, the median (line drawn from vertex
    to midpoint of opposite side) to any side is also the altitude.

11 In a parallelogram ABCD, $AD = AB$. Prove that the diagonals are right
    bisectors of each other.

C 12 In $\triangle ABC$, the midpoint S is found for AC. If $AS = BS$, prove that $\angle ABC$
       is a right angle.

Problem Solving

You often have to experiment to find a method of solving the problem.

Step 1: Draw any four points so that no three are collinear.
Step 2: Now, construct a square so that one of the points is on each
          of the sides of the squares.

6.5 Problem-Solving: Analyzing Deductions

Often, when trying to solve a problem, you cannot see, immediately, the steps for solving the problem. You can analyze the solution of a problem by asking the following questions:

I What am I asked to prove?   II What am I given?

For example, in the given diagram, you are asked to prove AB = DC. If you cannot see, immediately, the steps needed to prove this deduction, then you can organize your thinking by beginning with what you are asked

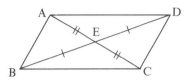

to prove and working backwards as follows, using the think titles of "I can" and "if I can".

	I can	if I can
Step 1	I can prove AB = CD	if I can prove △AEB ≅ △CED.
Step 2	I can prove △AEB ≅ △CED	if I can use one of ASA, SAS, SSS, HS.

In Step *2* you must now eliminate some of the congruence assumption choices you analyzed that you might be able to use.

ASA	SAS	SSS	HS
There are no equal angles given. *Probably* I cannot use ASA	Two sides are given I will likely be able to use SAS if I can prove $\angle$ AEB = $\angle$ CED.	Only two sides are given. I *cannot* use SSS.	There is no right angle. I *cannot* use HS.

You decide to try SAS with △AEB and △CED.

	I can	if I can
Step 3	I can prove △AEB ≅ △CED	if I can prove $\angle$ AEB = $\angle$ CED. But $\angle$ AEB = $\angle$ CED since the angles are vertically opposite angles.

Thus, you have analyzed the thought process starting with

Once you think the proof will work, you may then proceed to write up a complete proof as follows.

Once you have analyzed the problem, then you can write the proof.

Given: Quadrilateral ABCD, AE = CE, BE = DE

Required to prove: AB = DC.

Proof: AC and BD intersect at E. | *Reasons*

∠AEB = ∠CED	VOAT
In △AEB and △CED	
AE = CE	given S
∠AEB = ∠CED	proven A
BE = DE	given S
Thus △AEB ≅ △CED	SAS
AB = CD	congruent triangles
or AB = DC	

6.5 Exercise

A In this part of the exercise the diagrams are given. In part B, you will have to construct your own diagram. For each of the following questions, analyze the steps of the solution by using the headings

I can prove	if I can prove

Once the deduction has been analyzed, write the complete proof.

1 In △PSR and △PQR
∠RPS = ∠RPQ
∠PRS = ∠PRQ.
Prove that △PSQ
is isosceles.

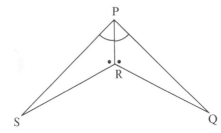

2 In the diagram,
QS = RS
∠PSQ = ∠PSR.
Prove that RP = QP.

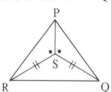

3 In the diagram,
AD = CD
AB = CB.
Prove that
AE = CE.

4 Two circles with centres O and P intersect at R and S. Prove that RS ⊥ OP.

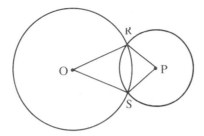

B For each of the following questions, sketch a diagram and record the information on it. Perform the following steps.

Step 1 Analyze each problem and record the information under the headings

I can prove	if I can prove

Step 2 Draw an accurate diagram to help plan your work.
Step 3 Then write a complete proof.

5 In △PQR, RP = QP and S is the midpoint of RQ. Prove that ∠PSR = 90°.

6 In quadrilateral STUV, ST = SV and VU = TU. Prove that ∠STU = ∠SVU.

7 Line segments SU and TV intersect at M so that MU = MS and ∠TSM = ∠VUM. Prove that M is the midpoint of VT.

8 Prove that, in an isosceles triangle, the bisector of the vertical angle is the right bisector of the base.

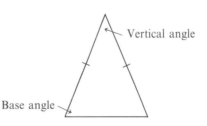

9 Given quadrilateral ABCD with AD = CB. BD is a diagonal. If ∠ADB = ∠CBD, prove AB = CD.

10 A rhombus is a quadrilateral with all sides equal. Prove that the diagonals are perpendicular to each other.

C 11 A quadrilateral has both pairs of opposite sides equal. Prove that the opposite angles are equal.

6.6 Using Language: Isosceles Triangle Theorem

Very often, when you study a diagram, you notice something that appears to be true (a conjecture). However, before you can rely on the information, you must prove that what appears to be true is in fact true. The process of mathematics, very often, is to examine a mathematical diagram, state a conjecture and then, deductively, attempt to prove your conjecture. For example, if you examine an isosceles triangle you can make a conjecture.

Conjecture: If a triangle is isosceles, then the angles opposite the equal sides are equal.

To prove the above conjecture, a particular isosceles triangle is drawn and labelled. However, on analyzing the diagram, you might find that, to prove two angles equal, you require two triangles. To obtain two triangles, a useful construction is performed in the example, as follows.

Given: △ABC, AB = AC

Required to prove: ∠B = ∠C

Proof: Construct the bisector of ∠A to meet BC at D.

In △ABD and △ACD	Reasons	
AB = AC	given	S
∠BAD = ∠CAD	constructed	A
AD = AD	common side	S
Thus △ABD ≅ △ACD	SAS	
Then ∠B = ∠C	congruent triangles	

Thus, you have developed an important result (a theorem).

Isosceles Triangle Theorem (ITT)

In an isosceles triangle, the angles opposite the equal sides are equal.

Using Language

Statements written in the form **if-then** are called **conditional statements**.

If a triangle	**then** the angles opposite the
↑ is isosceles	↑ equal sides are equal.
The hypothesis is the information given.	The conclusion is the information you are asked to prove.
II What do I know?	I What do I want to prove?

The **converse** of a statement is the statement obtained by reversing the hypothesis and conclusion. For example, for the isosceles triangle statement the converse is written

If the angles opposite	**then** the sides opposite
two sides in a	these angles are
triangle are equal	equal.

The converse of a statement may not be true, so you need to prove the converse.

Given: △PQS with ∠Q = ∠S.

Required to prove: PQ = PS

Proof: Construct PT ⊥ QS
as shown

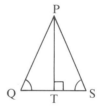

In △PQT and △PST	*Reasons*	
∠Q = ∠S	given	A
∠PTQ = ∠PTS	constructed	A
PT = PT	common side	S
Then △PQT ≅ △PST	AAS	
PQ = PS	congruent triangles	

You can use the *Isosceles Triangle Theorem* (*ITT*) and its converse to write **biconditional statements** that use "if and only if". Thus, the two previous statements are combined into one statement.

> In an isosceles triangle, the angles opposite two sides
> are equal if and only if the sides are equal.
> ↑

The symbol iff is used.

For example, if *p* then *q*
if *q* then *p* is written compactly as *p* iff *q*.

Now the theorem is applied when proving a deduction.

Example   In quadrilateral EFGH, if EF = EH and GF = GH prove that $\angle F = \angle H$.

Solution   Given: Quadrilateral EFGH,
EF = EH, GF = GH

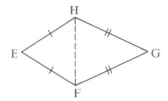

Required to prove: $\angle F = \angle H$

Proof: Join FH   Complete this construction.

		Reasons
In $\triangle$EFH, EF = EH		given
Then $\angle$EFH = $\angle$EHF ①		ITT
In $\triangle$FGH, GF = GH		given
Then $\angle$GFH = $\angle$GHF ②		ITT
From ① and ②		
$\angle$EFH + $\angle$GFH = $\angle$EHF + $\angle$GHF		addition of equal measures
Then $\angle F = \angle H$ as required.		

6.6   Exercise

A 1 Conditional statements also occur in algebra. Write the converse of each statement. Which converse is true (T)? Which is false (F)?

(a) If $2y - 1 = 9$, then $y = 5$.

(b) If an equilateral triangle has each side 5 cm, then the perimeter is 15 cm.

(c) If $a, b, > 0$ then $ab > 0$.   (d) If $x > 9$, then $x > 5$.

(e) If $y = mx + b$ then $m = \dfrac{y - b}{x}$.

(f) If x and y are odd then $x + y$ is even.

(g) If two triangles are congruent, then the corresponding angles are congruent.

(h) If $a > 0$, $b < 0$ then $ab < 0$.   (i) If m and n are even, then mn is even.

2 For each of the following
   • examine the diagram. Decide on a conjecture related to the diagram. List any information that you need to prove your conjecture.
   • then prove the deduction given in the question.

(a) In the diagram, D is the midpoint of BC, and $\angle$ADB = $\angle$ADC. Prove that $\triangle$ABC is isosceles.

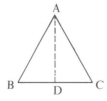

(b) From the given figure,
SA = SB and AU = BT
Prove that △STU is isosceles.

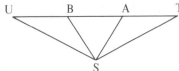

(c) Two circles with centres
O and P intersect at A
and B. Prove that
∠OAP = ∠OBP.

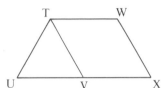

(d) For the diagram,
∠TUV = ∠WXU and
TU = TV = WX. Prove
that ∠TVU = ∠WXV.

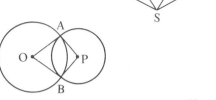

3 (a) Write the converse of the following. If the diagonals of a quadrilateral bisect each other then the quadrilateral is a rhombus.
   (b) Prove whether the converse in (a) is true or not.

4 (a) Write the converse of the following statement.
       If a point is on the right bisector of a line segment, then it is equidistant from the end points of the line segment.
       (This result is called the **Right Bisector Theorem RBT**.)
   (b) Show whether the converse is true or not.

B In the following questions, remember to analyze the given information. Try working backwards (previous section).

5 In quadrilateral RSTU, RS = RU and ∠UST = ∠TUS. Prove ∠S = ∠U.

6 In △QNP, A is a point on NP so that AN = AP. If ∠NAQ = ∠PAQ, prove that QN = QP.

7 In △TSR, P is the midpoint of SR. If TP ⊥ SR prove that △TSR is isosceles.

8 MNP is an isosceles triangle with MN = MP. A is the midpoint of NP. Prove MA ⊥ NP.

9 △SMP is isosceles with SM = SP. R is an interior point of △SMP such that SR bisects ∠MSP. Prove that ∠RMP = ∠RPM.

10 In rhombus RSTU, prove that the opposite angles are equal, that is, ∠S = ∠U and ∠R = ∠T.

C 11 Given △XYZ with point M in XY and point N in XZ so that MY = NZ and MZ = NY. Prove that △XZY is isosceles.

6.7 Parallel Lines

The process of a deductive system has often been referred to. Each time you introduce new vocabulary, skills, or concepts, you result in them being blended into the structure of deductive geometry.

Parallel lines: Two lines are parallel if they lie in the same plane and do not intersect.

A **transversal** is a line that intersects 2 or more lines in the plane. ST is a transversal for UK and VM. When a transversal intersects two or more lines certain angles are named.

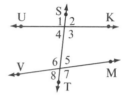

Corresponding Angles	Alternate (Interior) Angles	Interior Angles on the same side of the transversal
$\angle 1$, $\angle 6$	$\angle 4$, $\angle 5$	
$\angle 4$, $\angle 8$ etc.	$\angle 3$, $\angle 6$	$\angle 4$, $\angle 6$ $\angle 3$, $\angle 5$

If two lines are drawn parallel, and a transversal cuts them, certain pairs of angles appear to be equal.

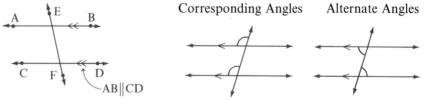

Corresponding Angles   Alternate Angles

AB∥CD

Since it appears more acceptable that corresponding angles are equal, the following postulate is accepted as true.

> **Parallel Postulate:** If two parallel lines are cut by a transversal then the corresponding angles are equal.

With the parallel postulate, you can prove the following theorem.

Example 1   Prove that if two parallel
lines are cut by a
transversal, then the
alternate angles are equal.

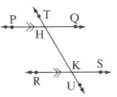

Solution   Given: PQ ∥ RS. TU is a transversal that cuts PQ and RS.

Required to prove: (a) ∠QHK = ∠HKR   (b) ∠PHK = ∠HKS

Proof: Since PQ ∥ RS, TU is a transversal, then

			Reasons
	∠THP = ∠HKR	①	parallel postulate
PQ, TU are intersecting lines at H.			
Thus	∠THP = ∠QHK	②	(VOAT)
From ① and ②	∠HKR = ∠QHK		Equality: If $a = b$,
Similarly,	∠PHK = ∠HKS		$c = b$, then $a = c$

Parallel Lines Theorem (PLT)

If a transversal intersects two parallel lines, then
- the corresponding angles are equal.
- the alternate angles are equal.
- the interior angles on the same side of the transversal are
  supplementary.

Example 2   In the diagram, PQ ∥ RS,
and PQ = RS. Prove
that M is the midpoint
of PS and QR.

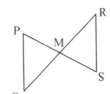

Solution   Given: PQ ∥ RS, PQ = RS

Required to prove: PM = SM, QM = RM

Proof: QR is a transversal for PQ ∥ RS.	*Reasons*
Then ∠PQR = ∠QRS.	PLT
PS is a transversal for PQ ∥ RS.	
Then ∠QPM = ∠MSR.	PLT
In △PQM and △SRM	
∠PQM = ∠SRM	proved A
PQ = SR	given S
∠QPM = ∠RSM	proved A
△PQM ≅ △SRM	ASA
PM = SM	congruent triangles
QM = RM	

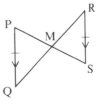

6.7 Exercise

A If you don't know the precise meaning of words, you will not be able to solve problems. After you complete part A, list the various words and their meanings.

1 For the diagram, name pairs of
   (a) corresponding angles.
   (b) alternate angles.
   (c) interior angles on the same side of the transversal.

2 For the diagram in Question 1, name pairs of angles that are
   (a) supplementary   (b) vertically opposite

3 In the diagram, PQ ∥ RS.
   (a) Prove that ∠PHK + ∠RKH = 180°.
   (b) Prove that ∠QHK + ∠SKH = 180°.
   (c) Why may the following be written as a theorem?
   If a transversal intersects two parallel lines, then the interior angles on the same side of the transversal are supplementary.

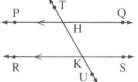

4 Which pairs of angles are equal for each diagram?
   (a)

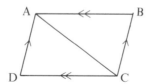

   (b)

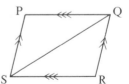

5 For each diagram, find the values of m and n. Give reasons for your results.
   (a)

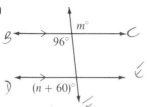

   (b)

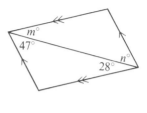

6 Calculate the values of m and n, if PQ = PR and QR ∥ ST.

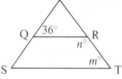

B   Copy each diagram and mark any given information on it.

7   For the diagram
PQ ∥ RS.
Prove that
∠ TUP = ∠ WVS.

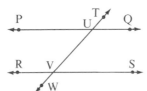

8   Prove that
∠ AME = ∠ BPF.

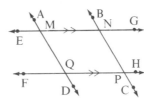

9   For the diagram, SV ∥ KL
and KV is the bisector of
∠ SKL. Prove that
KS = SV.

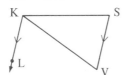

10   If TL ∥ RS and M is the
midpoint of LR, prove
that M is also the
midpoint of TS.

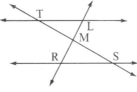

11   For the diagram, PQ ∥ BC.
(a) Why is ∠ 1 + ∠ 2 + ∠ 3 = 180°?
(b) For △ABC, prove that
∠ BAC + ∠ B + ∠ C = 180°.

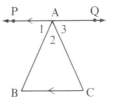

12   If RS and PQ are both perpendicular to AB prove that RS ∥ PQ.

13   In quadrilateral PQRS, PQ ∥ SR. If the diagonal QS bisects ∠ PSR, prove
that PS = PQ.

14   A rhombus is a quadrilateral with all sides equal. Prove that, in any rhombus,
the opposite angles are equal.

15   A parallelogram is a quadrilateral with opposite sides parallel. Prove the
following properties of a parallelogram.
(a) The opposite angles are equal.   (b) The opposite sides are equal.
(c) The diagonals bisect each other.

C 16   In square PQRS points M and N are chosen in PQ and QR respectively
so that PM = NR. Prove QS ⊥ MN and QS bisects MN.

6.8 Method of Indirect Proof

There are other methods of reasoning that you can use to prove deductions. One such method is the method of **indirect proof** based on the following observation.

There are two possibilities for a given statement. Either

> A: The statement is true.   or   B: The statement is not true.

The basis of indirect proof is to choose one of the possibilities. If the chosen possibility leads to a contradiction of a fact or theorem, (a fact or theorem being a statement which is true) then the alternative must be true. For example, if you assume a statement is not true, and this assumption leads to the contradiction of a fact, then your assumption is wrong and thus the statement must be true since a statement must be either true or false.

To prove some theorems, the indirect method is used, as shown in the following example. In this example, you are asked to prove the converse of the *Parallel Lines Theorem* (*PLT*).

Example 1 Prove that if a transversal intersects two lines and the alternate angles are equal, then the lines are parallel.

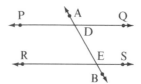

Solution Given: AB a transversal for PQ and RS. ∠PDE = ∠SED.

Required to prove: PQ ∥ RS

Think: Either the statement is true or it is not true.

Proof: Either PQ ∥ RS or PQ ∦ RS
Make the assumption that
PQ ∦ RS. Then PQ and RS
meet at the point M, as shown.

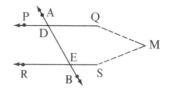

	Reasons
In △MDE, PDE is an exterior angle, ∠MED is a remote interior angle, and ∠MED = ∠PDE. This is impossible since, for △MDE, ∠MED < ∠PDE	given ← The alternate angles are given as equal. EAT

Thus, the assumption is incorrect and PQ ∥ RS.

Once you have proved the above theorem, you can show the corollaries.

Converse of Parallel Lines Theorem

If a transversal intersects two lines and if

- the alternate angles are equal or
- the corresponding angles are equal or
- the interior angles on the same side of the transversal are supplementary

then the lines are parallel.

As before, once you introduce new vocabulary and develop more theorems, you can apply them to the proof of other deductions, as shown in the following example.

Example 2   In $\triangle ABC$, $CA = CB$ and BC is produced to D. If exterior $\angle ACD$ is bisected by CE, prove that $CE \parallel BA$.

Translate the given information and construct a diagram to plan the solution.

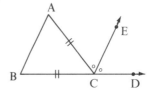

Solution   Given: In $\triangle ABC$, $CA = CB$,
$\qquad\qquad \angle ACE = \angle ECD$.

Required to prove: $BA \parallel CE$

	Reasons
Proof: In $\triangle ABC$, $AC = BC$	given
Then $\angle CAB = \angle CBA$	ITT
$\angle ACD = \angle ACE + \angle ECD$	
$= 2\angle ACE$ ①	construction
In $\triangle ABC$, $\angle ACD = \angle CAB + \angle CBA$	ASTT corollary
$= 2\angle CAB$ ②	$\angle CAB = \angle ABC$
From ① and ②	
$2\angle ACE = 2\angle CAB$	
$\angle ACE = \angle CAB$	
Since AC is a transversal for AB and	
CE and $\angle ACE = \angle CAB$,	proven
then $BA \parallel CE$.	PLT, alternate angles

Throughout the exercise, use the new theorems to prove further deductions.

6.8 Exercise

A 1   In the diagrams, which lines may be proven parallel?

(a) 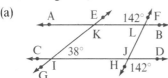   (b)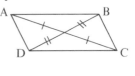

2   Prove that if AB ∥ CD and CD ∥ PQ, then AB ∥ PQ.

3   If ∠PQA = ∠SRD, prove AB ∥ CD.

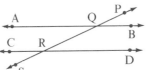

4   Use the diagram to prove that if two line segments PQ and RS bisect each other then PSQR is a parallelogram.

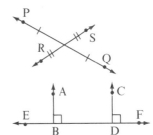

5   Use the indirect method of proof to prove that AB ∥ CD.

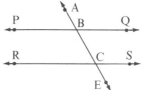

6   Prove the corollaries.

  (a) If a transversal intersects two lines and the corresponding angles are equal then the lines are parallel.

  (b) If a transversal intersects two lines and the interior angles on the same side of the transversal are supplementary, then the lines are parallel.

B   For the following questions, try to use, where possible, the method of indirect proof to prove the deductions.

7   Prove that if ∠QBC = ∠RCB then PQ ∥ RS.

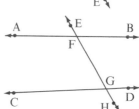

8   In the diagram, ∠BFG + ∠FGD ≠ 180°. Prove that AB ∦ CD.

9 (a) Prove that the equal angles of an isosceles triangle are acute.

   (b) $\triangle$PQR is a scalene triangle and PS $\perp$ QR where S is in QR. Prove PS is not a median.

10 In $\triangle$PQR, S is the midpoint of PR and T is the midpoint of QR. Prove PT and QS cannot bisect each other.

11 Prove that two lines perpendicular to the same line cannot intersect.

12 A quadrilateral PQRS is given so that $\angle$SRQ = $\angle$RSP. If RQ = SP prove RS $\parallel$ QP.

13 Prove that in any quadrilateral, if a pair of opposite sides are equal and parallel then the quadrilateral is a parallelogram.

14 (a) If the diagonals of a quadrilateral bisect each other, prove that the quadrilateral is a parallelogram.

   (b) If the diagonals of a quadrilateral are right bisectors of each other prove that the quadrilateral is a rhombus.

   (c) If the diagonals of a parallelogram bisect the angles, prove it is a rhombus.

C 15 $\triangle$TUS is constructed so that TL = LV and SK = KW. If K and L are midpoints of UT and SU respectively, prove points V, U and W are collinear.

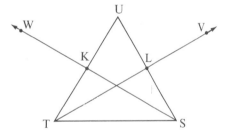

Computer Tip

The square root function is useful for solving problems about the horizon distance. For example,

The distance d, in metres, to the horizon from a building with height h, in metres, above sea level is given by the following formula.

$$d = \sqrt{2rh} \qquad r \text{ is the radius of the earth, in metres}$$

▶ How far is the horizon from a building, 95.85 m high? The radius of the earth is 12 759 km. (You will need a telescope to see the actual horizon).

▶ Write a computer program to find the distance d.

6.9 Inequalities for Triangles

Often, everyday problems may be solved using the properties of triangles.

For example, if you have a garden in the shape of $\triangle ABC$ and $\angle A < \angle B$ then you would need more fence for side AC than for side BC.

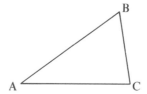

Triangle Inequality

The angle opposite the longest side of a triangle is the greatest angle; the angle opposite the shortest side of a triangle is the smallest angle.

Thus,
if AC > AB then $\angle B > \angle C$.

Given: $\triangle ABC$, AC > AB

Required to prove: $\angle ABC > \angle ACB$

Proof: On AC locate Q such that AQ = AB.

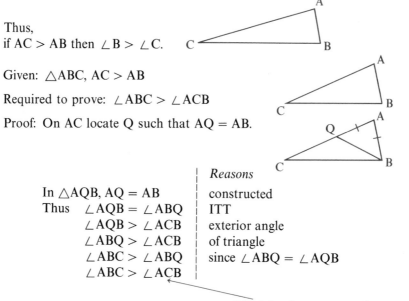

	Reasons
In $\triangle AQB$, AQ = AB	constructed
Thus $\angle AQB = \angle ABQ$	ITT
$\angle AQB > \angle ACB$	exterior angle
$\angle ABQ > \angle ACB$	of triangle
$\angle ABC > \angle ABQ$	since $\angle ABQ = \angle AQB$
$\angle ABC > \angle ACB$	

Which is what you wanted to prove.

You may ask whether the converse statement is true.

Converse of Triangle Inequality

The side opposite the larger angle of a triangle is greater in measure than the side opposite the smaller angle.

You can use the method of indirect proof to prove the converse of the *Triangle Inequality Theorem*.

6.9 Exercise

A 1 For △PQR, ∠Q > ∠P.
Prove that RP > QR.

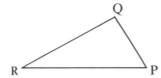

2 To prove the converse of the triangle inequality use the following diagram.

Given: △ABC, ∠C > ∠B.

Required to prove: AB > AC

Proof: One of AB < AC, AB = AC,
or AB > AC must be true.

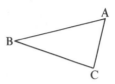

Use the method of indirect proof and show that
(a) the assumption AB = AC gives a contradiction.
(b) the assumption AB < AC gives a contradiction.

3 To prove that the sum of any two sides of a triangle is greater than the third, complete the following proof by writing the authorities (reasons) for each step.

Given: △PQR where PM bisects ∠QPR

Required to prove: PQ + PR > QR

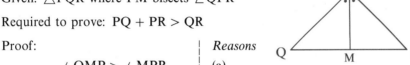

Proof:

	Reasons
∠QMP > ∠MPR	(a)
∠MPR = ∠MPQ	(b)
∠QMP > ∠MPQ	(c)
PQ > QM	(d)
∠PMR > ∠QPM	(e)
∠QPM = ∠MPR	(f)
∠PMR > ∠MPR	(g)
PR > MR	(h)
PQ + PR > QM + MR	(i)
QM + MR = QR	(j)
PQ + PR > QR	(k)

B Use the *Triangle Inequality Theorem* and its converse to prove the following deductions.

4 △ABC is a scalene triangle. D is any point in AC. Prove
AB + BC > AC.

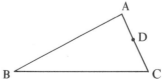

5 In quadrilateral OAFL prove that LA + OF > OL + AF.

6 Prove that the sum of any 3 sides of a quadrilateral is greater than the fourth side.

7 Given △EFG with EF > EG. If the bisectors of ∠F and ∠G intersect in T, then prove FT > TG.

8 Given △XYZ with YV the bisector of ∠Y. Prove that if XY > XV, then YZ > VZ.

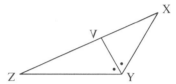

9 △XYZ is right-angled at Y. Prove that ∠X and ∠Z are acute.

10 ATEK is a quadrilateral in which ∠KTE = ∠E. Prove that TA + AK > KE.

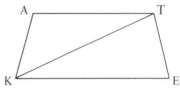

11 △PMN is isosceles with PM = PN. MN is produced to Q. Prove that PQ > PN.

C 12 Use the indirect method of proof to show that PS is the shortest segment to QR.

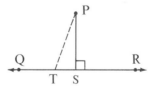

Problem Solving

To solve some problems, you can obtain an insight into the solution if you draw a sketch of the given information. Solve the following problem.

Points P(1, 2) and Q(−F, −2) are given and R is a point on the line $x = -1$. Find the co-ordinates of point R if PR + RQ is a minimum.

Try to remember this strategy throughout the study of the mathematics of this chapter.

6.10 Translations and Their Properties

Points on a plane are related by a correspondence as shown.

$A \rightarrow A'$ $B \rightarrow B'$ $C \rightarrow C'$
and so on

The correspondence $A \rightarrow A'$, $B \rightarrow B'$, $C \rightarrow C'$, and so on is a translation if and only if

$$AA' = BB' = CC' \quad \text{and} \quad AA' \parallel BB' \parallel CC'$$

The translation is shown by the notation

$$[A, B, C, \ldots] \rightarrow [A', B', C', \ldots]$$

pre-image   A is mapped onto A'.   translation image

As in plane geometry, you can use the properties of translations to prove deductions. To show the properties of translations, use the results you proved earlier in plane geometry.

Example 1   Show that if $[A', B']$ is the translation image of $[A, B]$ then $AB = A'B'$.

Solution   Given: $[A', B']$ is the translation image of $[A, B]$.

Required to prove: $AB = A'B'$

Proof: Join AB and A'B'.	*Reasons*
Since $[A, B] \rightarrow [A', B']$ then $AA' = BB'$ $AA' \parallel BB'$	property of translation
In quadrilateral AA'B'B $AA' = BB'$ $AA' \parallel BB'$ Then AA'B'B is a parallelogram. Thus $AB = A'B'$.	property of a parallelogram

From the above result, since a translation is a transformation that preserves length, it is referred to as a **congruence transformation** or **isometry**.

An **isometry** is a transformation that preserves length.

Example 2   If △A′B′C′ is the translation image of △ABC, then prove that
△ABC ≅ △A′B′C′.

Solution   Given: △A′B′C′ is the translation
image of △ABC.

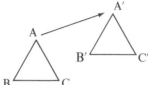

Required to prove: △ABC ≅ △A′B′C′

Proof: Since [A′, B′, C′] is the translation   *Reasons*
image of [A, B, C] then
[A, B] → [A′, B′]   property of translation
and AB = A′B′
[B, C] → [B′, C′]   property of translation
and BC = B′C′
[A, C] → [A′, C′]   property of translation
and AC = A′C′
In △ABC and △A′B′C′
AB = A′B′   proved S
BC = B′C′   proved S
AC = A′C′   proved S
Thus △ABC ≅ △A′B′C′   SSS

From Example 2, you observe the following properties of translations.

- The vertices of △ABC and △A′B′C′ have the same sense, namely counter-clockwise.
- AA′B′B is a parallelogram.

You may extend the results of Example 2 and prove these properties.

Translation Properties Theorem (TPT)

If [A′, B′, C′, . . .] is the translation image of [A, B, C, . . .] then:

- length is preserved.   • area is preserved
- slopes of lines are preserved   • the measures of angles are preserved
- the sense of the vertices of the congruent figures is preserved

---

A **direct isometry** is a translation in which lengths are preserved
and also the sense of the vertices are preserved.

You may now use the *Translation Properties Theorem* (*TPT*) for proving
deductions.

6.10 Exercise

A 1 On a grid, directed line segments represent translations. Which directed line segments represent the same translation?

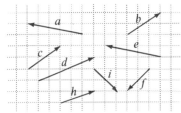
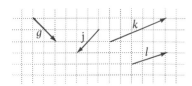

2 A directed line segment is shown. Find the translation image for each of the following.

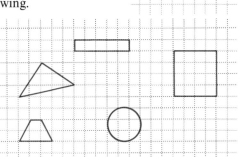

3 For the diagram,
PP′ ∥ RR′ ∥ QQ′ and
PP′ = RR′ = QQ′.
Prove △PQR ≅ △P′Q′R′.

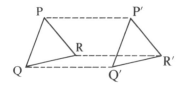

4 For the diagram, [P, Q, N, R] is the translation image of [A, B, M, C]
(a) Prove AM = PN and AM ∥ PN.
(b) Write the result in (a) as a general statement.

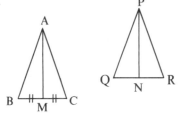

5 A translation is given by [P, Q, R] → [P′, Q′, R′]. Prove that
∠PQR = ∠P′Q′R′.

6 The translation image of [S, T, U, V] is given by [S′, T′, U′, V′]. Prove that STUV = S′T′U′V′.

7 Complete the proof of the following property. If A′ is the translation image of A, then the translation of any line through A is parallel to the translation image of the corresponding line through A′.

Given: [A′, B′] is the translation image of [A, B]. (Line s′ is the translation image of the line s.)

Required to prove: AB ∥ A′B′ (i.e. s ∥ s′).

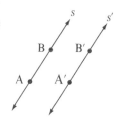

B 8 In the diagram, P′ is the translation image of P. l is a line parallel to PP′ as shown.

(a) Prove that for the same translation the translation image of the line l is itself.

(b) Write a general statement for the above result in (a).

9 If [S, T, U] is the translation image of [A, B, C], prove that

(a) the midpoint of AB is mapped onto the midpoint of ST.

(b) the perpendicular from A to BC is mapped onto the perpendicular from S to TU.

10 Prove that the translation image of a square is congruent to the pre-image.

11 (a) Prove that the translation image of a triangle is a triangle.

(b) Prove that the translation image of a straight line is a straight line.

12 The circle with centre P is translated. P → P′. Prove that the circle with centre P is congruent to the circle with centre P′.

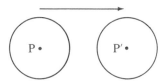

C 13 A mapping, in general, is defined for the Cartesian plane in the form

$$(x, y) \rightarrow (x + a, y + b) \qquad a, b \text{ real numbers.}$$

(a) Find the image of the given figure for each mapping.

Figure	Mapping
(i) A(2, 3), B(−1, 2)	$(x, y) \rightarrow (x − 3, y + 2)$
(ii) P(2, 2), Q(−3, −2), R(3, −3)	$(x, y) \rightarrow (x + 2, y − 4)$

(b) Show that the mappings in (a) define translations.

6.11 Reflections and Their Properties

Another transformation, called a **reflection**, is defined for points on a plane and involves a mirror line, *m*.

For the points shown, the mapping

$$A \to A'$$
$$B \to B'$$
$$C \to C'$$

is a reflection in the line *m*, if and only if the line *m* is the right bisector of AA', BB', and CC'.

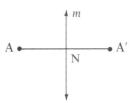

Line *m* is called the **mirror line** or **reflection line**. Thus from the definition

- if A and A' are corresponding points of a reflection in the mirror line *m*, then the mirror line *m* is the right bisector of AA'.

- if A and A' are two points such that a line *m* is the right bisector of AA' then A and A' are corresponding points of a reflection in the mirror line *m*.

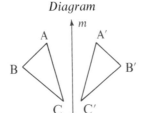

AN = NA'
AA' ⊥ mirror line *m*

The notation used to show a reflection is the same as that used to show a translation.

A reflection of △ABC is shown on the grid. △ABC is mapped onto △A'B'C'.

Diagram

Notation

Write
[A, B, C] → [A', B' C']

Called the reflection image of [A, B, C].

Just as your skills in plane geometry were used to develop the properties of a translation, so will the properties of reflections be developed.

Example [A', B'] is the reflection image of [A, B] in the mirror line *m*. Prove that AB = A'B'.

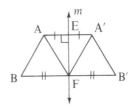

Solution   Given: [A, B] and its reflection image [A', B'].

Required to prove: AB = A'B'

Proof: Draw the lines as shown.

	Reasons
In △AEF and △A'EF, AE = A'E	definition of reflection
∠AEF = ∠A'EF	90°
EF = EF	common side
△AEF ≅ △A'EF	SAS
AF = A'F	congruent triangles
∠EFA = ∠EFA'	congruent triangles
Thus, ∠AFB = ∠A'FB'	CAT
In △ABF and △A'B'F, AF = A'F	proved
∠AFB = ∠A'FB'	proved
BF = B'F	definition of reflection
△ABF ≅ △A'B'F	SAS
Thus, AB = A'B'	Which is what you wanted to prove.

In the previous example, you proved that, for a reflection, since *length is preserved* then it is an isometry. By using your theorems of plane geometry, you can prove other properties of reflections. For example, you can use the result:

If [A, B] → [A', B'] then AB = A'B'.

to deduce the result:

if [A, B, C] → [A', B', C'] then △ABC ≅ △A'B'C'.

Properties of Reflections

For a reflection in a mirror line *m*

- if [A', B'] is the reflection image of [A, B] the mirror line *m* is the perpendicular bisector of AA' and BB'.

- length is invariant.

- measures of angles are invariant.

- sense of figures is not preserved.

- if *l* ∥ *p* then *l'* ∥ *p'*.

- if *l'* is the reflection image of *l*, then the mirror line bisects the angle between *l* and *l'*, *l*⫲*l'*

- any point, P, on the mirror line is its own reflection image P'. The point P is said to be an **invariant point** of the reflection.

6.11 Exercise

A 1 Use squared paper to find the reflection image of each shape in the mirror line *m*.

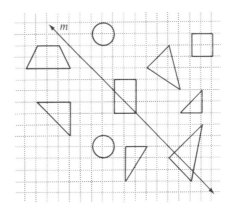

2 In the diagram, [A′, C′] is the reflection image of [A, C]. B is any point on AC. Use the diagram to prove that the reflection image of a straight line is a straight line.

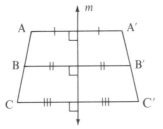

3 (a) Use the definition of reflection to prove that if ∠A′B′C′ is the reflection image of ∠ABC then ∠A′B′C′ = ∠ABC.

(b) Write a general statement for your result in (a).

4 (a) Prove that if [A′, B′, C′] is the reflection image of [A, B, C] then △ABC ≅ △A′B′C′.

(b) Write a general statement for your result in (a).

5 In the diagram, OF is the bisector of ∠GOH, the angle between lines *l* and *k*. Use the diagram to prove that if two lines intersect they are the reflection images of each other in the bisector of the angle between them.

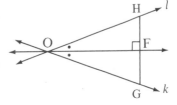

B 6 For a reflection, prove that

(a) length is invariant.   (b) the measure of an angle is invariant.

7 If two lines *l* and *p* are parallel and their reflection images are *l′* and *p′*, prove that *l′* ∥ *p′*.

8 If *l′* is the reflection image of *l*, then prove that the mirror line bisects the angle between *l* and *l′*. (Note *l* ∦ *l′*.)

9   Prove that a straight line, perpendicular to the mirror line of a reflection, is its own image.

10 Prove that if two circles are reflection images of each other then
    (a) they are congruent.
    (b) their centres are reflection images of each other.

11 Prove that two congruent circles are the reflection images of each other.

12 (a) For the diagram, the two circles are reflection images of each other. Prove that the mirror line is the right bisector of the segment PP'.
    (b) Write a general statement for your result in (a).

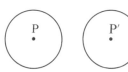

13 (a) Two congruent circles intersect in the common chord RS. Prove that the two circles are reflection images of each other in the line through RS.
    (b) Write a general statement for your result in (a).

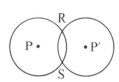

14 Prove that a circle is its own reflection image with the diameter as the mirror line.

15 (a) Show that the mapping $(x, y) \to (x, -y)$ is a reflection in the x-axis.
    (b) Show that the mapping $(x, y) \to (-x, y)$ is a reflection in the y-axis.

Problem-Solving

In the diagram P' is the image of P under reflection in the line $y = x$ on the plane.
• How are the co-ordinates of P and P' related for P, P' on the Cartesian plane?
• Prove that the graph of $x^2 + y^2 = 100$ is symmetrical with respect to the line $y = x$.

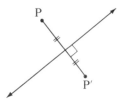

6.12 Rotations and Their Properties

A rotation is another correspondence between points on a plane. To specify a rotation you need to know

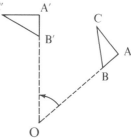

- **the rotation centre** or **turn centre**.
  For △ABC → △A'B'C', the rotation centre is O.

- the **rotation angle**.
  ∠ BOB' is the rotation angle.

For a rotation, △ABC → △A'B'C'. Then, in particular,

$$OB = OB', \qquad OA = OA', \qquad OC = OC',$$
$$\angle BOB' = \angle AOA' = \angle COC'.$$

In particular, a half-turn is defined by

$$\angle BOB' = \angle AOA' = \angle COC' = 180°.$$

In the exercise, some properties of half-turns will be explored.

Properties of Rotations

- A rotation is an isometry (length is invariant).
- The measures of angles are invariant.
- The sense is preserved.
- The centre of rotation lies on the perpendicular bisector of the segment joining a point and its image.

6.12 Exercise

A 1 Use squared paper to find the rotation image of each of the following for $\frac{1}{4}$-turn, $\frac{1}{2}$-turn about the turn centre P.

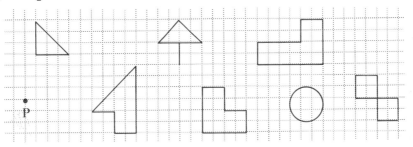

B 2   [A′, B′] is the rotation image of [A, B]. Prove that AB = A′B′.

3   [A′, B′, C′] is the rotation image of [A, B, C]. Show that
(a) △ABC ≅ △A′B′C′.   (b) the sense of △ABC is preserved.

4   For a rotation with turn centre O, A → A′. Prove that O lies on the right bisector of AA′.

5   [A′, B′] is the rotation image of [A, B] for a half-turn. Prove that AB ∥ A′B′.

6   Prove that the rotation image of a straight line is a straight line.

7   The line *l* passes through P. If P is the rotation centre for a half-turn, prove that the reflection image of *l* is itself.

8   Two parallel lines *p* and *q* are shown. T is a point half-way between the lines. Prove that, for a half-turn, *p* and *q* are rotation images of each other.

9   (a) The centre of a circle is O. RS is a diameter. Use the diagram to show that, for a half-turn, R and S are images of each other about O.

(b) Write a general statement about your result above.

10   (a) The circles with centres P and P′ are congruent. For a half-turn, prove that the circles are rotation images of each other about the midpoint of PP′.

(b) Write a general statement about your result in (a).

Math Tip

It is important to clearly understand the vocabulary of mathematics when solving problems. *You cannot speak the language of mathematics if you don't know the vocabulary.*

- Make a list of all the words you review and learn in this chapter.
- Provide a simple example to illustrate the meaning of each word.

6.13 Problem-Solving: Making Conjectures

In proving a deduction, you progress from one part of the solution to another, only after you have justified each step. Thus, if you are unable to justify a step, then very likely, the result in that step is not true. However, very often, after looking at a diagram, a conjecture can be made that may turn out to be true. Thus, you set out to prove that your conjecture is either true or not true. This process has resulted in the formation of many new ideas.

Look at each diagram shown on this page.

Step 1: Write a conjecture about the diagram, based on the information given.

Step 2: Prove your conjecture in Step 1.

Step 3: If you are not able to prove your conjecture, modify the diagram, so that the adjustment may allow you to complete the proof.

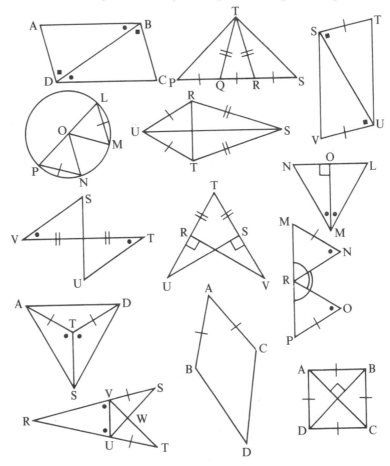

Practice and Problems: A Chapter Review

1 For each diagram, what fact is required so that the pairs of triangles are congruent?

(a)

(b)

2 Use the diagram to prove that PT = QT.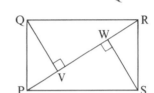

3 Use the diagram. If PQRS is a parallelogram, prove that △WSR ≅ △VQP.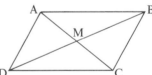

4 Use the headings to analyze each of the following deductions.

I can prove		If I can prove

Then write a proof for each deduction.

(a) For quadrilateral ABCD
   AD = CB   DC = BA.
   Prove that ∠B = ∠D.

(b) In quadrilateral RSTU
   ∠SRT = ∠UTR and ∠URT = ∠STR. Prove that ∠S = ∠U.

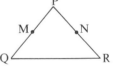

5 Use the method of indirect proof to prove the following deduction.

For △PQR, PQ = PR.
If MQ = NR prove that MN‖QR.

6 In △ABC, V and W are points in AB and AC respectively. If VW‖BC and AB = AC prove that △AVW is isosceles.

7 ABC is an isosceles triangle with AB = AC and BA is produced to D forming exterior ∠CAD. AP‖BC. Prove ∠DAP = ∠PAC.

8 Prove that the translation image of a triangle is congruent to the pre-image.

Test for Practice

1 Write 6 facts that are true if $\triangle PQS \cong \triangle ABC$.

2 In the diagram,
$\angle S = \angle Q$ and
$SP = QP$. Prove
that $QR = SR$.

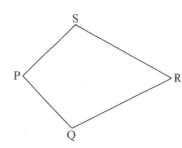

3 If $AB \parallel DC$
and $DA \parallel CB$,
prove that
$\angle A = \angle C$.

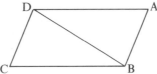

4 For the following deduction:
   • Analyze the steps of the solution using the headings

I can prove		If I can prove

   • Then write a complete solution

   Prove that the diagonals of a square are of equal length.

5 In $\triangle PQR$, QR is produced to T and RQ is produced to S so that $\angle PRT = \angle PQS$. Prove that $PQ = PR$.

6 If, in a triangle, the altitude from a vertex bisects the angle at the vertex, then prove that the triangle is isosceles.

7 $\triangle AXY$ is isosceles with $\angle X = \angle Y$. XY is produced to C and YX is produced to B so that $XB = YC$. Prove $\angle ABC = \angle ACB$.

8 The diagonals of quadrilateral PQRS intersect on N. Prove that $PQ + QR + RS + SP > QS + PR$.

9 Use transformations
to prove that $PQ = SR$.
O is the centre
of the circle.

10 Two circles with centres P and P′ are congruent. If P′ is the translation image of P, prove that the circle with centre P′ is the translation image of the circle with centre P.

Maintaining Skills

To do mathematics well you need to practise, regularly, skills in mathematics. You should review, not only skills you develop in each chapter, but also skills you need for developing more mathematics. You may wish to refer to this Maintaining Skills more than once to review these skills. Remember: try the review of skills in the Inventory of Essential Skills.

1   Calculate each of the following.
   (a) cos 30°   (b) csc 60°   (c) sec 45°   (d) sec 60°   (e) cot 60°
   (f) tan 30°   (g) sin 45° + cos 45° + cos 45° sin 45°

2   Calculate.
   (a) cos 225°   (b) sec (−30°)   (c) cot 330°
   (d) cos (−300°)   (e) csc 240°   (f) tan (−225°)

3   Calculate.
   (a) $\sec \dfrac{\pi}{6}$ (b) $\sin \dfrac{\pi}{4}$ (c) $\cot \dfrac{\pi}{3}$ (d) $\cot \dfrac{\pi}{4}$ (e) $\tan \dfrac{\pi}{6}$ (f) $\cos \dfrac{\pi}{3}$

4   Calculate.
   (a) $\csc \dfrac{3}{4}\pi$ (b) $\cot \dfrac{2}{3}\pi$ (c) $\cos \dfrac{5}{6}\pi$

   (d) $\tan \left(-\dfrac{\pi}{6}\right)$ (e) $\cos \left(-\dfrac{3}{4}\pi\right)$ (f) $\sec \left(-\dfrac{4}{3}\pi\right)$

5   Calculate.
   (a) cos 90°   (b) csc 90°   (c) cos 360°
   (d) sin (−90°)   (e) sec 270°   (f) tan (−180°)

6   Calculate.
   (a) $\cos \dfrac{\pi}{2}$ (b) $\tan \dfrac{3}{2}\pi$ (c) $\csc (-\pi)$

   (d) $\sec \left(-\dfrac{3}{2}\pi\right)$ (e) $\csc \dfrac{3}{2}\pi$ (f) $\tan \left(-\dfrac{3}{2}\pi\right)$

7   Solve each of the following triangles for the given information.
   (a) In △PQR ∠R = 90°, PQ = 14.2, ∠PQR = 29°.
   (b) In △ABC, AB = 15.1, CB = 34.9, ∠A = 90°.
   (c) In △ABC, ∠A = 29°, ∠C = 79°, and c = 96.3.
   (d) In △PQR, ∠P = 48°, ∠Q = 27°, r = 125.
   (e) In △ABC, b = 3, a = 5, ∠C = 77°.
   (f) In △PQR, r = 6, q = 3, p = 5.   (g) In △JKF, j = 4.6, k = 6.2, f = 5.6.

7 Similarity: Concepts and Skills

definition of similarity, similar triangles, similarity assumptions, SSS~, SAS~, proving deductions, applying skills with similarity, ratios of areas, analyzing diagrams, solving problems, strategies, problem-solving

Introduction

The process of solving deductions follows a set number of steps. For example, as you study mathematics, you acquire a variety of strategies and methods of exploring mathematics. Throughout the work, you apply the concepts, skills and strategies to solve problems. In particular, in geometry, you prove deductions. However, this process is based on a foundation which is illustrated in the following diagram.

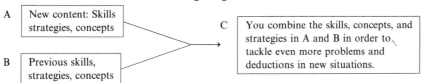

A | New content: Skills strategies, concepts

B | Previous skills, strategies, concepts

C | You combine the skills, concepts, and strategies in A and B in order to tackle even more problems and deductions in new situations.

In particular, the strategies, skills and concepts you learned about in geometry, in the previous chapter, are now combined with the concepts and skills of similarity in this chapter.

Since similar figures occur, readily, in many forms, you already have an idea about the skills and concepts of similarity. How are the following alike? How are they different?

7.1 Exploring Similar Figures

When a figure is enlarged, or reduced, in size, similar figures are obtained.

Polygons ABCD and PQRS are said to be similar if

- the corresponding sides are proportional.

$$\frac{AB}{PQ} = \frac{BC}{QR} = \frac{CD}{RS} = \frac{AD}{PS}$$

- corresponding angles are equal.

$$\angle A = \angle P, \quad \angle B = \angle Q, \quad \angle C = \angle R, \quad \angle D = \angle S$$

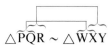

For quadrilaterals to be similar **both** conditions must be satisfied, as shown by the following.

▶ For quadrilaterals ABCD and PQRS, the ratios of corresponding sides are equal. But the quadrilaterals are not similar. For similarity, corresponding angles must also be equal.

▶ For quadrilateral, EFGH and PQRS, corresponding angles are equal. But the quadrilaterals are not similar. For similarity, corresponding sides also must be proportional.

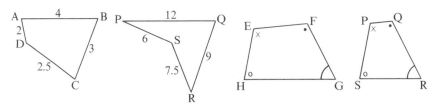

The symbol ~ is used to relate similar figures and means "is similar to." Thus, △ABC ~ △DEF means △ABC is similar to △DEF. The similarity relationship shows which angles and which sides correspond. Thus, for the similarity relationship,

These angles are related.

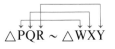

△PQR ~ △WXY

$\angle P = \angle W, \angle Q = \angle X, \angle R = \angle Y.$

Corresponding sides are related.

△P̂Q̂R ~ △ŴX̂Y

$$\frac{PQ}{WX} = \frac{QR}{XY} = \frac{PR}{WY}$$

Example Quadrilateral DEFG and quadrilateral KLMN are similar. Find the value of each variable.

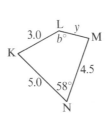

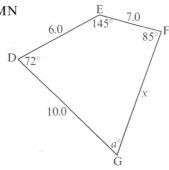

Solution Quadrilateral DEFG ~ quadrilateral KLMN.
Thus, $\angle D = \angle K$, $\angle E = \angle L$, $\angle F = \angle M$, $\angle G = \angle N$.
Since $\angle E = 145°$ then $\angle L = b° = 145°$.
Since $\angle N = 58°$ then $\angle G = a° = 58°$.

Also $\dfrac{DE}{KL} = \dfrac{EF}{LM} = \dfrac{FG}{MN} = \dfrac{DG}{KN}$

But, $\dfrac{DE}{KL} = \dfrac{6.0}{3.0} = \dfrac{2}{1}$ then $\dfrac{EF}{LM} = \dfrac{7.0}{y}$ and $\dfrac{FG}{MN} = \dfrac{x}{4.5}$

$\dfrac{7.0}{y} = \dfrac{2}{1}$ $\dfrac{x}{4.5} = \dfrac{2}{1}$

$y = 3.5$ $x = 9.0$

Thus, $x = 9.0$, $y = 3.5$, $a° = 58°$, and $b° = 145°$.

7.1 Exercise

A 1 For each of the following similarity relationships, name the corresponding angles and the corresponding sides. Write a proportion which relates corresponding sides.
(a) $\triangle JKL \sim \triangle STU$ (b) $\triangle ACE \sim \triangle BDF$
(c) quadrilateral GHJK ~ quadrilateral MNPQ

2 Which of the following pairs of figures is similar? Write a similarity relationship for each pair of similar figures.
(a)   (b)

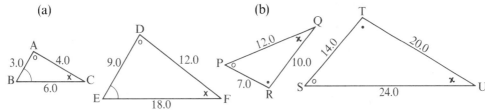

(c) 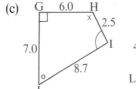   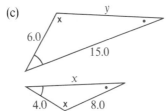 (d)

B 3 Each pair of polygons is similar. Find the missing values.

(a)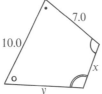

(b)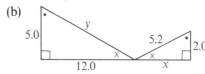

(c)

(d)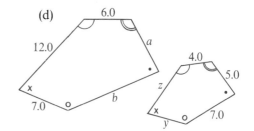

4 Use the diagram at right.
   (a) Name two triangles which are similar.
   (b) Write the similarity relationship.
   (c) Find the values of the missing variables.

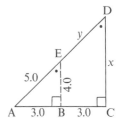

5 $\triangle ABC \sim \triangle DEF$, BC = 5.0, AC = 3.5, DE = 8.0, and EF = 10.0. Find AB and DF.

6 Quad (ABCD) $\sim$ quad (MNPQ), AB = 3.5, BC = 5.0, AD = 6.0, MN = 7.0, and PQ = 9.0. Find DC, NP, and MQ.

7 If figure (ABCDE) $\sim$ figure (MNPST) and $\angle A = 110°$, $\angle N = 90°$, $\angle C = 70°$, $\angle S = 120°$, find $\angle E$.

C 8 If two triangles are similar, prove that corresponding altitudes are proportional to corresponding sides.

7.2 Patterns of Study: Similarity Assumptions

In the study of mathematics, there are often patterns in developing skills and concepts. For example, in your study of congruence of triangles

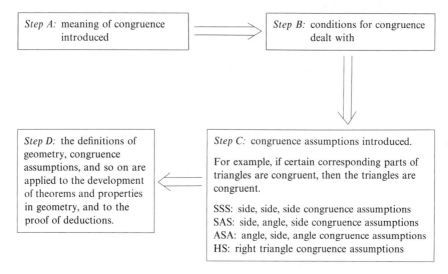

Step A: meaning of congruence introduced

Step B: conditions for congruence dealt with

Step C: congruence assumptions introduced.

For example, if certain corresponding parts of triangles are congruent, then the triangles are congruent.

SSS: side, side, side congruence assumptions
SAS: side, angle, side congruence assumptions
ASA: angle, side, angle congruence assumptions
HS: right triangle congruence assumptions

Step D: the definitions of geometry, congruence assumptions, and so on are applied to the development of theorems and properties in geometry, and to the proof of deductions.

In your study of similarity, a pattern will emerge which is the same as the pattern which emerged when you developed and applied the concepts and skills about congruence. In the previous section you learned that when triangles have equal corresponding angles and have corresponding sides in proportion, then they are similar (Steps A, B). However, the following set of conditions is sufficient for similarity (Step C).

Angle, Angle, Angle Similarity Assumption AAA ~

If the corresponding angles of two triangles are equal, then the triangles are similar.

For $\triangle ABC$ and $\triangle DEF$
if $\angle A = \angle D$, $\angle B = \angle E$
and $\angle C = \angle F$, then
$\triangle ABC \sim \triangle DEF$.

$$\triangle ABC \sim \triangle DEF$$

Use your skills with similar triangles (Step D), to solve the following example.

Example 1 Find the value of the variable. Justify your answer.

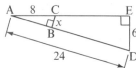

Solution

Statements	Reasons
In $\triangle ABC$ and $\triangle AED$ $\angle A$ is common.	
$\angle ABC = \angle AED$	both 90°
$\angle ACB = \angle ADE$	ASTT corollary
$\therefore \triangle ABC \sim \triangle AED$	(AAA $\sim$)

Since the triangles are similar, then

$$\frac{BC}{ED} = \frac{AC}{AD}$$

$$\frac{x}{6} = \frac{8}{24} \qquad \text{corresponding sides of similar triangles.}$$

$$x = 2$$

Based on the earlier property of a triangle that the sum of the angles of any triangle is 180°, then the following set of conditions is sufficient for similarity of triangles.

Similar Triangles (AA $\sim$)

If two angles of one triangle are equal to two corresponding angles of another triangle then the triangles are similar.

AA $\sim$ is now used to do the following deduction in Example 2.

Example 2 In $\triangle ABC$, D is on AB so that CD $\perp$ AB and E is on AC so that BE $\perp$ AC. If DC and BE intersect at P, prove that DP $\times$ EC = DB $\times$ EP.

Solution

An important first step in proving a deduction is to accurately sketch a diagram for the given information.

Statements	Reasons
In $\triangle PDB$, $\triangle PEC$,	
$\angle DPB = \angle EPC$.	vertically opposite angles
$\angle PDB = \angle PEC$	90° angle
$\triangle PDB \sim \triangle PEC$	AA $\sim$
Thus, $\dfrac{DP}{EP} = \dfrac{DB}{EC}$.	property of similar triangles

Thus, DP $\times$ EC = DB $\times$ EP.

7.2 Exercise

A Throughout A, give reasons for your statements.

(1) Why are the following pairs of triangles similar? Write the corresponding ratios for the sides.

(a)

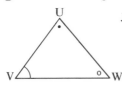

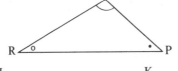

(b)

2 Refer to the diagram.

(a) Show that
△ABE ~ △DCE.

(b) Show that
AB:DC = AE:DE.

3 (a) Which pair of
triangles is similar?

(b) Write corresponding
equal ratios.

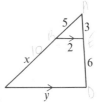

(4) Find the value of the variables.

(a)

(b)

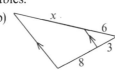

(c)

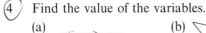

(d)

(e)

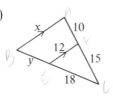

(f)

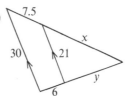

5 If DE ∥ BC, AE = 7.0, EC = 4.0,
DE = 4.0, find BC.

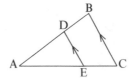

6   If PR = 8.0, RS = 6.2, QR = 10.0,
    $\angle$ PRQ = 90°, $\angle$ PSR = 90°, find SQ.

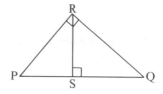

B   Remember: translate the given information, and record it on a diagram.

7   You are given $\triangle$XYZ with $\angle$ X = 90°. Point A is on XZ and point B is on
    YZ so that AB $\perp$ YZ. Prove that XY $\times$ ZB = AB $\times$ XZ.

8   In $\triangle$PQR, F, G, H are the midpoints of PQ, QR, and RP respectively.
    Prove that $\triangle$PQR $\sim$ $\triangle$GHF.

9   In $\triangle$PQR, QX $\perp$ PR, PY $\perp$ QR, where Y and X are points on QR and PR
    respectively. If QX and PY intersect at Z, prove that $\triangle$PZX $\sim$ $\triangle$PRY.

10 In $\triangle$XYZ, the points M and N are on XY and XZ respectively such that
    MN $\parallel$ YZ. Prove
    (a) XM $\times$ ZN = YM $\times$ XN   (b) XM $\times$ YZ = XY $\times$ MN

11 For two similar triangles, prove that the corresponding altitudes are
    proportional to corresponding sides.

12 A clubhouse has a roof whose slant height is 13.0 m and whose base is
    24.0 m. Two extra supports 2.0 m and 3.0 m from the peak need to be
    constructed. Find the length of each support.

13 Gene is 1.7 m above the ground and stands 1.3 m from a puddle. The
    puddle is 20.0 m from a pole in which the light is reflected. The reflection
    property of a puddle gives $\angle$ PDM = $\angle$ BDF. With this information find
    PM, the height of the pole.

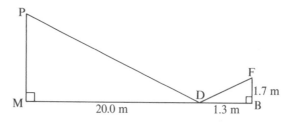

C 14 In parallelogram XYZW, A is a given on XW so that AZ bisects $\angle$ YZW.
    YX and ZA are produced to meet at B. Prove $\dfrac{WA}{AZ} = \dfrac{YB}{BZ}$.

7.3   Similar Triangles, SSS ~

You know that for similar triangles,

  I   corresponding angles are equal
  II corresponding sides are proportional

In the previous section, the similarity assumption for triangles, AAA ~, was applied to develop further properties about similar triangles as well as to prove deductions. The following assumption about similar triangles is applied to develop additional properties.

Side, Side, Side Similarity Assumption (SSS ~)

If the ratios of the corresponding sides of two triangles are equal, then the triangles are similar.

In △ABC and △DEF

if $\dfrac{a}{d} = \dfrac{b}{e} = \dfrac{c}{f}$,

then △ABC ~ △DEF.

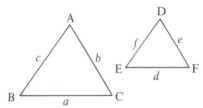

SSS ~ is applied in the following example.

Example   Find the values of a, b, and c.
Give reasons for your answers.

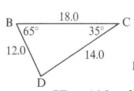 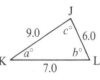

Solution   In △BCD and △JKL

$$\dfrac{BC}{JK} = \dfrac{18.0}{9.0} = \dfrac{2}{1}, \quad \dfrac{BD}{JL} = \dfrac{12.0}{6.0} = \dfrac{2}{1}, \quad \text{and} \quad \dfrac{CD}{KL} = \dfrac{14.0}{7.0} = \dfrac{2}{1}$$

Thus, since $\dfrac{BC}{JK} = \dfrac{BD}{JL} = \dfrac{CD}{KL} = \dfrac{2}{1}$ then

△BCD ~ △JKL	SSS ~
Thus ∠B = ∠J	Properties of
∠C = ∠K	similar triangles
∠D = ∠L	

Thus, since ∠B = 65° then ∠J = 65°
          ∠C = 35° then ∠K = 35°
          ∠D = 180° − (65° + 35°)
             = 80°
     then ∠L = 80° (Property of similar triangles)

7.3 Exercise

A Review the meanings of AAA~, AA~, SSS~.

1 The lengths of sides of triangles are shown in each column. Which corresponding triangles are similar?

	Column I	Column II
(a)	18, 12, 9	30, 20, 15
(b)	12, 8, 6	18, 12, 9
(c)	15, 20, 25	21, 28, 35
(d)	12, 24, 30	14, 28, 35
(e)	15, 20, 25	18, 24, 28

2 Which pairs of triangles are similar? Justify your answers.

(a) (b)

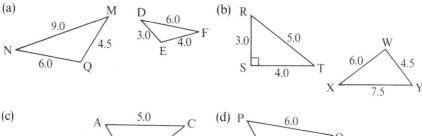

(c) (d)

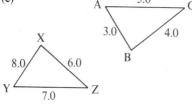
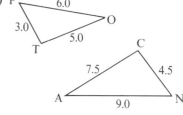

3 Name three pairs of equal angles for each of the following.

(a) (b)

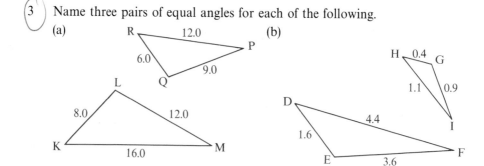

4 Use the diagram at the right to show
   (a) SW ∥ TR.   (b) VW ∥ UR.

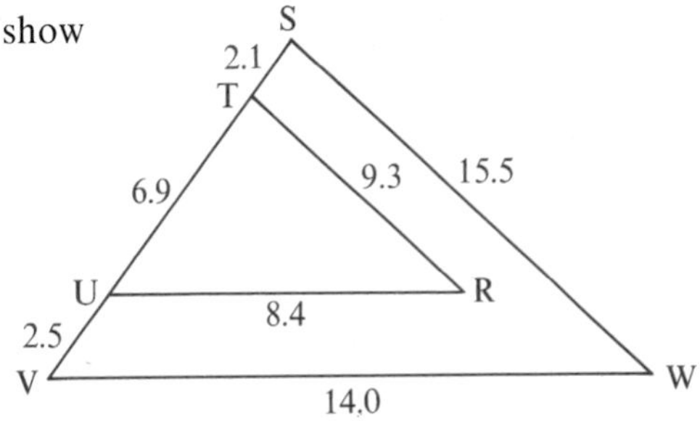

B Remember: sketch a diagram and record the given information on it.

5 △ABC is similar to △DEF and the sides of △ABC are 5.0 cm, 6.0 cm, 7.0 cm. The perimeter of △DEF is 360.0 cm. Find the lengths of the sides of △DEF.

6 △ABC has a perimeter of 45.0 cm. △TEC has sides 20.0 cm, 21.0 cm, and 25.0 cm. If △ABC ∼ △TEC, what are the lengths of the sides of △ABC (to 1 decimal place)?

7 The diagonals of quadrilateral MNOP intersect at S. If

$$MS:SN = OS:PS \quad \text{and} \quad MS:PS = OS:SN$$

prove that the quadrilateral is a parallelogram.

8 In △PQR, A, B, and C are the midpoints of sides PQ, QR, and RP respectively. Prove △PQR ∼ △ABC.

Problem Solving

Often, you are not able to prove something in mathematics because you have perhaps forgotten, or do not clearly understand, the meanings of the words given in the problem. For example, read the following.

> Prove that the sum of the squares of the lengths of the segments from any point P to two opposite vertices of a rectangle is equal to the sum of the squares of the lengths of the segments from P to the remaining vertices.

To solve the above problem you must understand clearly the meaning "sum of the squares", "opposite vertices" and "from any point P". Now solve the problem.

7.4 Similar Triangles, SAS ~

So far, you have used the following to develop properties and to prove deductions about similar triangles.

> AAA ~   If the corresponding angles of two triangles are equal, then the triangles are similar.
>
> AA ~   If two angles of one triangle are equal to the corresponding angles of another triangle, then the triangles are similar.
>
> SSS ~   If the ratios of the corresponding sides of two triangles are equal, then the triangles are similar.

If you compare the development of congruence and similarity for triangles, you will note that there is a counterpart for SAS for similar triangles, SAS ~.

> **Side, Angle, Side Similarity Assumption, SAS ~.**
>
> If two triangles have a pair of equal angles, and the corresponding sides containing the equal angles are proportional, then the triangles are similar.

In $\triangle ABC$ and $\triangle DEF$,

if $\angle A = \angle D$, and $\dfrac{AB}{DE} = \dfrac{AC}{DF}$,

then $\triangle ABC \sim \triangle DEF$.

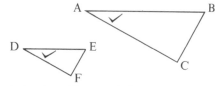

SAS ~ is used to prove the following deduction. The result in the following example is called a theorem and is used later to prove other deductions.

Example   If $\dfrac{AD}{AB} = \dfrac{AE}{AC}$ then $\triangle ABC \sim \triangle ADE$ and $DE \parallel BC$.

Solution

Statements	Reasons
In $\triangle ADE$ and $\triangle ABC$	
$\angle A = \angle A$	common
$\dfrac{AD}{AB} = \dfrac{AE}{AC}$	
Thus, $\triangle ADE \sim \triangle ABC$	SAS ~
Thus, $\angle ADE = \angle ABC$	property of similar triangles

But $\angle ADE$ and $\angle ABC$ are corresponding angles formed by lines DE and BC being cut by transversal AB. Thus DE $\parallel$ BC.

7.4 Exercise

A (1) Which pairs of triangles are similar? Give reasons for your answers.

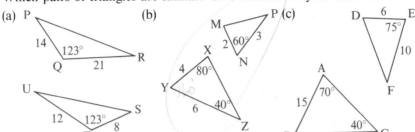

(a) (b) (c)

(2) Find the values of x and y. Give reasons for your answer.

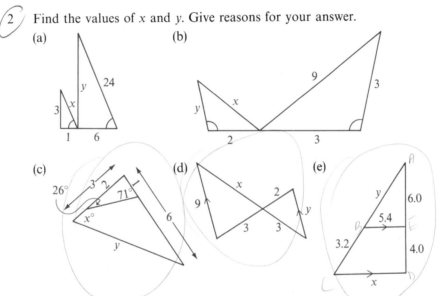

(a) (b) (c) (d) (e)

B Remember to organize your solution: show statements and reasons.

3 For the diagram, show that
YZ:AB = 5:1.

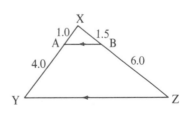

4 For the diagram, $\angle QMN = \angle QNP$
and $\dfrac{QM}{MN} = \dfrac{QN}{NP}$.
Prove that QN bisects $\angle MQP$.

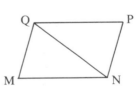

5   If the corresponding sides of two quadrilaterals are in the same ratio and a pair of corresponding angles are equal, prove that the quadrilaterals are similar.

6   In $\triangle ABC$, M and N are the midpoints of AB and AC respectively. Prove
(a) $MN \parallel BC$.   (b) $MN = \dfrac{1}{2}BC$.

7   A point P is chosen on side QT of parallelogram QRST so that PS bisects $\angle RST$. RQ and SP are produced to meet at M. Prove that $PT \times MS = PS \times RM$.

C 8   Quadrilateral EFGH has the property that $\angle F = \angle H$ and $EH:HG = GF:FE$. Prove $EH = FG$.

Problem Solving

Often in solving a problem, we suffer from a "mind set" characterized by thinking of only one way to solve a problem. Sometimes, this is called "tunnel vision."

To solve the following problem, approach it from different points of view.

Start with one of the points. Draw at most 4 lines segments to pass through each and all of the points once and only once. Once you start to draw you cannot lift your pen off the diagram.

Problem Solving

To solve some problems about geometric figures, we need to use our algebraic skills. Solve this problem.

The bisectors of the exterior angles at B and C of $\triangle ABC$ meet at D. Prove that $\angle BDC = 90^\circ - \dfrac{A}{2}$.

7.5 Problem Solving: Alternative Strategies

The more concepts and skills you learn in mathematics, the more alternative strategies you have for proving results. For example, in your earlier work in mathematics, you developed a proof of the Pythagorean Theorem.

Pythagorean Theorem: Right Triangle Property

For a right triangle, the square of the hypotenuse is equal to the sum of the squares of the two other sides.

Now you may use the similar triangle AA~ to prove the Pythagorean Theorem from a different point of view.

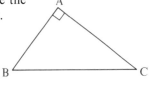

Given: $\triangle ABC$ with $\angle A = 90°$

Required to prove: $BC^2 = AB^2 + AC^2$

Proof: Do the construction. Draw a
perpendicular from A to meet BC at D.

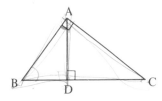

Statements	Reasons
In $\triangle ABC$ and $\triangle DBA$	
$\angle ABC = \angle DBA$	common
$\angle BAC = \angle BDA$	90° angles
$\triangle ABC \sim \triangle DBA$	AA~
Thus, $\dfrac{AB}{DB} = \dfrac{BC}{BA}$	property of similar triangles

Thus, $AB^2 = BC \times DB$ result ①

In $\triangle ABC$ and $\triangle DAC$	
$\angle ACB = \angle DCA$	common
$\angle BAC = \angle ADC$	90° angles
$\triangle ABC \sim \triangle DAC$	AA~
Thus, $\dfrac{AC}{DC} = \dfrac{BC}{AC}$	property of similar triangles

Thus, $AC^2 = BC \times DC$ result ②

Use results ① and ②

$$AB^2 + AC^2 = BC \times DB + BC \times DC$$
$$AB^2 + AC^2 = BC(DB + DC)$$
But, $DB + DC = BC$
Thus $AB^2 + AC^2 = BC \times BC$
$$AB^2 + AC^2 = BC^2$$ Which is what you wanted to prove.

7.5 Exercise

B 1 Find the value of the variables in each of the following to 1 decimal place.

(a)

y

32.0 24.0

x

(b)

4.00

2.25 5.00

y

x

(c)

1.5

0.9

y

2.5

x

2 Find the value of x in each of the following. Round your answer to 1 decimal place if necessary.

(a)

1.0 cm

(b)

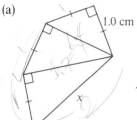

24.0 cm

x cm

7.0 cm

20.0 cm

(c)

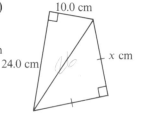

10.0 cm

24.0 cm

x cm

3 AG is a diagonal of the rectangular box shown. Prove that $AG^2 = AB^2 + BC^2 + CG^2$.

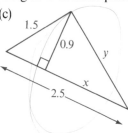

4 (a) Find the length of the diagonal of a rectangular box with dimensions 3 cm × 4 cm × 12 cm.

(b) Find the increase in the length of the diagonal when the dimensions of the box are doubled.

C 5 Use the measurements in the diagram to prove that B, C, and D are collinear.

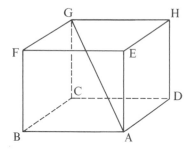

15.00 9.00 11.25

D   12.00   C   6.75   B

7.6 Proving Deductions: Making Decisions

An important skill in problem solving is to know which skills are needed to solve a problem. For example, with similarities, you have used the following

AAA$\sim$	angle-angle-angle similar triangles	To learn mathematics it is helpful to compare your work. For example, note the comparisons to your earlier work with the congruency.
AA$\sim$	angle-angle similar triangles	
SSS$\sim$	side-side-side similar triangles	
SAS$\sim$	side-angle-side similar triangles	

Along with your previous strategies for proving deductions (drawing diagrams, working backwards, using clues, and so on) you need to decide which properties of similar triangles are needed to prove deductions and develop theorems, as shown in the following example.

Example In $\triangle PQR$, $\angle P = 90°$, $PS \perp QR$ and S is a point on QR.

Prove that $\dfrac{QS}{SP} = \dfrac{SP}{SR}$.

Use a diagram to translate the problem.

Solution Given: $\triangle PQR$ with $\angle P = 90°$ and $PS \perp QR$.

Required to prove: $PS^2 = QS \times SR$

Proof:

Statements	Reasons
In $\triangle PSQ$, $\quad \angle PSQ = 90°$	given
Thus, $\angle SPQ + \angle SQP = 90°$	ASTT
In $\triangle QPR$, $\quad \angle QPR = \angle SPQ + \angle SPR = 90°$	
Thus, $\quad\quad\quad \angle SQP = \angle SPR$	angles complementary to the same angle are equal.
In $\triangle PSQ$ and $\triangle RSP$	
$\quad\quad\quad \angle PSQ = \angle RSP$	given
$\quad\quad\quad \angle SQP = \angle SPR$	proved
Thus, $\quad\quad \triangle PSQ \sim \triangle RSP$	AA$\sim$
Thus, $\quad\quad \dfrac{PS}{RS} = \dfrac{SQ}{SP}$	property of similar triangles
Thus, $\quad\quad PS \times SP = SQ \times RS$	
$\quad\quad\quad PS^2 = QS \times SR$	

The previous result is an important one in the study of geometry. It is called the **Mean Proportional Theorem (MPT)**.

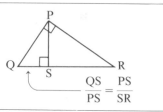

Mean Proportional Theorem

In a right-angled triangle, the altitude to the hypotenuse is the mean proportional between the segments of the hypotenuse.

$$\frac{QS}{PS} = \frac{PS}{SR}$$

You can now combine the similarity theorems with your other theorems in geometry to prove deductions, as well as to develop important theorems.

7.6 Exercise

A Review and list the properties of congruent and similar triangles.

1. Use the *mean proportional theorem* to find the missing variables.

(a)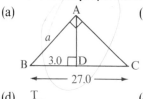

(b)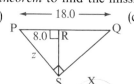

(c)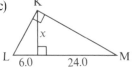

(d)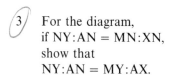

(e)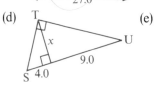

2. Use the diagram. Prove that
(a) $c^2 = ax$.
(b) $b^2 = ay$.

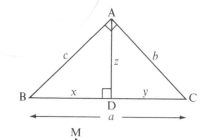

3. For the diagram, if $NY:AN = MN:XN$, show that $NY:AN = MY:AX$.

4. In the diagram, SQ bisects $\angle PQR$ and T is a point in SQ so that $\dfrac{PQ}{SQ} = \dfrac{PT}{SR}$. Show that $\angle RSQ = \angle TPQ$.

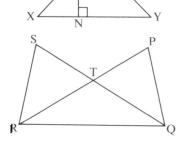

B   Remember: to prove deductions, you must know the answers clearly to the two questions:

I   What am I asked to prove?   II   What am I given?

5   Prove the converse of the mean proportional theorem, which states:

If $\dfrac{BD}{AD} = \dfrac{AD}{DC}$, then AD is the altitude from the right angle A to the hypotenuse, BC.

6   In $\triangle PQR$, S is a point on QR so that $PS \perp QR$. If $QS \times SR = PS^2$, prove that $\angle QPR$ is a right angle.

7   In $\triangle XYZ$, M and N are on XZ and XY respectively so that $\dfrac{XM}{XZ} = \dfrac{XN}{XY}$.

Prove $\dfrac{XM}{XZ} = \dfrac{MN}{ZY}$.

8   Prove that the lengths of corresponding altitudes of similar triangles are proportional to the lengths of the corresponding sides.

9   In $\triangle ABC$, P is a point on AB so that $AC^2 = AB \times AP$. Prove that $\angle ACP = \angle ABC$.

10   In $\triangle PQR$, A and B are points on PQ and PR respectively such that $\dfrac{PB}{PA} = \dfrac{PQ}{PR}$. Prove that $\angle PBA = \angle PQR$.

11   Prove that for any $\triangle ABC$, the triangle formed by vertices that are the midpoints of the sides is similar to $\triangle ABC$.

12   In quadrilateral PQRS, the diagonals intersect at O. If $\dfrac{PO}{OQ} = \dfrac{RO}{OS}$ and $\dfrac{PO}{OS} = \dfrac{RO}{OQ}$, prove that the quadrilateral is a parallelogram.

C 13   Prove that if the corresponding sides of two quadrilaterals are in the same ratio and a pair of corresponding angles are equal, then the quadrilaterals are similar.

14   In quadrilateral XYZW with $\angle Y = \angle W$ and $\dfrac{XW}{ZW} = \dfrac{YZ}{XY}$, prove that $XW = YZ$.

7.7 Applications: Similar Triangles

In the previous sections, you developed skills and concepts of similar triangles. Once you learned the theory, you applied it to the solution of problems and applications. For example, many devices have their designs based on the properties of similar triangles as will be seen in the exercise. Similar triangles can also be used to find distances that cannot be measured directly, as shown in the following example.

Example To calculate the height of a tower, PQ, a diagram is drawn with the following measurements. Find the height of the tower.

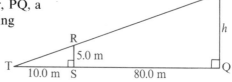

To write an equation to solve for h, you need to know which triangles are similar.

Statements	Reasons
In $\triangle RST$ and $\triangle PQT$	
$\angle RST = \angle PQT$	right angles
$\angle RTS = \angle PTQ$	common angles
Thus, $\triangle RST \sim \triangle PQT$	AA $\sim$
Thus, $\dfrac{RS}{ST} = \dfrac{PQ}{QT}$	property of similar triangles

Solution Let h, in metres, represent the height of the tower.

Thus, $\dfrac{5.0}{10.0} = \dfrac{h}{90.0}$

$10.0h = 450.0$

$h = 45.0$

Based on the above, you can use an equation to find the height of the tower.

Thus, the height of the tower is 45.0 m.

7.7 Exercise

B Throughout the exercise, know why the triangles are similar. Then solve the problem.

1 To determine the length of an irrigation pond, data are collected and recorded on the diagram. Find the length OP in the diagram.

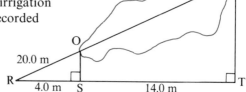

2 A jet climbs 25.0 m for every 100.0 m the plane travels horizontally.
   (a) If the plane is travelling at 250.0 km/h, how high will it be after 12 s?
   (b) How long will it take to reach 5.0 km?
   (c) What assumption did you make in solving (a) and (b)?

3 A pinhole camera works on the principle that light travels in straight lines through the pinhole. The image is inverted on the film in the camera.

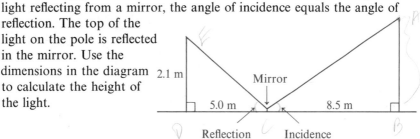

   If the image on the film is 1.8 cm high and the distance from the pinhole to the film is 4.5 cm, calculate the height of the object if the film is 280.0 cm from the object.

4 The following procedure can be used to calculate the height of objects. For light reflecting from a mirror, the angle of incidence equals the angle of reflection. The top of the light on the pole is reflected in the mirror. Use the dimensions in the diagram to calculate the height of the light.

   2.1 m

   Mirror

   5.0 m   8.5 m

   Reflection   Incidence

5 The top of a street light pole is reflected into a puddle 25.0 m away. A person 5.0 m away on the other side of the puddle sees the reflection. If the person's eyes are 2.1 m above street level, how high is the light?

6 In a lumberjack contest two poles are secured for a pole-climbing event. The poles are of different heights. The wires from the top of the poles to the ground are parallel as shown in the diagram. If the smaller pole is 8.0 m with its wire secured 9.2 m from the base, and the guide wire to the taller pole is 16.0 m, find the height of the taller pole.

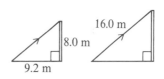

7 Two people attached to hand gliders jump from two different heights. Their descent paths are parallel. If Ian's cliff was 28.0 m high and he lands 52.8 m from his cliff, find the height of the cliff Shelly jumped from if she landed 120.0 m from her cliff.

8 The instrument shown is used to make an exact copy on a larger or smaller scale. The four rods are connected at A, B, C, D to form a parallelogram so that OD = DC and CB = BP. A tracing point is at P and a pencil at C. To obtain a larger copy, the tracing point is at C and a pencil at P. For a smaller copy, the points are reversed.

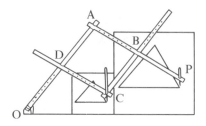

(a) Prove that △DOC ~ △BCP

(b) What ratio gives the magnification of the drawing?

9 The supports AC and DB open up so that the triangles formed are similar. From the picture, prove that

AB ∥ DC.

10 The longest run on the ski slope at Searchmount Valley Ski Resort is 1590.0 m. If the vertical drop corresponding to the run is 210.0 m, how much have you dropped vertically when you have skied 400.0 m down the slope?

In downhill racing, the skiers often reach speeds over 100 km/h.

11 In 600 B.C. the Greek mathematician, Thales, used the following method to find the height of the Great Pyramid. The shadow of the top of the pyramid of Cheops, Son of Snefru, was 280.0 m (to the centre of the pyramid). At the tip of the shadow a stick 2.2 m was placed having a shadow 4.2 m long. Calculate the height of the Great Pyramid to the nearest metre.

7.8 Similar Triangles: Ratios of Areas

The need to know the relationship among areas of similar figures on drawings is important in making decisions when designing buildings. From this point on, $\triangle ABC$, as well as meaning *triangle ABC*, will also mean *the area of triangle ABC*. Particular results such as that shown at the right suggest a generalization about the ratio of the areas of similar triangles, proved in the following example,

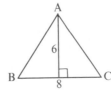

$$\frac{\triangle DEF}{\triangle ABC} = \frac{6}{24} = \frac{1}{4}$$

$$\frac{EF^2}{BC^2} = \frac{4^2}{8^2} = \frac{1}{4}$$

Example 1 Prove that if two triangles are similar, then the ratio of their areas is equal to the ratio of the squares of corresponding sides.

Solution Given: $\triangle ABC \sim \triangle PQR$

Required to prove: $\dfrac{\triangle ABC}{\triangle PQR} = \dfrac{AB^2}{PQ^2} = \dfrac{AC^2}{PR^2} = \dfrac{BC^2}{QR^2}$

Proof:

Statements	Reasons
Draw $AD \perp BC$, $PS \perp QR$.	
$\dfrac{\triangle ABC}{\triangle PQR} = \dfrac{\frac{1}{2}BC \times AD}{\frac{1}{2}QR \times PS} = \dfrac{BC}{QR} \times \dfrac{AD}{PS}$ ①	area of triangle
$\dfrac{AB}{PQ} = \dfrac{AC}{PR} = \dfrac{BC}{QR}$ ②	
In $\triangle ABD$ and $\triangle PQS$	
$\angle B = \angle Q$	$\triangle ABC \sim \triangle PQR$
$\angle ADB = \angle PSQ$	$\triangle ABC \sim \triangle PQR$ $90°$
Thus $\triangle ABD \sim \triangle PQS$	$AA \sim$
Thus $\dfrac{AD}{PS} = \dfrac{AB}{PQ}$ ③	property of similar triangles
From ② $\dfrac{BC}{QR} = \dfrac{AB}{PQ}$ ④	
Use results ③ and ④ in ①.	
$\dfrac{\triangle ABC}{\triangle PQR} = \dfrac{BC}{QR} \times \dfrac{AD}{PS}$	
$= \dfrac{AB}{PQ} \times \dfrac{AB}{PQ}$	from ③
$= \dfrac{AB^2}{PQ^2}$ ← Which is what you wanted to prove.	

Similarly, you can prove that $\dfrac{\triangle ABC}{\triangle PQR} = \dfrac{AC^2}{PR^2}$ and $\dfrac{\triangle ABC}{\triangle PQR} = \dfrac{BC^2}{QR^2}$.

In a similar way, the results for triangles could be extended to the ratios of areas of any pair of similar figures.

> If two polygons are similar, then the ratio of their areas is equal to the ratio of the squares of corresponding sides.

For example, if rectangles ABCD and PQRS are similar then $\dfrac{\text{area ABCD}}{\text{area PQRS}} = \dfrac{AD^2}{PS^2}$.

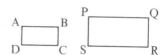

The result in Example 1 is used to prove deductions, such as that shown in the following example.

Example 2 In quadrilateral PQRS, PQ ∥ SR and diagonals QS and RP intersect at M. If $\triangle PQM : \triangle RSM = 16:25$ and PM = 8 cm, find the length of MR.

Solution Given: Quad PQRS, PQ ∥ SR, $\dfrac{\triangle PQM}{\triangle RSM} = \dfrac{16}{25}$, and PM = 8 cm

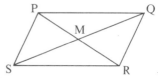

Required to find: the length of MR

Proof:

Statements	Reasons
PQ ∥ SR	given
then, ∠ SRM = ∠ QPM ①	property of
∠ RSM = ∠ PQM ②	parallel lines
In △RMS, △PMQ	
∠ SRM = ∠ QPM.	from ①
∠ MSR = ∠ MQP	from ②
∠ RMS = ∠ PMQ	vertically opposite angles
∴ △PMQ ∼ △RMS	AAA ∼
Thus $\dfrac{\triangle PMQ}{\triangle RMS} = \dfrac{PM^2}{RM^2}$	Ratio of areas of similar triangles
Use $\dfrac{\triangle PQM}{\triangle RSM} = \dfrac{16}{25}$	given
and PM = 8 cm.	
$\dfrac{16}{25} = \dfrac{(8)^2}{RM^2}$	
$RM^2 = 100$	
$RM = 10$	

Thus, the length of MR is 10 cm.

7.8 Exercise

A 1 The ratios of corresponding sides of similar triangles are given. What is the ratio of their areas?

(a) 3:4   (b) 5:2   (c) $3:\dfrac{1}{2}$ (d) $\sqrt{3}:4$ (e) $6:\sqrt{5}$

2 In $\triangle ABC$, $XY \parallel BC$.
If $\triangle AXY = 4.0$ cm^2, find
(a) $\triangle ABC$ (b) quad XYCB

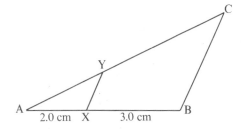

3 For the diagram,
$\triangle AXY = 4.0$ cm^2 and
trapezoid $XYCB = 12.0$ cm^2.
If $AX = 5.0$ cm, find XB.

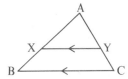

4 If $\triangle XYZ \sim \triangle MNQ$, $XY = 6$, $MN = 5$, find $\triangle XYZ:\triangle MNQ$.

5 Two triangles have the property that $\triangle OXB \sim \triangle TAF$, $OB = 16$, and $TF = 13$. Find $\triangle OXB:\triangle TAF$.

6 If $\triangle ABC \sim \triangle RST$ and $\dfrac{\triangle ABC}{\triangle RST} = \dfrac{16}{25}$, find AB if $RS = 10$.

7 For the diagram, $\dfrac{\triangle MNO}{\triangle OPQ} = \dfrac{9}{25}$
and $MO = 6.0$ cm. Find OP.

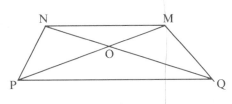

B Remember: Sketch a diagram and then record the given information accurately on your diagram.

8 Prove that if two triangles are similar then the ratio of their areas is equal to the ratio of the squares of the corresponding medians.

9 In $\triangle QRS$, $\angle R > \angle S$. If X is a point on QS so that $\angle QRX = \angle S$, prove that $\dfrac{QR}{QS} = \dfrac{RX}{RS}$.

10 In △MNP, AB ∥ NP, MA = 3.0 cm, AN = 4.0 cm, and △MAB = 18.0 cm². Find
(a) △MNP.   (b) quad (ABPN).

11 Quad ABCD ∼ quad MNPQ such that quad (ABCD):quad (MNPQ) = 16:9. If AB = 8.0 cm, find MN.

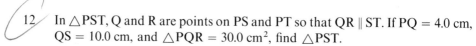

12 In △PST, Q and R are points on PS and PT so that QR ∥ ST. If PQ = 4.0 cm, QS = 10.0 cm, and △PQR = 30.0 cm², find △PST.

13 △ABC ∼ △DEF such that △ABC:△DEF = 16:49. Find the ratio *perimeter △ABC:perimeter △DEF*.

14 In △PQR, two medians PA and RB are drawn to intersect at C. What are the following ratios?
(a) △ABC:△CPR   (b) △ABC:△PQR

C 15 In a rectangle PQRS, PS > PQ. A and B are points in QS so that PB ⊥ QS and RA ⊥ QS, making QB = BA = AS. Prove that $QR^2 = 2PQ^2$.

Problem-Solving

Often in solving a problem you can only solve the problem if you know how to use facts at the right time.

For example, for △ABC you know the following

Fact 1: $\cos A = \dfrac{b^2 + c^2 - a^2}{2bc}$ *Fact 2:* $\sin^2 A + \cos^2 A = 1$ or

$$\sin A = \sqrt{1 - \cos^2 A}$$

Fact 3: Define $s = \dfrac{a + b + c}{2}$

Use the above facts to develop the formula below for the area of △ABC.

$$\triangle ABC = \sqrt{s(s - a)(s - b)(s - c)}$$

Math Tip

It is important to clearly understand the vocabulary of mathematics when solving problems.

• Make a list of all the words you have met in this chapter.
• Provide a sample example to illustrate each word.

Problem-Solving: Analyzing Diagrams

A lot of the work that is done in different fields of endeavour, such as engineering, astronomy, geography, are based on an accurate analysis of diagrams. A diagram contains much information in a compact form. To find the missing information in the diagram, you need to interpret the given information in the diagram.

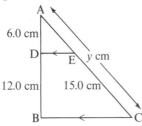

Think and analyze: Since DE ∥ BC, then
∠ADE = ∠ABC (Corresponding angles)
∠AED = ∠ACB (Corresponding angles)

In △ADE and △ABC, since all corresponding angles are equal, then
△ADE ~ △ABC (AAA ~)

In △ADE and △ABC, $\dfrac{AD}{AB} = \dfrac{AE}{AC}$ or $\dfrac{6}{18} = \dfrac{y-15}{y}$

$$6y = 18y - 270$$
$$270 = 12y \text{ or } y = 22.5$$

16 • Examine the given information in each diagram.
    • Then find the missing values.

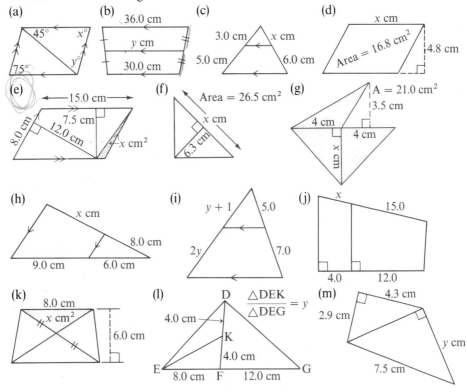

7.9 Problem-Solving: Extending Concepts

Earlier, you dealt with the ratios of the areas of similar polygons. The same strategies and skills can be applied to the exploration of the ratios of the volumes of similar polyhedra. Two polyhedra are similar if
- the vertices correspond one-to-one
- corresponding face angles are equal
- ratios of corresponding edges are equal

Examine the polyhedra at the right, which are similar. What can you deduce about the ratio of their volumes?

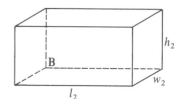

For rectangular solid A, $V_1 = l_1 w_1 h_1$

For rectangular solid B, $V_2 = l_2 w_2 h_2$

Thus, $\dfrac{V_1}{V_2} = \dfrac{l_1 w_1 h_1}{l_2 w_2 h_2} = \left(\dfrac{l_1}{l_2}\right)\left(\dfrac{w_1}{w_2}\right)\left(\dfrac{h_1}{h_2}\right)$

But, $\dfrac{l_1}{l_2} = \dfrac{w_1}{w_2} = \dfrac{h_1}{h_2}$ (property of similar polyhedra)

Thus, $\dfrac{V_1}{V_2} = \left(\dfrac{l_1}{l_2}\right)^3 = \dfrac{(l_1)^3}{(l_2)^3}$ or $\dfrac{V_1}{V_2} = \left(\dfrac{w_1}{w_2}\right)^3 = \dfrac{(w_1)^3}{(w_2)^3}$ or $\dfrac{V_1}{V_2} = \left(\dfrac{h_1}{h_2}\right)^3 = \dfrac{(h_1)^3}{(h_2)^3}$

> The ratio of the volumes of two similar polyhedra is equal to the ratio of the cubes of corresponding sides.

Thus, if the dimensions of a rectangular solid are doubled, then the volume of the resulting solid is 2^3 or 8 times greater than the original solid.

1 Two similar polyhedra have their sides in the following ratios. What is the ratio of the volumes for each of the following?
(a) 1:3   (b) 1:4   (c) 2:3   (d) 4:3

2 The ratio of the volumes of two similar polyhedra is given. What is the ratio of the sides for each of the following?
(a) $\dfrac{27}{8}$:1   (b) 64:125   (c) $8:\dfrac{27}{64}$ (d) 2:8

3 A box has dimensions 32.0 cm × 45.0 cm × 15.0 cm. If the dimensions of each side are increased by a factor $\dfrac{3}{5}$, how many times greater is the volume of the new box than the volume of the original box?

Practice and Problems: A Chapter Review

1   If ST ∥ VU, SW = 6, WV = 2.5,
    UW = 4, find TW.

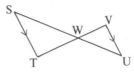

2   Find x.
    (a)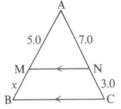

    (b)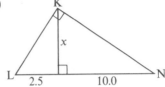

3   In △ABC, P, Q, and R are respectively the midpoints of the sides of
    △ABC. If △PQR = 25.0 cm², find △ABC.

4   Quadrilateral PQRS has the following properties.
    • ∠QPS and ∠SQR are congruent.
    • SP:PQ = SQ:QR
    Prove that QS is the bisector of ∠PSR.

5   In △XYZ, M is the midpoint of XY and N is the midpoint of XZ. If MZ
    and NY intersect at P, then find

    (a) $\dfrac{\triangle MPN}{\triangle YPZ}$ (b) $\dfrac{\triangle YPX}{\triangle XYZ}$ (c) $\dfrac{\triangle MPN}{\triangle XYZ}$

6   Prove that the corresponding medians of similar triangles are proportional
    to corresponding sides.

7   Given a quadrilateral XYZW with XZ and YW intersecting at A so that
    XA:ZA = YA:WA. Prove the quadrilateral is a trapezoid.

8   A radio antenna has two parallel wires attached to it at different points
    and secured to the ground. The first wire is attached at a height of 14.0 m
    from the ground and is anchored at 15.0 m from the base of the antenna.
    The second wire is anchored 25.7 m from the antenna. How high above
    the base of the antennae is the second wire attached?

Test for Practice

1   (a) What is the definition of similar triangles?

    (b) Show with an example that two quadrilaterals need not be similar if their corresponding sides are proportional and a pair of corresponding angles is congruent.

2   Find the missing values.

(a)

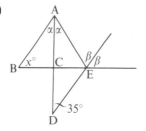

(b)

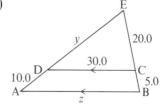

3   From the following diagram, find △ABC:△AXY.

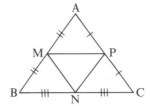

4   Use the diagram.

    (a) Prove △MNP ~ △ABC.

    (b) Find the ratio △MNP:△ABC.

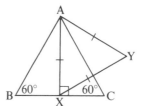

5   A ski tow rises 38.0 m for a horizontal distance of 100.0 m. If a skier is towed at the rate of 6.0 km/h, how far will the skier have risen after 9 s?

6   In right △ABC, ∠A = 90°. AT is an altitude drawn from A. Prove that BT × TC = AT × AT (i.e. AT^2).

7   △MOT is right-angled at T. If Q is on TM and R on OM so that QR ⊥ OM, prove TO:QR = TM:MR.

8   In △PST, Q and R are points in PS and PT respectively so that PR:PQ = PT:PS. Prove that QR ∥ ST.

8 Geometry of Circles

vocabulary of circles, properties of chords and angles, concyclic points and cyclic quadrilaterals, making conjectures, properties of tangents, tangent chord property, arcs, sectors, secants, solving problems, strategies, applications, problem-solving

Introduction

In your study of mathematics, you have seen a number of recurring themes.

- The results of the work of people in mathematics are often named after them in their honour. For this reason, you speak of Cartesian co-ordinates, Pascal's triangle, Napier's rods, Pythagorean Theorem, Fermat's Numbers, and so on.
- Often, a method, skill or strategy you develop for a particular skill can be extended and applied to solving more problems.
- The development of mathematics has often been based on asking useful questions such as: "What if . . . ?", or "Why?" or on making a conjecture.

The three themes above represent only a few of the recurring themes and steps in the process of developing mathematics. However, a very important consideration is the development of skills for organizing your work, and for leaving a record that others can understand. The *Steps for Solving Problems* apply not only to your work in geometry, but also to any branch of mathematics.

Steps for Solving Problems

Step A Read the problem carefully. Can you answer these two questions?
I What information are you asked to find (information you don't know)?
II What information are you given (information you know)?
Be sure to understand what it is you are to find.

Step B Translate from English to mathematics. (Draw a diagram; record the given information on the diagram.)

Step C Solve the problem. Do the deduction. Find the missing values of angles or line segments.

Step D Check the answers in the original problem.

Step E Write a final statement as the answer to the problem.

8.1   Geometry with Circles and Chords

You have often used the *Steps for Solving Problems* to help you organize your solutions. These steps are helpful, as well, in proving deductions.

Steps for Solving Problems

Step A   Read the problem carefully. Can you answer these two questions?
    I What information am I asked to find (information I don't know)?
    II What information am I given (information I know)?

Step B   Translate from English to mathematics. Draw a diagram to help you.

Step C   Do the work. Write the proof.

Step D   Check your work. Have you given reasons for the statements?

Step E   Write a final statement as the answer to the problem.
    Have you proven what you set out to prove?

In studying geometry, you must know clearly the various definitions you had learned about triangles to prove deductions. Similarly, to do mathematics involving circles, you need to know the vocabulary of the circle and understand the significance of each word.

Example 1   Prove that chords of equal length are equidistant from the centre of a circle.

Solution   Given: AB and CD are equal chords.
Required to prove: OE = OF.
Proof: Join OA, OB, OC, OD.

Translate the given information into a diagram. Label the diagram.

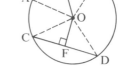

Statements	Reasons
In △OAB, △OCD	
OA = OC.	radii
OB = OD	radii
AB = CD	given
△OAB ≅ △OCD	SSS
Then, ∠OCF = ∠OAE ①	property of congruent triangles
In △OCF, △OAE,	
∠OCF = ∠OAE.	proved ①
∠OFC = ∠OEA	90°
OC = OA	radii
△OCF ≅ △OAE	ASA
Thus, OF = OE.	

Thus, chords of equal length are equidistant from the centre of a circle.

↖ Make a final statement.

The main properties of chords are listed in the following chart. You will prove these properties in the exercise.

Properties of Chords of a Circle

C1   A line from the centre, perpendicular to a chord, bisects the chord.

C2   The line segment drawn from the centre to the midpoint of the chord is perpendicular to the chord.

C3   Chords equidistant from the centre of a circle are equal.

C4   Equal chords are equidistant from the centre.

C5   The right bisector of a chord passes through the centre of a circle.

To find missing measures involving chords, you need to apply the properties of chords as well as the properties developed earlier for triangles.

Example 2   A circle has centre O and diameter 26 cm. A chord 24 cm in length is drawn in the circle. What is the perpendicular distance from the centre to the chord?

Solution $OE \perp AB$, $AE = EB$,
$AE = 12$ cm

Sketch a diagram
with the given information

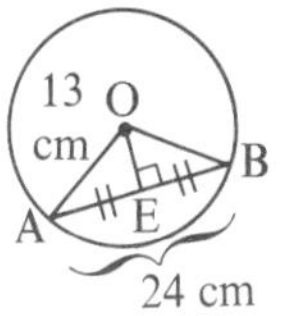

The diameter is 26 cm.
The radius is 13 cm.

In $\triangle OEA$, $OE^2 + AE^2 = OA^2$
$$OE^2 + 12^2 = 13^2 \qquad \text{Pythagorean Property}$$
$$OE^2 = 25$$
$$OE = 5 \qquad \text{Distance is positive.}$$

Thus, the perpendicular distance from the centre to the chord is 5 cm.

8.1 Exercise

A Throughout the exercise, the centres of circles are labelled by O.

1 For each diagram, what property of chords is shown?

(a) 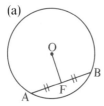 (b) 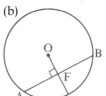 (c)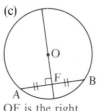

OF is the right
bisector of AB.

(d)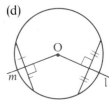

l and *m* are the
right bisectors of
the two chords.

(e) 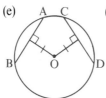 (f) 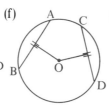 (g) 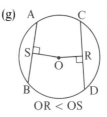 (h)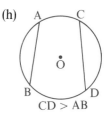

OR < OS   CD > AB

2 Find the missing values.

(a) 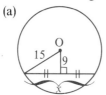 (b) 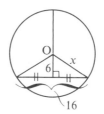   (c)

3 Refer to the diagram of the circle with
radius 30.0 cm. The chord PQ is
perpendicular to the diameter AOB. If
PQ is 48.0 cm, find the length of AS.

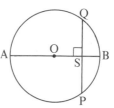

4 In a circle, a chord is 16 cm in length and
is 6 cm from the centre of the circle. Calculate the length of the diameter.

5 Two parallel chords of a circle, NM and PQ, on opposite sides of the
centre, have lengths of 12.0 cm and 16.0 cm. If the diameter of the circle is
20.0 cm, find the perpendicular distance between the chords.

6 A circle with centre O has a diameter 26.0 cm. A chord, PQ, of the circle is
24.0 cm. What is the distance of the centre to the chord?

B Organize your solutions. Refer to the *Steps for Solving Problems.*

7 Prove that the line segment that joins the centre of a circle to the midpoint of a chord is a perpendicular bisector of the chord.

8 Prove that the right bisectors of two chords of a circle intersect at the centre of the circle.

9 (a) Two chords, AB and CD, are equidistant from the centre of a circle. Prove AB = CD.

   (b) EF and GH are two chords of a circle. If EF = GH, prove they are equidistant from the centre of the circle.

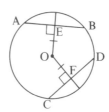

10 (a) Two chords AB and CD are drawn in a circle so that OE < OF. Prove AB > CD.

   (b) What is the converse of the statement in (a)? Prove the converse.

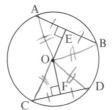

11 (a) Prove that, for any circle, the perpendicular bisector of any chord passes through the centre of the circle.

   (b) Three points, A, B, and C are given on the plane. Use the result in (a) to describe how a circle passing through A, B, and C is drawn.

   (c) Prove that only one circle can be drawn through three given points.

12 The circle in Question 11 is called the **circumcircle** of the triangle with vertices A, B, C.

   (a) Construct an isosceles triangle. Draw its circumcircle.

   (b) Construct an equilateral triangle. Draw its circumcircle. Label the centre O. Draw the three medians of the triangle. What do you notice about the point of intersection of the medians?

13 Two circles of unequal radii, intersect at S and T. ETF is a line drawn parallel to GSH meeting the circle in the points E, F, G, and H. Prove that EFHG is a parallelogram.

14 (a) Refer to Question 13. If the two circles have equal radii, make a conjecture about the point of intersection of the diagonals of EFHG.

   (b) Prove your conjecture in (a).

Applications: Designs and Chords

The properties of circles and their chords are often used in the making of designs.

15 To inscribe a regular hexagon in a circle, the value of the central angle is used.

(a) Why is $\angle AOB = 60°$?

(b) Inscribe a hexagon in a circle so that $AB = 3$ cm.

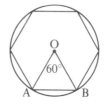

16 (a) What properties of the chords of a circle are used to construct this design?

(b) Construct a design of your own based on the same properties.

17 (a) What properties of the chords in a circle are used to construct this design?

(b) Construct a design of your own based on the same properties.

Problem-Solving

Very often, you can use skills which you have learned previously to draw a diagram. Sometimes some very interesting results occur. For example, in $\triangle PQR$, the altitudes are drawn so that PD, QE, and RF intersect at H. The points A, B, and C are the midpoints as shown. The points X, Y, and Z are the midpoints of PH, QH, and RH respectively.

Prove that

(a) $ZB \parallel PQ$.   (b) $AB \parallel QR$.

(c) $\angle PAB = \angle BCR$.

(d) A, F, B, Z, and C are concyclic points.

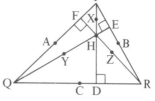

The circle passing through A, F, B, Z, and C also passes through E, X, Y and D. The circle is called the **nine-point circle** of $\triangle PQR$.

8.2 Angles in a Circle

In order to solve geometric problems, you must understand the meaning of the vocabulary used. A visual is a helpful reminder of the meaning of the words

∠ACB is said to be an **inscribed angle** of the circle, subtended by the arc ADB or the chord AB. The vertex, C, is on the circle.

∠AOB is called a **central angle** of the circle, subtended by the arc ADB or the chord AB. The vertex, O, is the centre of the circle.

subtended by   subtended by
the arc ADB   the chord AB

subtended by   subtended by
the arc ADB   the chord AB

In the following example, the relationship between the central angle and the inscribed angle is derived.

Example 1   In a circle with centre O, chord AB subtends ∠ACB at the circumference and a central angle, ∠AOB, at the centre. Prove that ∠AOB = 2∠ACB.

Solution   Given: ∠ACB, an inscribed angle on arc ADB; ∠AOB, a central angle on arc ADB.

Required to prove: ∠AOB = 2∠ACB.

Proof: Join CO and extend to D.

Statements	Reasons
In △AOC, OA = OC.	given radii
Then ∠OCA = ∠OAC,	isosceles triangle
and ∠AOD = 2∠ACO. ①	exterior angle
In △BOC, OB = OC.	radii
Then ∠OBC = ∠OCB,	isosceles triangle
and ∠BOD = 2∠BCO. ②	exterior angle

From ① and ②,

∠AOD + ∠BOD = 2∠ACO + 2∠BCO.

∠AOB = 2(∠ACO + ∠BCO)

∠AOB = 2∠ACB   This is what was required to be proved.

A proof, similar to the one in the previous example, can be obtained for ∠ACB and ∠AOB situated as shown in the diagram. The proof is left as an exercise.

Two important properties of inscribed angles can be derived from the above result.

- Angles inscribed on the same arc or chord are equal.

∠DAC and ∠DBC are angles inscribed on the same arc, DEC. Then ∠DAC = ∠DBC.

- Angles inscribed on the diameter of a circle are right angles.

∠AEB is inscribed on the diameter AOB. Then ∠AEB = 90°.

Inscribed Angle Theorem

- If a central angle and an inscribed angle of a circle are subtended on the same chord and on the same side of the chord, then the central angle is twice the inscribed angle.

- If two angles are inscribed on the same chord and on the same side of the chord, then they are equal.

- An inscribed angle subtended by the diameter is a right angle.

The results are applied to find numerical values of angles as shown in the following example.

Example 2 Find the missing values.
Give reasons for your steps.

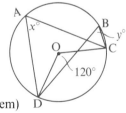

Solution ∠DOC is a central angle and ∠DAC is an inscribed angle on the same chord, DC. (given)

Thus, $∠DAC = \dfrac{1}{2} ∠DOC$. (Inscribed Angle Theorem)

$$x° = 60°$$
$$∠DAC = 60°$$

∠DAC, ∠DBC are inscribed angles on the same chord, DC, and on the same side. (given)

$$∠DAC = ∠DBC \quad \text{(Inscribed Angle Theorem)}$$
$$x° = y° \quad \text{or} \quad y° = 60° \quad ∠DBC = 60°$$

8.2 Exercise

A 1 A numerical example often suggests a general property of some geometric figure. For each diagram, show that $2a° + 2b° = c°$.

(a)   (b)

2 (a) Use the results in Question 1 and the diagram to show that $\angle QOR = 2\angle QPR$.

(b) Write the result in (a) as a theorem.

(c) Use the result in (a) to show that the angle inscribed in a semi-circle is a right angle.

(d) Use the result in (a) to show that angles inscribed on the same arc or chord are equal.

3 Find the value of the missing variables.

(a) 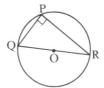   (b) 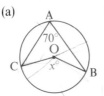   (c)    (d)

4 A chord subtends an angle at the circumference with the following measures. What is the measure of the corresponding central angle?

(a) 20°   (b) 50°   (c) 45°

5 A chord subtends an angle at the centre with the following measures. What is the measure of the corresponding inscribed angle?

(a) 50°   (b) 90°   (c) 120°

6 For the diagram, $\angle BP_1P_2 = 60°$ and $\angle AP_1O = 35°$.
Find the following.

(a) $\angle BOP_2$ (b) $\angle AOB$
(c) $\angle P_1AO$ (d) BP_1A

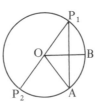

7 Find the values of the variables. Give reasons for your answer.

(a) 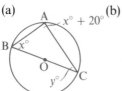 (b) 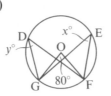 (c) 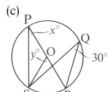 (d)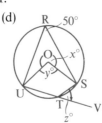

B As in your earlier work, sketch a diagram when necessary. Record the given information in the diagram.

8 (a) Prove that inscribed angles of a circle subtended by the same arc or chord are equal.

(b) Prove that an angle inscribed on the diameter of a circle is right.

9 Use each diagram to derive the result $\angle AOC = 2\angle ABC$.

(a) 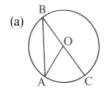 (b)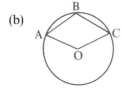

10 Three points on a circle are named P, Q, and R. If $\angle POQ = 86°$ and $\angle RPQ = 28°$, calculate the measure of the following.

(a) $\angle PRQ$ (b) $\angle PQR$

11 An inscribed angle of 80° is subtended by a chord 6.0 cm in length.

(a) Find the angle subtended at the centre by the chord.

(b) What is the radius of the circle?

12 Use the diagram to prove △CBM and △MDA are isosceles.

13 Refer to the diagram. Prove $\angle RQM = \angle SQN$.

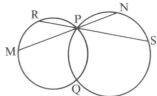

14 The circumcircle of △PQR is drawn. If PS is the bisector of $\angle P$ and intersects QR at T and the circle at S, prove △PQT ~ △PSR.

C 15 Use the diagram.

(a) Prove that if $\angle A = \angle B$ then points A, B, C, and D are on the same circle.

(b) What is the locus of all such points A and B, where $\angle A = \angle B$?

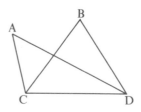

Applications: Sports and Inscribed Angles

A hockey player is at H. The angle at which the puck can be shot to score is shown. At what other positions on the ice would the same angle of shooting occur? Since angles inscribed on the same arc are equal, the diagram is used to locate the centre of the circle.

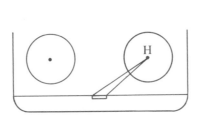 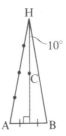

16   (a) Write a proof to show that C is the centre of the circle in which chord AB subtends ∠AHB = 20°.

     (b) Copy the diagram above. Locate all positions from which a hockey player could shoot at the net within the same angle.

17   A hockey player is situated at H. Copy the diagram. Construct the circle at which equal angles occur.

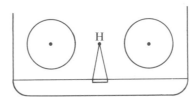

18   Copy each diagram. A hockey player is at H and the shooting angle is shown. Construct the circle on which angles equal to the shooting angle are found.

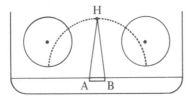 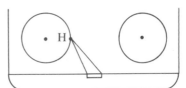

8.3 Concyclic Points and Cyclic Quadrilaterals

Once you know the meaning of the words *cyclic* and *concyclic*, you can develop the properties of concyclic points and cyclic quadrilaterals.

Points on the same circle are said to be **concyclic**. A, B, C, D, and E are concyclic points.

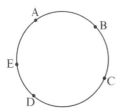

A quadrilateral whose vertices are concyclic is said to be a **cyclic quadrilateral**. ABCD is a cyclic quadrilateral

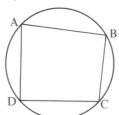

With the above definitions and the theorems developed in the previous section, theorems about the properties of cyclic quadrilaterals can be developed. This is shown in the next example.

Example   Prove that the opposite angles of a cyclic quadrilateral are supplementary.

Solution   Given: Quadrilateral PQRS.
              O is the centre of the circle.

Translate the given general information to a specific diagram.

Required to prove: $\angle P + \angle R = 180°$
                   $\angle Q + \angle S = 180°$

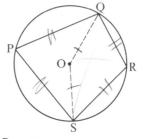

Proof: Construct the diagram as shown. Join OS and OQ. Since reflex $\angle SOQ$ is a central angle and $\angle SRQ$ is the corresponding inscribed angle, then

Statements			Reasons
reflex $\quad \angle SOQ = 2 \angle SRQ.$	①		Inscribed angle theorem
Similarly,			
obtuse $\quad \angle SOQ = 2 \angle SPQ.$	②		Inscribed angle theorem

From ① and ②,
$2 \angle SRQ + 2 \angle SPQ = $ reflex $\angle SOQ + $ obtuse $\angle SOQ.$
$2(\angle SRQ + \angle SPQ) = 360°$
$\angle SRQ + \angle SPQ = 180°$ ◀——This is what was required to be proved.
Similarly, by joining PO and OR it can be shown that
$\angle PSR + \angle PQR = 180°$

An important corollary of the above theorem is the following.

For a cyclic quadrilateral the exterior angle is equal to the interior and opposite angle. In the diagram, $\angle Q = \angle RST$.

$\angle Q$ is the interior and opposite angle of exterior $\angle RST$.

8.3 Exercise

A   Often a numerical example suggests a theorem about figures. This is shown in Questions 1 and 2.

1   (a) Use the diagram. Prove that $m° + n° = 180°$.

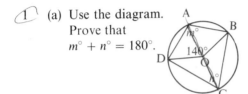

(b) Use the diagram. Prove that $m° + n° = 180°$.

2   (a) Use the diagram. Prove that $m° = n°$.

(b) Use the diagram. Prove that $m° = n°$.

3   PQRS is a cyclic quadrilateral. Find the missing measures.
(a) $\angle P = 70°$, $\angle Q = 85°$, $\angle R = ?$, and $\angle S = ?$
(b) $\angle P = ?$, $\angle Q = 27°$, $\angle R = 105°$, and $\angle S = ?$

4   Refer to the diagram.
(a) What is the measure of $\angle BDC$?
(b) Use your answer in (a). What is the measure of $\angle BAC$?

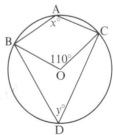

5   Refer to the diagram.
(a) What is the measure of $\angle BCD$?
(b) Use your result in (a). What is the measure of $\angle DAB$?
(c) Use your result in (b). What is the measure of $\angle BCA$?

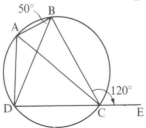

6 For each diagram, find the missing measures.

(a)

(b)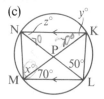

(c)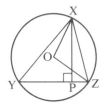

(d)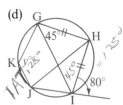

B Review the meaning of concyclic points and cyclic quadrilateral. You cannot solve problems if you do not know the meaning of the vocabulary.

7 Use the diagram.
If XP ⊥ YZ, prove
∠YXP = ∠OXZ.

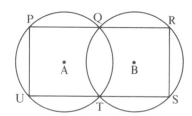

8 Circles with centres A and B intersect at Q and T. The lines PQR and UTS intersect the circles as shown. Prove PU ∥ RS.

9 Prove that an inscribed parallelogram of a circle is a rectangle.

10 L, M, N, and Q are concyclic points. If QL is a diameter of the circle and PL is perpendicular to MN or MN produced, prove that ∠PLM = ∠NLQ.

C 11 Prove that in any triangle, all the altitudes are concurrent.

Computer Tip

A calculator/computer is an important aid to any scientist. For example, the earth's path around the sun is not circular, but rather elliptical. The value of e given in the following formula shows a measure of how much the orbit deviates from a circular form, called the eccentricity. (See chapter 11.)

$$e = \frac{\sqrt{a^2 - b^2}}{a}.$$

▶ Find the approximate eccentricity if $a = 1.500\ 00 \times 10^5$ and $b = 1.499\ 81 \times 10^2$.
▶ Write a computer program to find the value of e.

8.4 Problem-Solving Strategy: What If . . . ?

Very often, mathematics can be developed by asking the question: "What if . . . ?" Previously, you learned that inscribed angles subtended on the same chord and on the same side of the chord are equal. What if a line segment, PQ, subtends equal angles, $\angle$ R and $\angle$ S on the same side of the line? Are the four points concyclic?

What if $\angle$ PSQ = $\angle$ PRQ? Then are P, Q, R, and S concyclic points?

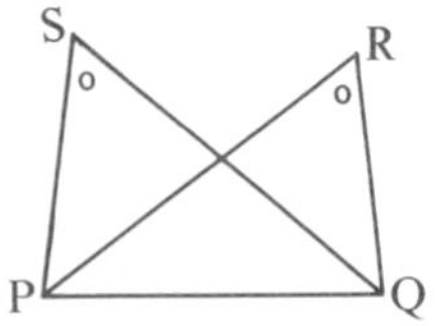

The method of indirect proof is used to show that the converse of the previous work on concyclic points is indeed true. Replace each Why? in the following proof with the correct reasons for the statement.

Given: $\angle$ PSQ = $\angle$ PRQ

Required to prove: P, Q, R, and S are concyclic points.

Proof: In the diagram, either Fact A or Fact B is true, but not both.
  Fact A: P, Q, R, and S are concyclic.
  Fact B: P, Q, R, and S are not concyclic.

Think. Refer to the diagram. Since any three points are concyclic, then P, Q, and S are concyclic, and R lies either *inside* or *outside* the circle containing P, Q, and S.

Suppose Fact B applies and R lies inside the circle. Then construct the circle as shown. $\angle$ PSQ and $\angle$ PTQ are inscribed on the same chord PQ.

$\angle$ PSQ = $\angle$ PTQ	①	Why?
But $\angle$ PSQ = $\angle$ PRQ.	②	Why?

Then from ① and ②,
  $\angle$ PTQ = $\angle$ PRQ.   Why?

In $\triangle$TQR,
exterior $\angle$ PRQ = interior $\angle$ RTQ. Why?
This is impossible.   Why?

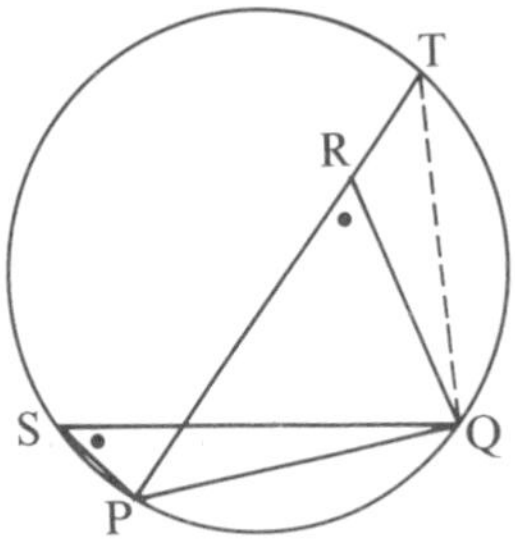

Similarly, the assumption that R lies outside the circle leads to a contradiction. Thus, R lies on the same circle as S, P and Q. Thus, P, Q, R, and S are concyclic points.
  — This is what was required to be proved.

Using similar methods, you can prove the following geometric facts.

- The circle constructed on the hypotenuse of a right triangle with the hypotenuse as diameter, passes through the vertex at the right angle.

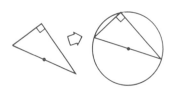

- If an exterior angle of a quadrilateral is equal to the corresponding interior and opposite angle, then the quadrilateral is cyclic.

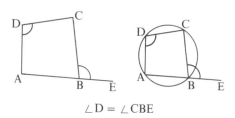

$$\angle D = \angle CBE$$

- If a pair of the opposite angles in a quadrilateral are supplementary, then the quadrilateral is cyclic.

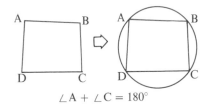

$$\angle A + \angle C = 180°$$

8.4 Exercise

B Remember: review the various strategies you know for proving deductions.

1 Prove X, Z, N, and M are concyclic if XY = ZW.

2 In the diagram, PT ⊥ RQ and RV ⊥ PQ. Prove that V, S, T, and Q, are concyclic points.

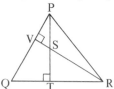

3 Prove each of the following.
   (a) The circle constructed on the hypotenuse of a right triangle as diameter passes through the vertex at the right angle.
   (b) If a pair of opposite angles in a quadrilateral are supplementary then the quadrilateral is cyclic.
   (c) If an exterior angle of a quadrilateral is equal to the corresponding interior and opposite angle, then the quadrilateral is a cyclic quadrilateral.

4   In the diagram, BN bisects ∠ABC and MC bisects ∠ACB. If AB = AC, prove M, B, C, and N are concyclic points.

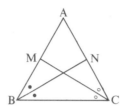

5   In the diagram, SP ∥ ON and MN ∥ RP. Prove that L, R, P, and S are concyclic points.

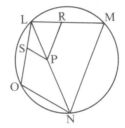

6   If PQ = RS and are drawn to bisect each other at B, prove that the end points are concyclic points.

7   J, K, L, and M are concyclic points. If Y and Z are points on JM and KL respectively so that YZ ∥ JK, then prove that LMYZ is a cyclic quadrilateral.

Problem Solving

Many theorems in mathematics are named after the original or subsequent investigator of the theorem. The following theorem is called the **Simson Line Theorem** after Robert Simson (1687–1768). △STU is inscribed in a circle. P is any point on the circle so that PY, PX, and PZ are perpendicular to the sides or sides produced of △STU as shown. Prove that X, Y, and Z are collinear.

Math Tip

It is important to understand clearly the vocabulary of mathematics when solving problems.
- Make a list of all the new words you meet in this chapter.
- Provide an example to illustrate each word.

Making Conjectures

In mathematics a theorem is a proposition that is proven, whereas a conjecture is a generalization that has yet to be proven. **A conjecture is a surmise made from observations.** Conjectures are worthwhile because, once proven, they may become theorems. Conjectures are also valuable because attempts to prove them may lead to the development of new skills and strategies. Use your skills with angles and chords to prove these conjectures.

8   (a) For the design, which subtended angles appear to be equal? Make a conjecture.

    (b) Prove your conjecture in (a).

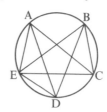

9   (a) For the design, which angles appear to be right angles? Make a conjecture.

    (b) Prove your conjecture in (a).

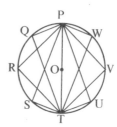

10 In the diagram, O is the centre of the circle and OS ⊥ XY.

    (a) Make a conjecture about which points in the diagram you believe to be concyclic.

    (b) Prove your conjecture in (a).

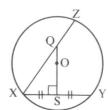

11 (a) For the diagram, make a conjecture about which points you believe to be concyclic.

    (b) Prove your conjecture in (a).

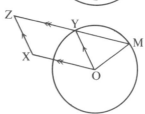

12 (a) For the diagram, make a conjecture about which points you believe to be concyclic.

    (b) Prove your conjecture in (a).

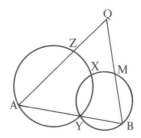

8.5 Properties of Tangents: Circles

A line can intersect a circle at one point, at two points, or at no points.

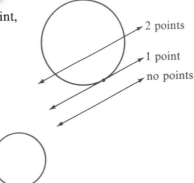

2 points

1 point

no points

A line which intersects a circle in one point is called a **tangent**. The line ST is a tangent to the circle. The point P is called the **point of tangency**.

To construct a tangent to a circle, you need to know the properties of tangents.

Example 1   Prove that the line drawn perpendicular to the end point of a radius of a circle is a tangent of the circle.

Solution   Given: Circle with centre O. OP ⊥ SQ.

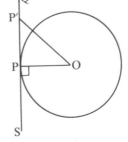

Required to prove: SPQ is
            a tangent of the circle.

Proof: Choose any point, P′, on SQ. Assume P′
        is in SQ and on the circle.

Statements	Reasons
In △POP′, ∠OPP′ = 90°.	given
Thus, ∠OPP′ > ∠OP′P	ASTT
OP′ > OP	The side opposite the greater angle is greater.
For all P′, OP′ > OP, and, thus, P′ lies outside of the circle.	OP is the radius.

Thus, P is the only point of intersection.

Thus, based on the above result:

> The line drawn perpendicular to the end point of a radius is a tangent to the circle.

Similarly, the result follows that:

> A line drawn perpendicular to a tangent at the point of contact with
> a circle passes through the centre of the circle.

To construct a tangent to a circle from a point outside the circle, you can
use the above result and a previously
developed property of angles drawn
on the diameter of a circle.
P is any point outside the circle with
centre O.

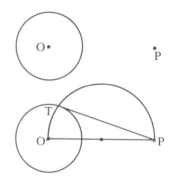

Use OP as diameter and draw a circle to
cut the given circle at T.
Since ∠OTP is subtended by a diameter,
then ∠OTP = 90°.
Thus, TP is tangent to the circle at T.

Tangents drawn from a point have certain
properties as shown in the next example.

Example 2   Tangents are drawn from P to the circle
with centre O to meet the circle at
S and T. Prove that PS = PT.

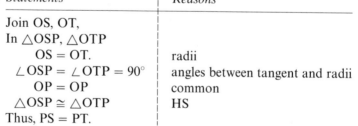

Solution   Given: Tangents PS and PT.
Required to prove: PS = PT.
Proof: *Statements* | *Reasons*

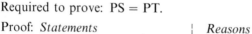

Statements	Reasons
Join OS, OT,	
In △OSP, △OTP	
OS = OT.	radii
∠OSP = ∠OTP = 90°	angles between tangent and radii
OP = OP	common
△OSP ≅ △OTP	HS
Thus, PS = PT.	

From the above result, you can deduce that, since △OSP ≅ △OTP, then
∠POS = ∠POT, and ∠OPS = ∠OPT.

> • Tangents drawn from a point outside the circle are equal in length.
>
> • Two tangents can always be drawn from a point outside of the circle.

8.5 Exercise

A 1 (a) In the diagram, $OP_1 = 1.0$ cm. Find (i) OA. (ii) $\angle P_1OP_2$.

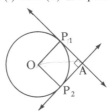

(b) PM and PN are tangents at M and N. Find $\angle MNP$.

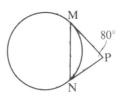

(c) If $OP_1 = 12.0$ cm, and $OA = 20.0$ cm, find P_1A.

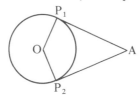

(d) YX is tangent to the circle at P. If OS ∥ YX, find $\angle SPX$.

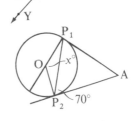

2 Find the value of each variable.

(a)

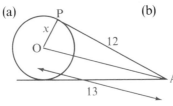

(b)

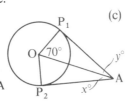

(c)

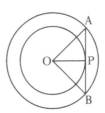

3 Two concentric circles are shown in the diagram. If $OP = 6.0$ cm and $OB = 10.0$ cm, find the length of AB.

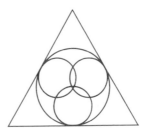

4 Draw a circle with radius 6.0 cm. Choose a point 4.0 cm from the circle.

(a) Construct a tangent to the circle.

(b) Find the length of the tangent.

5 (a) How are the properties of tangents used to construct the design?

(b) Make a copy of the design.

6   (a) Outline the steps needed to construct a tangent to a circle at a point
     on the circle.

     (b) Outline the steps needed to construct a tangent to a circle from a point
     outside the circle.

B Make a list of the definitions you have learned in this section. Also list the
     properties of tangents. You will need these properties to solve the following
     questions.

7   Two circles with centres P and Q touch at S (are tangent at S). Prove that
     P, S, and Q are collinear points.

8   Two tangents are drawn to a circle so that the points of contact are the end
     points of a diameter of the circle.
     Prove that the tangents are parallel.

9   PM and PN are tangents drawn
     from P to the circle with center O.
     Prove that PO is the right
     bisector of MN.

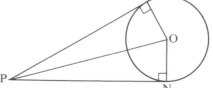

10   From P, tangents are drawn to a circle with centre O to meet the circle at
     M and N. If $\angle$ MPN = 90° and MO = 3.0 cm, find the length of MP.

11   Points R, P, and Q are concyclic and RP is a diameter with centre O. The
     tangents of the circle at P and Q intersect at S. Prove RQ $\parallel$ OS.

12   R, S, and T are concyclic points of a circle with centre O. Tangents are
     drawn at R and S to intersect at L. Tangents at S and T intersect at V.
     If VT $\parallel$ RL, prove $\angle$ LOV is a right angle.

C 13   L, M, and N are concyclic points. A line, QL, is drawn tangent to the circle
     at L. S and T are points in LN and LM, respectively, so that ST $\parallel$ QL. Prove
     that S, T, M, and N are concyclic points.

Problem Solving

It is helpful to sketch a diagram to visualize some problems. Sketch
a diagram and then solve this problem.

Find the equation of the line through the point (3, −2) which
meets the *x*-axis at the point (*a*, 0) and the *y*-axis at the point
(0, *b*), where 5*ab* = 1.

8.6 Tangent Chord Property

In the previous sections, you have learned properties about

• inscribed angles,   • tangents to circles.

These properties are used together to develop yet another property of circles, as shown in the following example.

Example 1   A tangent, BPR, meets the circle with centre O at P. A chord, PS, subtends the angle, T at the circumference. Prove $\angle SPB = \angle STP$.

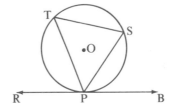

Solution   Given: chord PS, tangent BPR

Required to prove: $\angle SPB = \angle STP$

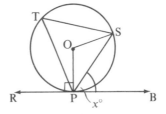

Proof:

Statements	Reasons
Join OS, OP.	
Use $\angle SPB = x°$. ①	
Since OP is a radius then $OP \perp RB$.	property of tangents
Thus $\angle OPS = 90° - x°$.	$\angle OPB = 90°$
In $\triangle OPS$, OP = OS.	radii
Then $\angle OPS = \angle OSP$.	isosceles triangle
$\angle POS = 180° - \angle OPS - \angle OSP$	sum of the angle of a triangle
$\angle POS = 180° - (90° - x°) - (90° - x°)$	
$\angle POS = 2x°$	
$\angle PTS$ is an inscribed angle and $\angle POS$ is a central angle on the same chord PS.	
Thus, $\angle PTS = x°$. ②	property of inscribed angles
From ① and ②, $\angle PTS = \angle SPB$. ←This is what was required to be proved.	

To obtain the above result, $\angle SPB$ was taken to be an acute angle. The result can also be proven if $\angle SPB$ is an obtuse angle or a right angle. (Refer to Question 9 in the exercise.)

The previous result is referred to as the **tangent chord property** of a circle.

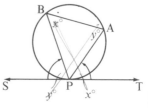

Tangent Chord Property of a Circle

The angle between a tangent and chord is equal to the angle subtended on the opposite side of the chord.

The tangent chord property of a circle is used to find numerical values.

Example 2   Calculate the measure of ∠RPS.

Solution   Join RP.

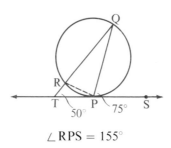

∠RPS = 155°

To find the measure of the angle, think through the following steps. Be sure you know why each step is true.

$$∠QPS = 75°$$
Thus,   ∠PRQ = ∠QPS = 75°
Thus,   ∠TRP = 105°
In △TRP, ∠TRP = 105°
             ∠RTP = 50°
Thus,   ∠TPR = 25°
Thus,   ∠RQP = ∠TPR = 25°
In △RPQ, ∠RQP = 25°
             ∠QRP = 75°
Thus,   ∠RPQ = 80°

Thus, ∠RPS = ∠RPQ + ∠QPS
      ∠RPS = 80° + 75°
      ∠RPS = 155°

8.6 Exercise

A   Throughout the exercise, justify each statement you use.

1   Find the missing values. Justify the steps for finding the missing values.

(a)   (b)   (c)   (d)

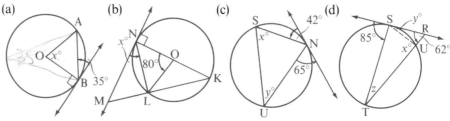

2   The sides of △ABC are tangents of
    the circle at X, Y, and Z. Calculate
    the measure of
    (a) ∠CAB.   (b) ∠ACB.
    (c) ∠ABC.

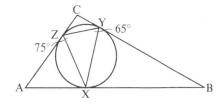

3   Sketch a diagram for each of the following. Record the given information
    on the diagram.
    (a) A tangent STR meets the circle with centre O at T. If DOA is a diameter
        of the circle and ∠RTA = 65°, then calculate the value of ∠DOT.
    (b) PS and PT are tangents to a circle at the points S and T. The chord
        TS subtends ∠TRS. R and P lie on opposite sides of chord TS. If
        ∠TPS = 36°, find the value of ∠TRS.
    (c) From T a tangent, TSR, is drawn to touch the circle at S. From T the
        line TPA meets the circle at P and A. If ∠RTA = 35° and ∠RSA = 75°,
        calculate ∠RSP.

4   A circle is inscribed in △ABC so that sides AB, BC, and CA touch the circle
    at X, Y, and Z respectively. If ∠A = 80° and ∠B = 40°, calculate the
    measures of
    (a) ∠ZYX.   (b) ∠YXZ.

B   Organize your work. Remember to refer to the *Steps for Solving Problems*.

5   In the diagram,
    EG bisects ∠DEF.
    Prove that
    △EGF is isosceles.

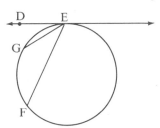

6   Refer to the diagram. RST is
    tangent to the circle at S.
    Prove AB ‖ RT.

7   In the diagram, ∠QPR = ∠RPS.
    TRV is tangent to the circle at R.
    Prove that TV ‖ QS.

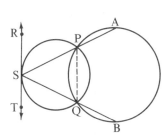

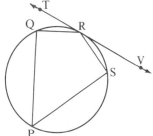

8   B, S, T are three points on a circle. If ST is parallel to the tangent at B, prove that △BST is isosceles.

9   Refer to the diagram for Example 1 in this section. A tangent, BPR, meets the circle with centre O at P. A chord PS subtends the angle T at the circumference.
(a) If ∠SPB is an obtuse angle, prove that ∠SPB = ∠STP.
(b) If ∠SPB is a right angle, prove that ∠SPB = ∠STP.

10 On a circle, the points P, Q, R are chosen and a tangent is drawn at P. S, and T are chosen on PQ and PR, respectively, so that ST is parallel to the tangent. Prove that ∠Q + ∠STR = 180°.

11 Two circles are tangent at the point P and have a common tangent, TPS. Through P a line is drawn to meet the circles respectively at M and N. Through P another line is drawn to meet the circles at U and V, respectively. Prove that MU ∥ VN.

12 In a circle, PQ and PR are equal chords and TPS is a tangent of the circle. The bisector of ∠PQR is drawn to intersect the tangent at I. Prove that ∠QPR = 2∠QIR.

C 13 From a point, P, on a circle, two chords, PZ and PX are drawn. If PY is the bisector of ∠XPZ, and PY is a chord prove that △XYZ is isosceles.

Problem Solving

A locus is a set of points that satisfies one or more conditions.

For example, a point P moves so that it is always the same distance from a fixed point. We already know this to be the locus of a circle.

However, if the point P moves so that the sum of the distances from two fixed points. A and B, is always the same, then what does the locus of P look like?

BP + AP
is a constant.

8.7 Secants, Tangents, and Chords: Properties

In developing mathematics, you often combine skills and concepts of one topic with the skills and concepts of another topic in order to develop more properties. For example:

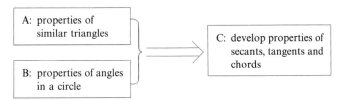

By combining the skills in A and B, you can prove the **chord theorem** for circles.

The Chord Theorem (CT)

If two chords, PQ and RS, intersect at T then

$PT \times TQ = RT \times TS$

From the diagram.

Statements	Reasons
In $\triangle PTR$, $\triangle STQ$,	
$\angle TPR = \angle TSQ$.	angles subtended by the same arc RQ
$\angle PRT = \angle SQT$	angles subtended by the same arc SP
$\angle PTR = \angle STQ$	vertically opposite angles
Thus, $\triangle PTR \sim \triangle STQ$.	AAA $\sim$

Thus, $\dfrac{TP}{TS} = \dfrac{TR}{TQ}$, or $TP \times TQ = TR \times TS$. This result is known as the Chord Theorem (CT).

A line that intersects a circle in two points is called a **secant** of the circle. PQ is a secant of the circle.

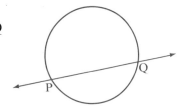

In the following example, the **Tangent Secant Theorem (TST)** is proven.

Example 1   The secant from point A intersects the circle at points B and C respectively. The tangent from point A meets the circle at D. Prove $AD^2 = AB \times AC$

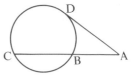

Solution   Given: External point A. Tangent AD and secant AC which also intersects the circle at B.

Required to prove: $AD^2 = AB \times AC$

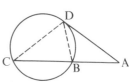

Proof: Join BD and DC

Statements	Reasons
In $\triangle ADB$ and $\triangle ACD$	
$\angle A = \angle A$	common
$\angle ADB = \angle ACD$	tangent chord property
$\angle ABD = \angle ADC$	ASTT corollary
Thus, $\triangle ADB \sim \triangle ACD$	AAA $\sim$
Thus, $\dfrac{AD}{AC} = \dfrac{AB}{AD}$	similar triangles
Thus, $AD^2 = AB \times AC$	

Tangent Secant Theorem (TST)

If a tangent from an external point A meets a circle at D, and a secant from the external point A meets the circle at B and C respectively, then $AD^2 = AB \times AC$.

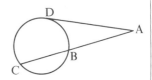

From the diagram, a corollary is obtained.

$AD^2 = AB \times AC$ (Tangent Secant Theorem)
$AD^2 = AE \times AF$ (Tangent Secant Theorem)

Thus, $AB \times AC = AE \times AF$.

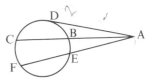

Corollary of Tangent Secant Theorem (TST)

If two secants, AC and AE, also cut the circle at B and D respectively, then $AC \times AB = AE \times AD$.

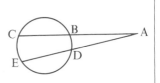

The properties of secants, tangents and chords are used to find missing numerical values in the following example.

Example 2 Find the value of the missing variable in each of the following.

(a)

(b)

(c)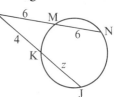

Solution

(a) $AE = 6$, $EB = 4$, $CE = 3$, $ED = x$.
Thus, $AE \times EB = CE \times ED$ (CT)

$6 \times 4 = 3 \times x$ or $x = \dfrac{6 \times 4}{3}$

$x = 8$

Thus $x = 8$.

Throughout the solution, can you give reasons for the steps?

(b) $PQ = 10$, $PR = 5$, $RS = y$.
Thus, $PS = y + 5$.
$PQ^2 = PR \times PS$ (TST)
$10^2 = 5 \times (y + 5)$
$100 = 5y + 25$
$5y = 75$
$y = 15$
Thus, $y = 15$.

(c) $LM = 6$, $MN = 6$. Thus, $LN = 12$.
$LK = 4$, $KJ = z$. Thus, $LJ = (z + 4)$
$LM \times LN = LK \times LJ$ (Corollary TST)
$6 \times 12 = 4 \times (z + 4)$
$72 = 4z + 16$
$4z = 56$
$z = 14$
Thus, $z = 14$.

8.7 Exercise

A   Give reasons for each step in your work.

1   Find the missing values in each of the following.

(a)

(b)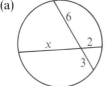

(c)

(d)

(e)

(f)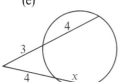

(g)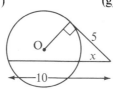

B Sketch a diagram of each of the following. Then translate the given information and record it on the diagram.

2 Q is a point outside the circle. The secant QBC meets the circle at B and C respectively. Secant QDE meets the circle at D and E respectively. If QB = 7.5 cm, QC = 12.0 cm and QD = 9.0 cm, find the length of DE.

3 Chords NM and RS intersect at P. If NM = 14.0 cm, NP = 8.0 cm and PS = 4.0 cm, find the length of RP.

4 A tangent from L meets the circle at M. Secant LNP meets the circle at N and P respectively. If LM = 13.0 cm, and LN = 6.5 cm, find PN.

5 K is the midpoint of chord FG of a circle. If chord JH passes through K such that JK = 9.0 cm and KH = 16.0 cm, find the length of chord FG.

6 Tangent AB meets the circle at B. Secant ACD meets the circle at C and D respectively. If BD is a diameter of a circle, find the length of CD and BD. AB = 8.0 cm, AC = 6.4 cm.

7 Two circles touch at T. P lies on the common tangent through T. Secant PQR meets one circle at Q and R respectively. Secant PSV meets the second circle at S and V respectively. If PQ = 8.0 cm, PS = 6.0 cm and SV = 10.0 cm, find the length of RQ and of PT to 1 decimal place.

Problem Solving

Mathematics abounds with interesting formulas. For example, the volume of a cone is given by

$$V = \frac{B + 4M + T}{6} H,$$

where B is the base cross-sectional area, M the mid cross-sectional area, T, the top cross-sectional area, and H the height.

- Check whether the formula indeed is true.
- Check whether the formula works for a cube, a sphere, a cylinder.

8.8 Arcs and Sectors

In your earlier work with trigonometry, you learned various skills with degrees and radians. To find the lengths of arcs of circles or the areas of sectors, you must use your earlier skills.

Length of Arc

The arc length, a, for a given circle with radius r is given by $a = r\theta$ where θ is the central angle measured in radians.

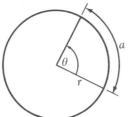

Area of Sector

A sector of a circle is shown by the shaded region.

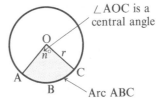

$\angle$ AOC is a central angle

Arc ABC

The area of the sector is given by

$$A = \frac{n}{360} \times \pi r^2$$

where n is the sector angle, in degrees.

Often in mathematics you can derive a different formula if the information is given in another way as shown by the following.

For the circle, the length of the arc ABC is a units. The area, A, of the sector, with radius r, is given by

$$A = \frac{1}{2} ar.$$

Also, since $a = r\theta$, then

$$A = \frac{1}{2}(r\theta)r$$

$$A = \frac{1}{2}r^2\theta \quad \text{where } \theta \text{ is the sector angle in radians.}$$

Example   (a) Find the area, to 1 decimal place, of the sector with central angle $200°$ in a circle of radius 11.4 cm. Use $\pi \doteq 3.14$.

(b) Find the length of the arc, to 1 decimal place, which subtends a central angle of $100°$ in a circle of radius 7.5 cm. Use $\pi \doteq 3.14$.

Solution   (a) Central angle $= 200°$, radius $= 11.4$ cm

Thus, $A = \dfrac{n}{360} \times \pi r^2$

$$\doteq \frac{200}{360} \times 3.14 \times (11.4)^2$$

$$= 226.7 \text{ (to 1 decimal place)}$$

Thus, the area of the sector is 226.7 cm^2 to 1 decimal place.

(b) Central angle = 100°, radius = 7.5 cm.

$$180° = \pi \text{ rad}$$

$$1° = \frac{\pi}{180} \text{ rad}$$

$$100° = \frac{\pi}{180} \times 100 \text{ rad} = \frac{5\pi}{9} \text{ rad}$$

$$a = r\theta$$

$$= 7.5 \times \frac{5\pi}{9}$$

$$\doteqdot \frac{7.5 \times 5 \times 3.14}{9}$$

$$= 13.1 \text{ (to 1 decimal place)}$$

Thus, the length of the arc is 13.1 cm to 1 decimal place.

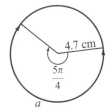

8.8 Exercise

A   Round all answers to 1 decimal place. Use $\pi \doteqdot 3.14$.

1   Find the length of the arc for each of the following.

(a) 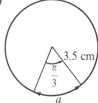   (b) 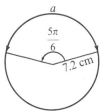   (c)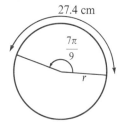

2   Find the radius of the circle for each of the following.

(a) 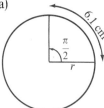   (b) 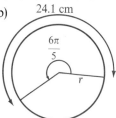   (c)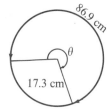

3   Find the sector angle for each of the following.

(a)   (b) 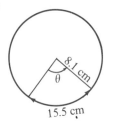   (c)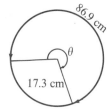

4 Find the area of each sector.

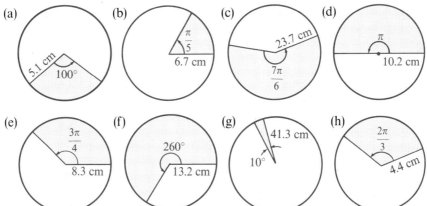

(a) 5.1 cm 100°

(b) $\frac{\pi}{5}$ 6.7 cm

(c) 23.7 cm $\frac{7\pi}{6}$

(d) π 10.2 cm

(e) $\frac{3\pi}{4}$ 8.3 cm

(f) 260° 13.2 cm

(g) 41.3 cm 10°

(h) $\frac{2\pi}{3}$ 4.4 cm

5 Find the radius of each circle for the given sector areas.

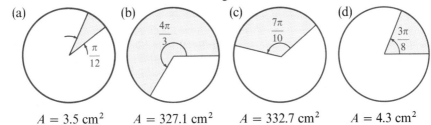

(a) $\frac{\pi}{12}$ $A = 3.5$ cm²

(b) $\frac{4\pi}{3}$ $A = 327.1$ cm²

(c) $\frac{7\pi}{10}$ $A = 332.7$ cm²

(d) $\frac{3\pi}{8}$ $A = 4.3$ cm²

6 Find the radius of each circle. The area of each sector is given.

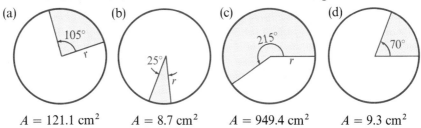

(a) 105° r $A = 121.1$ cm²

(b) 25° r $A = 8.7$ cm²

(c) 215° r $A = 949.4$ cm²

(d) 70° $A = 9.3$ cm²

7 Find the central angle for each of the following.

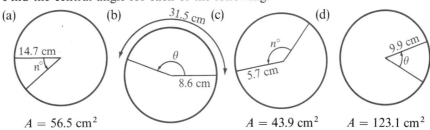

(a) 14.7 cm $n°$ $A = 56.5$ cm²

(b) 31.5 cm θ 8.6 cm

(c) $n°$ 5.7 cm $A = 43.9$ cm²

(d) 9.9 cm θ $A = 123.1$ cm²

B   Round your answers to 1 decimal place.

8   A circle has a diameter of 10.0 m and a central angle of 0.8 rad. Find the length of arc intercepted by the angle.

9   A pendulum with a length of 1.6 m swings through an arc of 20.0 cm. Find the measure of the angle through which the pendulum swings.

10   (a) What is the measure of the central angle for an arc of 2.5 m for a circle with radius 6.2 m?
     (b) Find the area of a sector if the central angle is 108° and the radius of the circle is 2.1 m.

11   Find the central angle, to the nearest degree, if the sector area is 38.2 cm² and the radius is 5.0 cm.

12   (a) Find the area of a sector if the arc length is 9.2 cm and the radius is 8.1 cm.
     (b) Find the length of the arc of a sector if the sector area is 10.2 cm² and the radius is 3.2 cm.

13   The length of the arc of a sector is 12.1 cm and the area of the sector is 27.8 cm². What is the radius?

C 14   Calculate the area of each shaded part.

(a)   (b)

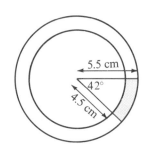

Calculator Tip

To do the calculation $\frac{49}{360} \times \pi r^2$ when $r = 3.65$ cm, use a calculator. Can you find a more efficient procedure than follows?

$$\boxed{c}\ 3.65\ \boxed{x^2}\ \boxed{\times}\ \boxed{\pi}\ \boxed{\times}\ 49\ \boxed{\div}\ 360\ \boxed{=}\ \overset{Output}{?}$$

Use your calculator to do the calculations in this section. Remember: be sure you have the calculator in the radian or degree mode for the appropriate formula.

Practice and Problems: A Chapter Review

1 Find the missing variables.

(a) 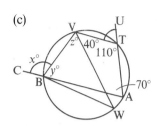 (b) (c)

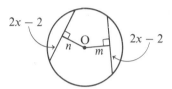

2 Refer to the figure shown. Show that

$$x = \frac{n^2 - m^2}{4}.$$

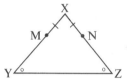

3 In the diagram, XN = XM and XY = XZ. Prove that M, Y, Z, and N are concyclic points.

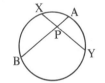

4 The figure RSTM is a cyclic quadrilateral. PTQ is tangent to the circle at T, so that $\angle STQ = 52°$ and $\angle MTP = 36°$. If RS ∥ MT then find the measures of

(a) $\angle RSM$ (b) $\angle SRM$.

5 For the diagram, prove that XP × PY = AP × PB.

6 From the diagram, circles with centre O_1, O_2 and O_3 intersect at P and Q. Prove that O_1 and O_2 and O_3 are collinear.

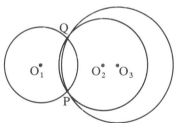

7 △XYZ is an inscribed triangle of a circle with centre C. If W is on YZ such that XW ⊥ YZ, prove $\angle YXW = \angle CXZ$.

8 Two circles with centres P and Q are drawn to intersect at A and B. The line PQ is extended to intersect the circle with centre Q at S so that PQ = PS. Prove that SA and SB are tangent to the circle with centre P.

Test for Practice

1 For each of the following diagrams, what angles are equal?

(a)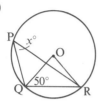

(b)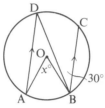

(c)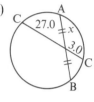

2 Find the value of the variables.

(a)

(b)

(c)

3 What is the value of each variable?

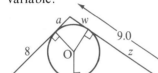

4 Use the diagram. Find the value of ∠BCD. SDR is tangent to the circle at D.

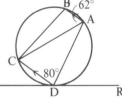

5 In a circle with centre O, a chord, PQ, is 16.0 cm in length. If T is a point of the chord such that OT ⊥ PQ and OT = 5.0 cm, calculate the radius.

6 (a) Find the length of the arc of a circle of radius 2.95 cm if the sector area subtended by the arc is 12.40 cm².
   (b) In a circle, the length of an arc is 11.9 cm and the area of the sector subtended by the arc is 30.2 cm². Find the measure of the radius.

7 For the diagram, use the given information to prove ∠A = 90°.

8 In the diagram, MN = MP. Prove NS = SP.

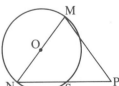

9 A, B, and C are points on a circle so that AB is a diameter. Y and Z are points on CB and AB respectively so that YZ ⊥ AB. Prove that A, C, Y, and Z are concyclic points.

9 Nature of Exponential Functions

concepts and skills with exponents, integral exponents, graphs of functions, exponential growth, exponential decay, solving exponential equations, strategies and solving problems, applications, problem-solving

Introduction

In the study of scientific phenomena, researchers look for patterns and similarities. The study of mathematics often provides a framework, within which the scientific researcher can use the principles of mathematics to report on the scientific findings. For example, mathematics can show relationships between and among phenomena that at first glance seem unrelated. For example, these phenomena have the same characteristics.
• the growth of money   • population growth   • cell growth

Each of these can be related by a graph that has similar properties, as you will see in this chapter.

To determine the age of a fossil, a scientist will use skills, involving exponential functions, which occur throughout the study of geology and archaeology. Similar concepts and skills occur in different situations.

To take another example, after the motor has been shut off, the speed of a large boat and that of a small cruiser exhibit the same characteristics.

Curves relating the speed of the vessels to the time elapsed are said to decay exponentially. In the chapter that follows, you will explore a variety of phenomena that can be described by similar curves.

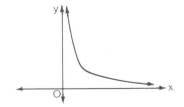

9.1 Process of Mathematics: Exponents

The process of mathematics is often illustrated by the following steps:

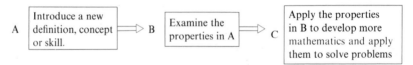

For example, in your earlier work in mathematics the definition of a power was introduced using exponents and bases.

A: **power** $a^n = \underbrace{a \times a \times a \times a \ldots \times a}_{n \text{ factors of } a}$ **exponent** n is the exponent in a^n.
base a is the base in a^n.

By using the definition of a power you developed the properties of exponents commonly called the Law of Exponents. The following chart summarizes the Laws of Exponents for $m, n \in N$.

B:

Laws of Exponents with $m, n \in N$	
product of powers	$a^m \times a^n = a^{m+n}$
quotient of powers	$a^m \div a^n = a^{m-n}, m > n, a \neq 0$
power of a power	$(a^m)^n = a^{mn}$
power of a product	$(ab)^m = a^m b^m$
power of a quotient	$\left(\dfrac{a}{b}\right)^m = \dfrac{a^m}{b^m}, b \neq 0$

C: In your subsequent work in this chapter you will apply your skills with exponents to develop further useful mathematics. However to do so you must clearly understand A and B.

In Example 1, the Law of Exponents are applied to simplify expressions that involve exponents.

Example 1   (a) Calculate $(2^2)^3 \times 4^2 \div 8$.   (b) Simplify $(2xy)^3 \times (3x^2)^2 \div \left(\dfrac{3x}{y}\right)^2$

Solution   (a) Express the powers with the same base.

$(2^2)^3 \times 4^2 \div 8$
$= (2^2)^3 \times (2^2)^2 \div 2^3$ ←
$= 2^6 \times 2^4 \div 2^3$
$= 2^{6+4-3}$ $4^2 = (2^2)^2$
$= 2^7$ or 128 $8 = 2^3$

(b) $(2xy)^3 \times (3x^2)^2 \div \left(\dfrac{3x}{y}\right)^2$

$= 8x^3y^3 \times 9x^4 \div \dfrac{9x^2}{y^2}$

$= 8x^3y^3 9x^4 \times \dfrac{y^2}{9x^2}$

$= \left(\dfrac{8 \times 9}{9}\right) \times (x^{3+4-2}) \times (y^{3+2})$

$= 8x^5y^5$

In the next example, an important technique is illustrated for organizing your work to prove relationships in mathematics.

Example 2   Prove that for all $n \in N$, $\dfrac{4^n \times 2^{2n+1} \times 8^n}{32^n} = \left(\dfrac{4^n \times 2^{n+4}}{8^{n+1}}\right) \times 2^{2n}$.

Solution   To prove the result, simplify each side.

$$LS = \frac{4^n \times 2^{2n+1} \times 8^n}{32^n}$$

$$= \frac{(2^2)^n \times 2^{2n+1} \times (2^3)^n}{(2^5)^n}$$

$$= \frac{2^{2n} \times 2^{2n+1} \times 2^{3n}}{2^{5n}}$$

$$= \frac{2^{2n+2n+1+3n}}{2^{5n}}$$

$$= \frac{2^{7n+1}}{2^{5n}}$$

$$= 2^{2n+1}$$

Express all powers with the same base.

$$RS = \left(\frac{4^n \times 2^{n+4}}{8^{n+1}}\right) \times 2^{2n}$$

$$= \frac{(2^2)^n \times 2^{n+4}}{(2^3)^{n+1}} \times 2^{2n}$$

$$= \frac{2^{2n} \times 2^{n+4} \times 2^{2n}}{2^{3n+3}}$$

$$= \frac{2^{5n+4}}{2^{3n+3}}$$

$$= 2^{2n+1}$$

Thus LS = RS. The expression is true for all $n \in N$.

9.1 Exercise

A   Questions 1 to 8 review various skills involving your earlier work with exponents.

1   Use the definition of a^n to prove each of the following. List any restrictions on your results.

(a) $a^m \times a^n = a^{m+n}$ (b) $a^m \div a^n = a^{m-n}$ (c) $(a^m)^n = a^{mn}$

(d) $(ab)^m = a^m b^m$ (e) $\left(\dfrac{a}{b}\right)^m = \dfrac{a^m}{b^m}$

2   Express
(a) 4^3 as a power of 2.   (b) 2^{12} as a power of 4.
(c) 9^8 as a power of 3.   (d) 3^{12} as a power of 9.

3   Express each of the following as a power of 3.
(a) $(3^3)^5$ (b) $(3^9)(3^3)$ (c) $(9)(3^7)$ (d) $(9)(27^2)$ (e) $(9)(3^4)$
(f) $(3^m)(3^n)$ (g) $(3^m)^n$ (h) 27^m (i) 9^n (j) $(3^m)(27)$

4   Evaluate.

(a) 2^5 (b) $6^6 \div 6^5$ (c) $(-4)^3$ (d) $(-1)^{10}$

(e) $(0.3)^2$ (f) $(2)(2^4)$ (g) $-(3^2)^3$ (h) $\dfrac{2^{11}}{2^8}$

(i) $\left(\dfrac{1}{3}\right)^8 \div \left(\dfrac{1}{3}\right)^6$ (j) $(1^2)^3 + (1^3)^2$

5   Simplify each of the following.

(a) $\dfrac{5^4 \times 5^5}{(5^3)^2}$ (b) $\dfrac{2^2 \times 2^4 \times 2^3}{(2^5)^2}$ (c) $\dfrac{27^3}{3^4}$ (d) $\dfrac{(125)^2}{25^2}$

6   a^n is the **exponential form** of a power. Write each of the following in exponential form.

(a) $(2^5)(2^7)$ (b) $(2^5)^2$ (c) $(3^3)(3^5)(3^6)$ (d) $2^7 \div 2^4$
(e) $(-3)^4(-3)^5$ (f) $(-4)^8 \div (-4)^2$ (g) $(3^4)^5 \div (3^2)^2$ (h) $(\pi^7)(\pi)$

7   Simplify each of the following.

(a) $(4x)^2$ (b) $(-3x)^3$ (c) $(x^3)^4$
(d) $(3x)(x^7)$ (e) $(5x)(4x^4)$ (f) $(-3x^2)(-6x^3)$
(g) $(10x^8) \div (2x^4)$ (h) $(-2x)^5(2x^2)$ (i) $(x^p)^3(x^3)^p$
(j) $(3x^4)^{3p} \div (2x^5)^{2p}$ (k) $(-x)^{5p-3}(x)^{2p+3}$ (l) $(x^7)(x^3) \div (x^3)^3$
(m) $\dfrac{(x^5)(x^2)}{(x^2)^6}$ (n) $\dfrac{(x^{15})(x^{24})(x^{25})}{(x^7)^8}$

8   Simplify.

(a) $3x^5 \times 2x^5$ (b) $3x^5 + 2x^5$ (c) $3x^5 \div 2x^5$ (d) $3x^5 - 2x^5$

B   List the Laws of Exponents, illustrating each law with examples of your own.

9   Simplify each of the following.

(a) $x^7 \times x^2 \div x^3$ (b) $y^a \times y^b \div y^c$ (c) $(a^3)^3 \div a^7$
(d) $x^{2a+b} \div x^{2a-b}$ (e) $a^{x+y} \times a^{2x+y} \div a^{2x-7}$ (f) $\dfrac{(ab)^{2x+y}}{a^x b^y}$

10   For each of the following, which is greater, A or B?

	A	B
(a)	$\dfrac{14^2}{7^2}$	$\dfrac{15^2}{3^2}$
(b)	$\dfrac{9^a \times 3^{a+3}}{27^a}$	$\dfrac{2^6 \times 64^2}{32^3}$

11. Simplify.

(a) $\left(\dfrac{a}{b}\right)^3\left(\dfrac{a}{c}\right)^2\left(\dfrac{c}{b}\right)^3$

(b) $\dfrac{(x^a)^4}{x^{3a+b}} \times \dfrac{(x^b)^4}{x^{a+3b}}$

(c) $\left(\dfrac{x}{y}\right)^{2a}\left(\dfrac{y}{x}\right)^{2b}(xy)^{2a+2b}$

12. Use $x = 1$, $a = -1$ and $b = 2$ to evaluate each of the following.

(a) $\dfrac{(6^{2x+y})(6^y)}{6^{2y}}$

(b) $\dfrac{(2^x)(2^{3x-1})(8^x)}{4^{x+1}}$

(c) $\dfrac{30a^4b^2}{5ab^4} \times \dfrac{15a^2b^5}{45a^3b^6}$

13. Prove.

(a) $2^k + 2^k = 2^{k+1}$

(b) $\dfrac{(3^a)(3^{a+1})(3^{a+2})}{27^a} = 27$

(c) $\dfrac{4}{2^n}(3^{n-2})(2^{n-2}) = \dfrac{1}{3^{2-n}}$

14. (a) Prove that $(2^3)^2 \neq 2^{3^2}$.

(b) For what values of m is $(m^3)^2 = m^{3^2}$?

C

15. Prove that for all $n \in N$, $\dfrac{5^{n+1} + 5^{2n}}{5 + 5^n} = 5^n$.

Problem Solving

You can use your calculator to explore the value of a^x when x takes on any real number value. To do so you use the $\boxed{y^x}$ key on the scientific calculator. To calculate a^x on the calculator use the procedure shown.

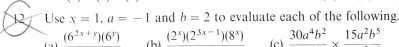

Value	input	output
a^x	$\boxed{C}$ $a\boxed{y^x}$ x $\boxed{=}$	?
$4^{0.5}$	$\boxed{C}$ $4\boxed{y^x}0.5\boxed{=}$	?
$27^{\frac{1}{2}}$	$\boxed{C}27\boxed{y^x}0.5\boxed{=}$	?

To calculate $27^{\frac{1}{2}}$ you need to use $27^{0.5}$.

A Remember: To calculate $3^{\frac{1}{2}}$ you need to use $3^{0.5}$. Calculate $2^{\frac{1}{2}}$, $3^{\frac{1}{2}}$, $4^{\frac{1}{2}}$, $5^{\frac{1}{2}}$, $6^{\frac{1}{2}}$, $7^{\frac{1}{2}}$, $8^{\frac{1}{2}}$, $9^{\frac{1}{2}}$.
What do you notice about your answers?

B Calculate $16^{\frac{1}{2}}$, $9^{\frac{1}{2}}$, $8^{\frac{1}{3}}$, $27^{\frac{1}{3}}$, $16^{\frac{1}{4}}$.
What do you notice about your answers?

C Calculate $2^{\frac{1}{2}}$, $2^{\frac{1}{3}}$, $2^{\frac{1}{4}}$, $2^{\frac{1}{5}}$, $2^{\frac{1}{6}}$.
What do you notice about your answers?

D Calculate $4^{\frac{1}{4}}$, $5^{\frac{1}{5}}$, $6^{\frac{1}{6}}$, $7^{\frac{1}{7}}$, $8^{\frac{1}{8}}$, $9^{\frac{1}{9}}$, $10^{\frac{1}{10}}$.
What do you notice about your answers?

E How can you interpret the value of $a^{\frac{1}{2}}$? of $a^{\frac{1}{3}}$? Find the value of $(6.83)^{\frac{1}{2}}$, $(6.83)^{\frac{1}{3}}$, $(12.95)^{\frac{1}{2}}$, $(12.95)^{\frac{1}{3}}$.

9.2 What if . . . Integral Exponents?

A question often asked in the study of mathematics is "What if . . . ?" In particular, when working with exponents, "What if the exponents include negative integers?". While answering this question and others like it beginning with "What if . . .", often the answer develops even further mathematics.

The following suggests how to define a^0 and a^{-1}.

The meaning of a^0:

Since $\dfrac{a^n}{a^n} = a^{n-n} = a^0$, $a \neq 0$ and since $\dfrac{a^n}{a^n} = 1$, $a \neq 0$ then it appears for the Laws of Exponents to include zero exponents, a^0 must be defined as:

$$a^0 = 1, \quad a \neq 0$$

The meaning of a^{-n}:

Since $a^n \times a^{-n} = a^{n-n}$, $n > 0$, and since $a^0 = 1$, $a \neq 0$, then you obtain $a^n \times a^{-n} = 1$. From this result, you apply your equation skills to obtain the results, A and B.

A: Divide both sides by a^{-n}.

$$\frac{a^n \times a^{-n}}{a^{-n}} = \frac{1}{a^{-n}}$$

$$a^n = \frac{1}{a^{-n}}$$

B: Divide both sides by a^n.

$$\frac{a^n \times a^{-n}}{a^n} = \frac{1}{a^n}$$

$$a^{-n} = \frac{1}{a^n}$$

Thus it appears that for the Laws of Exponents to include negative exponents, you need to define a^{-n} as follows:

$$\boxed{a^{-n} = \frac{1}{a^n}, \quad a \neq 0, n > 0} \qquad \boxed{a^n = \frac{1}{a^{-n}}, \quad a \neq 0, n > 0}$$

The definitions of a^0 and a^{-n} are applied in the following example.

Example Simplify $\dfrac{4xy^{-2}}{2x^{-2}y^{-1}}$. Express your answer with positive exponents.

Solution
$$\frac{4xy^{-2}}{2x^{-2}y^{-1}} = \frac{4}{2}x^{1-(-2)}y^{-2-(-1)}$$
$$= 2x^3y^{-1}$$
$$= \frac{2x^3}{y}$$

9.2 Exercise

A List the meanings of a^0 and a^{-n}. Provide an example of your own to illustrate their meanings.

1 Find the value of each of the following.
(a) 4^0 (b) 7^{-2} (c) $(0.03)^{-2}$ (d) -3^{-3} (e) $(-3)^{-3}$
(f) -3^3 (g) 3^{-3} (h) -100^0 (i) $(-0.1)^{-5}$ (j) 27^0

2 Evaluate.
(a) $\dfrac{3^{-1}}{2^{-1}}$ (b) $3^{-1} + 2^{-1}$ (c) $\dfrac{1}{3^{-1} + 4^{-1}}$ (d) $3^{-1} \times 3^2 \times 3^{-2}$

(e) $(4^{-3})^2$ (f) $[(-3)^2]^{-3}$ (g) $[(-2)^{-3}]^{-1}$ (h) $\left(\dfrac{1}{3}\right)^{-1}\left(\dfrac{3^{-3}}{3^{-2}}\right)$

(i) $-\dfrac{3^{-2}}{3^{-3}} - (0.3)^0 + \left(\dfrac{1}{5}\right)^{-1}$ (j) $2^2 + \dfrac{1}{2^{-1}} - 5 + 10^0$ (k) $\dfrac{\left(\dfrac{1}{4}\right)^{-2} + \left(\dfrac{1}{3}\right)^{-2}}{\left(\dfrac{1}{5}\right)^{-2}}$

(l) $\dfrac{\left(\dfrac{1}{2}\right)^{-2} \times \left(\dfrac{1}{2}\right)^{-1}}{2^{-1} + 3 \times 2^{-1}}$ (m) $\left(\dfrac{(0.7)^2(1.7)^{-3}}{0.7}\right)^0$ (n) $\left[\left(\dfrac{9^{-2}}{13^{-3}}\right)^{-1}\right]^0$

3 Simplify.
(a) $x^4 \times x^{-2} \times x^0$ (b) $y^{-3} \times y^3 \times y^2$ (c) $(x^{-4})^2$ (d) $(z^{-3})^{-2}$

(e) $\left(\dfrac{3^{-2}}{a^4}\right)^0$ (f) $\left(\dfrac{a}{2b^{-1}}\right)^{-5}$ (g) $\left(\dfrac{a^{-2}}{2^0 a^{-3}}\right)^{-1}$

4 Express each of the following with positive exponents.
(a) $x^3 y^{-2}$ (b) $\dfrac{3x^{-1}y}{p^{-1}}$ (c) $\dfrac{3xg^{-2}}{4g^{-2}}$

(d) $\dfrac{3^{-1}p^2 b^{-1}}{s^{-2}}$ (e) $\dfrac{(5p)^{-1}s}{4^{-2}}$ (f) $\dfrac{x^{-3}y^{-2}}{n^{-1}}$

B The skills you practise will be applied in your future work in exponents.

5 Express each of the following with positive exponents. Simplify each result.
(a) $p^{-3}q^2$ (b) $3x^{-2}$ (c) $\left(\dfrac{2m}{n}\right)^{-4}$ (d) $(3x)^{-2}$

(e) $9\left(\dfrac{3}{a}\right)^{-4}$ (f) $(3w)^4 \div (3w)^{-3}$ (g) $\dfrac{49}{7y^{-2}}$ (h) $(5w^{-3})^{-1}$

6 Simplify.

(a) $\dfrac{w^4 \times w^{-6}}{w^{-2}}$ (b) $v^{-3} \times v^{-5} \times v^2$ (c) $(3b)^{-5}b^7$

(d) $\dfrac{k^5 \div k^{-4}}{k^5 \times k^{-4}}$ (e) $\dfrac{4d^{-5} \times 12d^{-6}}{d^2 \div 3d}$ (f) $\dfrac{(0.5)^0 z^{-3} \times z}{(0.5)^{-1} z^{-1} \times z^4}$

(g) $\dfrac{(xy^3)^2 \times x^2 \times y^4}{3x^3 \times y^{-2}}$ (h) $\dfrac{7m^{-1}}{n^4}\left(\dfrac{m^2}{n^2}\right)^{-2}$ (i) $\left(\dfrac{3}{2s^{-2}}\right)^{-2}\left(\dfrac{1}{s^2}\right)^{-1}$

7 Express each of the following with a single base.

(a) $m^{x+y}m^x$ (b) $s^{2x-p} \div s^{p-x}$ (c) $(t^x)^2(t^3)^x$

(d) $q^{3x-1} \div (q^x)^2$ (e) $(y^m)(y^n) \div y^{2m-n}$ (f) $(k^2)^m(k^3)^{n+m}$

(g) $(g^{a+b})(g^{b+c})(g^{a-c})$ (h) $(q^{2a-b})(q^{b-2c}) \div (q^{a+b})$

8 Simplify each of the following. Record your answers with positive exponents.

(a) $3q^{-1}(q + q^{-3})$ (b) $(m + n)(m^{-1} + n^{-1})$ (c) $\dfrac{1 - 2p^{-1}}{1 + 2p^{-1}}$

9 Evaluate each of the following if $a = -4$ and $k = 1$.

(a) $(a^3)^k(a^k)^2(a^{2k-2})$ (b) $(a^{k+1})(a^{2k-1}) \div (a^{3k-1})$

(c) $(a^k)^2(a^2)^k \div (a^{2k-3})$

C 10 If $x = 2$, $y = -1$, and $z = -2$, find the value of each of the following expressions.

(a) $\dfrac{4xy^{-1} - 10xy^2}{20xy^{-3}}$ (b) $\dfrac{6x^{-2}y^{-1}z^{-1}}{7xy^{-2}}$ (c) $\left(\dfrac{y^{-1}}{x^2}\right)^{-2} - \left(\dfrac{3x^4}{y^{-2}}\right)$

(d) $\left(-\dfrac{y^2}{z}\right)^{-3} + \dfrac{5y^3}{z^6}$ (e) $\left(\dfrac{y^{-1} - z^{-1}}{y^{-2} - z^{-2}}\right)\left(\dfrac{1}{x}\right)^0$ (f) $\dfrac{x^{-1} + z^{-1}}{(x + z)^{-1}}$

Problem Solving

Estimate which of the following pipe systems carry more water?

A   5 pipes, each with a diameter of 10 cm.
B   10 pipes, each with a diameter of 5 cm.

Inventory: Rational Exponents

A lot of mathematics is developed when the question "What if . . ." is asked. As you have seen earlier, the use of integral exponents has been given a meaning. The definition of rational exponents, given below, can be used to develop more concepts about radicals and radical equations.

If $n \in N$, then $\sqrt[n]{b} = b^{\frac{1}{n}}$. If n is even then $b \geq 0$.	For $x \in I,\ y \in N,\ b^{\frac{x}{y}} = \sqrt[y]{b^x}$ or $(\sqrt[y]{b})^x$ If y is even then $b \geq 0$.

As the final step in the process, exponents are extended to include all real numbers.

The inventory that follows is based on your earlier skills with exponents. Refer to your earlier work with exponents to complete the inventory.

11 Write each of the following in radical form.

(a) $7^{\frac{1}{2}}$ (b) $x^{\frac{1}{5}}$ (c) $a^{\frac{1}{4}}$ (d) $8^{\frac{3}{5}}$ (e) $b^{\frac{3}{7}}$ (f) $5x^{\frac{2}{3}}$ (g) $x^{-\frac{1}{3}}$ (h) $a^{-\frac{2}{5}}$

12 Write each of the following in exponent form.

(a) $\sqrt{6}$ (b) $\sqrt{10}$ (c) $\sqrt[3]{7}$ (d) $(\sqrt[7]{12})^3$ (e) $(\sqrt[7]{8})^9$ (f) $\left(\dfrac{1}{\sqrt{7}}\right)^2$

13 Express the following with exponents.

(a) $\sqrt[5]{x}$ (b) $\dfrac{1}{\sqrt[3]{x}}$ (c) $\dfrac{1}{\sqrt{x^3}}$ (d) $\sqrt[6]{a^5}$ (e) $\dfrac{1}{\sqrt[4]{a^3}}$ (f) $\left(\dfrac{1}{\sqrt[3]{a}}\right)^{-2}$

14 Simplify each of the following.

(a) $3^{\frac{1}{2}} \times 3^{\frac{1}{4}}$ (b) $5^{\frac{1}{9}} \times 5^{\frac{1}{3}}$ (c) $(a^{\frac{1}{4}} \times b^{\frac{1}{2}})^2$ (d) $(a^{\frac{1}{3}}b^{\frac{1}{2}})^6$

(e) $(x^5y^{10}z^{15})^{\frac{1}{5}}$ (f) $(y^{\frac{1}{4}} + 2y^{\frac{3}{4}})y^{\frac{3}{4}}$ (g) $(16x^{12}y^8)^{\frac{1}{4}}$ (h) $\left(\dfrac{m^5}{2}\right)^{\frac{1}{6}} \times \left(\dfrac{4}{m}\right)^{\frac{5}{6}}$

15 Evaluate.

(a) $27^{\frac{1}{3}}$ (b) $8^{\frac{2}{3}}$ (c) $81^{\frac{3}{4}}$ (d) $32^{\frac{3}{5}}$ (e) $64^{\frac{5}{6}}$ (f) $4^{0.5}$ (g) $81^{0.75}$ (h) $(16^2)^{\frac{1}{4}}$

16 Evaluate.

(a) $32^{\frac{3}{5}}$ (b) $(27^2)^{-\frac{1}{3}}$ (c) $64^{-\frac{5}{6}}$ (d) $81^{-0.75}$ (e) $10\,000^{\frac{3}{4}}$

(f) $0^{1.78}$ (g) $625^{-\frac{1}{4}}$ (h) $\left(\dfrac{25}{4}\right)^{\frac{5}{2}}$ (i) $(0.04)^{-\frac{1}{2}}$ (j) $(\sqrt{32})^{0.4}$

17 Arrange the values of each of the following from smallest to greatest value.

A $\quad 27^{-\frac{2}{3}}$ B $\quad 216^{\frac{2}{3}}$ C $\quad 16^{-\frac{5}{4}}$ D $\quad 64^{-\frac{1}{6}}$

9.3 Graphing Exponential Functions

Exponential growth is illustrated by the example at the right. A person sends a letter to each of two people. They in turn each send a letter to two people. The number of letters, y, at each step grows exponentially and can be described by the equation $y = 2^x$, for $x = 1, 2, 3, 4 \ldots$.

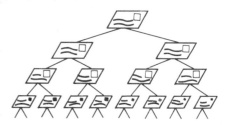

To draw the graph of the function described by $y = 2^x$, $x \in I$, begin with a table of values.

x	-3	-2	-1	0	1	2	3	4	5
2^x	$\frac{1}{8}$	$\frac{1}{4}$	$\frac{1}{2}$	1	2	4	8	16	32

Step 1

Then plot the points generated by the table of values.

Step 2

For all $x \in R$, the graph is obtained by drawing a smooth curve through the points that have been plotted.

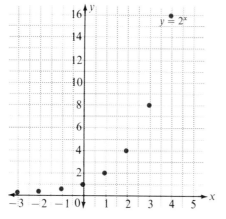

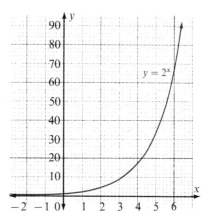

For each value of x there is exactly one value of y. Thus, the graph of $y = 2^x$, $x \in R$ is the graph of a function. From the graph, certain properties of the graph of $y = 2^x$, called the exponential function, can be seen.

Intercepts

- The y-intercept is 1.
- There is no x-intercept. (There is no value of x for which $2^x = 0$.)

Domain and range

- The domain is all real numbers.
- The range is all positive real numbers.
- 2^x is always greater than zero.

One-to-One

For each value of y, there corresponds a unique value of x. The function defined by $y = a^x$ or $f(x) = a^x$ is said to be *one-to-one*.

In industrial work or scientific work in a laboratory, the data obtained from an experiment often can not be related using algebra. Thus the data is recorded graphically and the solution to many problems can be read from the graph. Graphs can be used to obtain approximate values of the roots of equations. Refer to the graph of $y = 2^x$ as shown.

Example   Find the roots of each equation to 1 decimal place.

(a) $2^x = 3$ (b) $2^x = 14$

Solution   (a) Read the graph of $y = 2^x$. When $y = 3$, $x = 1.6$. The equation $2^x = 3$ is solved when $x = 1.6$. Thus, $x = 1.6$ is a root (to one decimal place).
(b) From the graph $y = 2^x$, $y = 14$ when $x = 3.8$. Thus, the root of $2^x = 14$ is $x = 3.8$ (to 1 decimal place).

For three different values of a, the graphs of the exponential functions defined by $y = a^x$, $x \in R$, are shown.

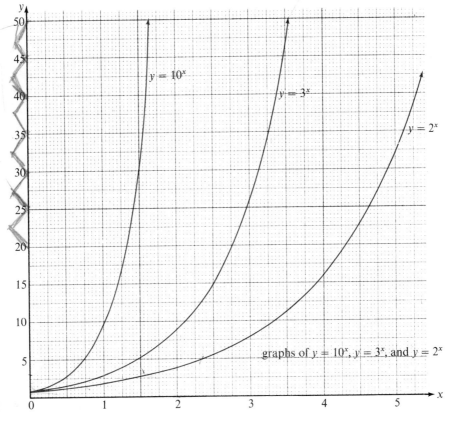

graphs of $y = 10^x$, $y = 3^x$, and $y = 2^x$

In general, the exponential function is defined by the equation,

$$y = a^x, \qquad a > 0 \text{ and } x \in R,$$
$$\text{or} \quad f(x) = a^x, \qquad a > 0 \text{ and } x \in R.$$

9.3 Exercise

A Use the graphs for $y = 2^x$, $y = 3^x$ and $y = 10^x$ as needed.

1 From the graph of $y = 2^x$; find the value of each of the following to one decimal place.

(a) $2^{1.3}$ (b) $2^{4.5}$ (c) $2^{3.8}$ (d) $2^{2.7}$ (e) $2^{3.3}$ (f) $2^{5.1}$

2 From the graph of $y = 3^x$, $x \in R$ find the value to 1 decimal place.

(a) $3^{2.1}$ (b) $3^{3.6}$ (c) $3^{1.5}$

3 From the graph of $y = 10^x$, $x \in R$, approximate each value to 1 decimal place.

(a) $10^{1.3}$ (b) $10^{1.8}$ (c) $10^{1.5}$

4 Solve each equation. Express the value of x to 1 decimal place.

(a) $2^x = 6$ (b) $3^x = 7$ (c) $10^x = 20$ (d) $10^x = 49$ (e) $2^x = 10$

5 Use the graph of $y = 10^x$, $x \in R$. Find the missing value. Express your answer to 1 decimal place.

(a) $(1.2, ?)$ (b) $(?, 50)$ (c) $(1.5, ?)$ (d) $(?, 60)$ (e) $(1.7, ?)$

B The skills for sketching graphs that you learned earlier are applied to the graphs of exponential functions.

6 (a) Construct a graph of $y = 4^x$, $x \in R$.

(b) Why is the graph in part (a) the graph of an exponential function?

(c) List the properties of the graph in part (a).

7 Use the same set of axes. Construct the graphs defined by

$$y = 2^x \text{ and } y = \left(\frac{1}{2}\right)^x.$$

(a) How are the graphs alike?   (b) How do the graphs differ?

(c) What transformation mapping of $y = 2^x$ will give $y = \left(\frac{1}{2}\right)^x$ as its image?

8   Use the same set of axes. Construct the graphs defined by
$y = 3^x$ and $y = \left(\dfrac{1}{3}\right)^x$.

(a) How are the graphs alike?   (b) How do the graphs differ?

(c) What transformation mapping of $y = 3^x$ will give $y = \left(\dfrac{1}{3}\right)^x$ as its image?

9   (a) Construct a graph of $y = 2^{-x}$.
(b) Did you construct the graph of a function?
(c) List the properties of the graph in part (a).

10   (a) On the same set of axes construct a graph of A: $y = 3^{-x}$ B: $y = 4^{-x}$.
(b) What properties do the graphs in part (a) have in common?
(c) How do the graphs in part (a) differ?

11   (a) Why is the term *curves of growth* appropriate to describe the graphs
of $y = 2^x$, $y = 3^x$, and so on?

(b) Why is the term *curves of decay* appropriate to describe the graphs of
$y = 2^{-x}$, $y = 3^{-x}$, and so on?

12   (a) Construct the graphs of

A $\boxed{y = \left(\dfrac{1}{2}\right)(2^x)}$ and   B $\boxed{y = 2^{x-1}}$

(b) Show algebraically why the graphs of A and B are the same.
(c) Construct the graphs of

C $\boxed{y = \left(\dfrac{1}{3}\right)(3^x)}$ and   D $\boxed{y = 3^{x-1}}$

(d) Show algebraically why the graphs of C and D are the same.

13   Use the graph of $y = 2^x$, $x \in R$ as a basis. On the same set of axes, sketch
graphs of each of the following.
(a) $y = 2^x + 2$ (b) $y = 2^x - 3$ (c) $y = 2^{x-2}$
(d) $y = 2^{x+2}$ (e) $y = 2^{x+1} + 3$ (f) $y = 2^{x-2} - 3$

14   Use the graph of $y = 3^x$, $x \in R$ as a basis. On the same set of axes sketch
the graph of each of the following.
(a) $y = 3^x - 2$ (b) $y = 3^{x-2}$ (c) $y = (2)(3^x)$ (d) $y = (2)(3^x) + 2$

C 15   Sketch the graph of each of the following.
(a) $y = (3)(2^{x-2}) - 1$ (b) $y = (2)(3^{x-3}) + 4$

Applications: Growth and Decay

To determine whether material is
radioactive, a Geiger Counter is used.
The Geiger Counter registers information
as *counts per minute*. The more counts
per minute the more radioactive the
material is.

*Hans Geiger is known for his work with Mueller
in the development of the Geiger-Mueller
Counter for the detection of atomic particles
and radiation. The piece of equipment used is
called a Geiger Counter.*

The following questions are based on the information in this chart. For
radioactive material, a chart is used to record the counts per minute over
a period of time.

Time elapsed in hours (t)	0	6	12	18	24
Counts per minute (c)	140	70	40	18	4.2

16 (a) Construct a graph of the data in the chart. Draw a smooth curve
    through the plotted points.
    (b) Which term best describes the graph, *growth* or *decay*?

17 You may use the graph to *interpolate* values from the graph. For what
    value of t are the following values of c obtained?
    (a) 100   (b) 80   (c) 30

18 For each value of t, what is the corresponding value of c?
    (a) 3   (b) 9   (c) 20

19 For an investment, the value, v, at certain times t is recorded.

t time elapsed in years	0	1	2	3	4
v value in dollars	350	402.50	462.88	532.31	612.15

(a) Construct a graph of the data in the chart. Draw a smooth curve
    through the plotted points.
(b) Which term best describes the graph in part (a), *growth* or *decay*?
(c) Use the graph to interpolate the values of v for each value of t in years.
    (i) 1.5   (ii) 2.4   (iii) 3.8
(d) How much time will it take the original investment v to reach the value
    (i) $500   (ii) $575?

9.4 Using Graphs

The skill of reading data from a graph is important in the study of many fields of science since often the only method of solving a problem is using data or results obtained from a graph. As you have seen, any number can be expressed as a power of 2. For example, you can read various values from the previous exponential graph, defined by $y = 2^x$.

$$6.5 = 2^{2.7} \qquad 13 = 2^{3.7}$$

In the development of mathematics, previously learned mathematics is examined in a new situation. For example, from the graphical data, any number can be expressed as a power with base 2. What if you examine this information in view of your earlier Laws of Exponents?

From the graphical data and the Laws of Exponents, you can combine the skills and illustrate how these powers may be used to find approximate values.

$$
\begin{aligned}
13 \times 6.5 &\doteq 2^{2.7} \times 2^{3.7} \\
&= 2^{2.7+3.7} \longleftarrow \text{Use the Laws of Exponents.} \\
&= 2^{6.4} \longleftarrow \text{Find the value of } 2^{6.4} \text{ from the graph. } 2^{6.4} \doteq 85.
\end{aligned}
$$

Thus $13 \times 6.5 \doteq 85$.

The above development illustrates how the process of mathematics expands into new ideas and additional skills which eventually paved the way for the current use and understanding of calculations using calculators.

Example   Use the graph of $y = 2^x$ to find the approximate value of

(a) $3.2 \times 2^{1.6}$ (b) $(6.5)^{\frac{1}{3}}$

Steps
1   Express each decimal number in exponential form.
2   Simplify exponents using the Laws of Exponents.
3   Express the power as a decimal number.

Solution   (a) From the graph $3.2 \doteq 2^{1.7}$.

$$
\begin{aligned}
3.2 \times 2^{1.6} &\doteq 2^{1.7} \times 2^{1.6} \\
&= 2^{1.7+1.6} \\
&= 2^{3.3} \\
3.2 \times 2^{1.6} &\doteq 9.8
\end{aligned}
$$

> Read the value of $2^{3.3}$ from the graph.

(b) From the graph $6.5 \doteq 2^{2.7}$.

$$
\begin{aligned}
(6.5)^{\frac{1}{3}} &\doteq (2^{2.7})^{\frac{1}{3}} \\
&= 2^{\frac{1}{3}(2.7)} \\
&= 2^{0.9} \\
&\doteq 1.9
\end{aligned}
$$

9.4 Exercise

A To perform calculations based on graphs and the Laws of Exponents, refer to the previous graphs of $y = 2^x$, $y = 3^x$, and $y = 10^x$. The graph you are to use will be stated in the question.

1 (a) Use the graph of $y = 2^x$ to calculate the approximate value of 3.6×2.8 to 1 decimal place.
   (b) Use the graph of $y = 3^x$. Calculate the approximate value of 3.6×2.8 to 1 decimal place.
   (c) What do you notice about your answers in parts (a) and (b)?

2 (a) Use the graph of $y = 2^x$ to calculate $\sqrt[3]{12}$.
   (b) Use the graph of $y = 3^x$. Calculate $\sqrt[3]{12}$ again.
   (c) What do you notice about your answers in parts (a) and (b)?

B 3 Use the graph of $y = 2^x$. Find the value of each of the following.
   (a) 5.6×3.9 (b) $18 \div 6.9$ (c) 4.9^2 (d) $\sqrt{6.8^3}$ (e) $(23)^{\frac{2}{3}}$

4 Use the graph of $y = 10^x$. Find the value of each of the following.
   (a) 5.6×3.9 (b) $18 \div 6.9$ (c) $(4.9)^2$ (d) $\sqrt{6.8^3}$ (e) $(23)^{\frac{2}{3}}$
   (f) Compare your answers to those obtained using $y = 2^x$ in the previous question. What do you notice?
   (g) Which approximate values are more accurate, A or B?

   A: those from the graph of $y = 2^x$?
   B: those from the graph of $y = 10^x$?

5 Use the graph of $y = 2^x$. Find an approximate value of each of the following.
   (a) $\sqrt{4.8 \times 12}$ (b) $\sqrt{30} \times 2\sqrt{20}$

6 (a) Use the graph of $y = 10^x$ to calculate again each of the values in the previous question.
   (b) Compare your answers to those obtained using $y = 2^x$. What do you notice?
   (c) Which values are more accurate, A or B?

   A: those from the graph of $y = 2^x$
   B: those from the graph of $y = 10^x$

7 Use the graph of $y = 10^x$ to calculate each of the following.
   (a) $\sqrt[5]{36} \times \sqrt[4]{44}$ (b) $(18)^{15} \div \sqrt[3]{20}$ (c) $(36)^{\frac{1}{4}} \times \sqrt[5]{3^3}$

9.5 Using Principles: Exponents

When you examine the face of a scientific calculator you will notice there is an exponential button of the form $\boxed{y^x}$. To use the calculator dealt with in the next section, you must understand the principles of exponents and their applications to calculators. To examine these principles, the exponential table for $y = 10^x$, $x \in R$, has been generated by a computer, part of which is reproduced here.

	0	1	2	3	4
0.68	4.786	4.797	4.808	4.819	4.831
0.69	4.898	4.909	4.920	4.932	4.943
0.70	5.012	5.023	5.035	5.047	5.058
0.71	5.129	5.140	5.152	5.164	5.176
0.72	5.248	5.260	5.272	5.284	5.297
0.73	5.370	5.383	5.395	5.408	5.420

From the table, numbers can be expressed as a power of 10.

$10^{0.680} = 4.786$ The degree of accuracy available from the table is
$10^{0.681} = 4.797$ far greater than that available from a graph but
$10^{0.682} = 4.808$ less than that obtained from computers or calculators.
$10^{0.683} = 4.819$
$10^{0.684} = 4.831$

The exponential table can be used to illustrate the principles for two kinds of evaluations. First, any number can be expressed as a power of 10.

Example 1   Express each number as a power of 10.   Refer to the complete
(a) 5.272   (b) 0.004 875   tables at the back of
the book.

Solution   From the table,
(a) $5.272 = 10^{0.722}$
(b) First write the number in scientific form.

$$0.004\ 875 = 4.875 \times 10^{-3}$$
$$= 10^{0.688} \times 10^{-3}$$
$$= 10^{0.688 + (-3)}$$
$$= 10^{-2.312}$$

Secondly the table illustrates how a power can be evaluated. To do this, the table must be read in reverse order.

Example 2 Evaluate. (a) $10^{0.694}$ (b) $10^{3.713}$ (c) $10^{-0.43}$

Solution From the table,

(a) $10^{0.694}$ (b) $10^{3.713} = 10^3 \times 10^{0.713}$ (c) $10^{-0.43} = 10^{-1} \times 10^{0.57}$
$ = 4.943$ $\phantom{(b) 10^{3.713} }= 10^3 \times 5.164$ $ -0.43 \uparrow = 10^{-1} \times 3.715$
$\phantom{(b) 10^{3.713} }= 5164$ $= -1 + 0.57 = 0.3715$

The exponential tables also illustrate the principles of calculating powers and roots, as shown in the following example. (Refer to the tables at the back of this text.)

Example 3 Calculate to 3 decimal places.

(a) $\sqrt[5]{5.623}$ (b) $\dfrac{\sqrt[4]{47.18}}{6.3}$

Solution (a) $\sqrt[5]{5.623} = (5.623)^{\frac{1}{5}}$ ◄——From the tables, $5.623 = 10^{0.750}$.
$\phantom{(a) \sqrt[5]{5.623} } = (10^{0.75})^{\frac{1}{5}}$
$\phantom{(a) \sqrt[5]{5.623} } = 10^{\frac{1}{5}(0.750)}$
$\phantom{(a) \sqrt[5]{5.623} } = 10^{0.150}$ ◄——From the tables, $10^{0.150} = 1.413$.

Thus, $\sqrt[5]{5.623} = 1.413$ to 3 decimal places.

A calculator check verifies this value.

(b) Use standard form to rewrite 47.18.

$47.18 = 10 \times 4.718$
$ = 10^1 \times 10^{0.674}$ ◄—— From the tables, $4.710 = 10^{0.673}$ and
$ = 10^{1 + 0.674}$ $ 4.721 = 10^{0.674}$.
$47.18 = 10^{1.674}$ i.e. $4.718 = 10^{0.674}$.

$\dfrac{\sqrt[4]{47.18}}{6.3} = \dfrac{(47.18)^{\frac{1}{4}}}{6.3}$

$\phantom{\dfrac{\sqrt[4]{47.18}}{6.3} } = \dfrac{(10^{1.674})^{\frac{1}{4}}}{10^{0.799}}$

$\phantom{\dfrac{\sqrt[4]{47.18}}{6.3} } = \dfrac{10^{0.419}}{10^{0.799}}$

$\phantom{\dfrac{\sqrt[4]{47.18}}{6.3} } = 10^{-0.380}$ ◄————————Express the negative number as a sum
$\phantom{\dfrac{\sqrt[4]{47.18}}{6.3} } = 10^{-1} \times 10^{0.620}$◄——of a negative integer and a positive
$\phantom{\dfrac{\sqrt[4]{47.18}}{6.3} } = 4.169 \times 10^{-1}$ rational number.
$\phantom{\dfrac{\sqrt[4]{47.18}}{6.3} } = 0.4169$

Thus, $\dfrac{\sqrt[4]{47.18}}{6.3} = 0.417$ (to 3 decimal places).

The previous examples illustrate the principles that were fundamental to the development of a branch of mathematics that is used in solving scientific problems, namely the study of logarithms. Furthermore, they provided part of the needed foundation for understanding the process required to develop technology to replace tedious calculations.

You will study logarithms
in the next chapter.

9.5 Exercise

A For the following exercises, use the exponential tables for $y = 10^x$ at the back of the book.

1 Evaluate each of the following.
(a) $10^{0.780}$ (b) $10^{0.783}$ (c) $10^{0.789}$ (d) $10^{0.613}$
(e) $10^{0.250}$ (f) $10^{0.251}$ (g) $10^{0.255}$ (h) $10^{0.289}$

2 Express each of the following as a power of 10.
(a) 4.467   (b) 1.552   (c) 8.892   (d) 6.855
(e) 1.795   (f) 5.058   (g) 3.097   (h) 2.547

3 Estimate each of the following as a power of 10.
(a) 5.463   (b) 3.693   (c) 1.608   (d) 2.870
(e) 2.253   (f) 6.337   (g) 7.753   (h) 9.718

4 Use the exponential tables. Express each number as a power of 10.
(a) 47.42   (b) 243.3   (c) 0.4018   (d) 84.33
(e) 315.5   (f) 0.6266   (g) 63.39   (h) 677.6

5 Evaluate each of the following.
(a) $10^{1.483}$ (b) $10^{2.691}$ (c) $10^{3.824}$ (d) $10^{0.0461}$ (e) $10^{0.069}$

B In the next section, skills will be explored for using a calculator and calculations involving exponents. Check your work.

6 Calculate each of the following.
(a) $10^{0.635} \times 10^{0.821}$ (b) $10^{0.125} \times 10^{0.809}$ (c) $10^{0.0135} \times 10^{1.469}$
(d) $10^{1.463} \times 10^{0.283}$ (e) $10^{0.931} \times 10^{1.893}$ (f) $10^{3.692} \times 10^{0.0468}$

7 Do the necessary calculations. Then evaluate each of the following.
(a) $10^{\frac{1}{2}(1.693)}$ (b) $10^{\frac{1}{3}(1.893)}$ (c) $10^{\frac{1}{3}(3.416)}$ (d) $10^{\frac{1}{2}(0.0162)}$

8   Calculate to 4 significant digits.
    (a) 4.693×8.215 (b) $8.962 \div 1.069$
    (c) $(2.613)^2 \div 1.895$ (d) $(6.312)^3 \div (1.261)^2$
    (e) $(46.89)^2 \div 18.32$ (f) $(0.01356) \div (0.02161)^2$

9   Calculate.
    (a) $\sqrt{8.312}\,\sqrt[3]{9.165}$ (b) $\sqrt[3]{2.613} \div \sqrt[4]{5.613}$ (c) $(\sqrt{48.62})^3$

Questions 10 to 12 are based on the following information.

The value, v, of each dollar invested in a particular fund is given by the formula $v = (1.16)^n$, where n represents the number of years.

10 Calculate the value of one dollar after
    (a) 3 years   (b) 8 years

11 Alan invested $350 in the fund. Calculate the value of his investment after four years.

C 12 Melanie has two choices for an investment.

   A   Invest $150 into a bond which returns $255 after 6 years.
   B   Invest $150 into the above fund.

   Which is the better investment? By how much?

Calculator Tip

A scientific calculator has an exponential feature. Refer to your manual provided with the calculator. How is the exponential feature used effectively?
In the next section, skills with a scientific calculator are explored. The principles with calculators are essentially the same, but each type of scientific calculator may have slightly different operational features. Always refer to your manual to check out the specific details and features.

9.6 Using Calculators Efficiently

Most scientific calculators have a $\boxed{y^x}$ button which can be used to calculate powers. Refer to the *Calculator Tip* on the previous page. For example, to calculate $3^{4.21}$ you follow these steps:

$$\text{Press} \quad \boxed{C} \; 3 \; \boxed{y^x} \; \underset{\text{Input}}{4.21} \; \boxed{=} \; \underset{\text{Display}}{102.01886}$$

Thus, $3^{4.21} = 102.02$ to 2 decimal places.

When calculating a power with a calculator, the base of the power must always be positive.

The buttons $\boxed{\text{INV}}$ and $\boxed{y^x}$ are used together to find the root of a number. For example, to find $\sqrt[1.2]{61}$ to 2 decimal places you follow these steps.

$$\text{Press} \quad \boxed{C} \; 61 \; \boxed{\text{INV}} \; \boxed{y^x} \; 1.2 \; \boxed{=} \; 30.745026$$

Thus, $\sqrt[1.2]{61} = 30.75$ to 2 decimal places.

Example 1   Calculate (a) $0.75^{1.85}$ (b) $\sqrt[-1.5]{176}$ (c) $2.7 \times 3.2^{1.5} + 4^{-0.75}$ to 2 decimal places.

Solution   (a) Press $\boxed{C}$ 0.75 $\boxed{y^x}$ 1.85 $\boxed{=}$ 0.5873045

Thus, $0.75^{1.85} = 0.59$ to 2 decimal places.

(b) Press $\boxed{C}$ 176 $\boxed{\text{INV}}$ $\boxed{y^x}$ 1.5 $\boxed{+/-}$ $\boxed{=}$ 0.0318414   — Changes the sign of the previous entry.

Thus, $\sqrt[-1.5]{176} = 0.03$ to 2 decimal places.

(c) Press $\boxed{C}$ 2.7 $\boxed{\times}$ 3.2 $\boxed{y^x}$ 1.5 $\boxed{+}$ 4 $\boxed{y^x}$ 0.75 $\boxed{+/-}$ $\boxed{=}$ 15.809255

Thus $2.7 \times 3.2^{1.5} + 4^{-0.75} = 15.81$ to 2 decimal places.

> Remember:
> Refer to your calculator manual for additional examples related to the operation of your particular calculator.

Skills with a scientific calculator allow you to solve problems that, otherwise, would be very tedious to calculate. For example:

What is the exponent x such that $13.6 = 3^x$?
Express your answer to 3 decimal places.

A method of successive estimates is used, in conjunction with the $\boxed{y^x}$ feature of your calculator. For example, to calculate what power of 3 the number 13.6 is, you follow these steps:

▶ First of all, you know from inspection that $3^2 = 9$ and that $3^3 = 27$. Since $9 < 13.6 < 27$ then you know $2 < x < 3$ where x is the required exponent.

Think	Calculator Steps and Display
First estimate	
$\dfrac{2 + 3}{2} = 2.5$	Press $\boxed{C}$ 3 $\boxed{y^x}$ 2.5 $\boxed{=}$ 15.588457 ◀——Too high.
Next estimate	
$\dfrac{2 + 2.5}{2} = 2.25$	Press $\boxed{C}$ 3 $\boxed{y^x}$ 2.25 $\boxed{=}$ 11.844666 ◀——Too low.
Next estimate	
$\dfrac{2.25 + 2.5}{2} = 2.375$	Press $\boxed{C}$ 3 $\boxed{y^x}$ 2.375 $\boxed{=}$ 13.588233 ◀—— Good estimate.

Thus $13.6 \doteq 3^{2.375}$. Thus, $x = 2.375$ to 3 decimal places.

9.6 Exercise

A   In the following exercise, use your scientific calculator to find the values.

1   Find each value to 3 decimal places.
(a) $3^{1.25}$ (b) $4^{3.5}$ (c) $71^{0.56}$ (d) $19^{-0.25}$ (e) $101^{-0.321}$
(f) $52^{2.1}$ (g) $14^{-0.02}$ (h) $42^{0.755}$ (i) $23^{0.022}$ (j) $81^{-0.32}$

2   Find each value to 1 decimal place.
(a) $25^{0.5}$ (b) $27^{0.333}$ (c) $32^{0.2}$ (d) $256^{0.25}$
(e) $27^{0.667}$ (f) $8^{1.33}$ (g) $64^{1.5}$ (h) $32^{0.833}$
What do you notice about your answers above?

3   Calculate to 2 decimal places.
(a) $\sqrt[3.2]{271}$ (b) $\sqrt[-0.5]{0.75}$ (c) $\sqrt[1.61]{38}$ (d) $\sqrt[0.91]{401}$
(e) $\sqrt[-7.5]{38.25}$ (f) $\sqrt[-8]{17.006}$ (g) $\sqrt[11.1]{91.3}$ (h) $\sqrt[1.02]{807.55}$
(i) $\sqrt[3.25]{0.0052}$ (j) $\sqrt[-1.15]{72\,431.1}$ (k) $\sqrt[-2.1]{0.022}$ (l) $\sqrt[13.55]{372.452}$

4   (a) What power of 7 is 15?   (b) What power of 11 is 3.2?
(c) What power of 6 is 1.5?   (d) What power of 0.5 is 2?
(e) What power of 2.75 is 0.5?

B You can use the calculator to make useful estimates.

5 Find a value for x which makes each of the following true. Round your answers to 2 decimal places.

(a) $9.1^x = 17$ (b) $2.9^x = 11$ (c) $5.2^x = 71$

(d) $0.25^x = 0.016$ (e) $1.11^x = 5.2$ (f) $12^x = 307$

6 Calculate to 2 decimal places.

(a) $7.321 \times 3^{1.5}$ (b) $13^{0.25} \times 25.26$ (c) $42^{0.48} \div 48.69^{0.789}$

(d) $0.08 \times 2^{4.2}$ (e) $1.82^{1.82} \div 3^{2.5}$ (f) 5.21×7.904^5

7 After 3 s the speed of a boat, v is given by the expression $v = 15 \times 10^{-0.12(3)}$ where the speed of the boat was 15 km/h when the engines were stopped. Find the speed v of the boat to 1 decimal place.

8 The number, N, of bacteria in a culture after 16 h is given by $N = 200 \times 10^{0.4(16)}$ where the original number of bacteria was 200. Find the number of bacteria, N, to the nearest whole number.

9 The amount, A, of money in dollars in a fund after 5 years is given by the value $A = \$100(1.14)^5$ where the inital amount of money invested was \$100 at 14%/a compounded annually. Calculate the amount of money, A.

10 The radius, r, of the nucleus of the zinc atom is given by the expression $r = 1.2 \times 10^{-13}(64)^{\frac{1}{3}}$. Calculate the value of r to 3 significant digits.

Calculator Tip

Questions 7 to 10 illustrate how a calculator can be used to solve problems that would involve tedious, if not impossible, calculations. Examine your calculator manual or handbook for additional uses and techniques for solving problems.

Math Tip

It is important to clearly understand the vocabulary of mathematics when solving problems.
- Make a list of all the words you have met in this chapter.
- Provide an example to illustrate your understanding of each word.

9.7 Applications With Exponential Growth

Yeast cells increase their numbers exponentially by a process called budding. They duplicate themselves about every half hour. The **doubling period** is thus said to be 0.5 h.

Yeast Cells
Yeast cells, which are microscopic plants, ferment sugars and starches to make alcohol and carbon dioxide. Yeast is used to make bread rise because of the gases produced by the fermentation.

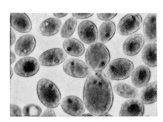

Here, the budding process is outlined. The initial number of yeast cells is N_0. After 0.5 h, the number of cells is $2N_0$. After the second 0.5 h, the number of cells is $2(2N_0)$ or $(2^2)N_0$. After the third 0.5 h, the number of cells is $2(2^2 N_0)$ or $(2^3)N_0$. Thus after n doubling periods, the number, N, of yeast cells is

$$N = N_0(2^n).$$

Some cells multiply at different rates than yeast. In general, if after time t, (in hours), an initial number of cells, N_0, has grown to N, then

$$N = N_0(2^{\frac{t}{d}}), \quad \text{where d is the time of the}$$
$$\text{doubling period in hours if } t \text{ is in hours}$$

Since N is a function of time, t, the formula above is often written as

$$N(t) = N_0(2^{\frac{t}{d}}).$$

Example A bacterial strain doubles its number every 3 min. If there are 1000 bacteria initially, how many will there be after 0.25 h?

Solution Use the formula $N(t) = N_0(2^{\frac{t}{d}})$ where $N_0 = 1000$, $t = 15$ min and $d = 3$ min.

$$N(t) = N_0(2^{\frac{t}{d}}) \qquad N(15) = 1000(2^{\frac{15}{3}}) \qquad \text{If } t \text{ is in minutes then } \frac{t}{d} \text{ is}$$
$$= 1000(2^5) \qquad \text{formed with } d \text{ in minutes.}$$
$$= 32\,000$$

Thus, there are 32 000 bacteria after 0.25 h.

9.7 Exercise

A 1 A bacterial culture has N(t) bacteria present after time, t (in seconds). If N(t) = 100(2^t), how many bacteria are present for each value of t?
(a) 0   (b) 1   (c) 5   (d) 10

2   A certain bacterial strain divides every hour producing two bacteria from every existing one. If there are 200 bacteria in a culture, how many will there be after

(a) 4 h?   (b) 8 h?   (c) n h?

3   A certain cell is allowed to divide and after 4 h it is estimated there are 10 000 cells. If the number of cells doubles every 2 h how many

(a) were there originally?   (b) would there be after 12 h?

B   4   For a biology experiment, the number of cells present is 1000. After 4 h the count is estimated to be 256 000. What is the doubling period?

5   A certain bacterial strain divides every 0.5 h. If 500 bacteria are present in a culture, how many will there be after

(a) 3 h?   (b) 6 h?   (c) n h?

6   During an experiment, it is recorded that there are 50 bacteria present originally in a culture. After 1 h there are 1600 bacteria present. What is the doubling period?

7   A culture of cells doubles in size every 0.5 min. After 2 min it is estimated that there are 160 000. How many

(a) were present originally?   (b) will there be after 4 min?

8   A certain bacterial strain doubles every minute in a culture. If there were 1000 bacteria in a culture originally, how many bacteria would there be after

(a) 4 min?   (b) 6 min?   (c) 10 min?

9   A certain strain of yeast cell doubles under certain conditions every 20 min. If there were 350 initially, how many will there be in 3 h?

10   Bacteria in a certain culture increase at a rate proportional to the number present. If the number doubles in 4 h, what percentage of the original number will be present at the end of one day?

Computer Tip

The computer program in BASIC provides the amount, A, for given values of P, the principal or initial deposit, I, interest rate, and T, time.

```
10 INPUT P, I, T
20 LET A = P * (I + 1) ↑ T
30 LET B = A − P
40 PRINT "THE AMOUNT IS" A
50 PRINT "THE INTEREST IS" B
60 END
```

Applications: Growing Money

The table shows what accumulates
when you deposit $1000 in a growth
fund that pays 15% per annum
compounded annually.

years	amount in fund
1	$1000 (1.15)
2	$1000 (1.15)^2
3	$1000 (1.15)^3

In other words, money grows exponentially to the amount $A(t)$ (in dollars)
given by $A(t) = A_0(1 + i)^t$ where i is the annual rate of interest,
A_0 is the initial deposit, and
t is the time in years

To calculate the amount after 3 years, the exponential tables, or your
calculator, can be used.

$$A(3) = 1000(1.15)^3 \longleftarrow 1.15 = 10^{0.061}$$
$$= 1000(10^{3(0.061)})$$
$$= 1000(10^{0.183}) \longleftarrow \text{ or } 10^3(10^{0.183})$$
$$= 1000(1.524) \longleftarrow = 10^3 \times 1.524$$
$$= 1524 \longleftarrow = 1524$$

Thus $1000 has grown to $1524 in 3 years when invested at 15% per annum
compounded annually.

11   Calculate the amount $A(t)$ for each investment for the time shown.
   (a) $500 invested at 12% per annum compounded yearly for 6 years.
   (b) $1000 invested in a second mortgage at 13.5% per annum compounded
      yearly for 4 years.
   (c) $800 invested in a real estate fund at 12.75% per annum compounded
      yearly for 7 years.

12   Pascal joined an investment club and was guaranteed 11.5% per annum
   compounded annually. The minimum period of time was 5 years.
   (a) Calculate the amount of money received at the end of 5 years if $1500 was
      invested in the club.
   (b) How much interest did Pascal actually receive for investing the money?

13   Lisa invested $800 in a municipal bond that pays 13% per annum
   compounded annually.
   (a) How much money will she receive in all at the end of 6 years?
   (b) How much interest was she actually paid?
   (c) What was the average amount of interest received each year?

9.8 Exponential Decay

Radioactive elements decay over time. For example, the data below, were recorded for a radioactive substance. The initial amount of radon was 12.6 g.

amount of radon (grams)	10.5	8.8	7.3	6.1	5.1	4.2	3.5	2.9	2.4	2.0
time (days)	1	2	3	4	5	6	7	8	9	10

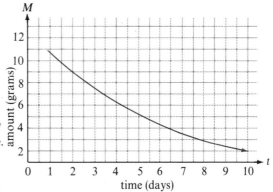

With the data displayed on a graph, you can see that the decay appears exponential.

The rate at which radioactive elements decay is described as the **half-life**. The half-life of a radioactive element is the time taken for the element to decay to one half of its initial amount. In general, for a radioactive element the amount of mass left after time, t, is given by

$$M(t) = M_0(2^{-\frac{t}{h}})$$ where M_0 is the initial mass and h is the half-life of the substance. The units for t and h must be the same. (ie. both minutes, years, days, and so on.)

Example 320 mg of Iodine 131 is stored in a laboratory for 40 d. At the end of this period only 10 mg of the element remain. What is the half-life of Iodine 131?

Solution Use $M(t) = M_0(2^{-\frac{t}{h}})$. $M(40) = 10$, $M_0 = 320$ and $t = 40$

t is time in days and h

$$10 = 320(2^{-\frac{40}{h}})$$ is the half-life in days.

Divide by 320. $\dfrac{1}{32} = 2^{-\frac{40}{h}}$

$$\frac{1}{2^5} = 2^{-\frac{40}{h}}$$

$$2^{-5} = 2^{-\frac{40}{h}}$$

$$-5 = \frac{-40}{h}$$

$$h = 8$$

The half-life of Iodine 131 is 8 d.

9.8 Exercise

A 1 The half-life of Strontium 90 is 25 years. For a 100 mg sample, how much would remain after 50 years?

2 At the end of 15 d, one eighth of a radioactive sample of Bismuth 210($^{210}_{83}$Bi) remains. What is the half-life of this element?

3 (a) How long will it take for a 1 g sample of Polonium 210($^{210}_{84}$Po) to lose $\frac{3}{4}$ of its radioactivity if its half-life is 140 d?

(b) How long will it take before it loses all but $\frac{1}{128}$ of its radioactivity?

B 4 In 1976 a research hospital bought half a gram of radium for cancer research. Assuming the hospital still exists, how much of this radium will the hospital have in the year 6836 if the half-life of radium is 1620 years?

5 Health officials found traces of Radium F beneath the local library. After 69 d they observed that a certain amount of the substance had decayed to $\frac{1}{\sqrt{2}}$ of its original mass. Determine the half-life of Radium F.

6 During the transportation of the isotope Thorium 243($^{243}_{1}$Th) to a nuclear waste facility, a spill occurred near a populated area. The area was evacuated and no one was to return home until the above isotope decayed to $\frac{1}{4}$ its original radioactivity. If the half-life of Thorium 243 is 24 d, for how long was the area evacuated?

The speed of a vehicle on water, in kilometres per hour, is described by the exponential decay equation given by $V = V_0 \times 10^{-kt}$, where
- V_0 is the speed at the time the engine is stopped and the boat coasts.
- t is the time, in seconds, after the engine is stopped.
- k is a constant, which is different for each vehicle.

7 A small power launch travels at a speed of 20 km/h. Find its speed 2.5 s after the engine has stopped if its initial speed, V_0, and final speed, V, are related by $V = V_0 \times 10^{-0.24t}$, where t is in seconds.

8 Sally Younger, at the age of 17 years became the world's fastest water skier, having reached a top speed of 168 km/h. Once she lets go of the tow-rope, her speed after time, t, in seconds, is given by $V = V_0 \times 10^{-0.16t}$. Find her speed 4.5 s after she lets go of the tow-rope.

Applications: Carbon Dating

How is it possible for *scientists* to pinpoint the age of the bones of animals that died so long ago? Carbon 14, ^{14}C, is a radioactive substance with a half-life of 5760 years. It can be used to determine how long ago a plant or animal lived. For example, suppose an animal, while living, absorbed Carbon 14. It is subsequently buried by volcanic ash, and thus the animal can no longer absorb ^{14}C. The ^{14}C is radioactive. After 5760 years half of the original ^{14}C absorbed by the animal will be left. If the scientist knows
- how much ^{14}C is left, $M(t)$, and
- how much ^{14}C there was to start with, M_0, then the scientist can use the exponential equation

$$M(t) = M_0(2^{-\frac{t}{5760}}) \longleftarrow \text{half-life: 5760 years}$$

to calculate the number of years ago, t, the animal died. This method of determining the age of living things is called **carbon dating**.

Archeological Find!

Scientists determine that animal bones found at a recent excavation are seventeen thousand two hundred eighty years old. With great excitement the scientists unfolded their discovery at a press conference upon arrival from their two-year exploration of

Express your answers to the nearest hundred years for each of the following.

9. The longest bone in the human body is the femur (or thigh) bone. At a site, a femur bone was found. If the amount of ^{14}C had decayed to $\frac{1}{4}$ of the original amount, how old is the bone?

10. In a recent dig, a human skeleton was unearthed. It was later found that the amount of ^{14}C in it had decayed to $\frac{1}{\sqrt{8}}$ of its original amount. If ^{14}C has a half-life of 5760 years, how old was the skeleton?

11. The remains of a flowering plant were found to have 0.36 g of ^{14}C. It was estimated that the plant originally had 2.9 g of ^{14}C. Calculate the age of the plant.

12. The oldest known wooden sled was found in Southern Finland. The original amount of ^{14}C was estimated to be 4.2 g. After conducting experiments, the present amount of carbon was found to be 1.5 g. How old is the oldest known sled?

13. A piece of fossilized wood was analyzed and it was found the original amount of ^{14}C was 8.5 g. If the amount of ^{14}C left is 0.75 g, calculate the age of the wood.

Practice and Problems: A Chapter Review

An important step for problem-solving is to decide which skills to use. For this reason, these questions are not placed in any special order. When you have finished the review, you might try the *Test for Practice* that follows.

1   Simplify.

   (a) $16^{\frac{1}{4}}$ (b) $8^{-\frac{2}{3}}$ (c) $\left(\dfrac{25}{9}\right)^{-\frac{1}{2}}$

2   If $a = -1$ and $b = 2$, find the value of $\dfrac{64a^3b^2}{2a^4b} \times \dfrac{32a^5b^3}{128a^5b^2}$.

3   Prove that $(a^{x^2+y^2})(a^{x+y})^{x-y} = a^{2x^2}$.

4   What is the value of each of the following? Express your answer to 1 decimal place.

   (a) $\sqrt[4]{32} \times \sqrt[3]{44}$ (b) $(28)^{\frac{1}{2}} \times \sqrt[3]{5^2}$

5   Solve and verify: $3^{2x} - 12(3^x) + 27 = 0$.

6   Jennifer invests $500 at 16%/a. After 8 years, the amount of money she has is given by the expression $A = 500(1.16)^8$. Use the exponential tables to calculate the value of a to the nearest cent.

7   There are 50 bacteria present initially in a culture. In 3 min the count is 204 800 bacteria. What is the doubling period?

8   The oldest rock known to data, found in the Minnesota River Valley, contains 2 g of carbon 14 (C^{14}). Using geological methods to estimate the original size of the rock, scientists estimate it to originally have contained 4.096 kg of C^{14}. The half life of C^{14} is 5760 years. What is the approximate age of the rock?

9   Land developers had to delay a housing development because the land had been previously used as a nuclear waste pit for Radium-D, an isotope. Health officials ordered them not to build until the isotope has decayed to $\dfrac{1}{\sqrt{2}}$ of its original mass. When could they build if Radium-D has a half life of 22 years?

Test for Practice

Try this test. Each *Test for Practice* is based on the mathematics you have learned in this chapter. Try this test later in the year as a review. Keep a record of those questions that you were not successful with, find out how to answer them, and review them periodically.

1 Simplify

(a) $\left(\dfrac{4}{3}\right)^2 \div \left(\dfrac{3}{2}\right)^3$ (b) $\dfrac{a^2b^2}{(ab)^2} \times \dfrac{a^5b^6}{(a^2b^2)^2}$

2 Simplify.

(a) $b^5 \div (b^2)^{-3}$ (b) $(a^{-3}) \div (a^{-3})$

(c) $3^{-1}(3^2 + 3^0)$ (d) $\dfrac{2^{-1} - 3^{-1}}{4^{-1} + 3^{-1}}$

3 Express each of the following with exponents.

(a) $\sqrt{a^3}$ (b) $\dfrac{1}{\sqrt[3]{a^4}}$ (c) $\dfrac{1}{\sqrt[4]{a^5}}$

4 Prove that $\dfrac{(4^{n+1})(16^{n+1})}{(4^3)^{n+1}} = 1.$

5 Find the solution set of $\left(\dfrac{1}{2}\right)^{x-1} = 2^{2x+4}.$

6 What is the value of each of the following? Express your answer to 3 significant digits.

(a) $\sqrt{4.631}$ (b) $\sqrt[3]{8.263}$ (c) $(\sqrt{43.92})^3$

7 Certain plant cells double every 40 min under controlled conditions. If there were 700 cells initially, how many will there be in 9 h?

8 In the game show "$64 000 Pyramid" contestants who answer the first question correctly win $125. If the contestants answer the next question correctly, they win $250. Each additional question doubles the amount of the prize. How many questions have to be correctly answered to climb to the top of the pyramid?

9 Jackie was doing research for her Ph.D. in nuclear physics when a 12 mg sample of radioactive material decayed to 3 mg in about 6 min. Determine the half life of this substance.

Looking Back: A Cumulative Review

1   (a) Find the zeroes of $g(x) = (x - 3)(x - 1) - 4(x + 2)$.
    (b) Use a calculator. Find the approximate values of the zeroes to 3 decimal places.

2   Find the value of k so that each polynomial is divisible by the binomial.
    (a) $x^3 - kx^2 + 2x - 8$; $x - 2$ (b) $x^3 - x^2 + 2x - k$; $x - 1$

3   Solve and draw the graph for each of the following:
    (a) $|x| + 3 \leq 9$ (b) $|2x| + 1 \geq 7$ (c) $2|x| - 1 < 10$
    (d) $23 > 3|x| + 5$ (e) $|3x| \leq 9$ (f) $3 < |x| - 2$

4   A chord of length k is drawn in a circle with radius r.
    (a) Prove the locus of midpoints of all chords of length k is a circle.
    (b) What is the radius of the circle in (a)?

5   The quadrilateral SOGN has $SN = OG$ and $\angle G = \angle N$. The midpoints of OS and GN are T and C respectively. Prove OS $\parallel$ NG.

6   Prove that in an isosceles triangle the circle drawn on one of the equal sides as diameter bisects the remaining side.

7   Use an example to illustrate the meaning of each type of function
    (a) constant   (b) increasing   (c) one-to-one

8   Prove.
    (a) $\sin^2 \theta \cot^2 \theta + \dfrac{\cos^2 \theta}{\cot^2 \theta} = 1$ (b) $\dfrac{1}{\cot \theta} + \sin \theta = \sin \theta \left(\dfrac{\cos \theta + 1}{\cos \theta} \right)$

9   Solve for $-\pi \leq \theta \leq \pi$.
    (a) $\cos^2 \theta + \cos \theta = 0$ (b) $3 \sin^2 \theta + 5 \sin \theta - 2 = 0$

10   The distance from Las Vegas to Brampton is 3100 km; from Las Vegas to Chicago is 2450 km; and from Chicago to Brampton is 700 km. What is the angle formed by the lines joining Brampton to Chicago and Brampton to Las Vegas?

11   Solve $x(x - 1)(x - 2)(x - 3) = 120$.

12   Determine the amplitude, period and phase shift of each of the following. Then draw the graph.
    (a) $y = 3 \cos \left(x - \dfrac{\pi}{4} \right)$ (b) $y = \tan \left(x + \dfrac{\pi}{6} \right)$ (c) $h(t) = \dfrac{3}{2} \sin \left(2\theta - \dfrac{\pi}{3} \right)$

10 Logarithmic Function

inverse of the exponential function, properties of logarithms, logarithmic function, decibels, laws of logarithms, Newton's law of cooling, solving logarithmic equations, strategies and solving problems, applications, problem-solving

Introduction

In the previous chapters you learned that the exponential function is very useful in describing, mathematically, many scientific phenomena. In the process of developing mathematics, the question "What if . . .?" is asked. For example, once the exponential function is drawn, what if the inverse is drawn? The result of asking this question is the graph of the logarithmic function, which has a rich history in the study of mathematics. Many people have played a role in the development of mathematics. One such person was John Napier (1550–1617) who was born in Scotland and invented logarithms. His invention reduced the countless hours of tedious calculations so that astronomers could more readily attend to the study of astronomy. With the advent of calculators, and computers, the application of logarithms and logarithmic functions became important in the study of earthquakes and other such geological phenomena and in the study of acoustics, medicine, and so on.

Seismologists have devised a scale to describe the intensity of an earthquake, called the Richter Scale. This scale was named after Charles Richter (1900–1985) and measures the energy released at the focus or centre of an earthquake. The scale is logarithmic with values that record earthquakes from 1 to 9.

Logarithmic functions are also used to compare the relative loudness of sounds. The unit used in such comparisons, the decibel, is described as 0.1 of a bel. The term, bel is derived from the name of the inventor of the telephone, Alexander Graham Bell. The loudness of the sound produced by the jet shown above is about 160 dB (decibels).

10.1 The Logarithmic Function: Inverses

You have already seen that the exponential function can be used to describe many scientific phenomena. In studying graphs, the researcher might ask "What if I drew the inverse of the exponential function?"

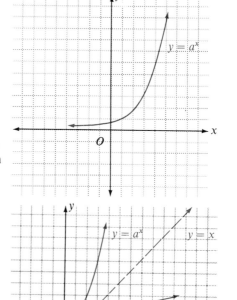

The graph of the exponential function, $y = a^x$, is reflected in the line $y = x$, $(x, y \in R)$. Thus, $y = a^x$ becomes $x = a^y$. It was noticed that this inverse function described certain scientific work.

From the graphs, various properties became evident.

- Both curves are one-to-one. The equations $y = a^x$ and $x = a^y$ both represent functions.
- The domain of $y = a^x$ is the range of $x = a^y$ and, conversely, the range of $y = a^x$ is the domain of $x = a^y$.

Graphs that are reflections of each other in the line $y = x$ describe inverse functions.

> $y = a^x$ is the inverse of $x = a^y$ and, conversely,
> $x = a^y$ is the inverse of $y = a^x$.

To refer to the new graph as the "inverse of the exponential function" is cumbersome. Furthermore, to deal with the new function in the form $x = a^y$ is again cumbersome, since y occurs as an exponent. For this reason, new vocabulary was invented. The *logarithmic* symbol was invented to describe this new graph and concept. The name **logarithm** was then used to define this new inverse function.

Exponential form **Logarithmic form**

$x = a^y$ is written as $y = \log_a x$

In the next section, the properties of $y = \log_a x$ are examined.

10.1 Exercise

B Use your calculator to obtain exponential values 2^x, 10^x, and so on.

1 (a) Sketch the graph of the function, f, given by $y = f(x) = 10^x$, $x \in R$,
    (b) Sketch the graph of the inverse f^{-1}.
    (c) Is the inverse a function? Give reasons for your answer.
    (d) What is the defining equation of the inverse in exponential form? In logarithmic form?

2 (a) Sketch the inverse for each of the following.

    (i) $y = 2^x$ (ii) $y = \left(\dfrac{1}{2}\right)^x$

    (b) Why is the inverse also a function?
    (c) Write the defining equation for the inverse of the graph in part (a)(i) and part (a)(ii).

3 For the function defined by $y = a^x$, what is the effect on y if
  (a) $a > 1$?   (b) $0 < a < 1$?   (c) $a = 1$?

Questions 4 to 6 are based on the graph of $y = \log_{10} x$ shown below. The graph of $y = \log_{10} x$ is obtained by reflecting $y = 10^x$ in the line $y = x$. Once you use the graph to answer the question, use your calculator to check. Only a partical graph of $y = \log_{10} x$ is shown.

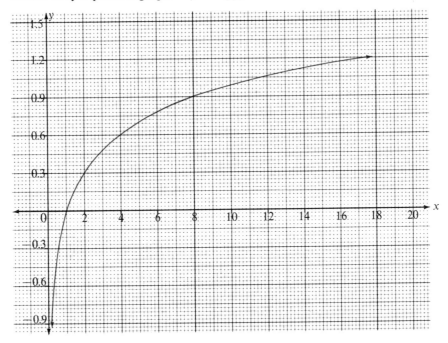

4   From the graph, find the value of y for the following values of x.
     (a) 2   (b) 5   (c) 8   (d) 10

5   From the graph, find an approximate value for
     (a) $\log_{10} 4$ (b) $\log_{10} 7$ (c) $\log_{10} 8.5$ (d) $\log_{10} 3.5$

6   What is the approximate value of x for each of the following?
     (a) $\log_{10} x = 0$ (b) $\log_{10} x = 1$ (c) $\log_{10} x = 0.7$
     (d) $\log_{10} x = 0.4$ (e) $\log_{10} x = 0.65$ (f) $\log_{10} x = 0.35$

Math Tip

In the early development of mathematics, three inventions had a significant influence on meeting the needs of calculating.
- Hindu Arabic notation
- decimal fractions
- logarithms

John Napier (1550–1617) made a number of contributions to the study of mathematics, one of which was the invention of *logarithms*.

The invention of logarithms provided the mathematical foundation for many labour-saving devices, such as calculators and today's powerful computers.

Problem Solving

Often to solve a problem, you need to know how the symbols are defined. Before you solve the problem, understand what $d(n)$ means. A chart is constructed for the first 100 counting numbers.

n	factors of n	$d(n)$
1	1	1
2	1, 2	2
3	1, 3	2
4	1, 2, 4,	3
5	1, 5	2
6	1, 2, 3, 6	4

What type of number is n for each of the following situations?

$$d(n) = 2 \qquad d(n) = 3 \qquad d(n) = 4$$

10.2 Concepts and Calculators: Logarithms

Once a concept is obtained, it is desirable to explore the properties and introduce suitable notations. For example, the equation $x = a^y$, is not a convenient equation to use. In an attempt to describe this function in terms of y, the logarithmic symbol was invented and the name logarithm was used to define the function.

$y = \log_a x$ which is read y is the logarithm of x to the base a.

When working with logarithms often you will find it helpful to refer to their inverses, namely the exponentials.

exponential form	logarithmic form	logarithmic functions
$x = a^y$	$y = \log_a x$	These are functions
Say, the base a is raised to the exponent y and the answer is x.	Say, y is the exponent to which you raise base a to get the answer x	defined by equations of the form $y = \log_a x$, where $a > 0$ and $a \neq 1$.

Example 1   Write each of the following in logarithmic form.

(a) $32 = 2^5$ (b) $2^{-5} = \dfrac{1}{32}$ (c) $x = 10^y$

Solution   Compare $x = a^y \leftrightarrow y = \log_a x$

(a) $32 = 2^5 \rightarrow 5 = \log_2 32$

Say, 5 is the exponent to which 2 is raised to obtain 32.

(b) $2^{-5} = \dfrac{1}{32} \rightarrow -5 = \log_2 \left(\dfrac{1}{32}\right)$

Say, -5 is the exponent to which 2 is raised to obtain $\dfrac{1}{32}$.

(c) $x = 10^y \rightarrow y = \log_{10} x$

When writing logarithms or evaluating expressions involving logarithms, you will find it useful to bear in mind the equivalent exponential form.

$$\boxed{x = a^y \leftrightarrow y = \log_a x}$$

These relationships are constantly referred to in the following examples and exercises, and are helpful when you use your calculator.

In order to work with the properties of the logarithmic function, examples such as the following establish important skills.

Example 2 Evaluate each of the following.
(a) $\log_{10} 100$ (b) $\log_2 64$

Solution (a) $\log_{10} (100)$
= $\log_{10} (10^2)$ Think, to what exponent
= 2 is the base 10 raised to
obtain 100?

(b) $\log_2 64$
= $\log_2 (2^6)$ Think, to what exponent
= 6 is the base 2 raised to
obtain 64?

Calculator Tip

On a calculator, use the key shown by $\boxed{\log}$ to base 10. To find $\log_{10} 100$, follow these steps.

INPUT	DISPLAY
$\boxed{c}$ 100 $\boxed{\log}$	2

Thus $\log_{10} 100 = 2$.

Another approach would be to write the equivalent form.

Let $y = \log_{10} 100$. Then $10^y = 100$.
$$10^y = 10^2$$
$$y = 2 \qquad \text{Thus } \log_{10} 100 = 2.$$

Skills with logarithms will be needed to solve equations involving logarithms. You must remember the meaning of the exponential and logarithmic forms. Refer to the section of your calculator manual that deals with logarithms and exponentials.

Example 3 Solve (a) $\log_3 m = 4$ (b) $\log_8 4 = y$

Solution (a)

Think: 4 is the exponent to which 3 is raised to obtain m.

$\log_3 m = 4$ Check ✓.
Then $m = 3^4$ $\log_3 81 = \log_3(3^4)$
$m = 81$ $= 4$

(b) $\log_8 4 = y$
means $4 = 8^y$
$2^2 = (2^3)^y$
$2^2 = 2^{3y}$
$3y = 2$
$$y = \frac{2}{3}$$

Thus, $\log_8 4 = \frac{2}{3}$.

10.2 Exercise

A Throughout the exercise, check your work with skills you have developed on your calculator.

1 Express each of the following in logarithmic form.
(a) $2^3 = 8$ (b) $3^4 = 81$ (c) $5^3 = 125$ (d) $7^3 = 343$
(e) $m^n = p$ (f) $10^0 = 1$ (g) $16^{\frac{1}{2}} = 4$ (h) $81^{\frac{3}{4}} = 27$

2   Express each logarithmic form in exponential form.

(a) $\log_3 27 = 3$ (b) $\log_{10} 1000 = 3$ (c) $\log_2 128 = 7$

(d) $\log_x y = z$ (e) $\log_5 1 = 0$ (f) $\log_{27} 3 = \dfrac{1}{3}$

(g) $\log_2\left(\dfrac{1}{8}\right) = -3$ (h) $\log_{49} 7 = \dfrac{1}{2}$ (i) $\log_{10} 0.001 = -3$

3   Express each of the following in an equivalent form, either exponential or logarithmic.

(a) $\log_2 16 = 4$ (b) $\log_6 6 = 1$ (c) $2^{-3} = \dfrac{1}{8}$

(d) $49^{\frac{1}{2}} = 7$ (e) $\log_{27} 3 = \dfrac{1}{3}$ (f) $\log_2 64 = 6$

(g) $12^0 = 1$ (h) $10^{-4} = 0.0001$ (i) $\log_7 1 = 0$

4   Find the value of each of the following.

(a) $\log_3 3^5$ (b) $\log_2 2^7$ (c) $\log_{10} 10^4$ (d) $\log_x x^3$
(e) $\log_x x^5$ (f) $\log_x x^{\frac{2}{3}}$ (g) $\log_x x^{\sqrt{3}}$ (h) $\log_x x^{-3}$

5   Calculate.

(a) $\log_{10} 10$ (b) $\log_{10} 100$ (c) $\log_{10} 1$ (d) $\log_{10} 1000$
(e) $\log_{10} 1.36$ (f) $\log_{10} 40.8$ (g) $\log_{10} 146$ (h) $\log_{10} 0.0302$

6   Properties of logarithms are examined in (a) and (b).

(a) Find the value of each of the following.

$$\log_2 2 \qquad \log_{10} 10 \qquad \log_4 4 \qquad \log_8 2^3 \qquad \log_6 (36)^{\frac{1}{2}}$$

Use the results. What is the value of $\log_a a$ for $a \in R$, $a > 0$?

(b) Find the value of each of the following.

$$\log_2 2^3 \qquad \log_{10} 10^2 \qquad \log_2 2^{-1} \qquad \log_{10} 10^{-3} \qquad \log_5 5^2$$

Use the above results. What is the value for all $a \in R$, of $\log_a a^x$, $a > 0$?

B   To work with logarithms, you must understand the vocabulary and meaning of symbols: exponential form, logarithmic form, $\log_a x$, and so on. Where needed, round to 1 decimal place.

7   Find the value of each logarithm.

(a) $\log_2 8$ (b) $\log_5 625$ (c) $\log_2 32$ (d) $\log_{10} \dfrac{1}{1000}$

(e) $\log_2\left(\dfrac{1}{32}\right)$ (f) $\log_{10} 1$ (g) $\log_5 \sqrt{5}$ (h) $\log_2 64$

(i) $\log_{27} 3$ (j) $\log_m m^x$ (k) $\log_a a$ (l) $\log_a 1$

8   Solve for x.

    (a) $\log_2 x = 3$ (b) $\log_2 32 = x$ (c) $\log_x 27 = 3$

    (d) $\log_3 \left(\dfrac{1}{27}\right) = x$ (e) $\log_5 x = -3$ (f) $\log_x 20 = 1$

    (g) $\log_x \dfrac{1}{8} = -3$ (h) $\log_x x^5 = 5$ (i) $x = \log_2 8\sqrt{2}$

9   You can use your calculator to solve for x to 4 decimal places. The first question is done for you. (Refer to your manual.)

    (a) $\log_{10} x = 4$

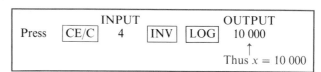

    Think: calculator steps

    | | INPUT | | | OUTPUT |
    Press CE/C 4 INV LOG 10 000

    Thus $x = 10\ 000$

    Check

    Press CE/C 10 000 LOG 4 ⟵——— Thus, $\log_{10} 10\ 000 = 4$

    (b) $\log_{10} x = 3$ (c) $\log_{10} x = 2.5$ (d) $\log_{10} x = 0.8$

    (e) $\log_{10} x = 3.61$ (f) $\log_{10} x = 0.542$ (g) $\log_{10} x = 0.0856$

10 Which of the following equations are equivalent? (Solve for the variable.)

    (a) $\log_{32} y = \dfrac{2}{5}$ (b) $\log_{16} 2 = y$ (c) $\log_y 81 = \dfrac{4}{3}$ (d) $\log_8 y = \dfrac{2}{3}$

11 (a) Find the value of $\log_2 64$.   (b) Find the value of $\log_2 128$.

    (c) What is the value of $\log_2 64 + \log_2 128$?

12 Find the value of each of the following.

    (a) $\log_3 81 + \log_3 27$ (b) $\log_{10} 1000 - \log_{10} 0.01$

    (c) $\log_3 \sqrt[4]{3} + \log_3 \sqrt[5]{81}$ (d) $\log_2 (4 \times \sqrt[5]{4}) - \log_2 (8^{\frac{1}{3}})$

13 Find the value of each of the following.

    (a) $2^{\log_2 4}$ (b) $10^{\log_{10} 100}$ (c) $5^{\log_5 25}$ (d) $3^{\log_3 \left(\frac{1}{27}\right)}$ (e) $4^{\log_4 (4^3)}$

    (f) Use the above results. What is the value for all $a \in R$, or $a^{\log_a m}$, $a > 0$?

14 Find the value of each expression. Use the results from the previous question.

    (a) $\log_2 73$ (b) $\log_6 (\log_2 64)$ (c) $3^{\log_3 27}$

    (d) $27^{\log_3 9}$ (e) $2^{\log_2 8 + \log_2 64}$ (f) $5^{(\log_5 8 - \log_5 2)}$

C 15 Simplify.

    (a) $3^{\log_3 27} + 10^{\log_{10} 1000}$ (b) $5^{\log_5 8} - 3^{\log_3 5 + \log_3 7}$

Applications: Decibels and Acoustics

To compare the loudness of sounds in acoustics, the unit commonly used is the decibel (dB), $\left(\dfrac{1}{10} \text{ of a bel}\right)$.

The loudness of sound is always given in reference to sound at the threshold of hearing. At this threshold, the level is said to be zero decibels or 0 dB. The formula used to compare sound is

$y = 10 \log_{10}\left(\dfrac{i}{i_R}\right)$, where
- i is the intensity of the sound being measured,
- i_R is the reference intensity, and
- y is the loudness in decibels.

Use the reference intensity, i_R, to be the intensity at the threshold of hearing. For a sound that is barely audible,

$$i = i_R, \text{ and } y = 10 \log_{10}\left(\dfrac{i_R}{i_R}\right) = 10(\log_{10} 1) = 10(0) = 0$$

For a sound 100 times more intense than i_R,

$$i = 100i_R \text{ and } y = 10 \log_{10}\left(\dfrac{100i_R}{i_R}\right) = 10(\log_{10} 100) = 10(2) = 20$$

16 (a) How many times more intense is a sound of 60 dB than a sound of 30 dB to the nearest 10?

(b) The sound at supper is measured at 62 dB. The sound at the library is measured at 26 dB. How many times more intense is the noise at supper than the noise at the library?

17 Jay finds that his car has a defective muffler. His friend Judy, who works at Can Tech Sound Laboratory, measures the sound at 118 dB. He has the muffler replaced. The new reading is 78 dB. How much more intense was the noise made by the muffler before the car was repaired?

18 The noise in a hospital is recorded at 46 dB. During an emergency it is recorded at 52 dB. By what factor does the intensity of the sound increase during the emergency?

19 During the busy part of the day the noise level near an airport was noted at 128 dB. After noise abatement procedures were introduced the noise level dropped to 112 dB. By what factor did the intensity of the noise decrease after noise abatement procedures were introduced at the airport?

20 Lesley records the average noise level in her house to be 68 dB. To cut down noise from outside, she installs certain insulation that reduces the average noise level inside to 56 dB. How many times as intense was the noise level in her house before she installed the insulation?

10.3 Problem-Solving: Exploring Properties

In the process of doing mathematics, often a suggestion, when pursued, will lead to a new branch of mathematics. For example, the graphs of the functions defined by $y = \log_{10} x$, $y = \log_2 x$ and $y = \log_3 x$ are given below. The graphs of these logarithmic functions can help to suggest properties about logarithms.

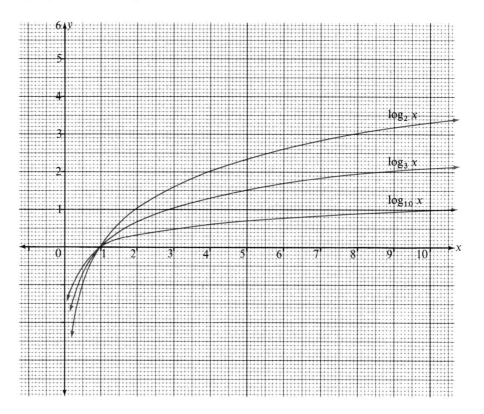

10.3 Exercise

B 1 (a) From the graph of the function defined by $\log_2 x$ find each value.

A: $\log_2 2$ B: $\log_2 3$

(b) Use your values in (a) to find a value for $\log_2 2 + \log_2 3$.

(c) Find a value for $\log_2 6$ from the graph of $\log_2 x$.

(d) How do your values in (b) and (c) compare?

(e) Find a relationship between the numbers 2 and 3 which results in the answer 6. What do you notice?

(f) Try some values of your own.

2   (a) Use the graph of $\log_3 x$ to find each value.

         A: $\log_3 9$ B: $\log_3 4.5$

   (b) Use your values in (a) to find a value for $\log_3 9 - \log_3 4.5$.

   (c) Find a value for $\log_3 2$. How does this value compare with the answer in (b)?

   (d) Find a relationship between the numbers 9 and 4.5 which results in the answer 2. What do you notice?

   (e) Try some values of your own.

3   (a) Find each of these values.

         A: $\log_{10} 2$ B: $\log_{10} 8$

   (b) Use the value of (a) A to find a value for $3 \log_{10} 2$.

   (c) What do you notice about the values of $3 \log_{10} 2$ and $\log_{10} 8$?

   (d) Calculate a value for 2^3.

   (e) Use the answers in (d) and (a) B to find a value for $\log_{10} 2^3$.

   (f) Compare your answers in (b) and (e). What do you notice?

   (g) Try some values of your own. What conclusion do you arrive at?

4   (a) Find each of these values from the graph of $\log_{10} x$.

         (i) $\log_{10} 8$ (ii) $\log_{10} 4$

   (b) Calculate a value for $\frac{2}{3} \log_{10} 8$.

   (c) Calculate a value for $8^{\frac{2}{3}}$

   (d) Find a value for $\log_{10} 8^{\frac{2}{3}}$

   (e) How do the values in (a), (b) and (d) compare?

   (f) Is the equation $\log_{10} 8^{\frac{2}{3}} = \frac{2}{3} \log_{10} 8$ true?

   (g) Try some values of your own.

Math Tip

Step A: The graphs of the logarithmic functions given by

$$y = \log_{10} x \qquad y = \log_2 x \qquad y = \log_3 x$$

have suggested certain properties about logarithms.

Step B: These steps are all part of the process of mathematics. You will prove these properties in the next section.

10.4 Laws of Logarithms: Products and Quotients

Part of the process of mathematics is that concepts and skills learned in one branch of mathematics, can often be transferred to another branch of mathematics. To find the transfer properties of logarithms, remember that the following statements are equivalent.

exponential form   **logarithmic form**

$x = a^y$ $y = \log_a x$

Product Law

The product law for exponents is used to develop the product law for logarithms. For this development,

$$\text{let } M = a^m \Rightarrow m = \log_a M, \text{ and}$$
$$\text{let } N = a^n \Rightarrow n = \log_a N.$$

$$\text{Now, } M \times N = a^m \times a^n$$
$$= a^{m+n}.$$

Take the logarithm to the base a of both sides of the equation.

$$\log_a (M \times N) = \log_a a^{m+n} \quad \longrightarrow \text{Remember:}$$
$$= m + n \quad\quad\quad \log_a a^k = k$$
$$= \log_a M + \log_a N$$

Product Law for Logarithms $\log_a (M \times N) = \log_a M + \log_a N$

Quotient Law

The quotient law for exponents leads to a quotient law for logarithms.

$$\text{Let } M = a^m \Rightarrow m = \log_a M.$$
$$\text{Let } N = a^n \Rightarrow n = \log_a N.$$

$$\text{Then, } \frac{M}{N} = \frac{a^m}{a^n}$$
$$= a^{m-n}$$

Take the logarithm of both sides of the equation to the base a.

$$\log_a \left(\frac{M}{N}\right) = \log_a a^{m-n}$$
$$= m - n$$
$$= \log_a M - \log_a N$$

Quotient Law for Logarithms $\log_a \left(\dfrac{M}{N}\right) = \log_a M - \log_a N$

At first glance, it appears an impossible task to evaluate

$$\log_3 54 + \log_3 \frac{3}{2}$$

because 54 cannot be expressed as a power of 3. However, the laws of logarithms allow you to evaluate this and other expressions, as shown in the example below.

Example Find the value of

(a) $\log_3 54 + \log_3 \left(\frac{3}{2}\right)$. (b) $\log_2 144 - \log_2 9$.

Solution (a) Use $\log_a (M \times N) = \log_a M + \log_a N$.

$$\log_3 54 + \log_3 \left(\frac{3}{2}\right) = \log_3 \left(54 \times \frac{3}{2}\right)$$
$$= \log_3 (81)$$
$$= \log_3 (3^4)$$
$$= 4$$

(b) Use $\log_a \left(\frac{M}{N}\right) = \log_a M - \log_a N$.

$$\log_2 144 - \log_2 9 = \log_2 \left(\frac{144}{9}\right)$$
$$= \log_2 (16)$$
$$= \log_2 (2^4)$$
$$= 4$$

10.4 Exercise

A Remember: Learn the Laws of Logarithms and apply them when you simplify expressions.

1 Express each of the following as a sum of logarithms.
(a) $\log_3 (13 \times 47)$ (b) $\log_2 (3.2 \times 78)$ (c) $\log_{10} (15.2 \times 33.8)$
(d) $\log_x (pg)$ (e) $\log_z (xy)$ (f) $\log_b (mn)$

2 Express each of the following as the logarithm of a product.
(a) $\log_7 13 + \log_7 41$ (b) $\log_2 28 + \log_2 36$ (c) $\log_7 43 + \log_7 81$
(d) $\log_{10} 22.7 + \log_{10} 36.3$ (e) $\log_a m + \log_a n$ (f) $\log_x a^2 b + \log_x ab^2$

3 Express each of the following as the logarithm of a quotient.
(a) $\log_3 37 - \log_3 22$ (b) $\log_2 85 - \log_2 74$ (c) $\log_{10} 222 - \log_{10} 75$
(d) $\log_x 71 - \log_x 17$ (e) $\log_b 33 - \log_b 11$ (f) $\log_a x^2 y - \log_a sy$

4 Express each of the following as a difference of logarithms.

(a) $\log_2\left(\dfrac{72}{35}\right)$ (b) $\log_7 (352 \div 19.3)$ (c) $\log_{10} (751 \div 82)$

(d) $\log_a\left(\dfrac{a}{b}\right)$ (e) $\log_x\left(\dfrac{52.5}{131}\right)$ (f) $\log_a\left(\dfrac{741}{337}\right)$

(g) $\log_b\left(\dfrac{842}{61.3}\right)$ (h) $\log_k\left(\dfrac{73.2}{13.7}\right)$

5 Use the properties of logarithms to write a sum or difference.

(a) $\log_4 (7 \times 6)$ (b) $\log_6 (0.28 \times 536)$ (c) $\log_7\left(\dfrac{421}{237}\right)$

(d) $\log_x\left(\dfrac{22.3}{481}\right)$ (e) $\log_{10} (7 \times 27 \times 361)$ (f) $\log_x \dfrac{pq}{mn}$

B The theory of logarithms is applied to the simplification of expressions that involve logarithms.

6 Evaluate.

(a) $\log_{10} 30 + \log_{10}\left(\dfrac{10}{3}\right)$ (b) $\log_5 2.5 + \log_5 10$

(c) $\log_3 18 + \log_3 1.5$ (d) $\log_2\left(\dfrac{4}{3}\right) + \log_2 24$

7 Evaluate.
(a) $\log_2 320 - \log_2 10$ (b) $\log_2 40 - \log_2 5$
(c) $\log_2 144 - \log_2 9$ (d) $\log_3 216 - \log_3 8$

8 Find the value of each of the following.
(a) $\log_2 5 + \log_2 25.6$ (b) $\log_3 63 - \log_3 7$ (c) $\log_3 108 - \log_3 4$
(d) $\log_3 81 - \log_3 27$ (e) $\log_2 16 + \log_3 9$ (f) $\log_5 100 - \log_5 4$

9 Which has the greater value, A or B?
A: $\log_3 72 - \log_3 8$ B: $\log_{10} 500 + \log_{10} 2$

10 Which of the following have the same value?
(a) $(\log_5 50 + \log_5 2) - \log_5 4$
(b) $\log_2 32 + \log_3 27 + \log_5 25$
(c) $(\log_6 12 + \log_6 9) - \log_6 3$

11 Evaluate.

(a) $\log_2 \sqrt{8}$

(b) $\log_3 27^3$

(c) $\log_2 (4 \times \sqrt[3]{16}) + \log_2 \sqrt[3]{16}$

(d) $\log_{\frac{2}{3}} \sqrt[4]{\frac{9}{4}}$

C 12 Express each of the following as a single logarithm.

(a) $\log_2 x + \log_2 y + \log_2 z - \log_2 a$

(b) $\log_3 x + \log_3 y^2 - \log_3 a^{\frac{1}{2}}$

(c) $\log_3 (x + y) + \log_3 (x - y) - (\log_3 x + \log_3 y)$

Calculator Tip

You can use your calculator to explore the laws of logarithms.

▶ Calculate A: $\log_{10} (4.5 \times 3.6)$ $\log_{10} 4.5 + \log_{10} 3.6$

B: $\log_{10} (8.6 \times 4.2)$ $\log_{10} 8.6 + \log_{10} 4.2$

To use your calculator efficiently, use the memory feature.

$\boxed{\text{MS}}$ memory store $\boxed{\text{MR}}$ memory recall
Do these steps. (Memory is clear.)

$\boxed{\text{C}}$ 4.5 $\boxed{\text{LOG}}$ $\boxed{\text{MS}}$ 3.6 $\boxed{\text{LOG}}$ $\boxed{+}$ $\boxed{\text{MR}}$ $\boxed{=}$ OUTPUT

What do you notice about your answers? Try other pairs of values.

▶ Calculate A: $\log_{10} \left(\dfrac{4.5}{3.6} \right)$ $\log_{10} 4.5 - \log_{10} 3.6$

B: $\log_{10} \left(\dfrac{8.6}{4.2} \right)$ $\log_{10} 8.6 - \log_{10} 4.2$

What do you notice about your answers? Try other pairs of values.

Problem Solving

What is wrong with the following?

$4 > 3$

$4(\log_{10} \tfrac{1}{3}) > 3(\log_{10} \tfrac{1}{3})$

$\log_{10} (\tfrac{1}{3})^4 > \log_{10} (\tfrac{1}{3})^3$

$\log_{10} (\tfrac{1}{81}) > \log_{10} (\tfrac{1}{27})$

$\tfrac{1}{81} > \tfrac{1}{27}$

10.5 Laws of Logarithms: Powers and Roots

Your earlier work with the Laws of Exponents is used again to develop additional theory about logarithms. Let $N = a^n$. Then $n = \log_a N$.

Now,
$$N^p = (a^n)^p.$$
$$N^p = a^{np} \longleftarrow \text{Take } \log_a \text{ of both sides of the equation.}$$
$$\log_a N^p = \log_a a^{np}$$
$$= np \longleftarrow \text{But } n = \log_a N.$$
$$= (\log_a N)p$$
Thus, $\log_a N^p = p(\log_a N)$.

Law of Logarithms for Powers $\log_a (N^p) = p \log_a N$, $N \in R$, $a > 0$

Since p can be a whole number or a fraction, this law can be expressed

---

Law of Logarithms for Roots $\log_a N^{\frac{p}{q}} = \dfrac{p}{q} \log_a N$, $N \in R$, $a > 0$

Example 1 Calculate the value of
(a) $\log_{10} \sqrt[4]{1000}$. (b) $\log_2 (32)^{\frac{1}{3}}$.

Solution (a) $\log_{10} \sqrt[4]{1000} = \log_{10} (1000)^{\frac{1}{4}}$ Think: $\log_{10} 1000 = \log_{10} 10^3$
$$= \frac{1}{4} \log_{10} 1000 \qquad\qquad\qquad\qquad = 3$$
$$= \frac{1}{4}(3) \text{ or } \frac{3}{4}$$

Use these steps on your calculator. Refer to your manual.

Press [CE/C] 1000 [INV] [10^x] [÷] 4 [=] [?]

(b) $\log_2 (32)^{\frac{1}{3}} = \dfrac{1}{3} \log_2 32$ Think: $\log_2 32 = \log_2 2^5$
$$= \frac{1}{3}(5) \qquad\qquad\qquad\qquad\qquad = 5$$
$$= \frac{5}{3}$$

The laws of logarithms for multiplication and division may be combined with the laws of logarithms for powers and roots.

Which property is used in each line in the solution of the following Example?

Example 2 Express each of the following as a single logarithm.

(a) $3 \log_3 2 + \log_3 4$ (b) $2 \log_2 3^2 + \log_2 6 - 3 \log_2 3$

Solution (a) $3 \log_3 2 + \log_3 4$
$$= \log_3 2^3 + \log_3 4$$
$$= \log_3 (2^3)(4)$$
$$= \log_3 32$$

(b) $2 \log_2 3^2 + \log_2 6 - 3 \log_2 3$
$$= \log_2 (3^2)^2 + \log_2 6 - \log_2 3^3$$
$$= \log_2 [(3^2)^2 \times 6] - \log_2 3^3$$
$$= \log_2 \frac{(3^2)^2 \times 6}{3^3}$$
$$= \log_2 18$$

10.5 Exercise

A Where possible check your results with a calculator.

1 Express each of the following in the form $k \log_a x$.
   (a) $\log_7 15^2$ (b) $\log_6 57^8$ (c) $\log_x a^4$
   (d) $\log_a y^{-2}$ (e) $\log_2 15^{\frac{1}{4}}$ (f) $\log_4 25^{\frac{4}{5}}$

2 Express each of the following in the form $\log_a k$.
   (a) $3 \log_8 7$ (b) $10 \log_6 5$ (c) $4 \log_x a$
   (d) $\frac{4}{5} \log_5 z$ (e) $-\frac{2}{3} \log_4 7$ (f) $-\frac{6}{7} \log_x a$

3 Express each of the following in the form $k \log_{10} p$.
   (a) $\log_{10} (1.68)^5$ (b) $\log_{10} (3.81)^{\frac{1}{2}}$ (c) $\log_{10} \sqrt{4.86}$
   (d) $\log_{10} \sqrt[3]{4.86^2}$ (e) $\log_{10} \sqrt[5]{4.26^2}$ (f) $\log_{10} (\sqrt{9.86})^3$

4 Find the value of each of the following
   (a) $\log_2 4^{25}$ (b) $\log_5 25^5$ (c) $\log_3 \sqrt[4]{27}$
   (d) $\log_{10} \sqrt[3]{10\,000}$ (e) $\log_{11} \sqrt[5]{11}$ (f) $\log_2 64 \sqrt[3]{2}$
   (g) $\log_2 \sqrt[5]{64}$ (h) $\log_3 \sqrt[5]{81}$ (i) $\log_2 \frac{1}{\sqrt{16}}$

5 Calculate to 4 decimal places. Use a calculator where needed.
   (a) $\log_{10} \sqrt[3]{1000}$ (b) $\log_{10} \sqrt[4]{10\,000}$ (c) $\log_{10} \sqrt{625}$
   (d) $\log_{10} \sqrt{48.6}$ (e) $\log_{10} (2.25)^{\frac{1}{3}}$ (f) $\log_{10} (46.8)^{\frac{3}{2}}$

B Summarize the laws of logarithms. You must understand them in order to apply them to do problems.

6 Which of the following are equal in value?

   A $\boxed{\log_3 27^4}$ B $\boxed{\log_7 \sqrt[4]{49^6}}$ C $\boxed{\log_2 8^4}$

7   Arrange the values of the following in order from least to greatest.

A $\log_{10} \sqrt[3]{10\ 000}$ B $\log_3 81^7$ C $\log_3 \sqrt[4]{27}$ D $\log_2 \sqrt[7]{8}$

8   Write each of the following as a single logarithm.
(a) $\log_2 6 + \log_2 10 + \log_2 3$ (b) $\log_6 4 + \log_6 10 - \log_6 8$
(c) $2 \log_4 3 + 3 \log_4 2$ (d) $3 \log_5 3 + \log_5 2 + \log_5 4$

9   If $\log_a x = m$ and $\log_a y = n$, express the value of each of the following in terms of m and n.

(a) $\log_a \left(\dfrac{x}{y}\right)$ (b) $\log_a (xy)^2$ (c) $\log_a \left(\dfrac{x^2}{y}\right)$ (d) $\log_a (xy)^3$

(e) $\log_a (xy)^{\frac{1}{2}}$ (f) $\log_a (xy)^{-\frac{2}{3}}$ (g) $\log_a \sqrt[3]{xy}$ (h) $\log_a \dfrac{1}{x^2 y}$

10   Evaluate.
(a) $\log_2 [(8)(\sqrt{32})] + \log_7 [(49)(\sqrt[4]{7})]$ (b) $\log_9 [(3)(\sqrt{27})] + \log_{81} 3$

C 11   Write as a single logarithm.

(a) $\dfrac{1}{2} \log_5 x + \dfrac{1}{3} \log_5 y - \dfrac{1}{4} \log_5 z$

(b) $\dfrac{1}{2} [\log_4 x + 3 \log_4 y] - 2[\log_4 a + \log_4 b]$

(c) $\dfrac{1}{3} [2 \log_2 x - 3 \log_2 y] - 4 \log_2 z$

Calculator Tip

You can use your calculator to explore the laws of logarithms.

▶ Calculate A: $\log_{10} (4.5)^2$ Use your calculator efficiently.

2 $\log_{10}$ 4.5   C 4.5 y^x 2 = LOG ?

C 4.5 LOG × 2 = ?

B: $\log_{10} (6.8)^{\frac{1}{2}}$ $\dfrac{1}{2} \log_{10} 6.8$

What do you notice about your answers? Try other values with your calculator.

10.6 Using Logarithms: Solving Equations

You will often find the logarithmic function occurring in scientific journals. The logarithmic function even occurs in everyday situations.

If you know the atmospheric pressure, P, in kiloPascals, then you can determine the approximate distance, d, in kilometres, above sea level by using the formula $d = \dfrac{500(\log P - 2)}{27}$.

When you work with equations involving logarithms, you need to use the Laws of Logarithms, which are summarized below.

Laws of Logarithms with $M, N \in R, a > 0$

logarithm of a product	$\log_a MN = \log_a M + \log_a N$
logarithm of a quotient	$\log_a \left(\dfrac{M}{N}\right) = \log_a M - \log_a N$
logarithm of a power	$\log_a M^p = p(\log_a M)$
logarithm of a root	$\log_a M^{\frac{p}{q}} = \dfrac{p}{q}(\log_a M)$

Example 1 Solve $\log_3 x - \log_3 4 = \log_3 12$.

Solution

$$\log_3 x - \log_3 4 = \log_3 12$$

$\log_a x$ is a function. $\log_3 x = \log_3 4 + \log_3 12$ Verification

If $\log_a x = \log_a y$, $\log_3 x = \log_3 48$ $LS = \log_3 x - \log_3 4$ $RS = \log_3 12$
  then $x = y$. $x = 48$ $= \log_3 48 - \log_3 4$
              $= \log_3 (48 \div 4)$
The root of the equation is 48. $= \log_3 12$
                  $LS = RS \checkmark$ Checks

The graph $y = \log_a x$ had $x > 0$. This restriction has been carried throughout your work on logarithms. It is important in the solution of logarithmic equations because it causes one root to be inadmissible, as shown in the following example.

Example 2 Solve $\log_6 (x + 3) + \log_6 (x - 2) = 1$, $x \in R$.

Solution

$\log_6 (x + 3) + \log_6 (x - 2) = 1$ $\log_6 6 = 1$ Restriction:
  $\log_6 (x + 3)(x - 2) = \log_6 6$ $x - 2 > 0$ and $x + 3 > 0$
                    $x > 2$ and $x > -3$
      $(x + 3)(x - 2) = 6$ Thus $x > 2$.
        $x^2 + x - 6 = 6$
        $x^2 + x - 12 = 0$
        $(x - 3)(x + 4) = 0$
$x - 3 = 0$ or $x + 4 = 0$
  $x = 3$ or $x = -4$ This is an inadmissible root since $x > 2$.

10.6 Exercise

A 1 (a) Solve $\log_2 72 = \log_2 x + \log_2 12$.

(b) Verify your answer in (a).

2 Find the solution set of each equation.

(a) $\log_2 6 + \log_2 9 = \log_2 x$ (b) $\log_{10} 3 = \log_{10} x - \log_{10} 5$

(c) $\log_8 x + \log_8 5 = \log_8 55$

3 Find the root of each equation.

(a) $\log_{10} 25 - \log_{10} 5 = \log_{10} x$ (b) $\log_3 x - \log_3 4 = \log_3 12$

(c) $-\log_5 1 = \log_5 7 - \log_5 x$

4 (a) Solve $\log_8 x = \dfrac{1}{4} \log_8 16$.   (b) Verify your answer in (a).

B 5 Solve each equation.

(a) $\log_2 x = 2 \log_2 8$ (b) $\log_{10} x = 4 \log_{10} 2$

6 Find the solution set.

(a) $4 \log_7 x = \log_7 625$ (b) $2 \log_8 x = \log_8 144$

7 Find the root of each of the following.

(a) $\log_7 x = 3 \log_7 4$ (b) $3 \log_7 x = \log_7 64$

8 (a) Solve $\log_7 (x - 2) = 1 - \log_7 (x + 4)$.

(b) What are the restrictions on x?

(c) Check your answer in (a).

9 Solve.

(a) $\log_4 (x + 2) + \log_4 (x - 1) = 1$

(b) $\log_6 (x + 3) + \log_6 (x - 2) = 1$

(c) $\log_8 x - \log_8 (x - 1) = 1$

10 Solve.

(a) $\log_8 (x - 4) = 1 - \log_8 (x + 3)$

(b) $\log_5 (7x + 1) - \log_5 (x - 1) = 2$

(c) $1 - \log_{10} (x - 4) = \log_{10} (x + 5)$

11 (a) Solve the logarithmic equation.

$\log_7 (2x + 2) - \log_7 (x - 1) = \log_7 (x + 1)$

(b) Verify your root in (a).

Applications: Atmospheric Pressure

The approximate distance above sea level, d, in kilometres, is given by

$$d = \frac{500(\log_{10} P - 2)}{27}$$

where P is the atmospheric pressure, in kiloPascals.

The reading on a barometer at sea level is 100 kPa. Use this value in the formula.

$$d = \frac{500(\log_{10} P - 2)}{27}$$

$$= \frac{500(\log_{10} 100 - 2)}{27}$$

$$= 0$$

If the pressure inside a jet liner changes drastically, the passengers receive a supply of oxygen that automatically is made available in the cabin of the jet.

Thus, the formula shows the distance above the earth at sea level is 0 km.

In Questions 12–16, round answers to 4 significant digits.

12 The highest inhabited buildings in the world are those in the Indian Tibetan border fort of Basisi. The atmospheric pressure of Basisi is 362 kPa. How far above sea level is the fort?

13 On May 29, 1953 Sir Edmund Hillary was the first person to reach the top of Mount Everest. The top of Mount Everest is 8850 m above sea level. Calculate the pressure loss the mountaineers underwent from sea level to the top of the mountain during this expedition.

14 On September 17, 1919 a skein of 17 Egyptian geese were photographed by an astronomer. The atmospheric pressure recorded at the same height as the geese were flying was 1088 kPa.
(a) Estimate the height at which the geese were flying.
(b) Calculate the height.

15 The greatest recorded vertical descent in parachute ski jumping is 1005 m achieved by Rick Sylvester. Assuming Rick started his jump at 2012 m above sea level, what pressure change did he undergo in his record breaking feat?

16 To ensure passenger safety, pressure within the cabin of a jet must remain constant at sea level pressure (100 kPa). What pressure difference must the jet withstand at 10.6 km above sea level?

10.7 Working with Proof

The process of proof occurs over and over again. The concepts and skills you have learned about logarithms are combined with the strategies for solving problems to solve any problem or prove any fact. In order to do this you must understand clearly

      A: what you are asked to prove.

      B: what information you are given.

Think of the above as you do the proof in the following example.

Example   If $\log_x y = m^2$ and $\log_y x = \dfrac{4}{m}$, prove that $m = \dfrac{1}{4}$.

Solution $\log_x y = m^2$ suggests $y = x^{m^2}$ ①   Think: What clues are given in the question?

$\log_y x = \dfrac{4}{m}$ suggests $x = y^{\frac{4}{m}}$ ②

From ① and ② you can obtain

$$x = y^{\frac{4}{m}}$$

$$x = (x^{m^2})^{\frac{4}{m}} \quad \text{since } y = x^{m^2}$$

$$x = x^{4m} \quad ③$$

From ③, since the bases are equal, then the exponents are equal.

$$1 = 4m \text{ or } m = \dfrac{1}{4}$$

which is what you wanted to prove.

10.7 Exercise

B 1   Given that $\log_m p = 2a^2$ and $a \log_p m = 3$. Prove that $a = \dfrac{1}{6}$.

2   Prove that if $\log_x y = am^2$, and for all a, b, m, $bam = 1$, then $\log_y x = \dfrac{b}{m}$.

3   Prove that for all m and n, $\log_m A = \dfrac{\log_n A}{\log_n m}$.

4   If $a^2 + b^2 = 23ab$, prove that $\log\left(\dfrac{a+b}{5}\right) = \dfrac{\log a + \log b}{2}$.

5   If $a(a - 7b) = -b^2$, then prove that $\log\left(\dfrac{a+b}{3}\right) = \dfrac{\log a + \log b}{2}$.

6   Prove each of the following for a, b, $\in R$, and $b > 0$.

    (a) $\log_p \dfrac{a}{b} + \log_p b = \log_p a$ (b) $\log_x \dfrac{a}{b} = \log_x a - \log_x b$

10.8 Alternate Strategies: Using Logarithms

You can use your skills with logarithms as an alternative strategy for solving equations involving exponents. The following example indicates another point of view and shows the process which can be applied to the solution of more complex equations.

Logarithms to the base 10 are called **common logarithms**. You write $\log_{10} 298$ as $\log 298$.

⎿Base 10 is assumed when no base is shown.

Example 1 Solve $x^{\frac{2}{3}} = 20.5$. Express your answer to 3 significant digits.

Solution $\log (x^{\frac{2}{3}}) = \log 20.5$ Think: To isolate the variable, take the common logarithm of both sides.

$$\frac{2}{3} \log x = \log 20.5.$$

$$\log x = \frac{3}{2} \log 20.5 \longleftarrow$$

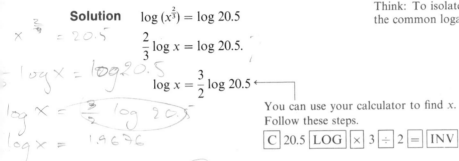

You can use your calculator to find x.
Follow these steps.

OUTPUT

| C | 20.5 | LOG | × | 3 | ÷ | 2 | = | INV | LOG | 92.817698

$\log x = 1.9676$ to 4 decimal places
$x = $ inv log 1.9676
$x = 92.8$ to 3 significant digits

In the next example you need to use additional skills with equations.

Example 2 Solve $4^{3x} = 55$ to 2 decimal places.

Solution

$\log (4^{3x}) = \log (55)$ Think: To isolate the variable, take the common logarithm of both sides.
$3x \log 4 = \log 55$

$$x = \frac{\log 55}{3 \log 4}$$

$$= \frac{1.7404}{3(0.6021)} \text{ (rounded)}$$

$$= \frac{1.7404}{1.8063}$$ Show the steps you would take to find the answer if you used a calculator.

$$= 0.9635 \text{ (rounded)}$$

Thus, $x = 0.96$ (to 2 decimal places).

10.8 Exercise

A   All logarithms in this exercise are to the base 10. Questions 1 to 4 require the use of the memory function on your calculator.

1   Calculate.
     (a) $\log 48 - \log 26$ (b) $\log 121 - 2 \log 16$
     (c) $3 \log 8 - \log 5$ (d) $5 \log 4 - 2 \log 3$

2   Calculate each of the following.
     (a) $\dfrac{\log 26}{\log 4}$ (b) $\dfrac{\log 96.3}{\log 28.2}$ (c) $\dfrac{2 \log 46}{\log 8}$ (d) $\dfrac{\log 181}{2 \log 65}$

3   Simplify each of the following.
     (a) $\dfrac{1}{2} \log 48.3$ (b) $\dfrac{1}{2} \log 0.413$ (c) $\dfrac{3}{2} \log 634$ (d) $\dfrac{2}{5} \log (0.163)$

4   Calculate.   (a) $\dfrac{\log 36 - \log 20}{\log 5}$ (b) $\dfrac{3 \log 12 - 2 \log 6}{\log 3}$

5   Find x to 2 decimal places in each of the following.
     (a) $\log x = 1.3698$ (b) $\log x = 3.692$
     (c) $\dfrac{2}{3} \log x = 0.8325$ (d) $\dfrac{3}{2} \log x = 0.046\ 29$

B   Use your calculator. Think through the procedure carefully before doing each step of the calculation.

6   Solve for the variable. Can you simplify before you do the work?
     (a) $\log m = \dfrac{\log 48 - \log 12}{2 \log 4}$ (b) $\log k = \dfrac{2 \log 36 - \log 5}{3}$

7   Solve for the variable.
     (a) $x^5 = 63$ (b) $y^7 = 92$ (c) $m^{\frac{2}{3}} = 28$ (d) $q^{\frac{3}{5}} = 126$

8   Solve each equation. Express your answer to one decimal place.
     (a) $3^x = 18$ (b) $8^x = 25$ (c) $4^{2x} = 90$
     (d) $5^{3x} = 63$ (e) $29^{\frac{x}{2}} = 76$ (f) $16^{\frac{3x}{5}} = 73$

9   Solve.   (a) $4^{2x} = 3^{x-1}$ (b) $6^{3x} = 2^{2x-3}$ (c) $(2^{2x})^3 = 4^{x+3}$

10   Solve for x.   (a) $(1.4)^x = (2.6)^{x+5}$ (b) $(1.93)^{2x} = 3(4.1)^x$
                  (c) $4(2^x) = 3^{x+1}$ (d) $(2.1)^{x+1} = 5(3.6)^{2x}$

10.9 Applications with Logarithms

In the process of developing concepts and skills in mathematics, often you devise different strategies for solving the same problem. It takes practice to be able to decide wisely on which strategy is needed to solve a problem. Sometimes the reason for your choice of strategy is based on personal experience. You may have found that one strategy is more efficient than another one. Thus, the more problems you solve and the more situations you are confronted with, the better you will be at choosing the correct strategy for solving a problem.

In the previous chapter, you used skills with the exponential function. You can also use your skills with logarithms and your calculator to solve problems that contain equations involving the variable in the exponent, as shown in the following example.

Example From experimental data, it was observed that 128 g of radioactive material decayed according to the equation

$$A = 128 \times 10^{-0.016t},$$

where A is the amount of material left, in grams,
t is the time, in hours.

How much time will have passed when 25 g of the material remain? Express your answer to 1 decimal place?

Solution

$A = 128 \times 10^{-0.016t}$ Always record the original equation.

Use $A = 25$.

$25 = 128 \times 10^{-0.016t}$ To isolate the variable, take the common logarithm of both sides.

$\log 25 = \log 128 + \log 10^{-0.016t}$

$\log 25 = \log 128 - 0.016t \log 10$

$\log 25 = \log 128 - 0.016t(1)$

$0.016t = \log 128 - \log 25$

$0.016t = 2.1072 - 1.3979$

$t = \dfrac{0.7093}{0.016}$

$t = 44.3$ to 1 decimal place

Thus, 25 g of material remain after 44.3 h.

When you solve problems, remember to organize your solution. Refer to the *Steps For Solving Problems.*

Steps For Solving Problems.

Step A Read the problem carefully. Can you answer these two questions?
  I What information am I asked to find (information I don't know)?
  II What information am I given (information I know)?
  Be sure you understand what it is you are to find.
Step B Decide on the method or strategy you need to use.
Step C Apply the method or strategy. Do the work. Obtain your answer.
Step D Check the answer in the original problem.
Step E Write a final statement as the answer to the problem.

10.9 Exercise

B Express all answers to 3 significant digits unless indicated otherwise.

1 The area, A, of an equilateral triangle with side m is given by $A = \dfrac{m^2(\sqrt{3})}{4}$.
Find the length of the sides if the area is 12.6 m^2.

2 An investment of \$100 is made in a term deposit that pays 9.5%/a compounded annually. The value, A, of the investment in dollars after time, t, in years, is given by A $= 100(1.095)^t$.
  (a) How long will it take the investment to be worth \$150?
  (b) How long will it take to double the value of the investment?

3 The value, V, of a coin appreciates according to the formula $V = 230(1.08)^t$, where V is the value of the coin, in dollars, after time, t, in years. How long will it take the coin to be worth \$400?

4 The volume, V, of a sphere is given by $V = \dfrac{4}{3}\pi r^3$, where r is the radius of the sphere. Use $\pi \doteq 3.1415$.
  (a) To one decimal place, calculate the radius of a sphere that has a volume of 69.60 cm^3.
  (b) A storage tank has a volume of 14.60 m^3. Calculate its radius to one decimal place.

5 The surface area, A, of a sphere of radius, r, is given by $A = 4\pi r^2$. The surface area of Mars is 1.45×10^8 km^2. Find the radius of Mars.

6 The number of cells in a culture grows according to the equation $A = 2500 \times 10^{\frac{t}{5}}$, where A is the number of cells after time, t, in seconds. How long will it take the number of cells to grow to 1.3×10^8?

7 A radioactive material decays according to the equation $A = A_0^{\frac{t}{14.5}}$, where A_0 is the original amount in grams and A is the amount remaining, in grams, after time t, in days. If 96.2 g of radioactive material is placed in the nuclear reactor, how long will it take the material to decay to 12.5 g?

8 When a projectile is dropped from an aircraft, the distance, s, in metres, that it has fallen, after time, t, in seconds, is given by $s = \dfrac{1}{2}gt^2$ where $g = 9.8 \text{ m/s}^2$. How long will it take a projectile to reach the earth if it is dropped from a height of 1150 m?

9 In deep water, the speed, V, in metres per second of a water wave is given by $V = \sqrt{\dfrac{g\lambda}{2\pi}}$, where g is acceleration due to gravity ($g = 9.8 \text{ m/s}^2$), and λ is the length of the wave, in metres.

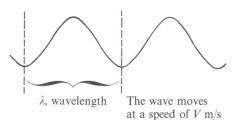

λ, wavelength   The wave moves at a speed of V m/s

(a) Calculate the speed of a wave of water for each length.
   (i) 1.2 m   (ii) 5.8 m   (iii) 20.5 m
(b) What is the length of a wave whose recorded speed is 9.6 m/s?

Applications: Earthquakes and Logarithmic Scales

You have probably seen the destructive nature of earthquakes. To record an actual measurement of the energy released by an earthquake would require large numbers. For this reason, a logarithmic scale is used for comparing intensities. Charles F. Richter (1900–1985) developed the Richter Scale, which compares the energy released at the centre of an earthquake. He defined the magnitude as

$$\log_{10}\frac{I}{I_0} = \text{ where } I \text{ is the intensity of the earthquake and}$$
I_0 is the intensity of a reference earthquake.

He assigned Richter Numbers to earthquakes, using a reference scale.

Richter Scale   3   4   5   6   7   8

magnitude of earthquake S,   magnitude of earthquake T.

$\log_{10}\dfrac{S}{I_0}$, is 5.1 ——————— 5.1   6.2 ——————— $\log_{10}\dfrac{T}{I_0}$ is 6.2

To compare the intensities of earthquakes S and T, you use your earlier skills with logarithms.

$$\log_{10}\left(\frac{T}{I_0}\right) - \log_{10}\left(\frac{S}{I_0}\right) = \log_{10} T - \log_{10} I_0 - (\log_{10} S - \log_{10} I_0)$$
$$= \log_{10} T - \log_{10} S$$
$$= \log_{10} \frac{T}{S}$$

Use the Richter Scale. $\quad \log_{10} \frac{T}{I_0} - \log_{10} \frac{S}{I_0} = 6.2 - 5.1$
$$= 1.1$$

Thus, $\quad \log_{10} \frac{T}{S} = 1.1$
$$\frac{T}{S} = 10^{1.1}$$
$$= 12.6$$

Thus, earthquake T is 12.6 times more intense than earthquake S.

10 Calculate the relative intensities of earthquake A and earthquake B. The Richter Scale readings are

	Earthquake A	Earthquake B
(a)	8.3	6.1
(b)	7.9	5.2

11 A great earthquake in India (August 15, 1950) had a Richter reading of 8.7. A slight tremor occurring in California had a reading of 2.5. How many times greater was the earthquake in India?

12 A magnitude of 8.9 was estimated for the earthquake off Northern Ecuador in 1906. An earthquake was registered as 6.5 in Desert Hot Springs, California in 1948. How many times greater was the earthquake in Ecuador?

Calculator Tip

To solve $4^x = 3$, you can use your calculator.

For example, $4^x = 3 \Rightarrow x \log 4 = \log 3$ or $x = \dfrac{\log 3}{\log 4}$

To calculate x, follow these steps on a calculator.

$$\boxed{3}\ \boxed{\text{LOG}}\ \boxed{\div}\ \boxed{4}\ \boxed{\text{LOG}}\ = \quad \textit{Output?}$$

Solve for x.
(a) $2^x = 6$ (b) $2^x = 6.8$ (c) $4.2^x = 5.6$
(d) $0.8^x = 3.2$ (e) $12.6^x = 65.3$ (f) $0.325^x = 0.862$

10.10 Problem Solving: Selecting a Strategy

To solve a problem, you often need to combine more than one strategy. For example, the following equation is written to solve a problem.

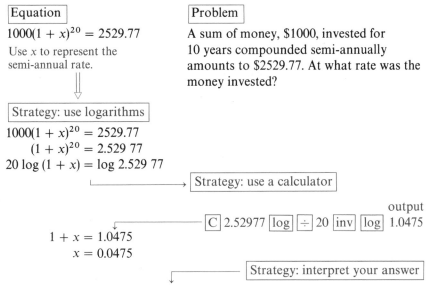

Equation
$1000(1 + x)^{20} = 2529.77$

Use x to represent the semi-annual rate.

Problem
A sum of money, $1000, invested for 10 years compounded semi-annually amounts to $2529.77. At what rate was the money invested?

Strategy: use logarithms

$$1000(1 + x)^{20} = 2529.77$$
$$(1 + x)^{20} = 2.529\ 77$$
$$20 \log (1 + x) = \log 2.529\ 77$$

Strategy: use a calculator

output

$\boxed{C}$ 2.52977 $\boxed{\log}$ $\boxed{\div}$ 20 $\boxed{\text{inv}}$ $\boxed{\log}$ 1.0475

$$1 + x = 1.0475$$
$$x = 0.0475$$

Strategy: interpret your answer

Thus, the rate of interest is $9\frac{1}{2}\%$ compounded semi-annually.

1 (a) At what rate, compounded annually, will an investment of $6000 accumulate to $12 861.53 in 8 years?

(b) A term deposit of $4950.00 earns 11% compounded semi-annually. If the deposit accumulated to $9410.98, how long was the money invested?

(c) Summer earnings of $969.86 are invested, compounded quarterly. If the amount of the investment in 4 years is $1384.59, what was the rate of interest?

2 Lori invested $3500 into a Registered Retirement Plan compounded quarterly for 5 years. If the amount in her plan after 5 years is $5735.16, what is the rate of interest?

3 To provide for a down payment on a recreation property, the Watsons invest their savings of $3832.82 at 9% compounded semi-annually.

(a) If the down payment required is $6500, how long does the money need to be invested?

(b) What assumption(s) did you make in obtaining your answer in (a)?

Practice and Problems: A Chapter Review

1   For each of the following, which is greater, A or B?

   <table>
   <tr><td></td><td>A</td><td>B</td></tr>
   </table>

	A	B
(a)	$\log_7 7^3$	$\log_7\left(\dfrac{1}{343}\right)$
(b)	$\log_3 \sqrt[4]{3}$	$\log_3 \sqrt[5]{81}$
(c)	$\log_{10}\left(\dfrac{1}{\sqrt[4]{1000}}\right)$	$\log_{10} 0.0001$

2   Use $\log_4 s = 3$. Find the value of the following.

   (a) $\log_4 16s^2$ (b) $\log_4 \dfrac{64}{\sqrt{s}}$

3   (a) If $\log x = m$ and $\log y = b$, find an expression for $\log\left(\dfrac{x^4}{y^2}\right)$.

   (b) If $\log K = 3 \log m + \log 5 - \log n$, express K in terms of m and n.

4   Solve.

   (a) $\log_2 x = \log_2 7 + \log_2 6$ (b) $\log_2(3x + 2) + \log_2 (x - 9) = 5$

5   Write as a single logarithm.

   (a) $\dfrac{1}{4}\left[2(\log_2 x + 3 \log_2 y) - 3 \log_2 z\right]$

   (b) $\dfrac{3}{5}\left[\dfrac{1}{2}(\log_2 x + 3 \log_2 y) - 2(\log_2 x - 4 \log_2 y)\right]$

6   If $a^2 + b^2 = 2ab$, prove $\log\left(\dfrac{a + b}{2}\right) = \dfrac{1}{2}(\log a + \log b)$.

7   (a) In an experiment, you need to solve the exponential equation $2.5^x = 6.8$

   for x. Show that $x = \dfrac{\log 6.8}{\log 2.5}$.

   (b) What is the value of x to 3 decimal places?

8   The amount of money, A, Sacha needs to invest today (compounded annually at 9%/a) in order to have $500 eight years from now is given by the solution to the equation

   $$8 \log 1.09 + \log A = \log 500.$$

   Find the amount Sacha would invest today.

Test for Practice

1   Prove that for all positive real numbers,

(a) $\log_a MN = \log_a M + \log_a N.$ (b) $\log_a \left(\dfrac{M}{N} \right) = \log_a M - \log_a N.$

2   Evaluate.

(a) $\log_3 81$ (b) $\log_2 2^{12}$ (c) $\log_a a^2$

(d) $\log_8 \sqrt{2}$ (e) $\log_2 (16 \times \sqrt[5]{8}) - \log_2 (16^{\frac{1}{4}})$

3   Solve for x.

(a) $\log_{10} x = 5$ (b) $\log_x 125 = 3$

(c) $\log_3 27 = x$ (d) $\log_{10} 0.0001 = x$

4   Solve for x to 2 decimal places.

(a) $5^x = 8$ (b) $2 \cdot 8^x = 9.3$

5   (a) If $\log_a x = m$, find the value of $\log_a x^3$ in terms of m.

(b) If $\log_a x = k$, find the value of $\log_a \sqrt[3]{x}$ in terms of k.

6   Which of the following are true (T)? Which are false (F)?

(a) $\log a + \log b \overset{?}{=} (\log a)(\log b)$

(b) $\log \sqrt[3]{a} \div \sqrt{b} \overset{?}{=} \dfrac{1}{3} \left(\log a - \dfrac{1}{2} \log b \right)$

(c) $\sqrt{xy} \overset{?}{=} \dfrac{1}{2} (\log x + \log y)$

7   If $\log_x y = m^4$ and $\log_y x = \dfrac{5}{m^3}$, prove that $m = \dfrac{1}{5}$.

8   (a) Solve $\log_{10} 9 = 2 \log_{10} x - \log_{10} 25$

(b) Verify your answer in (a).

9   Solve $\log_5 (2x + 1) + \log_5 (x - 1) = 1.$

10 A magnitude of 8.2 on the Richter Scale was recorded for the earthquake in Tangshan, China. An earthquake with Richter number 4.3 will cause a very noticeable vibration throughout the house, but no serious damage. How many times greater was the earthquake in China than one with a 4.3 reading?

Maintaining Skills

Skills related to the co-ordinate plane are used throughout your study of mathematics.

1 Show that the triangle with vertices $(0, \sqrt{6})$, $(\sqrt{2}, -\sqrt{6})$, and $(2\sqrt{2}, \sqrt{6})$ is isosceles.

2 A parabola given by $y = 2x^2 + x - 6$ intersects the x-axis. What are the co-ordinates of the points of intersection?

3 (a) Find the roots of $4x^2 - 6x - 1 = 0$
    (i) to 1 decimal place.   (ii) to 2 decimal places.
   (b) Verify your roots. Show why the roots in part (a)(i) provide a better approximation than those in part (a)(ii).

4 Find the zeroes of each function.
   (a) $f(x) = 2x^2 - 5x - 1$ (b) $g(x) = 2x^2 + 4x - 60$

5 The line $3x - 4y + 10 = 0$ is perpendicular to the line $Ax + By - 40 = 0$. If the lines intersect at $(2, 4)$, find A and B.

6 An iceberg has floated into the Gulf of St. Lawrence and lodged itself at map co-ordinates $(4, 1)$. A passenger liner is at map co-ordinates $(-2, 6)$ and heading for $(5, -1)$.
   (a) If the liner does not alter course, will it crash into the iceberg?
   (b) If the iceberg is not in the liner's path, how close does the liner pass by the iceberg?

7 (a) Write the equation of the family of lines with slope -0.2.
   (b) Select the member(s) of this family that form a triangle of area 10 square units with the axes.

8 Three ships patrol a triangular area whose vertices have co-ordinates $(7, -3)$, $(4, -5)$ and $(9, 6)$. Calculate the area that is patrolled to the nearest square unit.

9 A line defined by $3x - 4y = 12$ is translated by the mapping $(x, y) \rightarrow (x - 3, y + 2)$. Find the equation of the image line.

10 (a) Draw the region shown by $x^2 + y^2 < 25$, $x, y \in R$.
    (b) Which point is in the region, $A(3, 2)$ or $B(-3, -5)$?
    (c) Draw the region in (a) if the restriction $y \geq 0$ is given.

11 Equations and Applications: Conic Sections

nature of locus, writing equations, concepts and skills with circles, equations and properties, parabolas, ellipses, and hyperbolas, rectangular hyperbolas, eccentricity, solving systems of equations, strategies and solving problems, applications, problem-solving

Introduction

One of the main areas of study of an astronomer is the field of mathematics. In fact, Johannes Kepler (1571–1630) was influenced by his mathematics teacher and used mathematics to formulate the three famous Kepler Laws, named in his honour, which concern the motion of the planets. These laws played a very important role in the work of the physicist.

Sir Isaac Newton also used mathematics to formulate his findings. In the advanced study of astronomy, the mathematics you study is a beginning towards understanding the orbital paths of the moon as well as those of the artificial satellites launched from earth.

Astronomers can know the precise position of any planet, moon or satellite in the solar system at any given time. Computers translate all the conditions that affect these bodies into very precise equations. The path of Voyager 2 can be predicted to 1989 when it will pass Neptune on August 24, 1989, exactly.

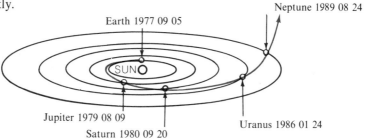

Neptune 1989 08 24

Earth 1977 09 05

SUN

Jupiter 1979 08 09

Saturn 1980 09 20

Uranus 1986 01 24

11.1 Working with Locus

Since a meteor or a comet in our solar system is affected by the pull of gravity of the sun and planets, its path can be predicted. The comet's path can be plotted when all the conditions acting on it have been considered. In a similar manner, the path or locus of points on the Cartesian plane, satisfying certain conditions, can be plotted.

A **locus** is a set of points that satisfy a given condition or conditions.

For example, each condition determines a locus.

Locus of points, P, equidistant from a point C.

Locus of points, P, equidistant from two lines.

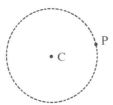

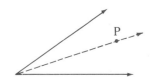

To find the locus of points, P, that satisfy a given condition or conditions, you need to find the locus of some of the points, and then sketch the locus as shown in the following example.

Example 1 Chords of equal length are drawn in a circle. Describe the locus of all midpoints of the chords.

Solution

Step 1
Draw a sketch to show a circle with chords of equal lengths.

Step 2
Mark the midpoints.

Step 3
Continue to mark equal chords with their midpoints until the locus can be sketched.

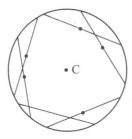

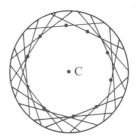

It seems, the locus of all the midpoints is a concentric circle with centre C.

You can find the equation of the locus of points occurring on the Cartesian plane as shown in the following example.

Example 2 Find the locus of all points, P, equidistant from the points A(-4, 1) and B(1, 3).

Solution *Step 1* Let P(x, y) be any point on the locus.

Draw a diagram to record the given information.

Step 2 P is equidistant from A and B.
Then PA = PB.

Step 3 PA = $\sqrt{(x + 4)^2 + (y - 1)^2}$
PB = $\sqrt{(x - 1)^2 + (y - 3)^2}$

But,   PA = PB.

$$\sqrt{(x + 4)^2 + (y - 1)^2} = \sqrt{(x - 1)^2 + (y - 3)^2}$$

Square both sides.
$$(x + 4)^2 + (y - 1)^2 = (x - 1)^2 + (y - 3)^2$$
$$x^2 + 8x + 16 + y^2 - 2y + 1 = x^2 - 2x + 1 + y^2 - 6y + 9$$
$$10x + 4y = -7$$

Thus, the locus of points equidistant from fixed points, A and B, has the equation $10x + 4y = -7$.

Think: To check, locate any point that satisfies the equation of the locus and verify that it is equidistant from A and B.

Follow these steps to find the equation of a locus.

Step 1 Draw a sketch. Let P(x, y) be any point on the locus.
Step 2 Write the geometric condition satisfied by P.
Step 3 Write the condition using skills from co-ordinate geometry.

11.1 Exercise

A 1 Describe the locus in each of the following situations.
(a) sitting in a seat on a turning ferris wheel
(b) going up an escalator
(c) taking an elevator
(d) walking from the corner of a room so that you are the same distance from a pair of adjacent walls.

2 Describe the locus of each of the following.
(a) a speck of dirt on a revolving record

(b) the needle of the record as the record is played

(c) the hub of a bicycle wheel going up a hill

(d) the tip of a motor boat propeller as the boat idles

(e) the tip of a motor boat propeller as the boat tows a water skier

3 What is the locus of each of the following? Use a sketch to help you picture each.

(a) all points 5 cm from a fixed point

(b) all points 4 cm from a line

(c) all points equidistant from two intersecting lines

(d) all points the same distance from two fixed points

(e) all points in a cubical box 1 cm from its surface

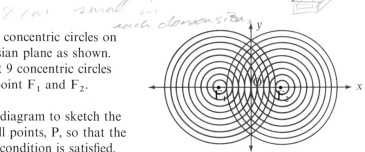

4 Construct concentric circles on the Cartesian plane as shown. Use about 9 concentric circles for each point F_1 and F_2.

Use your diagram to sketch the locus of all points, P, so that the following condition is satisfied.

co-ordinates | condition
$F_1(-4, 0), F_2(4, 0)$ | $PF_1 + PF_2 = 12$

5 The points, P_1 and P_2, shown in the diagram have the property that the difference of the distances from each point to F_1 and F_2 is a constant. That is,

$$PF_1 - PF_2 = 2 \text{ or } PF_2 - PF_1 = 2.$$

It is important that the distance be expressed as a positive number. Thus, the notation

$$|PF_1 - PF_2| = 2 \text{ is used.}$$

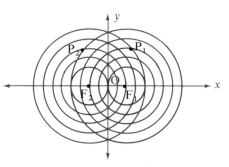

Use a diagram similar to the one above to do the following investigation. Find the locus of all points, P, that satisfy the condition.

fixed points | condition
$F_1(3, 0), F_2(-3, 0)$ | $|PF_1 - PF_2| = 2$

B Questions 6 to 10 are based on the following diagram.

A system of lines and circles has been drawn. The points, P, marked on the diagram have the property that they are equidistant from the fixed point, F, and the line, l_1.

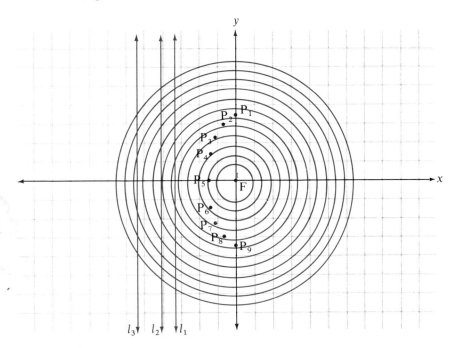

Make a working copy of the above diagram. Include at least five additional concentric circles. Then complete the following investigations.

6 (a) Mark all points that are equidistant from the point, F, and the line, l_1.
   (b) Draw a smooth curve through the points to show the locus of all points equidistant from the point, F, and the line, l_1.

7 Locate all points that are equidistant from the point, F, and the line, l_2. Draw a smooth curve through the points to show the locus.

8 Repeat the procedure in Question 6 for the point, F, and line, l_3.

9 Choose a line of your own on the diagram. Locate all the points that are equidistant from F and your line. Draw a smooth curve for the points to show the locus.

10 Use the results of your work in Questions 6 to 9. How are the loci (plural of locus) alike? How do they differ?

Writing Equations: Locus

In the previous questions you completed constructions for finding the locus of a point satisfying a given set of conditions. By first sketching the locus, you can gain useful information that is helpful in writing the equation of the locus.

11 Sketch the locus of each of the following on a set of co-ordinate axes.
(a) points six units above the line defined by $x - y = 8$
(b) points four units below the line defined by $x + 2y = 12$
(c) points equidistant from $(4, 6)$ and $(4, -8)$
(d) points 4 units from $(-1, 3)$

12 Sketch the locus of all points, $P(x, y)$, that have the following properties.
(a) The sum of the co-ordinates is 10.
(b) The difference of the co-ordinates is 6.
(c) The x co-ordinate is 4 less than the y co-ordinate.

13 Find the equation of the locus of all points
(a) 4 units to the right of the y-axis.
(b) 6 units below the x-axis.
(c) equidistant from $(4, 3)$ and $(-5, 3)$.
(d) equidistant from $(-2, 1)$ and $(-2, -6)$.
(e) 4 units from the origin.

14 A locus has the property that the slope of the line joining any point, P, of the locus and the point $(3, 2)$ is $\frac{1}{2}$. Find the equation of the locus.

15 Any point P, of a locus has the property that the distance from P to the line $2x - 3y - 5 = 0$ is 8. Find the equation of the locus.

16 A point, P, has the property that it is equidistant from the points $A(-1, -3)$ and $B(6, 4)$. Find the equation of the locus of all such points, P.

17 The right bisector of AB is the locus of all points equidistant from the points A and B.
(a) Find the equation of the right bisector of AB given the co-ordinates $A(6, 4)$ and $B(4, -2)$.
(b) Find the equation of the right bisector of BC given the co-ordinates $B(4, -2)$ and $C(-2, 0)$.
(c) Use the equation in (a) and (b) to find the co-ordinates of a point equidistant from A, B, and C.

11.2 Circles and Mappings: Thinking Visually

When an equation of a circle is given with the centre at the origin, the equation is said to be in **standard form**.

Circle with centre at the origin

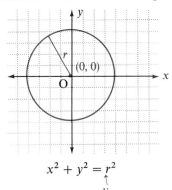

$$x^2 + y^2 = r^2$$
radius r

Circle with centre (a, b)

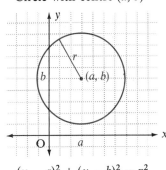

$$(x - a)^2 + (y - b)^2 = r^2$$
radius r

The equation is said to be in **standard form** (with its centre at the origin).

You obtained the equation of a circle with centre (a, b) as shown. In the following example, the converse question is asked. Find what mapping must be applied to a circle to obtain its equation in standard form.

Example   What mapping applied to the circle given by the equation

$$x^2 + y^2 - 4x + 2y - 11 = 0$$

will determine the equation in standard form?   Think: *Step* 1
To find the required mapping, write the given equation in the form
$(x - a)^2 + (y - b)^2 = r^2$.

Solution
$$x^2 + y^2 - 4x + 2y - 11 = 0$$
$$x^2 - 4x + 4 + y^2 + 2y + 1 = 11 + 5$$
Add 5 to each side.
$$(x - 2)^2 + (y + 1)^2 = 4^2$$

To find the mapping, think of the problem visually.

You have this circle.   You want this circle.

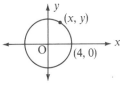

Choose a point (X, Y). Then complete the steps shown on the next page.

Let (X, Y) be any point on the circle given in standard form.

Use

$$X = x - 2 \quad \text{or} \quad x = X + 2 \qquad ①$$
$$Y = y + 1 \quad \text{or} \quad y = Y - 1 \qquad ②$$

Since (x, y) is a point on the original circle, then its co-ordinates satisfy the equation

$$x^2 + y^2 - 4x + 2y - 11 = 0. \qquad ③$$

Substitute ① and ② into ③.

$$(X + 2)^2 + (Y - 1)^2 - 4(X + 2) + 2(Y - 1) - 11 = 0$$
$$X^2 + 4X + 4 + Y^2 - 2Y + 1 - 4X - 8 + 2Y - 2 - 11 = 0$$
$$X^2 + Y^2 = 16$$

Thus, the required mapping is given by $(x, y) \rightarrow (x - 2, y + 1)$. ⟵⎤

The mapping yields a transformation which is a translation.

In general, the mapping $(x, y) \rightarrow (x - a, y - b)$ translates the circle given by $(x - a)^2 + (y - b)^2 = r^2$ to the origin. The translated circle has the equation $x^2 + y^2 = r^2$.

What if the mapping $(x, y) \rightarrow (3x, y)$ is applied to the circle $x^2 + y^2 = 1$? Thus, let (X, Y) be a point on the image of $x^2 + y^2 = 1$ under the transformation $(x, y) \rightarrow (3x, y)$. Then,

$$X = 3x \quad \text{or} \quad x = \frac{X}{3}$$

$$Y = y \quad \text{or} \quad y = Y$$

Substitute $x = \frac{X}{3}$, $y = Y$ in $x^2 + y^2 = 1$.

Then,

$$\left(\frac{X}{3}\right)^2 + Y^2 = 1$$

$$\frac{X^2}{9} + Y^2 = 1$$

$$X^2 + 9Y^2 = 9$$

Thus, the image of $x^2 + y^2 = 1$ under the transformation $(x, y) \rightarrow (3x, y)$ is $x^2 + 9y^2 = 9$.

When the graph of the equation $x^2 + 9y^2 = 9$ is drawn you see that the circle with equation $x^2 + y^2 = 1$ is transformed to an ellipse with equation $x^2 + 9y^2 = 9$.

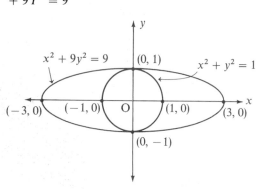

11.2 Exercise

1 For each circle, determine a mapping that will place the circle in standard position.

(a)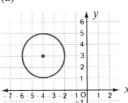

(b)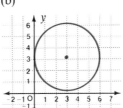

(c)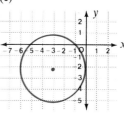

2 Write each of the following in mapping notation.
   (a) The x co-ordinate is translated 6 units to the left.
   (b) The y co-ordinate is translated 2 units downwards.
   (c) The x co-ordinate is translated 5 units to the right and the y co-ordinate is translated 7 units upwards.
   (d) a stretch of factor 2 parallel to the x-axis.
   (e) a stretch of factor 4 parallel to the y-axis.
   (f) a stretch of factor 3 parallel to the x-axis and of factor 2 parallel to the y-axis.

3 For the circle with centre $(-3, 1)$, translate the centre of the circle to the origin. What is the translation?

4 Find the translation for each of the following.

	Centre of pre-image circle	Centre of image circle
(a)	$(4, -3)$	$(0, 0)$
(b)	$(5, 2)$	$(0, 0)$
(c)	$(-6, -4)$	$(0, 0)$

B Remember: To find the mapping, think visually. Draw a sketch of both the original figure and the image figure in order to organize your steps.

5 (a) Find the equation of the image of the circle given by $x^2 + y^2 = 4$ under the mapping $(x, y) \rightarrow (2x, y)$. Sketch the graph of the image.
   (b) Find the equation of the image of the circle given by $x^2 + y^2 = 2$ under the mapping $(x, y) \rightarrow (x, 4y)$. Sketch the graph of the image.

6　Find the standard form of each equation.

(a) $(x-1)^2 + (y+2)^2 = 25$　　　　　　　What mapping did you use?

(b) $(3-x)^2 + (5+y)^2 = 100$

(c) $x^2 + y^2 + 8x - 6y - 11 = 0$

(d) $36x^2 + 36y^2 + 36x - 24y + 12 = 0$

(e) $x^2 + y^2 - 2\sqrt{2}x + 4\sqrt{2}y + 6 = 0$

7　(a) Prove that $x^2 + y^2 - 4x + 6y - 87 = 0$ is the equation of a circle.

(b) Find the centre and radius of the circle.

(c) $P(-6, 3)$ is a point on the given circle. What are the co-ordinates of P on the image circle with centre at the origin?

8　What mapping would transform the circle given by $x^2 + y^2 = 13$ into an ellipse given by the equation $36x^2 + y^2 = 117$?

9　What mapping maps the ellipse given by $8x^2 + 2y^2 = 7$ into the circle given by $x^2 + y^2 = 14$?

10　What transformation maps the ellipse given by $x^2 + 4y^2 = 72$ into the circle given by $x^2 + y^2 = 8$?

11　(a) What transformation maps the circle given by $x^2 + y^2 = 4$ into $(x-3)^2 + (y-2)^2 = 4$?

(b) What transformation maps the circle given by $x^2 + y^2 = 1$ into $x^2 + y^2 + 4x - 2y + 4 = 0$?

12　What transformation maps the circle given by $x^2 + y^2 = 1$ into

(a) $\left(\dfrac{x}{4}\right)^2 + \left(\dfrac{y}{3}\right)^2 = 1$　　(b) $9x^2 + 25y^2 = 225$

13　In the previous questions you have found various mappings that transform one graph into another. Use these results or test further examples to answer the following for real numbers $a, b, r > 0$. Identify the transformation that maps

the graph of	to the graph of
(a) $x^2 + y^2 = r^2$	$(x-a)^2 + (y-b)^2 = r^2$
(b) $x^2 + y^2 = r^2$	$(x+a)^2 + (y+b)^2 = r^2$
(c) $x^2 + y^2 = 1$	$\left(\dfrac{x}{a}\right)^2 + y^2 = 1$
(d) $x^2 + y^2 = 1$	$x^2 + \left(\dfrac{y}{b}\right)^2 = 1$
(e) $x^2 + y^2 = 1$	$\left(\dfrac{x}{a}\right)^2 + \left(\dfrac{y}{b}\right)^2 = 1$

11.3 Parabolas and Their Applications

The method used to find the equation of a circle is applied in a similar way to investigate other loci. For example:

What type of curve is obtained if a point P is equidistant from a fixed point, and a fixed line? In other words, what is the locus of P if PF = PD?

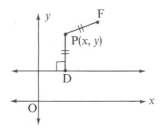

To find the equation, you need to sketch the given information on a diagram as shown in the following example.

Example 1   Find the equation of the locus of all points equidistant from a fixed point, F(1, 2), and a fixed line defined by $y = -4$.

Solution   Let P(x, y) be any point on the locus. Sketch the information on a diagram. The distances from P to F and P to D are equal.

Thus,   PF = PD.

$$\sqrt{(x - 1)^2 + (y - 2)^2} = \sqrt{(x - x)^2 + (y + 4)^2}$$
$$(x - 1)^2 + (y - 2)^2 = (y + 4)^2$$
$$(x - 1)^2 + y^2 - 4y + 4 = y^2 + 8y + 16$$
$$(x - 1)^2 - 12 = 12y$$

The equation becomes

$$12y = (x - 1)^2 - 12.$$
$$y = \frac{1}{12}(x - 1)^2 - 1$$

The graphs of such equations as in Example 1 are called **parabolas**. Thus, the *locus* of all points, equidistant from a fixed point F and a fixed line is a **parabola**.

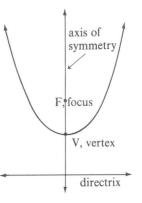

- The fixed point is called the **focus**.
- The fixed line is called the **directrix**.
- The **axis of symmetry** (or axis) is the line through the focus perpendicular to the directrix. The axis of the parabola shown is vertical.
- The **vertex** is the point at which the parabola intersects its axis of symmetry.

The definition of a parabola can be used to find the general equation of a parabola with axes that are either horizontal or vertical, as shown in the following example.

Example 2   A parabola has focus F(p, 0) and directrix defined by $x = -p$. Find its defining equation.

Solution   In particular, the locus passes through the origin. D has co-ordinates $(-p, y)$.

Sketch a diagram for the given information.

Step 1   Let P(x, y) be any point on the locus.

Step 2   P is equidistant from F and D. Thus
$$PF = PD.$$

Step 3 $\sqrt{(x - p)^2 + (y - 0)^2} = \sqrt{(x + p)^2}$
$$(x - p)^2 + y^2 = (x + p)^2$$
$$x^2 - 2px + p^2 + y^2 = x^2 + 2px + p^2$$
$$y^2 = 4px$$

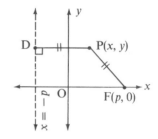

From Example 2, $y^2 = 4px$ is the equation of the parabola in standard position with focus F(p, 0) and directrix $x = -p$.

Similarly, $x^2 = 4py$ is the equation of a parabola in standard position with focus F(0, p) and directrix $y = -p$.

The vertex of each parabola is at the origin. The properties of the family of parabolas given in standard position are summarized in the chart.

	Axis of symmetry the x-axis.	Axis of symmetry the y-axis.
Equation	$y^2 = 4px$	$x^2 = 4py$
Focus	$(p, 0)$	$(0, p)$
Directrix	$x = -p$	$y = -p$
Graph	$p > 0 \qquad p < 0$	$p > 0 \qquad p < 0$

11.3 Exercise

A   Check your work. Choose a point. Does it satisfy the equation of the locus?

1   (a) Find the equation of the parabola with focus F(4, 0) and directrix $x = -4$.
    (b) Draw a sketch of the graph. Is the graph a function?

2   (a) Find the equation of the parabola with focus F(0, -3) and directrix
        $y = 3$.
    (b) Draw a sketch of the graph. Is the graph a function?

3   Parabolas are given by the following information. Find each equation.
    (a) focus F(2, 0); directrix $x = -2$.   (b) focus F(0, -3); directrix $y = 3$.
    (c) vertex V(0, 0); directrix $x = 4$.   (d) focus F(0, 5); vertex V(0, 0).

4   The equation of a parabola is given by $y^2 = 16x$.
    (a) Find the co-ordinates of the focus.
    (b) Find the equation of the directrix.
    (c) Draw a sketch. Locate the co-ordinates of two other points on the
        parabola.

5   For each parabola
    (i) draw a sketch.   (ii) label the co-ordinates of the focus, and vertex,
                                and the equation of the directrix.

    (a) $x^2 = 16y$ (b) $y^2 = -24x$ (c) $y^2 = 32x$ (d) $x^2 = -20y$

B   Review the meanings of the words *domain*, *range*, *intercepts*, *focus*, and *axis*.

6   A parabola has a horizontal axis with vertex at (0, 0). It passes through
    P(1, -10).
    (a) Find the equation of the parabola.
    (b) What are the co-ordinates of the focus?
    (c) What is the equation of the axis of the parabola?
    (d) What are its domain, range, and intercepts?

7   A parabola has a vertical axis with vertex at (0, 0). It passes through P(1, 8).
    (a) Find the equation of the parabola.
    (b) What are the co-ordinates of the focus?
    (c) What is the equation of the axis of the parabola?
    (d) What are its domain, range, and intercepts?

8   (a) Derive the equation of the parabola with focus $(0, p)$ and directrix
        $y = -p$.
    (b) Sketch the graph for   (i) $p > 0$ (ii) $p < 0$
    (c) What are the co-ordinates of the vertex?

9   Find the equation of the locus of points
    (a) equidistant from a fixed point $(0, 4)$ and a fixed line $y = -2$.
    (b) equidistant from a fixed point $(3, 0)$ and a fixed line $x = -5$.
    (c) What are the co-ordinates of the vertex in (a) and (b)?

10 Parabolas are given by the following information. What is the domain and
    range? Find each equation.
    (a) focus F(2, 0); directrix $x = -4$.   (b) focus F(0, 2); vertex V(0, -1).
    (c) vertex V(-2, 0); directrix $x = 3$.   (d) vertex V(2, 0); directrix $x = -3$.

11 For each parabola, a focus, F, and directrix are given. Find the defining
    equation and draw a sketch of it.
    (a) focus F(2, 4); directrix $y = -2$.   (b) focus F(-4, -6); directrix $y = 2$.
    (c) focus F(5, 2); directrix $x = -3$.   (d) focus F(-4, -3); directrix $x = 2$.

12 The arch of the bridge shown in the diagram
    is in the shape of a parabola.
    (a) Find its defining equation.
    (b) What are the co-ordinates of the focus?
        What is the equation of its axis?

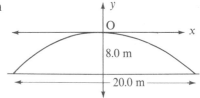

13 For each parabola find
    (i) the co-ordinates of the vertex.   (ii) the axis of symmetry.
    (a) $y = x^2 - 6x + 5$ (b) $x = y^2 - 2y + 3$ (c) $y = 2x^2 - 16x + 33$

14 A parabola passes through the points A(-1, 5), B(0, -1), and C(2, -1).
    (a) Find its defining equation.
    (b) Find the co-ordinates of the vertex.
    (c) Find the equation of its axis.

C 15   The diagram shows the *latus rectum* of a parabola,
       which is a line segment, ST, perpendicular to
       the axis and passing through the focus, F. Prove
       that the length of the latus rectum is $|4p|$ where $|p|$
       is the distance from the focus, F, to the vertex, V,
       of the parabola.

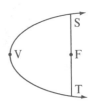

Applications: Parabolic Reflections

The parabolic shape is used in many ways to send out energy waves and collect energy waves. A property of the parabolic shape is shown in the following diagrams.

A   Waves travelling parallel to the axis of the parabola reflect to the focus.

B   Waves coming from the focus reflect parallel to the axis of the parabola.

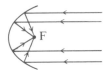
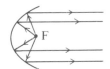

Thus, in situation A, parabolic reflectors are used to collect sound waves from centre field at a football game. Similarly, in situation B, a light bulb placed at the focus will give a beam of light (such as in a car headlight).

16   The headlight of a car has a reflector which is parabolic in shape. A cross sectional diagram is shown. The dimensions in the diagram represent lengths in centimetres.

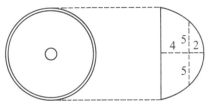

(a) Find the defining equation of the reflecting surface.

(b) What is the width of the headlight?

17   The cross section of a headlight reflector is in the shape of a parabola. The diameter of the reflector is 15.0 cm. The reflector is 10.0 cm deep. Find the distance from the vertex to the focus of the parabola.

18   At a sports event, a parabolic reflector is used to obtain the sound for television cameras. The opening of the reflector is 125.0 cm and the parabolic dish is 17.5 cm deep.

(a) Determine an equation for the reflecting surface.

(b) Find the position of the microphone placed at the focus.

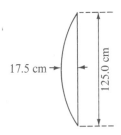

11.4 Ellipses and Their Applications

In your work thus far, the equation of a locus has been found by determining the geometric condition that the locus satisfies.

Circle: PF = constant   *Parabola:* PF = PD

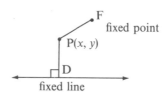

An **ellipse** is the *locus* of points, P, such that the sum of the distances from two fixed points to P is a constant.

Each of the fixed points, F_1 and F_2, is called a **focus** of the ellipse. PF_1 and PF_2 are called the **focal radii**.

Ellipse: $PF_1 + PF_2$ = constant

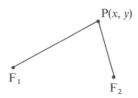

Based on its definition, the equation of an *ellipse* drawn on the Cartesian plane can be obtained by using the same procedure as that used for the circle and parabola.

Step 1 Draw a sketch. Let P(x, y) be any point on the locus.

Step 2 Write the geometric condition.

Step 3 Translate the geometric condition into a condition involving co-ordinates.

These steps are illustrated in the following example.

Example An ellipse has foci at $F_1(-3, 0)$ and $F_2(3, 0)$. If the sum of the focal radii is 10, find the equation of the ellipse.

Solution *Step 1* Let P(x, y) be any point on the locus.

Step 2 The sum of the distances from P to F_1 and F_2 is 10. Then $PF_1 + PF_2 = 10$.

Step 3 $PF_1 = \sqrt{(x + 3)^2 + (y - 0)^2}$

$PF_2 = \sqrt{(x - 3)^2 + (y - 0)^2}$

$$PF_1 + PF_2 = 10$$
$$\sqrt{(x + 3)^2 + y^2} + \sqrt{(x - 3)^2 + y^2} = 10$$
$$\sqrt{(x + 3)^2 + y^2} = 10 - \sqrt{(x - 3)^2 + y^2}$$

Square both sides and simplify.

$$(x + 3)^2 + y^2 = 100 - 20\sqrt{(x - 3)^2 + y^2} + (x - 3)^2 + y^2$$
$$x^2 + 6x + 9 + y^2 - x^2 + 6x - 9 - y^2 - 100 = -20\sqrt{(x - 3)^2 + y^2}$$
$$12x - 100 = -20\sqrt{(x - 3)^2 + y^2}$$
$$3x - 25 = -5\sqrt{(x - 3)^2 + y^2}$$

Square both sides and simplify.

$$9x^2 - 150x + 625 = 25(x^2 - 6x + 9 + y^2)$$
$$9x^2 - 150x + 625 = 25x^2 - 150x + 225 + 25y^2$$
$$16x^2 + 25y^2 = 400 \qquad \text{This form of the equation is referred to as the \textbf{standard form} of the equation.}$$

Other information can be obtained from the equation. The graph is drawn.

x-intercepts $\quad 16x^2 = 400$
Let $y = 0$. $\qquad x^2 = 25$
$$x = \pm 5$$

y-intercepts $\quad 25y^2 = 400$
Let $x = 0$. $\qquad y^2 = 16$
$$y = \pm 4$$

Ellipse with foci on x-axis

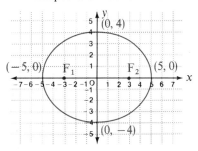

The ellipse has as two axes of symmetry, the x-axis and the y-axis.

The parts of the axes within the ellipse are referred to as the **major axis** and **minor axis** of the ellipse. The major axis contains the foci.

The points at the ends of the major axis are called the **vertices** of the ellipse.

The equation can be written in a form that makes the intercepts easy to identify.

$$16x^2 + 25y^2 = 400$$
$$\frac{x^2}{25} + \frac{y^2}{16} = 1 \qquad \text{Divide by 400.}$$
$$or \quad \frac{x^2}{(5)^2} + \frac{y^2}{(4)^2} = 1$$

The equation written in the above form is called the **intercept-form** of the equation.

$$\frac{x^2}{(x\text{-intercept})^2} + \frac{y^2}{(y\text{-intercept})^2} = 1$$

Problem-Solving Strategy: General Equations

The steps that were used to develop the equation of a specific ellipse involving numbers can also be used to find the *general* equation of an ellipse.

An ellipse has foci as shown at $F_1(-c, 0)$ and $F_2(c, 0)$. Find its equation.

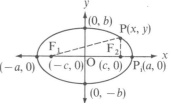

Step 1   Let $P(x, y)$ be any point on the ellipse.

Think: The sum PF_1 and PF_2 is a constant. Find the value for the constant.

Step 2   For $P_1(a, 0)$, a specific point,
$$P_1F_1 = a + c \quad \text{and} \quad P_1F_2 = a - c.$$
$$\text{Thus, } P_1F_1 + P_1F_2 = a + c + a - c$$
$$= 2a.$$
Thus for any point $P(x, y)$, $PF_1 + PF_2 = 2a$.
From the diagram
$$a^2 = b^2 + c^2 \quad \text{or} \quad b^2 = a^2 - c^2.$$

Step 3 $PF_1 = \sqrt{(x + c)^2 + (y - 0)^2}$ $PF_2 = \sqrt{(x - c)^2 + (y - 0)^2}$

$$PF_1 + PF_2 = 2a$$
$$\sqrt{(x + c)^2 + y^2} + \sqrt{(x - c)^2 + y^2} = 2a$$
$$\sqrt{(x + c)^2 + y^2} = 2a - \sqrt{(x - c)^2 + y^2}$$
$$(x + c)^2 + y^2 = 4a^2 - 4a\sqrt{(x - c)^2 + y^2} + (x - c)^2 + y^2$$
$$x^2 + 2xc + c^2 + y^2 = 4a^2 - 4a\sqrt{(x - c)^2 + y^2} + x^2 - 2xc + c^2 + y^2$$
$$4xc - 4a^2 = -4a\sqrt{(x - c)^2 + y^2}$$
$$xc - a^2 = -a\sqrt{(x - c)^2 + y^2}$$
$$x^2c^2 - 2xca^2 + a^4 = a^2(x^2 - 2xc + c^2 + y^2)$$
$$x^2c^2 - 2xca^2 + a^4 = a^2x^2 - 2xca^2 + a^2c^2 + a^2y^2$$
$$a^2x^2 - c^2x^2 + a^2y^2 = a^4 - a^2c^2$$
$$(a^2 - c^2)x^2 + a^2y^2 = a^2(a^2 - c^2) \longleftarrow \text{Use } b^2 = a^2 - c^2.$$
$$b^2x^2 + a^2y^2 = a^2b^2 \longleftarrow \text{Divide by } a^2b^2.$$
$$\frac{x^2}{a^2} + \frac{y^2}{b^2} = 1$$

Thus, for the ellipse given by $\dfrac{x^2}{a^2} + \dfrac{y^2}{b^2} = 1$, where $a^2 > b^2$, the vertices are given by $(a, 0)$, $(-a, 0)$ and the foci are given by $(c, 0)$ and $(-c, 0)$ where $c^2 = a^2 - b^2$.

Once you learn the properties of the above ellipse with the foci on the x-axis, you can transfer your skills to an ellipse with foci on the y-axis. The equation of the ellipse is

$$\frac{x^2}{b^2} + \frac{y^2}{a^2} = 1$$

The major axis again contains the foci. Thus the vertices are given by $(0, a)$ and $(0, -a)$ and the foci are given by $(0, c)$ and $(0, -c)$ where $c^2 = a^2 - b^2$.

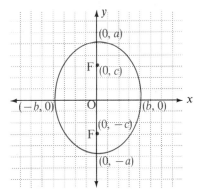

11.4 Exercise

A 1 For each ellipse, what are the
(i) lengths of the major and minor axes? (ii) co-ordinates of the vertices?
(iii) co-ordinates of the foci? (iv) intercepts?

(a) $\dfrac{x^2}{100} + y^2 = 1$ (b) $x^2 + \dfrac{y^2}{25} = 1$ (c) $\dfrac{x^2}{100} + \dfrac{y^2}{25} = 1$ (d) $\dfrac{x^2}{64} + \dfrac{y^2}{36} = 1$

(e) $\dfrac{x^2}{25} + \dfrac{y^2}{9} = 1$ (f) $\dfrac{x^2}{4} + \dfrac{y^2}{25} = 1$ (g) $\dfrac{x^2}{100} + \dfrac{y^2}{64} = 1$ (h) $\dfrac{x^2}{64} + \dfrac{y^2}{100} = 1$

2 For each graph find the
(i) co-ordinates of its vertices. (ii) co-ordinates of its foci.
(iii) length of the major axis and minor axis. (iv) equation of the ellipse.

(a) 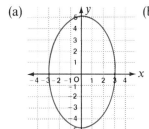 (b) 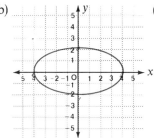 (c)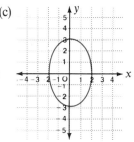

3 (a) Sketch the graph of the ellipse given by $\dfrac{x^2}{100} + \dfrac{y^2}{64} = 1$.

(b) Label the foci, vertices, major and minor axes.

B 4 The sum of the focal radii of an ellipse is 12. If the co-ordinates of the foci are $(4, 0)$ and $(-4, 0)$, find the equation of the ellipse.

5. The major axis of an ellipse is 20. If the co-ordinates of the foci are (0, 8) and (0, −8), find the equation of the ellipse.

6. Copy and complete the chart.

	Foci	Sum of focal radii	Equation of ellipse
(a)	(0, 5)(0, −5)	15	?
(b)	(−1, 0)(1, 0)	3	?
(c)	(0, m)(0, −m)	2k	?
(d)	(k, 0)(−k, 0)	2p	?

7. Each ellipse with centre (0, 0) has the following properties. Find its equation.
   (a) The x and y-intercepts are respectively ± 4 and ± 3.
   (b) The major axis is 8 and horizontal; the minor axis is 4.
   (c) One vertex is (−4, 0); the minor axis is 2.
   (d) The major axis is 10 and is vertical; the minor axis is 6.
   (e) One vertex is (6, 0); the minor axis is 4.
   (f) One vertex is (0, 5); one focus is (0, −4).

8. The equation of each ellipse is given. Sketch the graph. Label the foci, the vertices, and the major and minor axes.
   (a) $\dfrac{x^2}{9} + \dfrac{y^2}{4} = 1$ (b) $\dfrac{x^2}{4} + \dfrac{y^2}{9} = 1$ (c) $9x^2 + 25y^2 = 225$

9. The roof of a building is in the shape of an ellipse. Use the dimensions shown to find the equation of the ellipse.

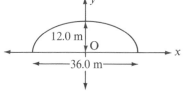

10. The dome of an arena is elliptical in shape. If the height of the dome is 30 m and has a span of 80 m, find an equation of the ellipse.

11. Each of the following equations defines an ellipse. Find the
    (i) co-ordinates of the foci.   (ii) co-ordinates of the vertices.
    (a) $3x^2 - 12 = -y^2$ (b) $2y^2 - 20 = -3x^2$ (c) $8y^2 = 10 - 4x^2$

12. (a) Show that the curve defined by $2(5x + 4) - y^2 = 2x(x + 5)$ is an ellipse.
    (b) What are the intercepts?   (c) Find the co-ordinates of the foci.

C 13. Sketch the region shown by each of the following.
    (a) $\dfrac{x^2}{25} + \dfrac{y^2}{4} \leq 1$ (b) $\dfrac{x^2}{4} + \dfrac{y^2}{16} \geq 1$ (c) $\dfrac{x^2}{64} + \dfrac{y^2}{100} > 1$

Applications: Orbits in the Solar System

Planets, comets, satellites often travel in elliptical orbits in our solar systems. For example, one famous comet called Halley's Comet travels in an elliptical orbit with the sun at one focus.

Comets are often named after their discoverers. Halley's Comet is named after Edmond Halley. The comet returns every 76 years and was last seen in 1910. When was Halley's Comet again visible on Earth?

14 The nearest distance that Halley's Comet approaches the sun is 88×10^6 km, while its farthest distance from the sun is 5.3×10^9 km.

(a) Draw a sketch of its orbit.

(b) Find the defining equation of the orbit of Halley's Comet.

15 The orbit of a satellite about the earth is often elliptical in shape, with the centre of the earth at one focus of the orbit. The point at which the satellite is nearest to earth is called the **perigee**. The **apogee** is the point at which the satellite is farthest from earth. The radius of the earth is 6.34×10^3 km. Use the diagram to determine an equation of the orbit of the satellite.

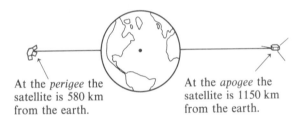

At the *perigee* the satellite is 580 km from the earth.

At the *apogee* the satellite is 1150 km from the earth.

16 The earth travels about the sun on an elliptical orbit with the sun at one focus of the ellipse. Use the information in the diagram to determine the equation for the orbit of the earth. (The centre of the elliptical orbit is placed at (0, 0).)

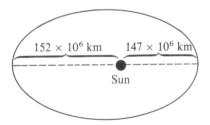

152×10^6 km 147×10^6 km

Sun

17 The closest distance of Mercury to the sun (called the **perihelion**) is 46.0×10^6 km and the most distant point of Mercury from the sun (called the **aphelion**) is 69.9×10^6 km.

(a) Determine an equation for the orbit of Mercury about the sun.

(b) List any assumptions you make in determining the orbit.

11.5 What If . . . ?: Hyperbolas

Many mathematicians have contributed to the development of new concepts in mathematics by asking the question, *What would happen if . . . ?*

For the definition of an ellipse,

$$PF_1 + PF_2 = 2a.$$

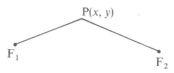

What would happen if the definition was altered so that the *difference* of the distances from two fixed points F_1 and F_2 was constant? That is, what is the locus of the point P which satisfies the condition $\left|PF_1 - PF_2\right| = 2a$?

You can now use your earlier steps to derive the equation of this locus, as shown in the following example.

Example Derive the equation of the locus of points, P, that satisfies the condition $\left|PF_1 - PF_2\right| = 6$ where the fixed points are $F_1(-5, 0)$ and $F_2(5, 0)$.

Solution *Step 1* Let $P(x, y)$ be any point on the locus.

Step 2 The difference of the distances from P to F_1 and F_2 is 6. Then $\left|PF_1 - PF_2\right| = 6$. Thus, $PF_1 - PF_2 = 6$ or $PF_1 - PF_2 = -6$. Without any loss of generality use $PF_1 - PF_2 = 6$.

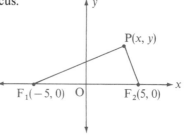

Step 3 $PF_1 = \sqrt{(x + 5)^2 + y^2}$ $PF_2 = \sqrt{(x - 5)^2 + y^2}$

Now, $PF_1 - PF_2 = 6.$

$$\sqrt{(x + 5)^2 + y^2} - \sqrt{(x - 5)^2 + y^2} = 6$$

$$\sqrt{(x + 5)^2 + y^2} = 6 + \sqrt{(x - 5)^2 + y^2}$$

$$(x + 5)^2 + y^2 = 36 + 12\sqrt{(x - 5)^2 + y^2} + (x - 5)^2 + y^2$$

$$x^2 + 10x + 25 + y^2 = 36 + 12\sqrt{(x - 5)^2 + y^2} + x^2 - 10x + 25 + y^2$$

$$20x - 36 = 12\sqrt{(x - 5)^2 + y^2}$$

$$5x - 9 = 3\sqrt{(x - 5)^2 + y^2}$$

$$25x^2 - 90x + 81 = 9(x^2 - 10x + 25 + y^2) \longleftarrow \text{Square both sides and simplify.}$$

$$25x^2 - 90x + 81 = 9x^2 - 90x + 225 + 9y^2$$

$$16x^2 - 9y^2 = 144 \longleftarrow \text{This form of the equation is referred to as the \textbf{standard form} of the equation.}$$

From the equation $16x^2 - 9y^2 = 144$ other information can be obtained.

x-intercepts
Let $y = 0$. $16x^2 = 144$
$$x^2 = 9$$
$$x = \pm 3$$

y-intercepts
Let $x = 0$. $-9y^2 = 144$
$$y^2 = -16$$

This result can be interpreted to mean that there are no y-intercepts.

If the equation is solved for y, then

$$16x^2 - 9y^2 = 144$$
$$9y^2 = 16x^2 - 144$$
$$y = \pm\frac{4}{3}\sqrt{x^2 - 9} \quad\longleftarrow\quad \begin{array}{l}\text{For all real } y, x^2 - 9 \geq 0, \\ \text{Thus, } x \geq 3 \text{ or } x \leq -3.\end{array}$$

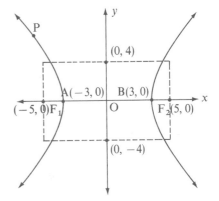

The previous information is used to sketch the graph of the locus. Other points needed to sketch the graph are plotted. The curve is called a **hyperbola**. A hyperbola is the *locus* of points, P, such that the difference of the distances from P to two fixed points, F_1 and F_2 is a constant.

$$\left|PF_1 - PF_2\right| = \text{constant}$$

The two fixed points, F_1 and F_2 are called the **foci** of the hyperbola.

For the graph, the hyperbola has the x-axis as its **axis of symmetry**.

The distance AB is called the **major** or **transverse axis**.

The points at the ends of the major axis are called the **vertices** of the hyperbola.

The **centre** of the hyperbola is $(0, 0)$.

The segment perpendicular to the transverse axis at its midpoint and with its endpoints $(0, 4)$ and $(0, -4)$ is called the **conjugate** or **minor axis**.

The equation can be rewritten in the following form.

$$16x^2 - 9y^2 = 144 \qquad \text{Divide by 144.}$$
$$\frac{x^2}{9} - \frac{y^2}{16} = 1$$

The method used above to develop the particular equation of the hyperbola involving numbers can be applied to find the general equation of a hyperbola with foci, F_1 and F_2, on the x-axis where $\left|PF_1 - PF_2\right| = 2a$.

The equation obtained is then

$$\sqrt{(x-c)^2 + y^2} - \sqrt{(x+c)^2 + y^2} = 2a \quad \text{or} \quad -2a,$$

which simplifies to

$$\frac{x^2}{a^2} - \frac{y^2}{b^2} = 1 \quad \text{where } a^2 + b^2 = c^2.$$

Write the equation

$$\frac{x^2}{a^2} - \frac{y^2}{b^2} = 1 \text{ in terms of } y.$$

$$\frac{y^2}{b^2} = \frac{x^2}{a^2} - 1 \quad \text{or} \quad y = \pm\frac{b}{a}\sqrt{x^2 - a^2}$$

The hyperbolas given by

$$\frac{x^2}{a^2} - \frac{y^2}{b^2} = 1 \quad \text{and} \quad \frac{x^2}{b^2} - \frac{y^2}{a^2} = -1$$

are called **conjugate hyperbolas**.

As x increases and becomes large, $x^2 - a^2$ approaches the value of x^2 and y approaches the value $\pm\frac{b}{a}x$.

The lines defined by $y = +\frac{b}{a}x$ and $y = -\frac{b}{a}x$ are called the **asymptotes** of the hyperbola. The asymptotes of the hyperbola are the lines that the branches of the hyperbola seem to approach as x increases in value in a positive or negative direction.

In the development for ellipses, two forms of this equation were obtained. Similarly, the equation of the hyperbola with foci on the y-axis is given by

$$\frac{y^2}{a^2} - \frac{x^2}{b^2} = 1 \quad \text{or} \quad \frac{x^2}{b^2} - \frac{y^2}{a^2} = -1.$$

The vertices of the hyperbola at right are $(0, a)$ and $(0, -a)$. The asymptotes are given by

$$y = \pm\frac{a}{b}x.$$

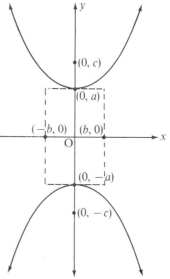

11.5 Exercise

A 1 For each hyperbola, find the values of a, b, and c.

(a) $\dfrac{x^2}{20} - \dfrac{y^2}{5} = 1$
(b) $\dfrac{x^2}{36} - \dfrac{y^2}{10} = -1$
(c) $3x^2 - 8y^2 = 24$

2 For each hyperbola, find
(i) the length of the major or transverse axis. Is the axis vertical or horizontal?
(ii) the co-ordinates of the vertices.
(iii) the intercepts.
(iv) the co-ordinates of the foci.

(a) $\dfrac{x^2}{25} - \dfrac{y^2}{16} = 1$
(b) $\dfrac{x^2}{4} - \dfrac{y^2}{9} = 1$
(c) $\dfrac{x^2}{25} - \dfrac{y^2}{9} = 1$

(d) $\dfrac{x^2}{4} - y^2 = 1$
(e) $\dfrac{x^2}{100} - \dfrac{y^2}{64} = -1$
(f) $\dfrac{x^2}{36} - \dfrac{y^2}{64} = -1$

(g) $\dfrac{x^2}{9} - \dfrac{y^2}{25} = -1$
(h) $\dfrac{y^2}{64} - \dfrac{x^2}{36} = -1$
(i) $\dfrac{y^2}{144} - \dfrac{x^2}{25} = -1$

3 Use the hyperbola in the diagram.
(a) What are the co-ordinates of the vertices?
(b) What are the co-ordinates of the foci?
(c) What is the length of the major or transverse axis?
(d) What is the equation of the hyperbola?

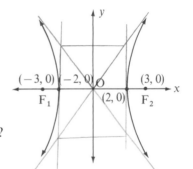

4 Find the defining equation of each of the following hyperbolas.

(a)

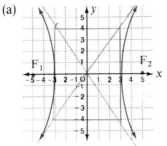

(b)

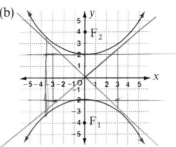

5 What are the equations of the asymptotes of each of the following hyperbolas?

(a) $\dfrac{x^2}{4} - y^2 = 1$
(b) $\dfrac{x^2}{9} - \dfrac{y^2}{16} = -1$
(c) $\dfrac{y^2}{16} - \dfrac{x^2}{9} = 1$

6   What are the slopes of the asymptotes of each of the following hyperbolas?

(a) $\dfrac{x^2}{100} - y^2 = 1$ (b) $x^2 - \dfrac{y^2}{25} = -1$ (c) $\dfrac{x^2}{36} - \dfrac{y^2}{64} = 1$

B   To work with hyperbolas, learn the meanings of *foci*, *vertices*, *transverse axis* and *conjugate axis*.

7   (a) Sketch the graph of the hyperbola given by $\dfrac{x^2}{10} - \dfrac{y^2}{9} = 1$.

(b) Label the foci, vertices, transverse axis and conjugate axis.

(c) Draw its asymptotes.

8   For each hyperbola, the equation is given.
  (i) Sketch the graph.
  (ii) Label the foci, vertices, transverse axis, and conjugate axis.
      Draw the asymptotes.
  (iii) What are the intercepts?

(a) $\dfrac{x^2}{16} - \dfrac{y^2}{9} = 1$ (b) $\dfrac{x^2}{9} - \dfrac{y^2}{16} = 1$ (c) $\dfrac{x^2}{40} - \dfrac{y^2}{10} = -1$

(d) $x^2 - 4y^2 = 36$ (e) $25y^2 - x^2 = 25$ (f) $4y^2 - 9x^2 = -36$

9   Sketch the graphs of each of the following hyperbolas on different pairs of axes.

(a) $\dfrac{x^2}{144} - \dfrac{y^2}{25} = 1$ (b) $\dfrac{x^2}{25} - \dfrac{y^2}{144} = 1$ (c) $\dfrac{x^2}{25} - \dfrac{y^2}{144} = -1$

(d) How are the graphs alike? How do they differ?

10   The difference of the focal radii of a certain hyperbola is 10. If the co-ordinates of the foci are $(6, 0)$ and $(-6, 0)$, find the equation of the hyperbola with its centre the origin.

11   Each hyperbola below has its centre at the origin. Find the equations of hyperbolas with the following properties.

(a) a vertex at $(3, 0)$; a focus at $(4, 0)$ $C^2 = A + B^2$

(b) a focus at $(-6, 0)$; transverse axis of 8 units

(c) a vertex at $(0, 5)$; a focus at $(0, 8)$

(d) a vertex at $(8, 0)$; an asymptote given by $y = 4x$.

(e) conjugate axis 6; a focus at $(0, -8)$

12   For each of the following hyperbolas draw the graph and the graph of the conjugate on the same set of axes.

(a) $\dfrac{x^2}{16} - \dfrac{y^2}{9} = -1$ (b) $\dfrac{x^2}{144} - \dfrac{y^2}{25} = 1$

4

Applications: LORAN—A System of Navigation

Have you ever wondered how a ship at sea can determine its position? A hyperbolic navigation system is used and in part is shown in the diagram.

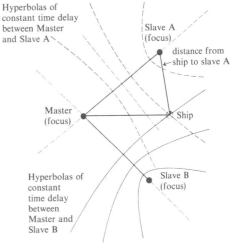

In order to determine positions at sea, or in the air, a system of navigation called

LORAN

(LOng Range Air Navigation)

is used.

Electronic signals are sent from two different LORAN stations M and Y_1 shown in the diagram at the right.

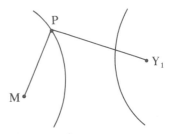

Since the signals come from different stations, there is a time difference. The navigator on the aircraft or ship at P converts the time difference into a distance difference $PM - PY_1$. This distance is coded on a map showing systems of hyperbolic curves and the aircraft or ship is determined to be on a branch of a hyperbola.

To find its exact position, the navigator tunes in on another pair of stations, M and Y_2, and finds the corresponding hyperbolic path on the map for this pair of stations.

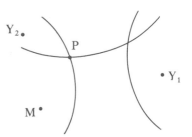

The point P, at which the curves intersect, is the location of the aircraft or ship.

13 For the hyperbolic navigation station above, one station, M, is designated as a master. The signals received from other stations are synchronized with the master signal. An aircraft at P is 30 km closer to station M than it is to station Y_1. The stations are 150 km apart. Find an equation for the hyperbola.

14 The master, M, and slave station, Y, are 100 km apart. A ship is 120 km from M and 95 km from Y. Find the equation of the hyperbola.

11.6 Rectangular Hyperbolas

You know that the general equation of a hyperbola, centre (0, 0) is

$$\frac{x^2}{a^2} - \frac{y^2}{b^2} = 1$$

Hyperbolas that have the property $a = b$ have a general equation

$$\frac{x^2}{a^2} - \frac{y^2}{a^2} = 1 \quad \text{or} \quad x^2 - y^2 = a^2$$

The asymptotes of such hyperbolas are $y = x$ and $y = -x$.

The graph at the right is the graph of the relation given by

$$x^2 - y^2 = 4$$

with asymptotes $y = x$ and $y = -x$.
Notice that the graph of $x^2 - y^2 = 4$ is similar to the graphs of relations of the form

$$\frac{x^2}{a^2} - \frac{y^2}{b^2} = 1$$

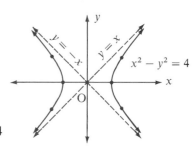

except that the asymptotes of $x^2 - y^2 = 4$ are $y = x$ and $y = -x$ and are perpendicular to each other.

Equations of the form $xy = k^2$, $k \in R$, also represent hyperbolas.

A sketch of the graph of $xy = 1$ is shown at the right. Notice that the asymptotes are the co-ordinate axes ($x = 0$, $y = 0$), which, as you know, are perpendicular to each other. Also the transverse and conjugate axes are equal. The vertices are (k, k) and $(-k, -k)$.

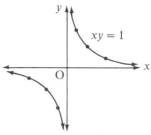

Relations whose equations represent a hyperbola with perpendicular asymptotes and equal major and minor axes are called **rectangular hyperbolas** or **equilateral hyperbolas**.

Thus, equations of the form $\quad \dfrac{x^2}{a^2} - \dfrac{y^2}{a^2} = 1 \quad \text{or} \quad x^2 - y^2 = a^2$

and equations of the form $\quad xy = k^2, \quad k \in R$

all represent rectangular hyperbolas.

11.6 Exercise

B 1 (a) Sketch the graphs of the following hyperbolas.
   (i) $x^2 - y^2 = 16$ (ii) $y^2 - x^2 = 16$
(b) How are they alike? How are they different?

2 (a) Sketch the graphs of the following hyperbolas.
   (i) $xy = 9$ (ii) $xy = -9$
(b) How are they alike? How are they different?

3 For the hyperbola given by $x^2 - y^2 = 36$ give
(a) the lengths of the transverse and conjugate axes.
(b) the equations of the asymptotes.

4 (a) Sketch the graph given by $x^2 - y^2 = 16$.
(b) On the graph in (a) indicate the vertices, foci and asymptotes.

5 (a) Sketch the graph given by $\dfrac{x^2}{9} - \dfrac{y^2}{9} = 1$.

(b) On the graph in (a) indicate the vertices, foci and asymptotes.

6 Sketch the graphs of each of the following. Mark any important points, and lines related to the graphs.
(a) $xy = 4$ (b) $xy = -4$ (c) $x^2 - y^2 = 16$ (d) $y^2 - x^2 = 16$

C 7 P is any point on a rectangular hyperbola with vertices v_1 and v_2. Show that the bisectors of the angles between the lines joining P to the vertices are parallel to the asymptotes.

Problem Solving

The strategies you use to do mathematics occur over and over again, but you need to decide which strategy to use. Refer to your work in Section 11.2. For real numbers $a, b, > 0$ identify each transformation that maps the graph of $x^2 - y^2 = 1$ into the graphs given by

A $\left(\dfrac{x}{a}\right)^2 - y^2 = 1$ B $x^2 - \left(\dfrac{y}{b}\right)^2 = 1$ C $\left(\dfrac{x}{a}\right)^2 - \left(\dfrac{y}{b}\right)^2 = 1$

D $\left(\dfrac{x}{a}\right)^2 - y^2 = -1$ E $x^2 - \left(\dfrac{y}{b}\right)^2 = -1$ F $\left(\dfrac{x}{a}\right)^2 - \left(\dfrac{y}{b}\right)^2 = -1$

Identify the graphs shown by the above.

11.7 Equations of Inverse Variation

Often, the study of the *same* graph occurs in *different* situations. For example, to illustrate that the older a car is, the less money it is worth, can be done using a graph. The value, C, in dollars, is said to vary inversely with the age, A, in years, of the car. This inverse relationship can be shown by the formula $C \times A = k$ where k is a constant and is called the constant of proportionality.

You say C is inversely proportional to A
or C varies inversely with A.

An equation such as this represents an inverse variation and can be illustrated on a graph as follows.

After 18 months, the value of a car is $9360. Thus, when $A = 1.5$ years (18 months) then $C = \$9360$.
Thus, $9360 \times 1.5 = k$
$14\,040 = k$

Age, in years	Value, in dollars
1.5	9360
2	7020
4	3510
6	2340
8	1755
10	1404

The formula $CA = 14\,040$ can be used to complete the chart. As you can see from the chart,

- as the age of the car increases its value decreases.
- as the age of the car decreases, its value increases.

You can draw a graph of
$$C \times A = 14\,040$$
to show the graphical relationship. Since, in this case, money and time are positive, then the domain for C and A are positive real numbers. The graph of the relationship is a hyperbola shown by the branch of the hyperbola in the first quadrant.

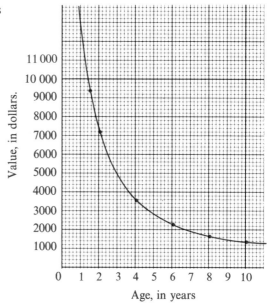

Choose two points on the graph given by (A_1, C_1) and (A_2, C_2). If (A_1, C_1) and (A_2, C_2) are two sets of values which satisfy the proportion then, since

$C_1A_1 = k$ and since $C_2A_2 = k$ then you can say

$$C_1A_1 = C_2A_2, \quad \text{or rearranged as} \quad \frac{C_1}{C_2} = \frac{A_2}{A_1} \quad ①$$

The equations shown in ① are needed to solve problems about inverse variation.

Example a varies indirectly with b. If $a = 72$ when $b = 5$, find b when $a = 120$.

Solution a varies indirectly with b. Thus, $a_1b_1 = a_2b_2$.
Given $a_1 = 72$, $b_1 = 5$, $a_2 = 120$, $b_2 = ?$.

Thus, $72 \times 5 = 120 \times b_2$
$$b_2 = \frac{72 \times 5}{120}$$
$$b_2 = \frac{360}{120}$$
$$b_2 = 3$$

Thus, when $a = 120$, $b = 3$.

11.7 Exercise

A To solve problems, you must know the meaning of the words *inversely*, *inversely proportional* and *constant of proportionality*.

1 M varies inversely with N. If $M = 1.5$ when $N = 10$, find N when $M = 2$.

2 B is inversely proportional to S.
(a) Write a statement to express the relationship.
(b) If $B = 100$ when $S = 8$, find the constant of proportionality.
(c) If $B = 250$, find S.
(d) If $S = 10$, find B.

3 An inverse variation is given by $GH = $ constant. If $G = \frac{1}{2}$, $H = 4$, find H when $G = \frac{1}{4}$.

4 An inverse proportion is given as $xy = k$.
(a) Find k when $x = 1.5$ and $y = 20$.
(b) Find y when $x = 7.5$.
(c) Find x when $y = 3.75$.

5 w is inversely proportional to v. If $w = 15$ when $v = 65$, find v when $w = 25$.

6   Use the inverse proportion $c_1d_1 = c_2d_2$ to find
     (a) d_2 when $c_1 = 100$, $d_1 = 2$, and $c_2 = 125$.
     (b) c_1 when $d_1 = 80$, $c_2 = 160$, and $d_2 = 1.25$.
     (c) d_1 when $c_1 = 20$, $c_2 = \frac{1}{2}$, and $d_2 = 400$.
     (d) c_2 when $d_1 = 250$, $d_2 = 0.2$, and $c_1 = 0.8$.

B   Refer to the *Steps For Solving Problems* to organize your work.

7   The number of slices from one pizza any one person can have is inversely
     proportional to the number of people sharing the pizza.
     (a) If 3 people each have 5 slices, how many slices does the pizza have?
     (b) How many slices each can 5 people have?
     (c) A group of people each have 1.5 slices of the pizza. How many people
         are in the group?

8   The number of bricklayers required to build a wall is inversely proportional
     to the length of time it takes to build the wall. It takes 4 bricklayers 3 days
     to build a wall.
     (a) How long would it have taken 6 bricklayers to build the wall?
     (b) It is decided that the wall must be built in 1.5 days. How many
         bricklayers must be employed to build the wall?

9   In order to balance a seesaw the distance, in centimetres, a person sits from
     the point of support (fulcrum) of the seesaw varies inversely with the mass,
     in kilograms, of the person. A 25-kg student sitting 120.0 cm from the
     fulcrum balances a 24-kg student sitting 125.0 cm from the fulcrum. How
     far from the fulcrum should a student with mass 30 kg sit in order to
     balance a student of mass 48 kg sitting 62.5 cm from the fulcrum?

10   The height of a triangle is inversely proportional to the base of the triangle
     for a triangle of constant area.
     (a) When the height of the triangle is 12.5 cm its base is 14.0 cm. What is
         the height of the triangle if its base is to be 10.0 cm?
     (b) What is the area of the triangle?

11   The time needed to fill an aquarium tank varies inversely with the square
     of the diameter of the hose used to fill the tank.
     (a) If it takes 10 min to fill the tank with a hose of diameter 2.5 cm, how
         long, to 1 decimal place, will it take to fill the tank with a hose of diameter
         3.8 cm?
     (b) If it takes 2 min to fill the tank with a piece of hose, what is the diameter
         of the hose, to 1 decimal place?

11.8 Conic Sections and Eccentricity

The shapes of the conic sections vary depending on the angle of the slicing plane.

| The slicing plane is horizontal. | The slicing plane is at an angle. | The slicing plane is parallel to the surface of the cone. | The slicing plane is parallel to the axis of the cone. |

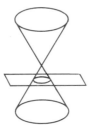 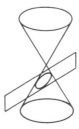 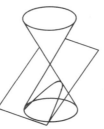 

Since the various curves can be related to cross sections of the cone they are often referred to as **conic sections**. As the slicing plane changes in direction the ellipses become elongated. As a measure of this elongation of the conic sections, a number called **eccentricity**, e, is defined.

$$\text{eccentricity} = e = \frac{c}{a} \quad \begin{array}{l} \leftarrow \text{distance from centre to focus} \\ \leftarrow \text{distance from centre to vertex} \end{array}$$

The conic sections are then related as follows.

circle	ellipse	parabola	hyperbola
$e = 0$	$0 < e < 1$	$e = 1$	$e > 1$

The eccentricity of a conic section can be obtained from its equation, as shown in the following example.

Example 1 Find the eccentricity of the ellipse defined by $\dfrac{x^2}{100} + \dfrac{y^2}{36} = 1$.

Solution

$\dfrac{x^2}{100} + \dfrac{y^2}{36} = 1$ ⟵ Think: Compare the equation to

$a^2 = 100 \qquad b^2 = 36$ the intercept form. $\dfrac{x^2}{a^2} + \dfrac{y^2}{b^2} = 1$

$a = \pm 10 \qquad b = \pm 6$

Use $c^2 = a^2 - b^2$.   ⟵ Remember, for the ellipse,

$c^2 = 100 - 36$ $a^2 = b^2 + c^2$.

$c^2 = 64 \qquad e = \dfrac{c}{a}$

$c = \pm 8$

$e = \dfrac{8}{10} \quad \text{or} \quad 0.8$

Thus, the eccentricity of this ellipse is 0.8.

If the eccentricity is known, the equation of the conic section can be found, as shown in the following.

Example 2   Find the equation of a hyperbola with eccentricity $e = \dfrac{3}{2}$ and one focus at $(4, 0)$.

Solution   Since the eccentricity is $\dfrac{3}{2}$, $e = \dfrac{3}{2}$ or $\dfrac{c}{a} = \dfrac{3}{2}$.

Co-ordinates of one focus are $(4, 0)$, Then, $c = 4$.

$$\frac{4}{a} = \frac{3}{2} \qquad a = \frac{8}{3}$$

Also, $b^2 = c^2 - a^2.$ ⟵ Remember, for the hyperbola, $a^2 + b^2 = c^2$.

$$= (4)^2 - \left(\frac{8}{3}\right)^2$$

$$= \frac{80}{9}$$

The equation of the hyperbola is given by

$$\frac{x^2}{a^2} - \frac{y^2}{b^2} = 1. \qquad \longrightarrow \qquad \frac{x}{\left(\dfrac{8}{3}\right)^2} - \frac{y^2}{\dfrac{80}{9}} = 1$$

(foci on x-axis)

$$\frac{9x^2}{64} - \frac{9y^2}{80} = 1 \qquad \text{which is the equation wanted}$$

11.8   Exercise

A 1   For each of the following:

▶ Sketch the conics obtained for each of the slicing planes A, B, and C.
▶ How are the conics alike? How are they different?

(a) 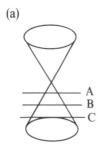   (b) 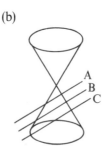   (c) 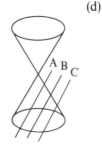   (d)

2   Which conic section is described by each eccentricity?

(a) $e = 1$ (b) $e = \dfrac{c}{a}, c > a$ (c) $e = 0$ (d) $e = \dfrac{c}{a}, c < a$

3   What is the eccentricity of each of the following conic sections?
    (a) ellipse with major axis 12; minor axis 8.
    (b) circle with centre $(-4, 1)$; radius 6.
    (c) hyperbola with focus $(-6, 0)$; vertex $(4, 0)$.
    (d) ellipse with one focus at $(0, 6)$; one vertex at $(0, -8)$.
    (e) parabola with focus at $(-4, 0)$; vertex at $(2, 0)$.
    (f) hyperbola with major axis 6; one focus at $(-6, 0)$.

4   Find the eccentricity of each ellipse. Express your answer as needed to one
    decimal place.
    (a) $\dfrac{x^2}{16} + \dfrac{y^2}{9} = 1$ (b) $\dfrac{x^2}{25} + \dfrac{y^2}{36} = 1$ (c) $25x^2 + 9y^2 = 225$

5   For each hyperbola, find the eccentricity.
    (a) $\dfrac{x^2}{9} - \dfrac{y^2}{4} = 1$ (b) $\dfrac{x^2}{100} - \dfrac{y^2}{36} = 1$ (c) $25x^2 - 16y^2 = -400$

B   To solve problems, you must know the meaning of the words *eccentricity*,
    major axis, *minor axis*, and *intercept*.

6   Find the equation of the ellipse with eccentricity 0.5 and vertex at $(-6, 0)$.

7   What is the equation of a hyperbola with its centre at the origin, eccentricity
    $\dfrac{3}{2}$ and major axis 12 units on the y-axis?

8   Find the equation of each conic section with centre at $(0, 0)$ given its
    properties as follows.
    (a) major axis 12; minor axis 8; $0 < e < 1$.
    (b) horizontal major axis 8; $e = \dfrac{3}{2}$.   (c) a focus at $(4, 0)$; major axis 10.
    (d) one focus at $(-4, 0)$; $e = 2$.   (e) x- and y-intercepts ± 2 and ± 4.
    (f) major axis 10; one focus at $(4, 0)$; $0 < e < 1$.

Problem Solving

A parabola is given by the equation $y = ax^2 + bx + c$.

What relationship must be true among the co-efficients of a, b, and c
for the graph of the parabola
- to touch the x-axis?
- to intersect the x-axis in two points?

Questions 9 to 11 are based on the following definition.

The conic sections can all be defined in terms of a directrix line, l and a fixed point, F, the focus, using the following definition.

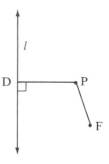

> **A conic** is defined as the set of points for which the ratio
>
> $$\frac{PF}{PD} = \text{constant.}$$

9   Draw a fixed line, l_3, and a fixed point, F_3, on a page.

(a) Find all points that satisfy the condition $\dfrac{PF_3}{PD} = \dfrac{1}{2}$.

(b) What type of curve is constructed in (a)?

10   Draw a fixed line, l_5, and a fixed point, F_5, on a page.

(a) Find all points that satisfy the condition $\dfrac{PF_5}{PD} = 2$.

(b) What type of curve is constructed in (a)?

11   For any conic the ratio given by $\dfrac{PF}{PD}$ is defined to be the eccentricity of the conic. Different conics are drawn below for a fixed line, l, the directrix, and a fixed point, F, the focus. Calculate the eccentricity $\dfrac{PF}{PD}$ for each conic. Use the points, P, shown on each conic.

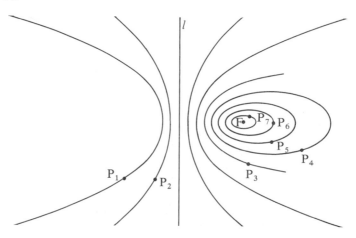

11.9 Solving Systems of Equations

The method used to locate points at which the paths of solar bodies intersect can be quite involved. The actual conic equations defining the paths of solar bodies are complex and are solved with the aid of computers. But those principles used to solve the system of conic equations for the solar bodies are the same as the principles used to solve each system of equations in this section. In the following example, the system consists of a linear equation and a quadratic equation. Such a system is referred to as a **linear-quadratic system**. You can interpret the solution visually by drawing the graphs.

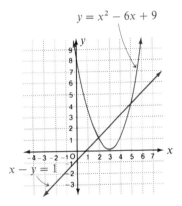

The graph at the right shows a graphical solution of $x - y = 1$ and $y = x^2 - 6x + 9$. It shows the intersection of a parabola and a straight line. Thus, the solution consists of two points, as shown. To find the solution algebraically, you use your earlier skills in solving equations.

Example 1 Solve the linear-quadratic system defined by the equations

$$y = x^2 - 6x + 9 \quad \text{and} \quad x - y = 1.$$

Solution

$x - y = 1 \qquad \text{①}$
$y = x^2 - 6x + 9 \qquad \text{②}$

From ① use $y = x - 1$. Substitute $y = x - 1$ into ②. The equation becomes

$$x - 1 = x^2 - 6x + 9$$
$$x^2 - 7x + 10 = 0$$
$$(x - 5)(x - 2) = 0$$
$$x - 5 = 0 \quad \text{or} \quad x - 2 = 0$$
$$x = 5 \quad \text{or} \qquad x = 2$$

Substitute the value of x in ①.

When $x = 5$	When $x = 2$
$5 - y = 1$	$2 - y = 1$
$5 - 1 = y$	$2 - 1 = y$
$y = 4$	$y = 1$

Thus, the solution of the system is given by $(5, 4)$ and $(2, 1)$.

Be sure to check your results in the original equations.

A similar procedure is used to solve a
system of quadratic equations as shown
in the next example. The system in the
following example is called a
quadratic-quadratic system. On the graph,
the solution is shown at the intersection
points S and Q.

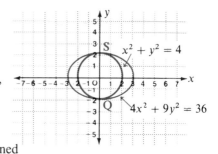

Example 2 Solve the quadratic-quadratic system defined

by $\quad 4x^2 + 9y^2 = 36 \quad$ and $\quad x^2 + y^2 = 4$.

Solution
$$4x^2 + 9y^2 = 36 \qquad ①$$
$$x^2 + y^2 = 4 \qquad ②$$

From ②, $x^2 = 4 - y^2$.
Substitute in ①.

Remember: Before you solve the system
think of the possible intersections
that might occur with the system.

$$4(4 - y^2) + 9y^2 = 36$$
$$16 - 4y^2 + 9y^2 = 36$$
$$5y^2 = 20$$
$$y^2 = 4$$
$$y = \pm 2$$

$y = +2$	or	$y = -2$

Substitute in ②. $\qquad\qquad$ Substitute in ②.

$$x^2 + (2)^2 = 4 \qquad\qquad x^2 + (-2)^2 = 4$$
$$x^2 + 4 = 4 \qquad\qquad\quad x^2 = 0$$
$$x^2 = 0 \qquad\qquad\qquad\quad x = 0$$
$$x = 0$$

Check ✓ You must check
whether the value satisfies
both equations.

The solution of the quadratic-quadratic system is given by (0, 2) and (0, − 2).

To solve quadratic-quadratic systems, it is often helpful to visually interpret
the possible solutions before you actually solve the system. For example, a
few situations are shown.

Ellipse and Parabola $\qquad\qquad\qquad$ *Hyperbola and Ellipse*

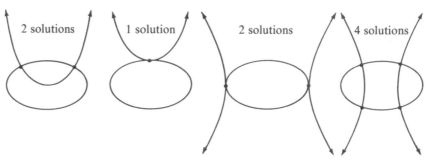

2 solutions $\qquad$ 1 solution $\qquad\qquad$ 2 solutions $\qquad$ 4 solutions

11.9 Exercise

A 1 What is the solution set for each system?

(a)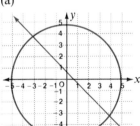

(b)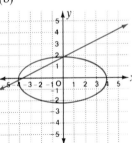

(c)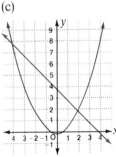

2 A line and a circle can have no points, 1 point, or 2 points of intersection, as shown.

No points of intersection   1 point of intersection   2 points of intersection

Sketch diagrams to show possible intersections of the following systems:
(a) line and ellipse   (b) line and hyperbola   (c) ellipse and circle
(d) circle and hyperbola   (e) ellipse and hyperbola
(f) parabola and ellipse   (g) parabola and circle

3 (a) Draw the graph of $x^2 + y^2 = 65$ and $x + y - 3 = 0$.
(b) What are the co-ordinates of the intersection points?
(c) What is the solution set for the linear-quadratic system of equations, $x^2 + y^2 = 65$ and $x + y - 3 = 0$?

4 (a) Sketch the graphs of $x^2 + y^2 = 25$ and $\dfrac{x^2}{25} + \dfrac{y^2}{9} = 1$.
(b) What is the solution set of the system of equations in (a)?

B In the questions that follow, any irrational number solutions can be left in radical form.

5 A line given by $x + 7y = 50$ intersects the conic section defined by the equation $x^2 + y^2 = 100$.
(a) Solve the linear-quadratic system.
(b) Interpret your solution in (a) graphically.

6   (a) Solve $x^2 + y^2 = 20$ and $y = -x + 6$.
    (b) Interpret your solution in (a) graphically.

7   (a) Solve the system $x^2 + 4y^2 = 36$ and $x = 6 + 2y$.
    (b) Verify your solution in (a).

8   Find the points of intersection of each of the following systems.
    (a) $x + 2 = y$ (b) $y = 4 - 2x$ (c) $y + 3x = 2$
        $y^2 - x^2 = 8$ $x^2 + y^2 = 9$ $y = x^2 - 2$

9   Find the solution set for each system.
    (a) $x - y = -2$ (b) $x = 8 - 2y$ (c) $x - y = -1$
        $x^2 + y^2 = 10$ $x^2 + 4y^2 = 40$ $(x - 1)^2 + (y + 2)^2 = 9$

    (d) $4x^2 - y^2 = 7$ (e) $x^2 + y^2 = 25$
        $2x^2 + 5y^2 = 9$ $xy = 12$

    (f) $\dfrac{x^2}{16} + \dfrac{y^2}{9} = 1$ and $\dfrac{x^2}{9} + \dfrac{y^2}{16} = 1$

10 Often a sketch may suggest an approach to solving a problem.
    (a) Sketch the system given by

    $$x^2 + y^2 - 6x + 8y - 9 = 0 \quad \text{and}$$
    $$x^2 + y^2 - 6x - 4y - 21 = 0.$$

    (b) Find the solution set of the system given in (a).

11 Prove that the centre of the circle given by $x^2 + y^2 - 10x - 10y + 25 = 0$ lies on the right bisector of the chord with equation $x + 3y - 5 = 0$.

12 The telecommunications satellite Anuk 3 follows an elliptical course described by $3x^2 + 7y^2 = 55$. Determine the co-ordinates of the points of reception of a light laser beam directed along the path $3x + y - 11 = 0$.

13 The path of a sonic boom on the earth's surface is described by the equation $x^2 - y^2 = -45$. The path of a highway is described by the equation $5x - 2y = 19$. What are the co-ordinates of points on the highway at which the sonic boom will be heard?

C 14 What are the conditions on a, b, and m, so that the curves given by $ax + y = b$ and $x^2 + y^2 = m^2$
    (a) intersect in 2 points?
    (b) intersect in 1 point?
    (c) do not intersect?

Problem-Solving Strategy: Generalizing Skills

The same skills can be applied to many different situations. In your earlier work in analytic geometry, you drew the graphs of regions defined by linear inequations. These same skills can be extended to the drawing of graphs of quadratic regions.

To draw the graph of the quadratic region defined by a system of inequations, you need to find the intersection points.

Graph of the region

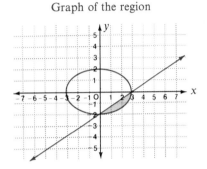

System of Inequations
$$4x^2 + 9y^2 \le 36$$
$$2x - 3y \ge 6$$

15   (a) Solve the system $x + y + 2 = 0$ and $x^2 + y^2 = 25$.
   (b) Sketch the region defined by $x + y + 2 \ge 0$ and $x^2 + y^2 \le 25$.

16   Sketch the region given by each system of inequations.
   (a) $x + y \le 9$ (b) $x + y \ge 9$ (c) $x + y > 9$ (d) $x + y < 9$
       $x^2 + y^2 \le 41$ $x^2 + y^2 \ge 41$ $x^2 + y^2 < 41$ $x^2 + y^2 > 41$
   (e) How are the regions in (a) to (d) alike? How do they differ?

17   Each region is defined by a linear-quadratic system of inequations. Draw the graph.
   (a) $2x + 1 \le 3y$ (b) $5x \ge 13 + y$ (c) $y > 5 - x$
       $x^2 + y^2 \le 25$ $x^2 + y^2 \ge 13$ $2x^2 + 3y^2 < 35$

18   Each region is defined by a quadratic-quadratic system of inequations. Draw the graph.
   (a) $y \le x^2$ (b) $x^2 + y^2 \ge 5$ (c) $x^2 + y^2 < 25, y > 0$
       $x^2 + y^2 \le 20$ $y \ge x^2 - 5$ $16x^2 - 20y^2 > 320$

Problem Solving

The word *locus* is a Latin word meaning *place* or *location*. Problems about locus involve finding the equation of points that satisfy certain conditions. Find the equation of the locus of a point P which satisfies $PA^2 - PB^2 = AB^2$ for A(1, 3) and B(2, 5).

11.10 The Process of Mathematics: Conics

When you solve problems, you accumulate skills for solving more advanced problems. For example, you have learned the skills and strategies for solving problems about conics. In so doing, you have also learned new vocabulary. Thus, conics with centre $(0, 0)$ are said to be in **standard position**. Conics not in standard position have defining equations such as

$$9x^2 + 25y^2 - 36x - 100y - 89 = 0 \qquad \text{\textcircled{1}}$$

which are more complex in form. An attempt to learn more about the properties of the curve given in the form ① would be more cumbersome than necessary. For this reason, the reverse process is explored first. Thus, the question is asked: "How did this equation occur in the first place?"

In the diagram, an ellipse is shown to have the property

$$PF_1 + PF_2 = 10.$$

The conditions of the problem are translated into the form of an equation given by

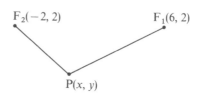

$$\sqrt{(x + 2)^2 + (y - 2)^2} + \sqrt{(x - 6)^2 + (y - 2)^2} = 10$$
$$\sqrt{(x + 2)^2 + (y - 2)^2} = 10 - \sqrt{(x - 6)^2 + (y - 2)^2}$$
$$(x + 2)^2 + (y - 2)^2 = 100 - 20\sqrt{(x - 6)^2 + (y - 2)^2} + (x - 6)^2 + (y - 2)^2$$
$$x^2 + 4x + 4 + y^2 - 4y + 4 = 100 - 20\sqrt{(x - 6)^2 + (y - 2)^2} + x^2 - 12x + 36 + y^2 - 4y + 4$$
$$16x - 132 = -20\sqrt{(x - 6)^2 + (y - 2)^2}$$
$$4x - 33 = -5\sqrt{(x - 6)^2 + (y - 2)^2}$$
$$16x^2 - 264x + 1089 = 25(x^2 - 12x + 36 + y^2 - 4y + 4)$$
$$16x^2 - 264x + 1089 = 25x^2 - 300x + 900 + 25y^2 - 100y + 100$$
$$9x^2 + 25y^2 - 36x - 100y = 89$$

which is an equation of the form ①. This equation represents an ellipse with its centre not at the origin. Thus, to simplify your study, your task is to rewrite the equation in a form which will provide information.

$$9x^2 + 25y^2 - 36x - 100y = 89$$
$$9(x^2 - 4x + 4) + 25(y^2 - 4y + 4) = 89 + 136 \quad \leftarrow \text{Add the same value to}$$
$$\underbrace{\qquad}_{36} \qquad \underbrace{\qquad}_{100} \qquad \qquad \underset{136}{} \qquad \text{both sides of the equation.}$$
$$9(x - 2)^2 + 25(y - 2)^2 = 225$$
$$\frac{(x - 2)^2}{25} + \frac{(y - 2)^2}{9} = 1$$

From the equation given in this form, you can list the properties of the ellipse

- centre (2, 2) semi-minor axis 3 units, and
- semi-major axis 5 units, foci $F_2(-2, 2)$, $F_1(6, 2)$.

In order to find the equation of the conic in standard position, we translate the conic with centre (2, 2) to the conic with centre (0, 0) using the mapping

$$(x, y) \rightarrow (x - 2, y - 2).$$

This gives the equation $\dfrac{x^2}{25} + \dfrac{y^2}{9} = 1$ with its centre at the origin.

ellipse

The equations of an ellipse, with its centre (h, k), and the sum of the focal radii $2a$ units are

$$\frac{(x - h)^2}{a^2} + \frac{(y - k)^2}{b^2} = 1, \qquad \longleftarrow \quad \text{The major axis is parallel to the } x\text{-axis.}$$

and $\quad \dfrac{(x - h)^2}{b^2} + \dfrac{(y - k)^2}{a^2} = 1. \qquad \longleftarrow \quad \text{The major axis is parallel to the } y\text{-axis.}$

hyperbola

The equations of a hyperbola, with its centre (h, k) and the difference of the focal distances $2a$ are

$$\frac{(x - h)^2}{a^2} - \frac{(y - k)^2}{b^2} = 1, \qquad \longleftarrow \quad \text{The transverse axis is parallel to the } x\text{-axis.}$$

and $\quad \dfrac{(y - k)^2}{a^2} - \dfrac{(x - h)^2}{b^2} = 1, \quad$ or $\quad \dfrac{(x - h)^2}{b^2} - \dfrac{(y - k)^2}{a^2} = -1.$

$$\qquad\qquad\qquad\qquad\qquad\qquad\qquad \underset{\text{The transverse axis is parallel to the } y\text{-axis.}}{\llcorner}$$

parabola

The equations of a parabola with vertex (h, k) and with $|p|$ as the distance from the focus to the vertex, are given by

$$(y - k)^2 = 4p(x - h), \qquad \longleftarrow \quad \text{The axis of the parabola is parallel to the } x\text{-axis.}$$

and $\quad (x - h)^2 = 4p(y - k). \qquad \longleftarrow \quad \text{The axis of the parabola is parallel to the } y\text{-axis.}$

Math Tip

Before you begin the exercise, list the various equations of the ellipse, hyperbola, and parabola. Also list the various properties of each conic.

11.10 Exercise

A   Through the exercise, express eccentricities in radical form where required.

1   For each of the following   ▶ identify the conic.
                                      ▶ list the co-ordinates of the centre.

(a) $(y - 3)^2 = 9(x - 5)$ (b) $\dfrac{(x - 2)^2}{4} + \dfrac{(y + 1)^2}{9} = 1$ (c) $\dfrac{(x + 3)^2}{16} - \dfrac{(y - 1)^2}{9} = 1$

(d) $(x + 2)^2 = 16(y - 3)$ (e) $\dfrac{(x - 1)^2}{4} + \dfrac{(y - 2)^2}{4} = 1$ (f) $\dfrac{(x + 4)^2}{16} + \dfrac{(y - 1)^2}{9} = 1$

2   For each conic given by the following equations
      ▶ identify the conic.
      ▶ list the co-ordinates of the centre.
      ▶ tell whether the foci are on a line   parallel to the x- or y-axis.

(a) $9(x - 1)^2 + 4(y + 1)^2 = 36$ (b) $4(x - 2)^2 - 9(y + 3)^2 = 36$
(c) $25(x + 1)^2 + 36(y - 2)^2 = 900$ (d) $25(x + 3)^2 - 16(y - 3)^2 = 400$

3   What is the eccentricity of each conic?

(a) $(x - 3)^2 = 16(y - 2)$ (b) $\dfrac{(x - 2)^2}{9} - \dfrac{(y - 1)^2}{4} = 1$ (c) $\dfrac{(x - 3)^2}{25} + \dfrac{(y + 1)^2}{9} = 1$

(d) $(y + 1)^2 = 8(x - 3)$ (e) $\dfrac{(x - 4)^2}{9} + \dfrac{(y + 2)^2}{16} = 1$ (f) $\dfrac{(x + 2)^2}{64} + \dfrac{(y - 1)^2}{16} = 1$

4   Express the equation of each conic in standard form.
      (a) $x^2 - 4x + 4y + 8 = 0$ (b) $4x^2 - 9y^2 - 8x + 36y - 68 = 0$
      (c) $y^2 + 6y - 12x - 39 = 0$ (d) $4x^2 + 9y^2 - 16x - 18y - 11 = 0$

B 5   A conic is given by the equation $(x - 3)^2 = 6(y + 1)$.
      (a) Identify the conic.
      (b) What are the co-ordinates of the focus and of the vertex?
      (c) What translation will place the conic at the origin? What is the defining
          equation of the conic with centre at $(0, 0)$?

6   A conic is given by the equation $\dfrac{(x - 3)^2}{36} + \dfrac{(y + 1)^2}{64} = 1$.

      (a) Identify the conic.
      (b) What are the co-ordinates of the foci and of the vertices?
      (c) Where is its centre?
      (d) What mapping will translate the conic to the origin as its centre? What
          is the defining equation of the conic with centre $(0, 0)$?

7   A conic is given by the equation $\dfrac{(x-1)^2}{16} + \dfrac{(y+2)^2}{9} = 1.$

   (a) Identify the conic.
   (b) What are the co-ordinates of the foci and the vertices?
   (c) Where is its centre?
   (d) What mapping will translate the conic to a conic with the origin as its centre? What is the defining equation of the conic with centre $(0, 0)$?

8   A conic is given by the equation $\dfrac{(x-1)^2}{16} - \dfrac{(y+3)^2}{9} = 1.$

   (a) Identify the conic.
   (b) What are the co-ordinates of the foci and the vertices?
   (c) Where is its centre?
   (d) What mapping will translate the conic to a conic with the origin as its centre? What is the defining equation of the conic with centre $(0, 0)$?

9   For each conic

   • sketch the graph.
   • sketch the graph of the image with centre at $(0, 0)$.
   • define the transformation used to translate the conic to the origin.

   (a) $y^2 - 4y - 9x + 13 = 0$ (b) $x^2 + 6x - 36y + 153 = 0$
   (c) $4x^2 + 9y^2 - 8x - 18y - 23 = 0$
   (d) $9x^2 - 4y^2 - 54x - 36y + 81 = 0$

10   What is the equation of each conic for the given properties?
   (a) The foci are $(-2, 0)$, $(4, 0)$; the sum of the focal radii is 10.
   (b) The foci are at $(3, 0)$, $(-5, 0)$; the difference of the focal radii is 6.

11   For each hyperbola, certain conditions are given. Find the defining equation.

   (a) The foci are at $(-6, 3)$, $(2, 3)$; the eccentricity $= \dfrac{4}{3}$.

   (b) The vertices are at $(5, 4)$, $(-7, 4)$; the eccentricity $= \dfrac{5}{3}$.

C 12   A conic is given by the equation $25x^2 + 4y^2 + 100x - 16y + 16 = 0.$
   (a) Identify the conic.
   (b) What are the co-ordinates of its centre?
   (c) Find the co-ordinates of the focus (or foci) and vertex (or vertices).
   (d) What is the length of its major and minor axis?
   (e) What is its eccentricity?

Practice and Problems: A Chapter Review

1 What is the locus of each of the following? Use a sketch to help you picture each.

   (a) all points equidistant from three fixed points

   (b) all points that are the midpoints of chords 2 cm in length, drawn in a circle with radius 6 cm

2 (a) Find the equation of a circle with centre $(0, 0)$ and radius 8 units.

   (b) Find its domain, range, intercepts and symmetry.

3 Find the equation of the image of the circle $x^2 + y^2 = 9$ under the mapping $(x, y) \rightarrow (2x, 3y)$. Sketch the graph of the image.

4 For each equation of a parabola
   (i) draw a sketch.
   (ii) label the vertex, focus, directrix, axis of symmetry.
   (a) $y^2 = 4px, p > 0$ (b) $y^2 = 4px, p < 0$
   (c) $x^2 = 4py, p > 0$ (d) $x^2 = 4py, p < 0$

5 The foci of an ellipse are given by $F_1(6, 0)$ and $F_2(-6, 0)$. If $P(x, y)$ is any point on the ellipse, derive an equation in x and y that shows the condition $PF_1 + PF_2 = 16$.

6 (a) Derive the equation of the locus which satisfies the condition that the difference of the distances from a point to $(0, 6)$ and $(0, -6)$ is 9.

   (b) What type of conic is it? What are its properties?

7 A hyperbola has its centre at the origin with foci $(5, 0)$ and $(-5, 0)$. If the transverse axis has length 8, find the equation of the hyperbola.

8 Solve each system. Illustrate your solution by sketching the system.
   (a) $y = -x^2 + 5x, y = 3x + 1$ (b) $8 + y = x, xy = -12$
   (c) $x^2 + 4y^2 = 20, x + 4y = 10$ (d) $x^2 + \dfrac{y^2}{2} = 6, x + \dfrac{y}{2} = 1$

9 Draw the graph of the region defined by the following linear-quadratic system of inequations.
$$x \leq 6 + 3y, \qquad xy \geq 24, \quad y \leq 0$$

10 Prove that the intersection points of $4x^2 + 16y^2 = 64$ and $x - 2y = 4$ lie on the conic defined by $x^2 + y^2 - 8x + 10y + 16 = 0$.

Test for Practice

1   A locus of points $P(x, y)$ has the property that the sum of the distances from P to $A(3, 0)$ and $B(-3, 0)$ is 10. Sketch the locus.

2   A circle has centre $(3, 3)$ and radius $\sqrt{10}$.
    (a) Find its equation.   (b) What are its domain, range and intercepts?

3   What mapping would transform the circle given by $x^2 + y^2 = 3$ into an ellipse given by the equation $x^2 + 49y^2 = 147$?

4   (a) Find the equation of a parabola with vertex $(0, 0)$ and focus $(0, -5)$.
    (b) Draw a sketch of the curve. Label the vertex, focus, directrix, and axis on the sketch.

5   The equation of an ellipse is $16x^2 + 4y^2 = 64$.
    (a) Sketch the graph.
    (b) Label the foci, the vertices, and the major and minor axes.

6   A curve is given by $\dfrac{x^2}{25} - \dfrac{y^2}{16} = 1$.

    (a) What type of curve is it?   (b) Find the intercepts and eccentricity.
    (c) What are the co-ordinates of the foci?
    (d) Write the equation of the asymptotes.   (e) Sketch the curve.

7   For each ellipse, calculate the eccentricity.
    (a) 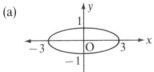   (b)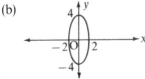

    (c) 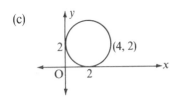   (d)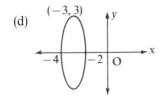

8   (a) Solve $x^2 + y^2 = 10$ and $2x = 5 - y$.
    (b) Sketch a graph to illustrate your answer in (a).

9   Write the equation of the conic defined by
    $9x^2 + 4y^2 - 36x + 16y + 16 = 0$ in standard position.

Looking Back: A Cumulative Review

1   Simplify.
(a) $(3\sqrt{2} - 3\sqrt{3})(2\sqrt{2} - 5\sqrt{3})$ (b) $(2\sqrt{5} - 3)^2$
(c) $(5\sqrt{3} - 2\sqrt{2})^2 - 3(\sqrt{3} - \sqrt{2})(\sqrt{3} + \sqrt{2})$

2   Solve for x: (a) $x^{\frac{1}{2}} = 5$ (b) $x^{\frac{3}{5}} = 8$ (c) $x^{\frac{1}{3}} = \dfrac{32}{\sqrt{x}}$

3   Solve $(x^2 + 2x)^2 - 2(x^2 + 2x) - 3 = 0$

4   A triangular sail for a racing boat must have an area of 22.5 m². If the height must be 4 m longer than the base, to what length must the base be cut?

5   (a) Solve $\triangle PQR$: $\angle P = 43°$, $\angle Q = 63°$, $p = 112$.
(b) Solve $\triangle PQR$: $\angle P = 62°$, $\angle Q = 52°$, $r = 15.2$.

6   Prove.   (a) $\dfrac{\sec \theta}{\cot \theta} = \dfrac{\sin \theta}{\cos^2 \theta}$ (b) $\dfrac{1 - \tan \theta}{\tan \theta} = \dfrac{\cos \theta - \sin \theta}{\sin \theta}$

7   The ratio of distances from the top of a hill to the ground on opposite sides of the hill is $3:\sqrt{5}$. If the angle of elevation to the top of the hill on the longer side is 37°, find the angle of elevation on the other side.

8   For the lines given by $2x + y = 1$, $3x + 2y = 12$, what
(a) are the slopes and intercepts?   (b) is the intersection point?
(c) If $3x + 2ky = -1$ is perpendicular to $2x + y = 1$, find k.

9   In $\triangle ABD$, $\angle B = 2\angle D$. C is the perpendicular bisector of AD. Prove AC = AB.

10   Points P, Q, R, and S are named in order on the circle such that $\angle PSR = 70°$, $\angle PQS = 60°$, and $\angle QPS = 75°$. Find the following.
(a) $\angle PRS$ (b) $\angle PRQ$ (c) $\angle QRS$

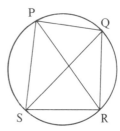

11   In $\triangle VWX$, VW = VX. If VW and VX are produced to Y and Z respectively so that WY = XZ, prove that the figure WYZX is cyclic.

12 Process of Statistics: Concepts and Skills

steps of statistics, sampling, random digits, frequency distributions, histograms, measures of central tendency, measures of dispersion, standard deviation, normal distribution, scatter diagrams, line of least fit, correlation coefficients, strategies and solving problems, applications, problem-solving

Introduction

You are constantly confronted with data in your everyday activities.

Economic Barometers	
Real GNP growth	+6.6%
Retail sales	$10.96 billion
Unemployment rate	10.2%
Trade balance	+$1 820 million
Consumer Price Index	+0.3%

results

CENTRAL JUNIOR A

	W	L	T	F	A	PTS
Pembroke	19	8	1	156	112	41
Brockville	17	9	1	155	141	36
Ottawa	16	13	0	161	126	35
Nepean	17	9	0	134	103	34
Gloucester	16	11	0	152	115	33
Hawkesbury	8	22	0	105	189	16
Smith F.ll.	4	25	0	110		

Weather forecast

Cloudy with chance of a flurry, says Environment Canada, and a high near 3. Cloudy with scattered fl....

The study of statistics is of immense value in today's world. For example, in the business of insurance, much of the economic theory, whether it be life insurance, car insurance, accidental insurance, and so on, is based on the mathematics of statistics. In the manufacturing business, skills with statistics are used to make business predictions which will minimize risk and increase profitability. These skills also provide a foundation for marketing techniques.

There are many studies in which the process of statistics plays a major role: noise and river pollution, traffic and transportation, communications, etc. Statistics is a branch of mathematics which involves

A: collecting data

B: organizing and analyzing the data

C: interpreting and making inferences, predictions and decisions about data.

The constant improvement in air travel and in other forms of transportation is in no small part due to the analysis of data, resulting in improved efficiency, more competitive rates and a high degree of safety.

12.1 Steps of Statistics: Sampling

Were you to ask a number of people what statistics means, you would get a variety of answers. People who work in business, industry, government, sports, recreation, entertainment, or research might think of the many uses of statistics in their jobs. Other people might think of the increasingly common use of statistics in the media: newspapers, magazines and television. In order to work with statistics or to understand the use of statistics in the media, it is necessary to understand the process of statistics. Statistics proceeds in three steps.

Steps of Statistics

Step A   Carefully collect and record the data.

Step B   Organize and analyze the data.

Step C   Make inferences, predictions and decisions based on the data.

The process of statistics can be seen in quality control of the manufacture of ballpoint pens. The following example describes the process of statistics and introduces useful vocabulary.

Step A   The cost of checking each and every pen would be prohibitive. The manufacturer selects a **sample** of the pens, performs experiments, collects data and records the data. It is important that

- the sample be representative of the complete shipment of pens.
- the pens in the sample be selected at **random**. In a random sample, each member of the population has an equal chance of being selected.

In this example, all the pens, taken together, are referred to as the **population**. The pens selected at random are called a **sample** from the population.

Step B   The manufacturer organizes and analyzes the recorded results of the experiments performed on the sample. The organization and analysis may include tables, charts or graphs. There may be a breakdown of defects by type. The number of defects will be compared to the number of pens produced for shipment.

Step C   The manufacturer uses the tables, charts or graphs to decide whether the shipment meets quality control specifications. If it does, the pens are shipped. If it does not, the data will help the manufacturer to isolate and correct production problems.

In either case, the manufacturer will have made a decision about the **production** of pens based on the **sample** studied.

In the previous example, the manufacturer selected a sample because it would not be economical to test the entire population. The sample, however, had to be representative of the population, so the manufacturer selected a **random sample**.

The relationship of a sample to a population is crucial in statistics. This is shown in three further examples of sampling: **clustered**, **stratified** and **destructive** sampling.

Clustered Sampling

A manufacturer of tools for mechanics has a choice of three magazines in which to advertise. The manufacturer does not sell tools to the general public. Rather, the sales are almost entirely to mechanics. In this case, the manufacturer would survey a sample of the general population. This sample would be made up of mechanics, who would be asked which of the three magazines they read. When a sample is taken from a particular segment of the population, the sample is said to be **clustered**.

Stratified Sampling

The manager of a singer wants to book concerts for a nation-wide tour. In order to decide where to hold concerts, the manager surveys a sample of people in each province. However, she should not survey the same number of people in each province because the total population varies from province to province. Rather, she should survey a number of people in each province that is proportional to the population of that province. The proportion, not the number of people surveyed, should be the same from province to province. In this case, the population of Canada is divided into **strata** or classes. The strata are the provinces. A choice of the number of concerts in each stratum (or province) is based on data obtained from the sample in proportion to the population from which it is taken. Such a sample is said to be **stratified**.

Destructive Sampling

To test quality control on a production run of tires, a manufacturer chooses tires at random from the assembly line. The tests are very rigorous, so that after testing, the tires are no longer usable. This type of sampling is said to be **destructive** because the sample cannot be reincorporated into the population.

12.1 Exercise

A 1   For each situation, why can the entire population not be surveyed?

(a) testing the taste quality of oranges

(b) checking cars for defective horns   continued →

(c) establishing the recreational habits of Canadians

(d) finding the average life of a ballpoint pen

2   Often it is neither possible nor desirable to use just a sample from a population to make a decision. Rather, the entire population is tested. For each of the following, decide whether to use a sample or the population to make a decision.

(a) to test the accuracy of the brake system of cars on a production line

(b) to check the quality of sausages processed in a meat plant

(c) to determine the quality of parachutes

(d) to check the effectiveness of a headache pill

(e) to test the quality of packaged pizza

(f) to determine the quality of an electrocardiograph (a machine that checks the operation of the human heart)

3   The quality of motorcycle tires is tested by first choosing a tire randomly from the production line, and then performing tests on it in the laboratory. The rubber in the tested tire is then used to construct new tires.

(a) What type of sample is obtained?

(b) What are some advantages of the above method?

(c) What are some disadvantages of the above method?

4   A theatre chain wants to determine the suitability of a suburban plaza for a theatre. A team of interviewers conducts a poll at the plaza. They collect data to help head office make a decision.

(a) What type of sample is obtained?

(b) What are some advantages of the above approach?

(c) What are some disadvantages of the above approach?

B 5   For each of the following decide whether destructive (D) or non-destructive (ND) sampling is required.

(a) to test the life length of toasters.

(b) to test the efficiency of pencil sharpeners.

(c) to test the strength of a metal support.

(d) to test the quality of a soft drink.

(e) to poll athletes to predict who will win an election for team representative.

(f) to test the quality of pork chops.

(g) to check the amount of silver in coins.

(h) to poll the popularity of a Canadian television personality.

6 Which of the following samples are good examples (G) of clustered sampling? Which are bad (B) examples of clustered sampling?
   (a) asking mechanics about a new engine repair procedure.
   (b) polling campers about the favourite parks in Canada.
   (c) asking baseball managers about the quality of baseball bats.
   (d) asking football players about the quality of football pads.
   (e) asking owners of large cars about energy conservation.
   (f) asking scientists their opinion of the metric system.
   (g) asking store owners their opinion of the metric system.

7 For each of the following, use descriptions I, II, III, IV, or V to indicate the type of data collection required. (You may choose more than one description.)

   I   Obtain a random sample.   II   Check the whole population.
   III Use a destructive sample.   IV   Use a stratified sample.
   V   Use a clustered sample.

   (a) to test a river for pollution.
   (b) to poll the popularity of a political leader.
   (c) to check the mass of a bag of packaged coffee.
   (d) to test the radiators of cars on a production lines.
   (e) to check the quality of a harvest of grapes.
   (f) to test the safety of seat belts.
   (g) to decide whether a shopping mall is a suitable location for a convenience store.
   (h) to determine Canada's favourite sport.
   (i) to check the accuracy of a safety device on a jet aircraft.
   (j) to determine the most popular song among teenagers.
   (k) to test the crispness factor for soda crackers.

Math Tip

Look up the meaning of each of these words in the dictionary:

statistics, data, random, sample.

Using Random Digits: Random Samples

To obtain a sample representative of the population, the sample must be chosen at random. That is, each member of the population must have an equal chance of being selected. If this is not the case, then the procedure of choosing the sample is *biased*.

In a company of 512 people, how would you choose a sample of 25 persons, randomly, to answer the following question?

Should the company offer staggered hours for quitting time?
☐ yes ☐ no

One procedure for obtaining a random sample is to get a computer to generate random digits from 000 to 999. A part of the table is shown below. Then proceed as follows.

Assign each person in the company a number from 000 to 511. Then choose a digit from the table by pointing to it with your eyes closed. You must choose the next 24 digits in an ordered fashion. That is, you must move left, right, up, down, or diagonally from your starting point to get the other 24 people. Here you moved right.

Start here.
The person assigned 143 is interviewed.

There is no person assigned the number 823. The next number to the right is chosen, namely 127.

The person assigned this number 138 is interviewed next.

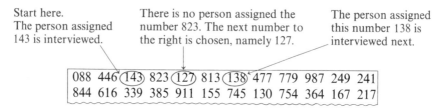

```
088  446 (143) 823 (127) 813 (138) 477  779  987  249  241
844  616  339  385  911  155  745  130  754  364  167  217
```

The complete table of random digits is given on the next page.

8   From 100 persons, a sample of 10 people are to be interviewed.
(a) How can the table of random digits be used to obtain the 10 persons?
(b) Choose a 3-digit number at random from the table of random digits.
(c) List the digits of the 10 persons who will be interviewed.

9   From your class, you are to choose a person at random to take the free trip to the Barbados.
(a) How would you use the table to select the person?
(b) Who won the trip based on the procedure in (a)?
(c) Who is the runner up and would take the trip if the first chosen person were sick?

10 For a lottery, 1000 tickets are printed.

(a) To obtain a winner, describe how you might use a pair of dice to determine the first number to be chosen from the table of random digits.

$$\begin{pmatrix} 707 & 913 & 658 \\ 850 & 199 & 667 \\ 604 & 791 & 859 \\ 073 & 781 & 168 \\ 599 & 745 & 161 \\ 678 & 720 & 021 \end{pmatrix}$$

(b) The winner of the contest is the fifth random set of digits occurring in the same column below the number in (a). What is the winning number?

11 Refer to the previous question.

(a) List two other ways of locating the first random digit for the above lottery.

(b) List two other different ways of deciding on a winner of the lottery.

Table of Random Digits: 000 to 999

```
302 068 416 505 346 808 242 349 956 892 265 546 092 488 336 201 057 728 343 640
809 229 534 531 633 874 682 353 794 607 039 713 764 623 563 527 794 604 069 799
186 408 090 103 644 774 892 279 486 409 124 305 294 429 903 019 884 456 332 049
806 288 827 206 422 754 358 536 443 239 557 307 438 468 847 699 863 930 558 362
561 451 088 502 255 677 218 380 672 059 585 703 955 914 203 172 855 871 751 277
296 316 999 001 673 088 446 143 823 127 813 138 477 779 987 249 241 394 580 874
221 135 036 015 710 844 616 339 385 911 155 745 130 754 364 167 217 550 050 439
430 842 575 965 692 648 655 436 558 359 446 124 353 779 993 137 282 748 196 546
503 267 339 408 077 922 397 456 309 538 363 219 371 222 088 567 664 162 373 300
690 599 249 213 883 471 946 899 126 718 609 101 208 604 067 056 378 576 968 645
078 998 022 103 730 287 923 410 168 418 591 504 309 563 533 573 862 948 933 626
711 796 541 265 500 140 801 419 617 350 978 444 204 287 951 998 038 981 383 818
411 225 172 869 786 814 095 987 251 378 566 619 428 951 992 157 907 064 933 622
999 007 478 742 047 998 025 258 954 941 793 637 992 160 607 025 588 592 470 924
149 041 487 362 290 698 826 199 210 812 144 745 128 920 333 065 536 458 418 560
941 803 385 926 480 667 090 042 646 721 540 313 826 200 106 767 700 872 736 123
615 285 952 978 427 986 278 465 689 568 692 653 499 104 882 503 277 364 189 704
321 695 739 031 609 100 017 494 079 630 787 765 654 490 242 316 977 462 560 408
632 824 147 265 515 749 243 219 357 583 790 693 680 304 186 414 392 711 800 450
200 126 688 537 408 095 966 671 046 805 328 287 934 643 824 144 697 781 919 316
646 731 247 079 565 585 704 983 331 089 341 511 595 362 281 674 118 584 737 092
954 929 546 095 971 570 767 720 560 424 874 676 176 634 903 025 314 868 829 268
691 612 211 948 938 737 091 201 067 018 430 848 725 407 033 263 665 129 825 160
229 532 601 165 011 926 466 746 161 487 368 029 354 723 485 471 969 612 194 779
718 606 002 809 224 094 775 906 046 650 615 288 889 348 919 321 687 505 369 072
895 202 076 619 431 443 283 754 354 725 526 816 046 626 637 984 306 322 609 097
480 655 454 224 163 096 829 270 129 882 432 766 677 204 298 281 670 003 723 473
041 294 453 190 852 066 188 585 699 847 413 323 549 014 078 269 193 821 078 936
302 114 600 193 879 745 125 508 483 546 213 840 516 795 588 592 464 676 193 894
363 227 400 302 089 641 878 595 382 790 088 397 449 001 566 977 450 051 700 872
690 595 366 060 061 013 289 767 711 802 046 541 252 511 600 601 167 295 407 033
996 071 799 476 805
```

12.2 Frequency Distributions and Histograms

Data given in an unorganized manner cannot be used to detect any patterns nor to aid in making a decision. For example, examine the following data.

At a supermarket, 40 food baskets are chosen at random. The time taken to check out each of the 40 baskets is recorded to the nearest minute. Each amount of time is referred to as an **outcome**.

8	3	7	6	2	9	6	7	4	5	6	7	9	1	7	10	5	8
6	8	8	4	5	7	9	4	7	8	7	9	10	3	7		5	8	6
11	5	12	6														

The above information or raw data are unorganized and as a result no pattern is seen. Learning to organize data in such a way as to show patterns is an important part of the study of **descriptive statistics**.

The data can be recorded in a table that shows the frequency of each amount of time. The table is referred to as a frequency distribution table. Any patterns that exist in the data will reveal themselves in the table.

Frequency Distribution Table

Outcome	Tally	Frequency
1			1
2			1
3				2
4					3
5							5
6								6
7										8
8								6
9						4
10				2
11			1
12			1

The number of times each outcome occurs.

The sum is 40
The total frequency is 40.

If each outcome is shown by x, then the frequency distribution is a function given by $f(x)$.

$$f(4) = 3 \qquad f(7) = 8$$

outcome — frequency

To show how the frequencies are distributed in a graphical form, a **histogram** is used as shown on the next page.

A line graph can be obtained from the histogram by connecting the midpoint of each bar. This line graph is called a **frequency polygon** and it too represents the frequency distribution, $f(x)$.

Histogram

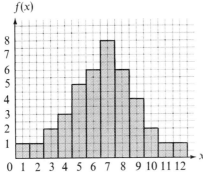

$f(x)$

Each bar of the histogram
is centred on the outcome.

Frequency Polygon

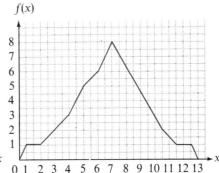

$f(x)$

Notice how the polygon
is completed at each side.

The area of each bar is proportional to the frequency.

From the histogram or the frequency polygon, patterns in the data can be noted and information read. Calculations can be made.
- The data seem to cluster around 7 min.
- The least amount of time is 1 min.
- The greatest amount of time is 12 min.
- About 5 per cent of customers waited more than 10 minutes.

12.2 Exercise

A Review the meaning of *histogram, frequency polygon, outcome frequency* and *frequency distribution.*

Questions 1 to 4 are based on the histogram given below.

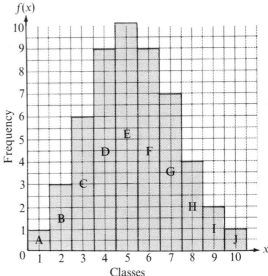

1 (a) How many outcomes are there in all?
   (b) What per cent of the outcomes in the sample is in class C? E? I?

2 (a) Calculate (area C) ÷ (area E).
   (b) Calculate the quotient of the frequencies $f(3) \div f(5)$.
   (c) What do you notice about your answers in (a) and (b)? Why is this so?

3 Calculate each of the following.
   (a) $\dfrac{\text{area F}}{\text{area B}}$ and $\dfrac{f(6)}{f(2)}$ (b) $\dfrac{\text{area C}}{\text{area F}}$ and $\dfrac{f(3)}{f(6)}$
   (c) What do you notice about your answers in (a) and (b)? Why is this so?

4 What are the advantages of using a histogram to show numerical data?

Questions 5 and 6 are based on the frequency polygon shown.

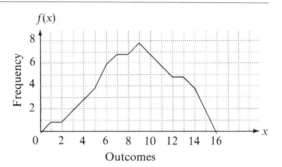

5 What value is each of the following?
   (a) $f(7)$ (b) $f(12)$ (c) $f(1)$ (d) $f(15)$ (e) $f(9)$ (f) $f(4)$

6 (a) What is the domain of f?   (b) What is the range of f?
   (c) What is the greatest value (maximum) of the frequency?

B Questions 7 to 11 are based on the following information.

   In an investigation of the pattern in rolling two dice, the following results were obtained for 300 tosses.

Sum	2	3	4	5	6	7	8	9	10	11	12
Frequency	8	16	27	36	44	48	43	34	23	17	4

7 (a) Construct a histogram for the data.
   (b) Construct a frequency polygon.

8 The function for the above frequency distribution is given by f. Find each of the following.
   (a) $f(6)$ (b) $f(12)$ (c) $f(7)$ (d) $f(3)$ (e) $f(4)$ (f) $f(9)$

9 What per cent of the tosses give
   (a) a sum of 2?   (b) a sum of 12?
   (c) a sum less than 7?   (d) a sum greater than 7?

10   Based on the information,
   (a) which sum seems most likely to turn up when rolling two dice?
   (b) which sums have about the same chance of occurring?

11   In a thousand tosses of a pair of dice, how many of each of the following sums would you expect to turn up (based on the above results)?
   (a) 5   (b) 9   (c) 7   (d) 2   (e) 12   (f) 1

12   The various heights (in centimetres) of a new strain of barley were recorded during its growth.

   18.3   18.5   18.8   18.7   18.5   18.3   18.5   18.4   18.8   18.1
   18.8   18.7   18.4   18.0   18.7   18.8   18.6   18.6   18.8   18.4
   18.2   18.4   18.9   18.5   18.7   18.8   18.3   18.5   18.5   18.5
   18.8   18.7   18.2   18.1   18.6   18.4   18.3   18.9   18.8   18.4

   (a) Construct a frequency table summarizing outcomes and draw the corresponding histogram.
   (b) Based on previous results, the strain is considered successful if at least 50% of the barley exceed a height of 18.6 cm. Is this batch successful?
   (c) Use the graph in (a). Write a problem based on the information and solve the problem.

13   Some computer chips were tested in each batch produced. The number of defective chips in each batch was recorded.

   4   1   3   2   0   6   2   1   2   4   4   6   0   2   0   0   3
   0   2   1   0   2   1   3   0   2   0   3   0   1   3   1   5   3
   8   2   0   1   9   0   5   0   3   7   2   0   1   2   4   3

   (a) Construct a frequency table showing outcomes of the number of defective chips.
   (b) Use the data in (a). Construct a histogram.
   (c) What per cent of the batches have six or more defective chips?
   (d) For purposes of quality control, the per cent of batches having four or more defective chips cannot exceed 50%. How would you describe the above sample?

Math Tip

It is important to understand clearly the vocabulary of mathematics when solving problems.
- Make a list of all the new words you meet in this chapter
- Provide an example to illustrate each word.

Displaying Data Graphically

There are different methods that can be used to display data. A **stem-and-leaf** diagram is another useful way of organizing data. The data below are taken from a list of points scored in basketball games.

63 48 74 92 51 60 81 32 63 87 95 64 59 54
56 43 49 88 60 91 57 36 59 75 85 64 78 65

A stem-and-leaf diagram is constructed for the above data as follows.
The organization of the data in the stem-and-leaf plot allows you to interpret data visually. For example, you can quickly answer:
How many scores are there?
Highest score? Lowest score?
How many scores are 70 and above? Less than 50?

The stem is formed by the tens digit.

The units digit forms the leaves.

Tens	Units						
3	2	6					
4	8	3	9				
5	1	9	4	6	7	9	
6	3	0	3	4	0	4	5
7	4	5	8				
8	1	7	8	5			
9	2	5	1				

The data shown in this row is 92 95 91

The need to display data occurs in weather forecasting, business, the study of economics, news reporting and so. You have already used a line graph to show data.

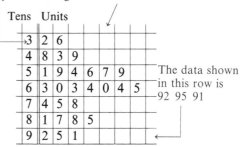

CRUDE OIL AND NATURAL GAS LIQUIDS PRODUCTION
(000's Barrels Per Day)

A quick glance at any newspaper illustrates the variety of ways in which people use graphical methods to display data. A few of these are shown.

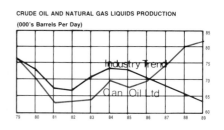

Pictograph

Number of Jeans Sold

Bar Graph

Circle Graph

14   The following data for pulse rates were obtained from a sampling of people attending a concert.

66 71 86 53 79 90 61 81 64 67 58 67 51 82
67 93 59 67 91 73 66 87 86 68 73 76 86

(a) Construct a stem and leaf plot for the data.

(b) How many people had a pulse rate less than 60? pulse rate 90 or more?

(c) Predict the average pulse rate from the graph. Calculate the average pulse rate. How does your calculation compare to your prediction?

(d) How many persons have a pulse rate above the average? below the average?

(e) List the advantages of displaying data on stem-leaf plot.

15 Refer to the graphical display on the previous page.

(a) For the data displayed in each graph, create a question based on the information.

(b) Answer the question you have created in (a).

(c) List the advantages and the disadvantages of each type of graphical display.

16 For the data shown in each of the following.

Step 1 Decide which type of graphical display best presents the data.

Step 2 Then construct the graphical display.

Step 3 Create one question based on your display. Write a solution for the question you have created.

(a) the height, in metres, reached by various members of a basketball team

1.79	1.93	1.46	2.03	1.62	1.85	1.54	1.88	1.92
2.11	1.58	1.78	1.73	1.83	1.63	1.84	1.81	2.08

(b) the average temperature in Puerto Rico for 20 consecutive weeks

20.8°C 21.3°C 21.4°C 21.9°C 21.0°C 22.9°C 23.2°C 24.5°C 25.6°C 26.8°C
27.3°C 26.9°C 27.2°C 26.8°C 25.0°C 25.4°C 25.8°C 25.5°C 26.3°C 25.4°C

(c) the percent of water in each of the tissues of the human body

Blood 80%, Bone 25%, Connective Tissue 60%, Fat 20%, Kidney 80%, Liver 70%, Muscle 75%, Nervous Tissue 78%, Skin 70%

(d) In 100 g of each food, a comparison is made of the nutritive value.

	Proteins	Carbohydrates	Fats
Peanuts	26.2 g	58.2	15.6
Popcorn	14.2 g	76.6 g	9.2 g

17 (a) Choose a newspaper each day for a week. Collect the various methods used in the newspaper to display data graphically. List reasons why the method chosen is effective or not.

(b) Obtain a copy of the financial report of a Canadian company. List the various methods used to display data graphically. List reasons why the method chosen is effective or not.

(c) Obtain a copy of a text in economics, geography, health studies, etc. List the various methods used to display data graphically. List reasons why the method chosen is effective or not.

12.3 Grouping Data

When an array of large numbers is given, the data are grouped before recording. For example, the number of days of sunshine on a south sea island each year for the last 50 years is shown in the table.

272	236	251	271	277	251	260	265	272	274
293	281	254	282	267	225	248	265	252	279
262	233	259	218	274	245	271	251	284	265
231	248	241	282	225	292	275	282	236	274
241	265	261	265	265	202	252	268	245	265

The outcomes are compressed into a convenient number of distinct classes.

Step 1: Calculate the *range*. This is the difference between the greatest and least value in the sample. Range: $293 - 202 = 91$

Step 2: Decide on the number of classes. For most of the work done, 10 classes will well represent the data. To approximate the width of each class, divide the range by 10.

Width (approximate): $\dfrac{91}{10} = 9.1$

Step 3: Use the number from Step 2 to determine the width of each class. Usually the number is rounded to a multiple of 5. $(9.1 \simeq 10)$

Choose class widths of ten units. The **class width** is the difference between two consecutive lower bounds.

When organizing grouped data, the outcome column of the table is represented by two columns and the columns are labeled *Class number* and *Classes*. The **class limits** are the extreme values in each class. The first class has limits 200 and 210, the second 210 and 220, and so on. You must decide on an *endpoint convention*, namely, class intervals include the left endpoint, but not the right endpoint.

Class number	Classes (Days)
1	200–210
2	210–220
3	

The frequency distribution table above is completed by tallying the sample data. The data are grouped as shown on the next page and a histogram is drawn as follows.

The *midpoint* of each interval is used to label the horizontal axis as follows. The **midpoint** is calculated as the average of the class limits for each interval. Thus, the midpoints are used to label the classes.

For class 1:

$$\text{midpoint} = \frac{200 + 210}{2} = 205.$$

For class 5:

$$\text{midpoint} = \frac{240 + 250}{2} = 245.$$

Frequency Distribution Table

Class number	Classes (Days)	Tally	Frequency
1	200–210	Class 1	1
2	210–220	excludes 210.	1
3	220–230	Class 2	2
4	230–240	includes 210.	4
5	240–250	|||	6
6	250–260	|||| ||	7
7	260–270	|||| |||| ||	12
8	270–280	|||| ||||	10
9	280–290	||||	5
10	290–300	||	2

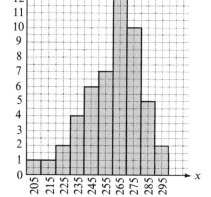

$f(x)$ Histogram (for grouped data)

For the grouped data, the values in the class are taken to be evenly spread.

Once the data are displayed and patterns are observed in the frequency distribution, problems about the data can be solved, as you will see in subsequent sections.

12.3 Exercise

A   Questions 1 to 5 are based on the following information.

A tourist firm polled international travellers to investigate their spending habits. People were asked how much money, on the average, they spent on each day of their vacations.

1   From the histogram,
   (a) how many classes are there?
   (b) what are the class limits?
   (c) how many persons were interviewed?

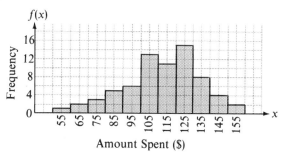

2   The frequency distribution is given by $f(m)$, where m is the value of the midpoint. Read and record each value.

(a) $f(95)$ (b) $f(115)$ (c) $f(125)$

(d) $f(155)$ (e) $f(85)$ (f) $f(55)$

3   (a) What is the greatest frequency?

(b) What is the least frequency?

4   What per cent of the people spent each amount?

(a) $90 − $100   (b) $130 − $140

(c) less than $90   (d) more than $130

5   If you design an advertising package to appeal to those vacationers that spend more than $100 per day, how much would you spend out of a total advertising budget of $500 000?

B   Review the meaning of range, class width, class number and class limit.

6   The number of phone requests recorded each day by Information Incorporated is shown for 50 consecutive days.

44	48	52	60	42	44	56	52	48	50	56	52	46	52	48
52	54	46	54	58	42	46	56	50	46	44	52	48	54	54
48	48	50	52	50	54	50	54	50	46	52	50	52	50	48
50	48	50	46	50										

(a) Construct a frequency table with five classes.

(b) Construct a histogram to represent the sample.

(c) What per cent of the days had fifty or more phone requests?

(d) If the per cent of days having fifty or more calls is more than 58.5%, then extra staff are hired for the following weeks. Based on the data will more staff be hired?

For questions 7 to 10 use the following data.

The table records the number of shots at the basket taken in 36 basketball games by the Lions.

74	79	93	91	73	85	81	73	89	77	88	76	83	66	62
76	85	69	76	72	77	86	84	83	76	92	77	81	88	72
66	63	84	58	87	93									

7   (a) Divide the data into classes of width 5. Set up a frequency table.

(b) Draw the corresponding histogram.

(c) In which class do the fewest data occur? the most data occur?

8 (a) Divide the data into classes of width 10. Set up a frequency table.
(b) Draw the corresponding histogram.
(c) In which way are the histograms of questions 7 and 8 different? the same?

9 Draw the frequency polygon for each of the histograms in questions 7 and 8.

10 Based on the results, what per cent of the games had
(a) more than 80 shots?   (b) fewer than 65 shots?

Questions 11 to 13 are based on the following information.

In an English course, the following grades (out of 100) were given this semester.

| 72 | 83 | 63 | 88 | 22 | 66 | 74 | 36 | 54 | 64 | 93 | 63 | 41 | 66 | 75 |
| 52 | 25 | 69 | 45 | 55 | 52 | 29 | 55 | 45 | 72 | 53 | 64 | 32 | 43 | |

11 (a) Choose an appropriate class interval. Construct a histogram.
(b) Based on your graph, which class has the greatest frequency?

12 What per cent of the class received
(a) first class honours (80 marks or greater)?
(b) a failing mark (less than 50 marks)?

13 The marks awarded in the same English course for the previous semester are shown.

| 61 | 53 | 84 | 44 | 66 | 71 | 38 | 46 | 63 | 75 | 73 | 67 | 15 | 58 | 35 |
| 88 | 62 | 50 | 42 | 59 | 55 | 41 | ·65 | 42 | 66 | 52 | 71 | 69 | 32 | |

(a) Compare the histograms for each semester? How are they different? the same?
(b) What explanation might you have to account for the differences in the histograms?
(c) What information do the differences indicate to the instructor of the course?

Using Percentiles

Whether you write an exam, play a sport, or dance, you are constantly making comparisons. Have you ever wondered how you did on an exam compared to others?

A piece of information used by many educational institutions is a measure of an individual's standing or rank within a group, called a **percentile**. A percentile rank indicates the percentage of people or students who achieved below a particular score or standard. For example, suppose in a bake-off contest you were given a score of 186. By itself a score of 186 is meaningless, but if you were told that the score you obtained has a percentile rank of 92, then you know that 92% of the people in the contest had a score less than your score of 186. Similarly, a percentile rank of 92 means that 8% of the contest group had a score greater than yours.

In effect, percentiles group data into 100 parts, and based on this grouping, a ranking is then made of any individual piece of data.

14 On a test marked out of 100, the following scores were obtained.
   43 61 78 88 36 70 60 89 54 93 42 84 86 72 63 74 59
   38 53 63 91 58 23 76 93 71 64 39 65 89 68 79 52
   85 82 96 72 80 95 51 59 75 73 56 49 98 62 95 94
   (a) Mark received 76 on the test. What is his percentile rank?
   (b) Louise received 89 on the test. What is her percentile rank?
   (c) If you are in the 80th percentile, what mark would you expect?

15 To enter a legal course, Stephen wrote a qualifying test. His score was 36th out of 653 people who wrote the test. Find what percentile rank Stephen achieved.

16 Peter received a score of 53 out of 70 on a history test. Of the 36 students in the class, Peter stood 22nd. What is his percentile score?

17 Tony was the twelfth highest scorer of the students who wrote the final exam. He was in the 94 percentile. How many students wrote the exam?

18 Applicants for an accountancy position were asked to write an Accountancy Awareness test. The results out of 75 are shown for the applicants.
   34 46 31 32 40 36 24 12 16 40 32 19 50 26 35 34 42
   54 10 27 30 51 47 22 31 28 19 8 36 42
   (a) Calculate the percentile rank for the applicant who received scores of
       (i) 35, (ii) 50, (iii) 27.
   (b) Which person would you hire from the applicants who wrote the test? List the assumptions you made in making your decision.
   (c) Create a question based on the data. Write a solution to the question.

12.4 Measures of Central Tendency

To show any patterns in raw data, frequency distribution tables and histograms can be used. From the graphical representations the data can be analyzed. For example, for two sets of data, histograms are drawn. The outcomes appear clustered around a value that is distinct for each sample.

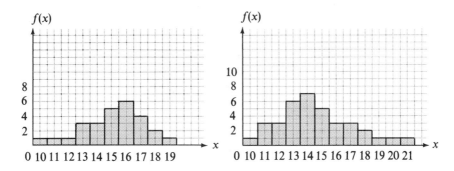

There are calculations that can be done to determine how data *cluster*, without constructing a graph. Three such representatives of data that are used to provide information about the location of the centre of the data are

- the mean.
- the median.
- the mode.

Mean

The arithmetic mean includes all the data. It is commonly referred to in everyday conversation as the average. The mean, $\bar{x}$, is calculated as follows for the given data.   62, 75, 72, 68, 54, 82, 62, 62, 66

$$\bar{x} = \frac{62 + 75 + 72 + 68 + 54 + 82 + 62 + 62 + 66}{9} \longleftarrow \text{the sum of the results}$$

$\longleftarrow$ the number of results

$\bar{x} = 67$ $\longleftarrow$ the mean is 67

Note that there is no score which is 67. That is, the mean, 67, is not necessarily a member of the sample or data collected.

Median

The median is the middle member when the data are arranged from smallest to greatest.   54   62   62   62   66   68   72   75   82

smallest   middle number   greatest

The median is an easy representative of data to obtain. It is not influenced by the value of the greatest and least numbers. For example, if the top mark was 100, the median would remain *unchanged*. However, the mean would change (increase in this case).

If there is an even number of test results, the median is the average of the two middle numbers.

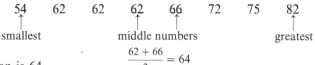

$$\frac{62 + 66}{2} = 64$$

The median is 64.

Mode

The mode is the representative of the data which occurs most frequently. Like the median, the mode does not take into consideration the values of all the data. For the previous sample, the most frequent mark is 62. Thus the mode is 62.

For the following results, there are 2 modes. Data such as this are said to be **bimodal**.

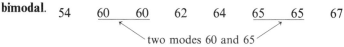

Example For a traffic count, the number of persons in commuter cars was recorded in a table.

	x_1	x_2	x_3	x_4	x_5	x_6
Outcome (number of persons in the car) x_i	1	2	3	4	5	6
Frequency (number of cars) $f(x_i)$	18	14	10	26	9	3

Calculate the (arithmetic) mean number of people in each car. Express your answer to the nearest whole number.

Solution

Think: From the table, there was 1 person in each of 18 cars, 2 people in each of 14 cars, 3 people in each of 10 cars, and so on.

Step 1
The total number of people commuting is given by

$$1(18) + 2(14) + 3(10) + 4(26) + 5(9) + 6(3)$$
$$= 243$$

The result is written in a compact form using the Σ, sigma, symbol.

$$\sum_{i=1}^{6} x_i f(x_i) = 243 \text{ means}$$

$$x_1 f(x_1) + x_2 f(x_2) + x_3 f(x_3) + \cdots + x_6 f(x_6) = 243$$

Step 2
The total number of cars in the sample is given as

$$18 + 14 + 10 + 26 + 9 + 3$$
$$= 80$$

The result shown in a compact form

$$\sum_{i=1}^{6} f(x_i) = 80 \text{ means}$$

$$f(x_1) + f(x_2) + \cdots + f(x_6) = 80.$$

Step 3 The mean, or $\bar{x}$, is calculated.

$$\bar{x} = \frac{\text{total number of people}}{\text{number of cars}}$$

$$= \frac{243}{80}$$

$$= 3.0375$$

Thus, to the nearest whole number, the (arithmetic) mean is 3.

This means that, on the average, there ⌐
are about 3 people in each car.

In general, for a set of data in a table with values $x_1, x_2,$ $x_3, \ldots, x_n$ and corresponding frequencies, $f(x_1), f(x_2),$ $f(x_3), \ldots, f(x_n)$, the arithmetic mean, $\bar{x}$, is given by the compact symbol shown on the right.

The Mean

$$\bar{x} = \frac{\displaystyle\sum_{i=1}^{n} x_i f(x_i)}{\displaystyle\sum_{i=1}^{n} f(x_i)}$$

A statistic is a value obtained from the sample data. Thus, the mean, median, and mode are examples of *statistics* which show the location of the centre of the raw data. These three statistics are referred to as **measures of central tendency**.

12.4 Exercise

A Use your calculator to help you find the measures of central tendency. Round your answers to 1 decimal place.

1 For each set of data find the mean, median, and mode.
(a) 20, 24, 28, 18, 26, 24, 12, 16, 20
(b) 5, 9, 13, 12, 2, 4, 0, 1, 7, 14, 12

2 Calculate each statistic for the following data.

12.5, 12.4, 12.2, 12.7, 12.9, 12.2, 12.3, 12.2, 12.6, 12.8, 12.1

(a) mean   (b) median   (c) mode
(d) Which statistic(s) show(s) the *centre* of the data distribution?

3 (a) The mean for a set of data is 21. Illustrate with a counter-example that 21 does not have to be one of the data values.

   (b) The mode for a set of data is 21. Illustrate with an example that 21 is a data value or show with a counter-example that 21 is not necessarily a data value.

   (c) The median for a set of data is 21. Illustrate with an example that 21 is a data value or show with a counter-example that it need not be.

4 Calculate the mean in each frequency distribution.

(a)

x_i	$f(x_i)$
2	3
4	5
6	7
8	6
10	2

(b)

x_i	$f(x_i)$
20	3
21	4
22	7
23	8
24	5
25	1

5 For the set of data shown in the frequency table, calculate

(a) $\sum_{i=1}^{5} f(x_i)$.   (b) $\sum_{i=1}^{5} x_i f(x_i)$.

(c) What is the mean, $\bar{x}$, of the data?

Outcome x_i	Frequency $f(x_i)$
x_1 3	13
x_2 4	14
x_3 5	17
x_4 6	16
x_5 7	12

For Table A,

6 (a) calculate the mean to 1 decimal place.
   (b) what is the median of the data?

Table A

Outcome x_i	10	11	12	13	14	15
Frequency $f(x_i)$	2	5	8	9	8	3

7 Additional data are collected and placed with the data from Table A to construct Table B.

Table B

Outcome x_i	10	11	12	13	14	15	25
Frequency $f(x_i)$	2	5	8	9	8	5	2

   (a) How are tables A and B alike? How do they differ?

   (b) Calculate the mean to 1 decimal place. How does it compare to the mean for Table A?

   (c) What is median of the data in Table B?

8 Use your results for Tables A and B above.
   (a) Why are the medians the same?   (b) Why do the means differ?
   (c) Use an example of your own to show why the following statement is true: *The median of a set of data is not affected by the greatest and smallest values, but the mean of the same data is affected by these values.*

B To do statistics, you must have a clear understanding of the words: mean, median, mode, . . .

9 From collected samples, a farm magazine made the following statement.

   Most farmers use Earth Brand fertilizer.

   Which of the following statistics were probably used to come to the above conclusion?   • the mean   • the mode   • the median
   Give reasons for your answer.

10 On which statistic: mean, median, or mode is each statement probably based?
   (a) The average cost of groceries per week is $103.65.
   (b) Most water-skiing accidents occur at the dock.
   (c) The country's favourite song is *Moonshine*.
   (d) The mean mass of the gymnasts is 62.5 kg.
   (e) The average speed of an aircraft on take-off is 212 km/h.
   (f) More people go to football games than go to baseball games.
   (g) The hourly rate for part-time workers is about $4.50/h.

11 The batting averages for 22 baseball players are recorded below.

0.267	0.251	0.265	0.265	0.279	0.305	0.202	0.329
0.252	0.221	0.376	0.265	0.236	0.265	0.392	
0.233	0.401	0.320	0.271	0.208	0.251	0.225	

   What is the mean batting average?

12 During August, the temperatures of 20 cities, chosen randomly from the same province are shown.

26°C	33°C	29°C	32°C	30°C	22°C	28°C	22°C	25°C	32°C
28°C	27°C	24°C	22°C	20°C	28°C	20°C	27°C	33°C	28°C

   (a) Find the mean, median, and mode for the data.
   (b) Which of the answers in (a) is useful in describing the location of the *centre* of the data.
   (c) Would you say there was a *best* measure of central tendency? Why or why not?
   (d) Which statistic was the hardest to calculate: mean, median, or mode?

13 For a test marked out of 10, the following scores were recorded.

Score x_i	0	1	2	3	4	5	6	7	8	9	10
Frequency $f(x_i)$	1	1	0	1	2	2	5	7	4	2	3

(a) How many students wrote the test?

(b) Find the mean score to 1 decimal place?

(c) Find the median.

14 For boxes of biscuits the mass is printed as 500 g. Fifteen samples are selected and the masses are recorded.

$$491 \quad 516 \quad 493 \quad 505 \quad 496 \quad 503 \quad 476 \quad 512$$
$$491 \quad 505 \quad 480 \quad 480 \quad 507 \quad 496 \quad 498$$

(a) If the mode is less than 500 g, then another sample will be taken. Will there be another sample taken? Give reasons for your answer.

(b) If the arithmetic mean is in the range $495 \leq x < 505$, then the amount in each box is acceptable. For the above sample, is the amount acceptable?

Computer Tip

To calculate the mean, M, of a set of data, a computer program written in BASIC language is used, as follows.

```
10  INPUT N
15  DIM X(100)
20  LET S = 0
30  FOR I = 1 TO N
40  INPUT X(I)
50  LET S = S + X(I)
60  NEXT I
70  LET M = S/N
80  PRINT "THE MEAN IS", M
90  END
```

Math Tip

It is important to understand clearly the vocabulary of mathematics when solving problems about statistics. Bring your list of new words up to date. Provide an example to illustrate the meaning of each word. Continue to add to your vocabulary list.

12.5 Finding the Mean: Grouped Data

When the data are grouped, it is not possible to determine the mean exactly. It is assumed that the data in each interval are distributed uniformly. The midpoint of the interval is used to calculate the mean score.

Example Data were collected and stored in the table. For the grouped data, find the value of the mean to 1 decimal place.

Interval	Frequency $f(x_i)$
0–5	2
5–10	5
10–15	7
15–20	6
20–25	3
25–30	1

Solution Complete the table.

$$\text{midpoint} = \frac{0 + 5}{2} = 2.5$$

Interval	Midpoint x_i	Frequency $f(x_i)$	$x_i \times f(x_i)$
0–5	2.5	2	5.0
5–10	7.5	5	37.5
10–15	12.5	7	87.5
15–20	17.5	6	105.0
20–25	22.5	3	67.5
25–30	27.5	1	27.5

← The score 2.5 occurred 2 times for a total of 5

Do the calculations.

Total frequency is given by $\sum\limits_{i=1}^{6} f(x_i) = 24$.

Total midpoint times frequency is given by $\sum\limits_{i=1}^{6} x_i f(x_i) = 330$.

The mean is given by $\bar{x} = \dfrac{\sum\limits_{i=1}^{6} x_1 f(x_i)}{\sum\limits_{i=1}^{6} f(x_i)}$. ← This is the total of the number of times a score appeared times its frequency.

$$= \frac{330}{24}$$

$$= 13.8 \text{ (to 1 decimal place)}$$

12.5 Exercise

A Use your calculator.

1 Use the data given in the histogram.
   (a) Use the histogram to estimate the mean.
   (b) Calculate the mean for the data given in the histogram. Set up a table first.

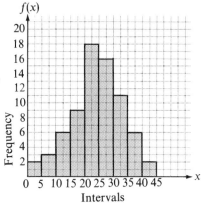

2 (a) Estimate the mean given for the data in the table.

Interval	0–3	3–6	6–9	9–12	12–15	15–18	18–21
Frequency	3	6	8	7	4	2	1

   (b) Calculate the mean for the data in (a).

3 Find the mean to 1 decimal place for the data given in the table.

Midpoint
$$= \frac{6 + 9.5}{2}$$
$$= 7.75$$

Interval	Frequency
6.0–9.5	6
9.5–13.0	12
13.0–16.5	16
16.5–20.0	14
20.0–23.5	8
23.5–27.0	2

B Skills with your calculator are useful in doing mathematics. Read the manual provided with your calculator to identify additional techniques.

4 Out of eighty-five people interviewed, the table shows the answers to the question, *At what age did you own your first car?*

23	32	25	29	26	32	40	27	31	22	27	18	31	30	35
24	28	37	24	38	36	25	20	27	34	39	26	35	28	22
29	34	20	27	25	17	30	21	33	28	25	30	33	32	30
26	34	39	21	31	26	23	19	28	34	42	22	26	27	36
29	37	23	31	43	45	29	21	30	24	33	24	33	25	27
30	39	26	37	27	25	37	41	31	22					

   (a) Construct a frequency distribution for the data. Do not group the data.
   (b) Calculate the mean for the data to one decimal place.

5   (a) Construct a frequency distribution for the data in Question 4. Use the intervals 17–20, 20–23, 23–26, and so on.
    (b) Compute the mean for the data. Compare your answer to that in question 4(b).

Questions 6 to 9 are based on the following data.

For a group of 50 people, pulse rates are recorded and given as raw data.

81	93	78	62	82	96	72	77	81	73	92	68	66
72	79	81	84	82	86	71	61	76	76	56	86	87
56	83	79	73	53	76	71	78	82	79	67	62	
67	83	88	76	80	67	90	76	67	71	64	91	

6   (a) Choose at random 10 values from the chart. Calculate the mean $\bar{x}_1$, for the sample.
    (b) Choose at random another 10 values from the chart. Calculate the mean $\bar{x}_2$ for the sample.
    (c) Calculate $\bar{x} = \dfrac{\bar{x}_1 + \bar{x}_2}{2}$.
    (d) Calculate the mean, μ, of the complete set of data.
    (e) Compare the values of $\bar{x}_1$, $\bar{x}_2$, and $\bar{x}$ with that of μ. What do you notice?

7   (a) For the above data, what is the median?
    (b) Use the value of μ and the median to predict where the data are clustered on a histogram.

8   Use the previous data.
    (a) Construct the histogram. Decide on suitable class intervals.
    (b) Calculate the mean of the data you have grouped.
    (c) Where are the data clustered on your histogram? Compare your answer to your predicted answer in 7(b).

9   (a) Based on the sample results, out of a population of 1000 people, what percentage would have a pulse rate of
        (i) 75?   (ii) 80?   (iii) 90?
    (b) What would you expect the average (mean) pulse rate to be?

Math Tip

It is important to clearly understand the vocabulary of mathematics when solving problems. Make a list of all the new words you have met in this chapter. Provide a simple example to illustrate each word.

Using Calculators: Weighted Means

For three consecutive weeks Jackson bought chicken wings at these prices, $2.29/kg, $2.96/kg and $2.53/kg, for his get togethers. For the same quantity at each price the mean A is obtained by the following calculation.

$$A = \frac{2.29 + 2.96 + 2.53}{3} \qquad \text{Use your calculator.}$$

$$= \$2.59 \longleftarrow \text{The average or mean price is \$2.59/kg.}$$

However, Jackson purchased 7.2 kg at $2.29/kg, 2.3 kg at $2.96/kg, and 5.9 kg at $2.53/kg. The mean, A, obtained previously is misleading since different quantities were purchased at different prices. For this reason a more accurate representative of the data is the *weighted mean* obtained by the following calculation.

$$W = \frac{(7.2)(2.29) + (2.3)(2.96) + (5.9)(2.53)}{7.2 + 2.3 + 5.9}$$

$$= \$2.48 \longleftarrow \text{Thus the weighted mean price, in this case, is \$2.48/kg.}$$

Can you find a more efficient way on your calculator of doing the calculation above?

$$\boxed{C}\,7.2\,\boxed{\times}\,2.29\,\boxed{=}\,\boxed{M+}\,2.3\,\boxed{\times}\,2.96\,\boxed{=}\,\boxed{M+}\,5.9\,\boxed{\times}\,2.53\,\boxed{=}$$

$$\boxed{M+}\,7.2\,\boxed{+}\,2.3\,\boxed{+}\,5.9\,\boxed{=}\,\boxed{\tfrac{1}{x}}\,\boxed{\times}\,\boxed{MR}\,\boxed{=}\,\underset{\textit{Output}}{2.4820105}$$

10 During October the prices of Mini Sizzlers fluctuated. Sarah purchased 9.1 kg at $2.15/kg, 5.5 kg at $2.49/kg, and 12.6 kg at $2.06/kg. Calculate the weighted mean.

11 (a) Tomatoes fluctuate in price depending on the season. What is the average price for a year based on the following information on purchases for Henley's Hamburg Haven?
26.5 kg at $1.86/kg,   28.8 kg at $1.93/kg,   24.3 kg at $2.29/kg
16.9 kg at $3.06/kg,   14.5 kg at $4.39/kg,   19.9 kg at $4.18/kg
16.3 kg at $3.96/kg,   19.5 kg at $3.12/kg,   21.3 kg at $2.96/kg
26.4 kg at $2.75/kg,   22.3 kg at $2.43/kg,   21.9 kg at $1.99/kg
(b) Use the data in (a). Create your own question and write the solution.

12 Over a 6-month period, the promotion department of Canadian Motors Limited wanted to know the average price paid for football tickets.
(a) Use the following data to determine the average price for tickets.
26 at $12.50   52 at $9.95   36 at $16.25   42 at $18.95
83 at $11.75   48 at $14.50   39 at $13.90   12 at $56.50
(b) Based on (a), which ticket would classify as medium priced, high priced?
(c) In order to budget next year's promotion give aways, the department expects an increase of 9.6% in prices. Estimate the average price next year. Calculate the average price. How accurate was your estimate?
(d) Refer to the data. Create your own question and write a solution.

12.6 Interpreting Data: Deviation

How does a coach often know which player to use in an intense crucial part of a game? Often data are collected and analyzed in professional sports so that the results of the analysis are readily available to a coach and a better decision can be made. Once you obtain data, you develop techniques for interpreting and analyzing the data, often to make important decisions. For example, the points scored by two players on a team are shown for 8 games.

One poor decision can decrease the earnings of players of post season play by thousands of dollars.

| Tank | 9 | 12 | 2 | 21 |
| (nickname) | 18 | 26 | 34 | 22 |

| Fridge | 15 | 22 | 2 | 18 |
| (nickname) | 34 | 17 | 16 | 20 |

Given the data in the above form, you can do some simple calculations for each player. For example, you could calculate the mean, or even the difference between the greatest and least number of points scored, called the **range**.

Tank $\text{Mean} = \dfrac{144}{8}$
              $= 18$
        $\text{Range} = 32$

Fridge $\text{Mean} = \dfrac{144}{8}$
               $= 18$
         $\text{Range} = 32$

At first glance, there is no noticeable difference in the two calculations. However, the scores for Fridge seem to deviate from the mean less than the scores for Tank. To analyze the data from this point of view, you are interested in how much each score deviates from the mean. Thus, the absolute value symbol is used. With the absolute value of the deviations known, you can now calculate the **mean deviation**.

Number of Points Tank scored	Deviation from the Mean	
	d	|d|
34	16	16
26	8	8
22	4	4
21	3	3
18	0	0
12	−6	6
9	−9	9
2	−16	16
Total		62

Number of Points Fridge scored	Deviation from the Mean	
	d	|d|
34	16	16
22	4	4
20	2	2
18	0	0
17	−1	1
16	−2	2
15	−3	3
2	−16	16
Total		44

The mean of the deviations for Tank is calculated

$$\text{Mean deviation} = \frac{62}{8}$$
$$= 7.8$$

The mean of the deviations for Fridge is calculated.

$$\text{Mean deviation} = \frac{44}{8}$$
$$= 5.5$$

To analyze the data, a diagram is helpful.

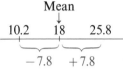

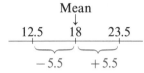

For a game, Tank's total points scored are more likely to occur between 10.2 and 25.8.

For a game, Fridge's total points scored is more likely to occur between 12.5 and 23.5.

Thus, based on the interpretation of the data Fridge appears to be the more consistent scorer. Thus, with this information, a coach, under pressure, can use the analysis of data to make a better decision.

12.6 Exercise

A Use a calculator. Where needed round your answers to 1 decimal place.

1 The points awarded in a contest are recorded.

47   50   52   42   73   53   61   43   51

(a) Calculate the mean.   (b) Calculate the mean deviation.

2 The errors in measurement, shown in centimetres, is given as follows.

3.6   7.5   4.2   6.5   4.8   6.3   2.8   5.6   4.3   2.2   5.3

(a) Calculate the mean   (b) Calculate the mean deviation

3 The number of sightings obtained from various stations is shown.

318   618   246   526   569   169   429   513   601   123

(a) Calculate the mean   (b) Calculate the mean deviation

B Review the meaning of mean, mean deviation and range.

4 The number of batteries found defective in 8 samplings by inspector A are shown by the data

83   156   74   92   78   9   101   69

(a) Calculate the mean deviation.
(b) Use your calculation in (a). Estimate the number of defective batteries inspector A would expect in the next sample.

5   Inspector B tested an identical set of samples as in the previous
    question and obtained the following results. The number of batteries
    found defective in 8 samplings in this case are shown by the following
    data.

    119   97   156   41   101   9   83   55

    (a) Calculate the mean deviation.

    (b) Use your calculation in (a). Estimate the number of defective
        batteries inspector B would expect in the next sample.

    (c) Compare the work of inspectors A and B. Based on the data and
        your calculations, which inspector appears to be more reliable?

6   The number of points scored by two players is shown.

| Myrna | 35 | 13 | 34 | 45 | 29 | 14 | 42 | 19 |
| Eileen | 29 | 45 | 27 | 35 | 13 | 32 | 24 | 26 |

    (a) For each player, calculate the mean; the mean deviation.

    (b) Based on your calculations, which player would you use in a crucial
        situation, in which you *must* score points?

7   The marks obtained by students for the same exams are shown in the
    table.

| | Session 1 | | | | Session 2 | | | | Session 3 | | | |
Exam	1	2	3	4	1	2	3	4	1	2	3	4
David	89	81	99	69	98	55	65	92	68	84	96	59
Lesley	76	94	58	41	89	87	72	73	84	66	79	81
Kevin	53	67	59	71	56	83	62	48	39	62	71	49
Andrew	41	32	48	29	52	53	30	39	51	32	36	39
Joelle	58	23	78	94	13	32	41	66	32	20	58	31

    (a) For each student calculate the mean; the mean deviation.

    (b) Based on your calculations, which student is most consistent?
        List any conditions for which this is not desirable?

    (c) List any additional information that you can obtain from these exam
        marks.

    (d) Based on the data, create a question of your own. Provide a
        solution for the problem you have created.

12.7 Measures of Dispersion

Two sets of data may have the same mean and same median, but can vary quite significantly in how the data is distributed.

For example, the following two samples have the same mean and median, but are dispersed (distributed) quite differently.

A: 2, 10, 12, 12, 12, 14, 22
   Mean 12   Median 12

B: 6, 8, 8, 12, 16, 16, 18
   Mean 12   Median 12

To indicate how widely the data of a sample are spread, other representatives of the data are needed. One representative of the data used earlier is the **range**.

The range of a set of data is the difference between the greatest and smallest data obtained. For example, for the above samples A and B

Sample A	*Sample B*
Greatest data = 22	Greatest data = 18
Smallest data = 2	Smallest data = 6
Range = 22 − 2	Range = 18 − 6
= 20	= 12

For a sample, the mean, median and mode give some useful information about where the data are *clustered*. The range will give us some information as to how the data are *dispersed*. Two sets of data, M and N, are compared.

Sample M: 1, 4, 3, 4, 5, 2, 4, 4, 6, 3,
            7, 3, 4, 5, 5
                 Range 7 − 1 = 6

Sample N: 6, 5, 4, 1, 3, 1, 6, 2, 7, 3,
            5, 1, 2, 6, 7
                    Range 7 − 1 = 6

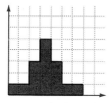
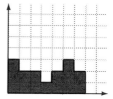

Although the data of the samples M and N are dispersed quite differently, the range is the same and thus may provide misleading information when comparing the two samples. Like the median and mode, the range does not

consider how the data may be clustered with respect to the rest of the data. For this reason, the concept of **standard deviation** is introduced, which provides further information about the dispersion.

Standard Deviation

To describe the data for M and N given previously, you might say

M: the data are clustered, in this case, about the mean.

N: the data are dispersed

As a *measure* of the amount of dispersion, a representative of the data, called the **standard deviation** is calculated.

Sample M

For the data in Sample M, x_1 and x_2 are the same "absolute distance" from the mean ($\bar{x} = 4$).

For x_1
$$x_1 - \bar{x} = -3$$

For x_2
$$x_2 - \bar{x} = 3$$

Note difference in sign.

To avoid negative results in calculating a measure of the dispersion, the deviations are squared, the mean of the squares is found and the square root of the last result is taken.

For a set of data values $x_1, x_2, x_3, \ldots, x_n$, the following steps are followed to calculate the standard deviation.

Step 1: Calculate the mean of the data.

$$\bar{x} = \frac{x_1 + x_2 + \cdots + x_n}{n}$$

Step 2: Calculate the deviation (difference) between each data value and the mean.

$$x_1 - \bar{x}, x_2 - \bar{x}, \ldots, x_n - \bar{x}$$

Step 3: Square each result in Step 2.

$$(x_1 - \bar{x})^2, (x_2 - \bar{x})^2, \ldots, (x_n - \bar{x})^2$$

Step 4: Calculate the mean of the squares in Step 3.

$$\frac{(x_1 - \bar{x})^2 + (x_2 - \bar{x})^2 + \cdots + (x_n - \bar{x})^2}{n}$$

Step 5: Calculate the square root of the mean in Step 4.

$$S_x = \sqrt{\frac{(x_1 - \bar{x})^2 + (x_2 - \bar{x})^2 + \cdots + (x_n - \bar{x})^2}{n}}$$

The $\sum$ notation is used to write the result for standard deviation in a compact form.

$$S_x = \sqrt{\frac{\sum_{i=1}^{n} (x_i - \bar{x})^2}{n}}$$

To check quality control in industry and manufacturing, the standard deviation is calculated as shown in the following example.

Example Packages of raisins are chosen randomly from a production line to check their contents. Each package should contain 60 g.

Sample Masses

61 g   59 g   62 g   60 g   60 g   58 g   63 g   60 g   61 g   61 g   56 g   59 g

The standard deviation must be less than 1.5 g in order for production to continue. Will production continue?

Solution *Step 1*

Think: To calculate the standard deviation, first find the value of the mean of the data.

$$\bar{x} = \frac{61 + 59 + 62 + 60 + 60 + 58 + \cdots + 56 + 59}{12}$$

$$= \frac{720}{12}$$

$$= 60$$

Steps 4 and 5

Think: Calculate the standard deviation. Use the formula.

$$S_x = \sqrt{\frac{\sum_{i=1}^{n} (x_i - \bar{x})^2}{n}}$$

$$= \sqrt{\frac{38}{12}}$$

$$= 1.78 \text{ (to 2 decimal places)}$$

		Step 2	*Step 3*
x_i	$\bar{x}$	$x_i - \bar{x}$	$(x_i - \bar{x})^2$
61	60	1	1
59	60	−1	1
62	60	2	4
60	60	0	0
60	60	0	0
58	60	−2	4
63	60	3	9
60	60	0	0
61	60	1	1
61	60	1	1
56	60	−4	16
59	60	−1	1

Sum: 38

$$\sum_{i=1}^{n} (x_i - \bar{x})^2.$$

From the calculations, the standard deviation, S_x, of the sample is 1.78. Since the standard deviation for quality control must be less than 1.5 g, production is stopped (until the situation is corrected).

For the above data, you say,
▶ *one standard deviation for the data is 1.78 g.*
▶ *two standard deviations for the data are 2 × 1.78 g or 3.56 g.*

12.7 Exercise

A Unless indicated otherwise, record answers accurate to 1 decimal place.

Questions 1 to 5 are based on the following information.

Samples A, B, C, are chosen from a production line, shown in the table.

	x_1	x_2	x_3	x_4	x_5	x_6	x_7	x_8	x_9
A	13	16	14	15	14	16	17	15	15
B	14	12	11	15	13	17	18	16	14
C	12	16	18	10	22	14	24	20	16

1 Calculate the range for
(a) sample A.   (b) sample B.   (c) sample C.

2 Calculate the mean $\bar{x}$ for
(a) sample A.   (b) sample B.   (c) sample C.

3 Calculate the standard deviation S_x for
(a) sample A.   (b) sample B.   (c) sample C.

4 Which sample has the
(a) greatest deviation?   (b) least deviation?

5 (a) Use the information in the above question. Predict the shape of the histogram for each set of data.
(b) Draw a histogram for each set of data. How accurate was your sketch to predict the histogram for each sample?

6 If $x_1, x_2, \ldots, x_6$ represent 4, 8, 12, 12, 16, 20 and $\bar{x}$ is the mean of the data, calculate

(a) $\sum_{i=1}^{6} (x_i - \bar{x})$.   (b) $\sum_{i=1}^{6} (x_i - \bar{x})^2$.

7 If $x_1, x_2, \ldots, x_8$ represent 19, 21, 21, 25, 25, 27, 31, 31 and $\bar{x}$ is the mean of the data then calculate

(a) $\sum_{i=1}^{8} (x_i - \bar{x})$.   (b) $\sum_{i=1}^{8} (x_i - \bar{x})^2$.

B 8 For the set of data given by   P: 22, 16, 23, 18, 19, 21, 14

(a) calculate the mean $\bar{x}$.

(b) calculate $x_i - \bar{x}$ where $x_1, x_2, \ldots, x_7$ represent the data from least to greatest.

(c) calculate the mean of the sum of the deviations, given by $\sum_{i=1}^{7} (x_i - \bar{x})$.

(d) calculate the mean of the sum of the squares of the deviations, given by $\sum_{i=1}^{7} (x_i - \bar{x})^2$.

9 For the set of data given by   Q: 7, 15, 8, 16, 9, 16, 6 calculate

(a) the mean $\bar{x}$.

(b) the mean of the sum of the deviations as shown by $\sum_{i=1}^{7} (x_i - \bar{x})$.

(c) the mean of the sum of the squares of the deviations $\sum_{i=1}^{7} (x_i - \bar{x})^2$.

10 Use your results in questions 8 and 9.

(a) Compare the ranges for the samples P and Q. Why is the range not a satisfactory statistic for samples P and Q?

(b) Compare the values of the sum of the deviations for samples P and Q. Why is the sum of the deviations for samples P and Q not a satisfactory statistic?

(c) Compare the values of the sum of the squares of the deviations for samples P and Q.

Why is the sum of the *squares of the deviations* used in the calculation of the standard deviation, rather than the sum of only the deviations?

(d) What is the standard deviation for sample P?

(e) What is the standard deviation for sample Q?

11 A set of data $x_1, x_2, \ldots, x_6$ is recorded in the table. The mean, for the data is 44. $\bar{x} = 44$.

(a) What is the range for the data?

(b) Copy and complete the table.

(c) Use the results in (b) to calculate the standard deviation of the data.

x_i	$x_i - \bar{x}$	$(x_i - \bar{x})^2$
39	?	?
42	?	?
43	?	?
45	?	?
46	?	?
49	?	?

12 Another set of data $x_1, x_2, \ldots, x_6$ is recorded in a chart. The mean for the data is $\bar{x} = 44$.

(a) What is the range for the data?

(b) Copy and complete the table.

(c) Use the results in (b) to calculate the standard deviation of the data.

x_i	$x_i - \bar{x}$	$(x_i - \bar{x})^2$
34	?	?
38	?	?
42	?	?
46	?	?
50	?	?
54	?	?

13 For the data in questions 11 and 12 the means are the same ($\bar{x} = 44$), but the standard deviations are different. Why is this so?

14 A sample of masses is taken from a production line and the data are recorded.

64 g   68 g   72 g   74 g   76 g   80 g   84 g

(a) Calculate the mean.

(b) Calculate the standard deviation.

(c) Is the standard deviation within 5 g?

15 The number of hours on different days that a machine is in operation is given below.

12 h   18 h   15 h   6 h   4 h   17 h   10 h
13 h   10 h   7 h   16 h   11 h   4 h

(a) Find the range. Is the range a useful representative of the above data?

(b) Calculate the mean.

(c) Find the standard deviation.

(d) Is the standard deviation within 1.5 h?

Computer Tip

Use the following BASIC program to calculate the standard deviation of a set data.

```
10 INPUT N                80 LET S2 = 0
15 DIM X(100)             90 FOR I = 1 TO N
20 LET S1 = 0             100 FOR S2 = S2 + (X(I) − M)↑2
30 FOR I = 1 TO N         110 NEXT I
40 INPUT X(I)             120 LET S3 = S2/N
50 LET S1 = S1 + X(I)     130 LET D = SQR(S3)
60 NEXT I                 140 PRINT "THE STANDARD DEVIATION IS". D
70 LET M = S1/N           150 END
```

12.8 Standard Deviation: Using Frequency Tables

Frequency tables have earlier been used to organize data in a compact form. You use your previous skills to calculate the standard deviation of data that is given in a frequency table, as well as for grouped data.

Test results for 30 students were recorded in a frequency table. The test was marked out of 5. What is the standard deviation to 2 decimal places?

Class number	1	2	3	4	5	6
x_i	0	1	2	3	4	5
$f(x_i)$	3	7	8	6	4	2

Do the calculations, as follows, to calculate the standard deviation. Copy and complete the table.

Class number	x_i	$f(x_i)$	$x_i \times f(x_i)$	$(x_i - \bar{x})^2$	$(x_i - \bar{x})^2 \times f(x_i)$
1	0	3	0	4.9729	14.9187
2	1	7	7	1.5129	10.5903
3	2	8	16	0.0529	0.4232
4	3	6	18	0.5929	3.5574
5	4	4	16	3.1329	12.5316
6	5	2	10	7.6729	15.3458

$$\sum_{i=1}^{6} f(x_i) = 3 + 7 + 8 + 6 + 4 + 2$$
$$= 30$$

$$\sum_{i=1}^{6} x_i \times f(x_i)$$
$$= 0 + 7 + 16 + 18 + 16 + 10$$
$$= 67$$

$$\bar{x} = \frac{\sum_{i=1}^{6} x_i \times f(x_i)}{\sum_{i=1}^{6} f(x_i)} = \frac{67}{30}$$

$$= 2.23 \text{ (to 2 decimal places)}$$

$$\sum_{i=1}^{6} f(x_i) \times (x_i - \bar{x})^2 = 57.367.$$

Use the results to calculate the standard deviation.

$$S_x = \sqrt{\frac{\sum_{i=1}^{6} f(x_i) \times (x_i - \bar{x})^2}{\sum_{i=1}^{6} f(x_i)}} = \sqrt{\frac{57.367}{30}}$$

$S_x = 1.38$ to 2 decimal places

Thus, the standard deviation for the data is 1.38, to 2 decimal places.

12.8 Exercise

B 1 Calculate • the mean $\bar{x}$ • the standard deviation S_x
for each frequency distribution below.

(a)

x_i	$f(x_i)$
0	3
1	7
2	11
3	8
4	6
5	2

(b)

x_i	$f(x_i)$
10	4
11	6
12	8
13	12
14	7
15	2

(c)

x_i	$f(x_i)$
40	3
41	5
42	8
43	9
44	2
45	1

2 For data that are grouped, the
midpoints of the intervals are used
for the values of x_i. Otherwise the
procedure is the same as outlined
earlier. The table shows grouped
data.

Classes	Midpoint m_i	Frequency $f(m_i)$
0 to 5	?	1
5 to 10	?	3
10 to 15	?	8
15 to 20	?	6
20 to 25	?	2

(a) Copy and complete the table.

(b) Calculate $\sum\limits_{i=1}^{5} f(m_i)$ and $\sum\limits_{i=1}^{5} m_i \times f(m_i)$.

(c) Calculate $\bar{x} = \dfrac{\sum\limits_{i=1}^{5} m_i \times f(m_i)}{\sum\limits_{i=1}^{5} f(m_i)}$.

(d) Calculate the standard deviation $S_x = \sqrt{\dfrac{\sum\limits_{i=1}^{5} (m_i - \bar{x})^2 \times f(m_i)}{\sum\limits_{i=1}^{5} f(m_i)}}$

3 For each table showing grouped data
• find the midpoint of each class.
• find the mean $\bar{x}$.
• find the standard deviation S_x.

(a)

Interval	Frequency
10–20	2
20–30	3
30–40	4
40–50	6
50–60	2
60–70	1

(b)

Interval	Frequency
4.0–4.5	1
4.5–5.0	2
5.0–5.5	4
5.5–6.0	6
6.0–6.5	3
6.5–7.0	2

4   The lengths of time (to the nearest minute) required by people in an office pool to type letters are given.

4	7	3	9	8	10	5	10	5	6	8	12	7	9	10
8	3	7	1	6	9	8	5	11	8	6	9	4	7	2

    (a) Group the data in a frequency table.
    (b) Calculate the mean $\bar{x}$, the standard deviation S_x and the range.
    (c) Predict the shape of the histogram. Then construct the histogram for the given data.

5   The time, to the nearest minute, needed to check out groceries at a supermarket is recorded. Twenty persons are sampled.

5	2	7	11	10	1	5	6	4	5	4	5	1	3	9	6	1	8	1	1

    (a) Group the data in a frequency table.
    (b) Calculate the mean $\bar{x}$, the standard deviation S_x and the range.
    (c) Predict the shape of the histogram. Then construct the histogram for the given data.

Math Tip

Very often the mean and standard deviation cannot be computed for the entire population. You will find that the mean, $\bar{x}$, and the standard deviation, S_x, *for a sample* do not exactly have the same value as the mean, μ, and the standard deviation, σ, for *the entire population*. There are many statistical problems in which it is impossible to calculate μ and σ. However there is a connection between x, S_x, and μ and σ. Later, you will explore this connection.

Problem Solving

Often a problem may not appear to have enough information to solve it. Does the following puzzle have enough information?

"George's age, at the time of his death, was $\dfrac{1}{29}$ of the year of his birth. He saw action in the First World War but died before the Great Stock Market Crash. When was George born?"

Applications: Quality Control

In the manufacture or production of merchandise, it is important to check quality at frequent intervals.

6 The heights in centimetres of 12 plants selected at random are shown.

121	117	125	121
118	122	124	113
120	119	122	118

The actual sampling procedures used and the mathematics involved are advanced. However, the following questions illustrate the basic principles involved.

In order that they can be shipped to fill an order, the standard deviation must not be greater than 3.1 cm. What is the decision?

7 Boards from a rail shipment of wood are selected at random and measured. The following results are obtained. (The lengths are in metres.)

4.01   3.96   4.05   3.98   3.92   3.95   4.08   4.03   4.03   3.98

To pass inspection, the standard deviation of the above data must not exceed 4.8 cm. Does the shipment pass inspection?

8 For the shipment in the previous question another sample was taken of 5 boards. (They were measured in metres.)

3.99   4.06   4.01   3.93   4.07

Combine the above data with that given in the previous question. To pass inspection, the standard deviation of the above data must not exceed 4.8 cm. Does the shipment pass inspection?

9 Chocolate covered clusters are chosen randomly and their masses are checked. A sample of 25 clusters have the following masses (in grams).

24.8   25.3   24.8   25.2   24.7   25.8   25.2   25.1   24.5   24.5   24.9   24.6   25.0
24.7   24.3   24.7   25.4   24.9   25.1   24.0   24.6   25.1   24.4   24.8   25.0

(a) Construct a histogram. Use the intervals 24.0–24.2, 24.2–24.4, and so on. The axis should be labelled, masses (in grams).

(b) Calculate the standard deviation.

(c) The standard deviation must be less than 1.75 g in order that production continue. What decision is made for the batch of clusters that were tested in the above sample?

12.9 Interpreting Data: Scatter diagrams

In previous sections, you have displayed data and then used various techniques to analyze the data. Based on your analysis you interpret the data and use the results to help you make better decisions. The more techniques you acquire in analyzing the data, the better you are at interpreting the data and thus making decisions. For example, the coach of the Red Wings invented a sequence of skating drills and assigned various points for the successful completion of each part. Data were collected and recorded as shown for 10 players.

Player	1	2	3	4	5	6	7	8	9	10
Points earned on drills	40	15	28	5	48	18	29	36	43	49
Number of Goals	35	18	20	8	37	10	27	31	30	42

The numbers appear to be related in some way, but are not precisely related. To analyze the data, a scatter diagram can be constructed.

Step 1
Data are recorded as ordered pairs.
(40, 35), (15, 18), (28, 20), (5, 8), (48, 37), (18, 10), (29, 27), (36, 31), (43, 30), (49, 42)

points for drill ⌐ ⌐ goals scored

By plotting the points, the graph shows a simple relationship. A line is drawn through the origin which best fits the data shown by the points. To do so, you estimate the position of the line so that the points lie near or on the line and visually displays the relationship. The data appear to be related linearly. The line is called the **line of best fit**.

Step 2
The ordered pairs are plotted on a graph with axes labelled as shown.

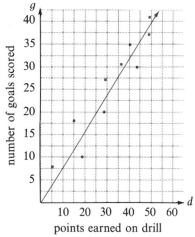

From the graph, an estimate of the slope can be obtained by choosing two points on the line. For example, chose the points (0, 0) and (50, 40).

$$\text{Slope} = \frac{40}{50} \text{ or } 0.8$$

If you use a number of the points on the graph you can use the calculator.

$$\text{Slope} = \frac{\text{Sum of second components}}{\text{Sum of first components}} = \frac{35 + 18 + 20 + 8 + 37 + 10 + 27 + 31 + 30 + }{40 + 15 + 28 + 5 + 48 + 18 + 29 + 36 + 43 + }$$

$$= \frac{258}{311} \text{ or } 0.83 \text{ (to 2 decimal places)}$$

Thus, the variables d and g are related by the equation $g = 0.83d$.

Based on the graph, it appears that a player that scores more points on the drills will score more goals in the season. However, you must be careful in interpreting the data. There are probably a lot of skaters who could score better on the drills, but who would not score goals. There are other factors that play a role in how well a player performs in scoring goals in a game. The scatter diagram displays *only one* of these factors.

Positive (or Direct) Correlation: From the scatter diagram on the previous page, note that as the number of points earned on the skating drills increases, so does the number of goals scored. The data are said to have a positive correlation.

Negative (or Inverse) Correlation: The scatter diagram at the right shows the data related in such a way that when the variable P increases the variable q decreases. Thus the data are said to have a negative correlation.

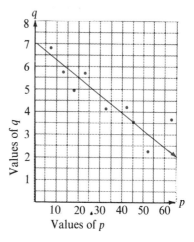

12.9 Exercise

A Review the meaning of scatter diagram, line of best fit. Remember, use a calculator.

1 Refer to the scatter diagram illustrating positive correlation. Estimate the number of goals scored by a player with the following points awarded for the drills.
   (a) 18   (b) 36   (c) 12   (d) 60
   What assumptions did you make in obtaining the answers (a) to (d)?

2 Refer to the scatter diagram illustrating negative correlation.
   (a) Estimate the slope of the line.
   (b) Calculate the slope of the line using the data shown on the diagram.
   (c) Estimate the missing components of A(20, ?)   B(?, 4)   (c) (55, ?).

B Be careful of how you interpret the information provided by a scatter diagram. Questions 3 to 7 are based on the data in the chart.

A coach recorded the total number of points scored and the amount of time that a player was in the game. The data are given in the chart.

Player	3	7	11	15	19	21	27	33	66	99	00
Playing Time (min)	106	74	151	93	172	20	160	127	55	181	200
Points Scored	31	24	37	22	38	10	49	36	11	54	53

3 Construct a scatter diagram to display the data.

4 (a) Sketch a line that best fits the data. Estimate the slope of the line.
(b) Calculate the slope of the line. How does the slope obtained by calculation compare with your answer in (a)?

5 (a) Describe the correlation of the data as positive or negative.
(b) Describe any properties of the relationship you can interpret from the scatter diagram.
(c) List any assumptions you used to answer (b).

6 Use your results to find the missing entries in the table.

	Player	Playing Time	Points Scored
(a)	Chico	123 min	?
(b)	Ashley	?	29
(c)	Jeremy	36 min	?
(d)	Ranch	?	45
(e)	Stenko	185 min	?

7 (a) Create a question based on the information in the chart or scatter diagram.
(b) Write a solution to answer the question in (a). List any assumptions you make in obtaining your answer.

Questions 8 to 10 are based on the following data.

A manufacturer's recorded data regarding the length of employment and the number of production errors that have occurred as follows.

Employee	A	B	C	D	E	F	G	H	I	J
Length of Employment (weeks)	26	13	49	60	5	72	78	46	35	89
Production Errors (class A)	16	22	12	15	27	12	8	17	21	8

8 Construct a scatter diagram to display the data.

9 (a) Sketch a line of best fit. Estimate the slope of the line.
(b) Calculate the slope of the line. How does the slope obtained by calculation compare to your answer in (a).

10 (a) Describe the correlation of the data as direct (positive) or inverse (negative).
(b) Describe any observations you can make based on the display shown by the scatter diagram.
(c) List any assumptions you used to answer (b).

12.10 Problem Solving: *The* Line of Best Fit

Often when you are drawing a line of best fit, it appears that by estimation there is more than one line that could be reasonably drawn. These questions could be asked: "What if I want to fit the best possible line to the data? How can I mathematically decide that the line is indeed the line of "best fit"? Suppose the line of best fit has an equation given as

$$y = mx + b \quad \text{or} \quad y - (mx + b) = 0$$

To establish the following, only 2 points are dealt with, but the procedure applies to any number of points. Points (x_1, y_1) and (x_2, y_2) are not on the line and

$$y_1 - (mx_1 + b) \neq 0,$$
$$y_2 - (mx_2 + b) \neq 0$$

Thus some error, E, occurs since the co-ordinates do not satisfy the equation.

$$E_1 = y_1 - (mx_1 + b), \qquad E_2 = y_2 - (mx_2 + b)$$

In determining the line of best fit, you would like to choose m and b, so that E_1 and E_2 are minimal. To do so, the average of the squared errors is used.

$$E^2 = \frac{1}{2}(E_1{}^2 + E_2{}^2) \quad \text{or in general} \quad E^2 = \sum_{i=1}^{n} \frac{E_i{}^2}{n} \quad \text{where } E_i = y_i - (mx_i + b)$$

The smaller the value of E^2, the better your choice of the line. When E^2 is as small as possible (a minimum value) you have obtained *the* line of best fit defined by $y = Mx + B$, called the **least squares line**.

The previous steps provide some insight into the actual process used to calculate the values of M and B to determine the line of best fit which is as follows.

Step 1 Calculate the mean of the data, in this case the co-ordinates.
$$X = \sum_{i=1}^{n} \frac{x_i}{n}, \quad Y = \sum_{i=1}^{n} \frac{y_i}{n}$$

Step 2 Calculate the mean of the products of the deviation given by

$$D = \frac{(X - x_1)(Y - y_1) + (X - x_2)(Y - y_2) + \cdots + (X - x_n)(Y - y_n)}{n}$$

$$= \sum_{i=1}^{n} \frac{(X - x_i)(Y - y_i)}{n}$$

Step 3 Calculate the square root of the mean of the sums of the squares of the deviations.

$$S_x = \sqrt{\frac{(X - x_1)^2 + (X - x_2)^2 + \cdots + (X - x_n)^2}{n}} = \sqrt{\sum_{i=1}^{n} \frac{(X - x_i)^2}{n}}$$

$$S_y = \sqrt{\frac{(Y - y_1)^2 + (Y - y_2)^2 + \cdots + (Y - y_n)^2}{n}} = \sqrt{\sum_{i=1}^{n} \frac{(Y - y_i)^2}{n}}$$

Step 4 Calculate the value of r, the **correlation coefficient**.
$$r = \frac{D}{S_x S_y}$$

The correlation coefficient, r, is used to calculate the values of M and B.

$$M = r\frac{S_y}{S_x} \quad \text{and} \quad B = Y - MX$$

The value of the correlation coefficient, r, between 2 sets of data lies between 1 and -1 and indicates the degree of linear relationship. It is defined concisely as

$$r = \frac{\sum\limits_{i=1}^{n}(X - x_i)(Y - y_i)}{\sqrt{\sum\limits_{i=1}^{n}(X - x_i)^2 \sum\limits_{i=1}^{n}(Y - y_i)^2}} \quad \text{where} \quad \begin{aligned} X &= \sum\limits_{i=1}^{n}\frac{x_i}{n}, \\ Y &= \sum\limits_{i=1}^{n}\frac{y_i}{n} \end{aligned}$$

The correlation coefficient measures how close the points with co-ordinates $(x_1, y_1), (x_2, y_2), \ldots, (x_n, y_n)$ is to a straight line. If $r = 1$, then the points lie on the line and you can say the correlation is perfect.

12.10 Exercise

A 1 Refer to the data in the scatter diagram.

(a) Estimate a line of best fit. What is the slope of the line?

(b) Calculate the correlation coefficient for the data?

(c) Use your results in (a). What is the equation of the line of best fit? How does your answer compare to that in (a)?

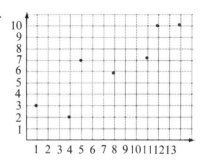

B 2 In an earlier exercise you determined by estimation that the data (d, g) given in the chart were related by the linear relationship, $g = 0.83d$.

d	40	15	28	5	48	18	29	36	43	49
g	35	18	20	8	37	10	27	31	30	42

(a) Calculate the correlation coefficient for the above data. Would you describe the data as having a high, medium or low degree of correlation?

(b) Determine the equation of the line of best fit.

(c) Complete (i)A(35, ?) (ii)B(?, 12) (iii)C(53, ?) for the ordered pairs (d, g).

3 From laboratory research, data were collected for typical rats as shown.

mass of rat (in grams)	122	154	126	119	123	108	112	113	97	101
mass of heart (in milligrams)	882	937	861	1010	791	722	844	689	790	685

(a) Construct a scatter diagram. Label the axes to correspond to the data as shown.

(b) Sketch a line of best fit. Write its equation.

(c) Calculate the correlation coefficient. Interpret this result.

(d) Determine the line of best fit.

(e) Predict A(135, ?)   B(?, 915)   C(105, ?) for the laboratory research.

$$(122, 882)$$
$$\uparrow \qquad \uparrow$$
$$\text{mass of rat} \quad \text{mass of heart}$$

Test for Practice

1 (a) Use an example to illustrate the three main steps when dealing with statistical information.
(b) Use an example to show the meaning of *sample* and *population.*

2 Use the table of random digits at the back of the book. Suggest a method of choosing a captain.

3 List 2 methods of collecting data. Indicate the advantages and disadvantages of each method.

4 The temperatures of 18 randomly chosen cities from the Atlantic Provinces are given.

26°C 20°C 20°C 22°C 26°C 28°C 23°C 26°C 24°C
25°C 31°C 30°C 26°C 25°C 27°C 30°C 18°C 31°C

(a) Find the mean, median, and mode.
(b) Which of the answers in (a) is the best measure of central tendency for the data? Give reasons for your answer.

5 The data for the life length in days, of a fibre, randomly sampled were

222 234 248 264 254 238 272 238 236 218

(a) Calculate the mean, $\bar{x}$.   (b) Calculate the standard deviation (in days).
(c) How many fibres had a life length within 1 standard deviation of the mean?
(d) How many fibres out of 10 000 would be within 1 standard deviation of the mean?

6 The Acme rent-a-car company has a hundred cars to service and it estimates that it costs a mean of 8.8¢/km (with a standard deviation of 1.2¢) to service each car. Assuming the costs are normally distributed, find the expected number of cars whose servicing per kilometre costs
(a) between 8¢ and 9¢.   (b) between 6¢ and 7¢.
(c) less than 6¢.   (d) more than 10¢.

7 In a medical experiment, the time taken in hours to complete two medical procedures was recorded to the nearest hour for 12 samples.

Procedure A	9	7	18	14	10	12	16	12	16	8,	12	12
Procedure B	10	4	15	11	6	10	13	7	7	5	8	6

(a) Construct a scatter diagram.   (b) Sketch a line of best fit.
(c) Calculate the correlation coefficient.   (d) Interpret your result in (c).

Looking Ahead: Applications with Normal Distribution

To make predictions and solve problems based on data, the data are grouped by various means: tables, charts, frequency distributions, graphs, and so on.

Each of the following frequency distributions has compressed the information by classifying the data so that predictions based on the data can be made more easily. Each of the following sets of data seems to "cluster" at some central point

- How are these three frequency distributions alike?

- How are these three frequency distributions different?

Graph A

Graph B

Graph C

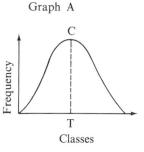

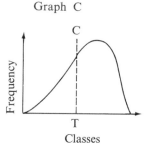

Graph A: This frequency distribution curve is symmetrical. (CT seems to be an axis of symmetry.) Note how the data are evenly distributed. This type of distribution is often called a **normal distribution** and the curve is often called a **bell curve**.

Graph B: Compared to Graph A, this frequency distribution curve is not symmetrical. Such a distribution is said to be "skewed to the left". Most of the data are below or to the left of CT.

Graph C: Compared to Graph A, this frequency distribution curve is not symmetrical. Such a distribution is said to be "skewed to the right". Most of the data are above or to the right of CT.

To test the effectiveness of new packaging of a product a manufacturer wants to determine the distribution of the ages of the persons that now use the product. To do so, some samples are taken from the population. The mean, $\bar{x}$, and standard deviation, S_x, are calculated for each sample. The population has its own mean, denoted by μ, and standard deviation, denoted by σ.

	Sample	Population	
Mean	$\bar{x}$	μ	called *mu*
Standard Deviation	S_x	σ	called *sigma*

A histogram and a frequency polygon or curve can be drawn for the data from any sample. If the results of more random samples are taken and the intervals made smaller, the graph approaches a shape called a bell curve, which is symmetrical about its mean μ. The type of distribution shown by a bell curve is named the **normal distribution**.

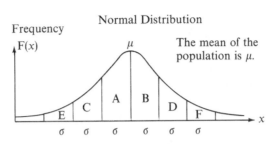

Normal Distribution

Frequency

The mean of the population is μ.

Each σ represents one standard deviation.

The normal distribution has certain properties, which are given as follows:

The values in section A and B are *within one standard deviation* (σ) *from the mean* μ. The values in sections C and D are *greater than one but less than two standard deviations from the mean.*

For a normal distribution, the data are distributed as follows.

- 68% of the data are within *one standard deviation* of the mean. In other words, values from A and B represent 68% of the population.

- 95% of the data are within *two standard deviations* of the mean. In other words, values from A, B, C, and D represent 95% of the population.

- 99.7% of the data are within *three standard deviations* of the mean. In other words, values from A, B, C, D, E, and F represent 99.7% of the population.

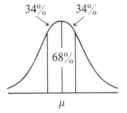

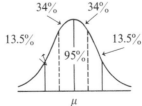

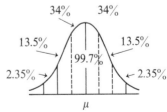

Economic problems in business and industry are tackled sometimes by examining their statistical properties. Answers to questions such as, "Do the data form a normal distribution?" "What are the means and standard deviations?", help trouble-shooters to attack the problems.

Example   The amounts of soap used daily by a car wash terminal yield a normal distribution with a mean of 150 kg and a standard deviation of 20 kg. On what per cent of the days will a terminal use

(a) less than 130 kg of soap?   (b) more than 110 kg of soap?

Solution   Draw the normal distribution curve and label it.

Plan the construction of the normal distribution carefully.

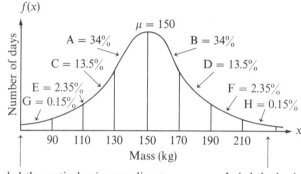

Per cent distributions are rounded to 2 decimal places.

Label the vertical axis according to the normal distribution of data.

Label the horizontal axis using $\mu = 150$ and $\sigma = 20$.

Answer the question.

(a) The per cent of days the car wash uses less than 130 kg of soap is given by the values in C, E, and G on the graph.

The required per cent is $13.5\% + 2.35\% + 0.15\% = 16\%$

Thus on 16% of the days, less than 130 kg of soap are used.

(b) The per cent of days the car wash uses more than 110 kg of soap is given by values in C, A, B, D, F, and H. The required per cent is

$$13.5\% + 34\% + 34\% + 13.5\% + 2.35\% + 0.15\% = 97.5\%$$

Thus, on 97.5% of the days, more than 110 kg of soap are used.

The mean and standard deviation completely specify a normal distribution. Because populations have different means and standard deviations, their curves will not be exactly the same. However, all normal distribution curves are bell-shaped.

Exercise

A 1 Use the curve of normal distribution, with mean 75 min and standard deviation 4.5 min to answer the following questions.

Think: This means $\sigma = 4.5$ min.

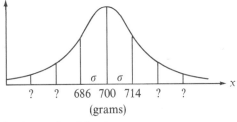

(a) What are the missing values on the horizontal axis?

(b) What per cent of the data is within 1 standard deviation of the mean?

(c) What per cent of the data are greater than 66 min?

2 A sample is taken. A curve of normal distribution shows the data for the masses of boxes of biscuits.

(a) What per cent of the boxes of biscuits have a mass within 1 standard deviation of the mean μ?

(b) What is the standard deviation?

(c) What are the missing values on the horizontal axis?

(d) The quality control regulations require that not more than 3% of the boxes have a mass of less than 672 g. Do the above boxes meet the required regulations?

3 A population has a normal distribution with mean μ and standard deviation σ. What per cent of the population lie in the intervals with the limits shown?

(a) $\mu, \mu + \sigma$ 　　(b) $\mu - 2\sigma, \mu$ 　　(c) $\mu - \sigma, \mu + \sigma$

(d) $\mu - 2\sigma, \mu + 2\sigma$ 　(e) $\mu - 3\sigma, \mu + 3\sigma$ 　(f) $\mu - \sigma, \mu + 2\sigma$

(g) $\mu + \sigma, \mu + 2\sigma$ 　(h) $\mu - 3\sigma, \mu$ 　　(i) $\mu - 2\sigma, \mu + \sigma$

B Questions 4 to 6 are based on the following information.

For fifty days the pollution counts of an industrial centre were recorded.

11	15	7	16	19	14	11	20	15	22	26	22	14	15	16	10	18
17	12	19	24	13	19	16	17	16	13	10	15	16	14	17	15	18
18	20	17	18	21	17	13	21	19	25	18	12	9	14	16	23	

4 (a) Decide on a method of choosing a sample, A, of ten numbers randomly from the population in the table.

(b) Calculate the mean, $\bar{x}$, and the standard deviation, S_x, for the sample.

5 (a) Choose another sample, B, of ten numbers randomly from the given population.
   (b) Calculate the mean, $\bar{x}$, for the sample.
   (c) Calculate the standard deviation, S_x, for the sample.

6 (a) Calculate the mean, μ, of the given population.
   (b) Compare the value of μ with the means, $\bar{x}$, calculated for the samples. What do you notice?
   (c) Calculate the standard deviation, σ.
   (d) Compare the value σ with the standard deviations calculated for the samples. What do you notice?

7 To determine whether to harvest the wheat, 50 samples are taken in the field. The height (in centimetres) is measured and recorded.

72	185	159	86	119	96	57	108	53	125	83	125	98
116	71	98	168	144	138	24	117	68	155	138	98	125
53	119	150	144	156	146	95	116	180	53	117	72	
108	110	72	98	164	74	59	143	159	189	27	107	

   (a) For the above sample, the mean is close to the mean, μ, of the population. What is the value of μ?
   (b) Calculate the standard deviation of the sample.
   (c) Of 1000 samples, how many would be within two standard deviations of the mean?

8 For packaged chocolates, the mean mass of each piece is to be 40 g. The standard deviation is 1.5 g.
   (a) If for a production run, 10 000 chocolates were produced, how many should be within 1.5 g of the required mass?
   (b) How many chocolates would be rejected if the chocolates must have a mass between 37 g and 43 g?

9 From past records, it is found that the life of a hair dryer used every day is about 6.5 years. The data are normally distributed with a standard deviation of 1.5 years. A retail store bought 5000 hair dryers and offered a two-year guarantee. How many dryers will the store expect to replace?

10 The mean mass of game fish in a certain locale is determined to be 2.5 kg and the standard deviation is 0.75 kg. There are about 1000 game fish available.
   (a) How many are between 1 kg and 4 kg?
   (b) Suppose a fish with a mass of less than 1.75 kg must be thrown back. How many of the above fish cannot be kept if caught?

Year-End Review

1. β is an angle and has its terminal arm in the first quadrant. If $\sec \beta = \frac{5}{3}$, find the value of $\sec \beta + \sin \beta + \cos \beta - \tan \beta$.

2. (a) For the function defined by $f(x) = 3(x - 2)^2 + 5$, find each of the following: $f(2 - h)$, $f(2 + h)$.
   (b) Use the result in (a) to sketch the graph of f.

3. Solve.
   (a) $x^3 - 3x^2 - x + 3 = 0$ (b) $x^3 + 5x^2 + 7x + 3 = 0$
   (c) $x^3 - 4x^2 - 9x + 36 = 0$ (d) $2x^3 + 5x^2 - 4x - 3 = 0$

4. From the diagram, prove $\angle A = 60°$.

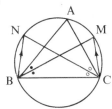

5. For each of the following, describe the method or type of sampling you might use when obtaining a sample.
   (a) testing the quality of grapefruit
   (b) deciding how many people would purchase from a discount store
   (c) deciding how many people would be interested in installing a swimming pool
   (d) obtaining an opinion about recent car repair prices
   (e) determining the most popular TV star
   (f) checking the quality of tires for a car

6. For each triangle, find the measure of the parts indicated.
   (a) 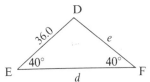   (b)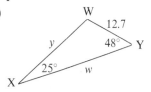

7. Find the equation of the conic section with centre at $(0, 0)$ given its properties as follows: one vertex at $(0, 4)$ and one focus at $(0, 5)$.

8. For $\triangle TAC$, $TA = TC$. On TA there is a point Z and on TC point Q so that $ZQ \parallel AC$. Prove $\triangle TZQ$ is isosceles.

9   For the diagram, $\triangle PQR$ is isosceles with $PQ = PR$. If $QR \parallel ST$ prove that $QS = RT$.

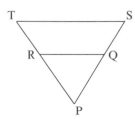

10 (a) In $\triangle ABC$, $a = 5$, $b = 11$, and $c = 13$. Find cos B.
    (b) In $\triangle PQR$, $p = 7$, $q = 5$, and $r = 3$. Find $\angle P$.

11 (a) A point is in the first quadrant and has the property that the product of its co-ordinates is 24. Draw the locus of all such points.
    (b) Find the locus of all points on the Cartesian plane that have the property that the product of their co-ordinates is 24.

12 Sketch the graph of each of the following. Indicate the period, amplitude and phase shift.
    (a) $y = \dfrac{1}{2} \sin 2\left(\theta - \dfrac{\pi}{3}\right)$ (b) $y = 2 \cos \left(2\theta - \dfrac{\pi}{2}\right)$ (c) $y = \tan \left(2\theta - \dfrac{\pi}{3}\right)$

13 A small plane takes off from Edmonton, Alberta with a heading of N32°E. Another plane takes off from North Battleford, Saskatchewan with a heading of N36°W. The paths of the planes cross at point A. If Edmonton is 350.0 km due west of North Battleford, how far is the crossing point from North Battleford?

14 Solve.
    (a) $3^{2x} - 30(3^x) = -81$ (b) $(2x^2 - x)^2 - 7(2x^2 - x) + 6 = 0$

15 Simplify.
    (a) $4y\sqrt{20} - 3y\sqrt{45} + 2x\sqrt{125}$ (b) $-4x\sqrt{27} - 2y\sqrt{12} + 2x\sqrt{48} + 3y\sqrt{75}$

16 The number of hours of practice by various teams are listed below.
        Barons 14.0   Rams   9.1   Raiders 9.5   Warriors 8.5
        Braves 9.1   Hawks 7.8   Spartans 8.3   Bulls   8.3
    (a) Find the mode. Why is the mode not a good measure of central tendency for the above data?
    (b) Which team is not represented well by the mean?
    (c) What is the median?
    (d) What is more representative of the centre of the data: the mean, the median, or the mode?

17 Simplify.

(a) $\dfrac{8^{\frac{2}{3}} + 8^{\frac{2}{3}}}{8^{\frac{1}{3}} - 8^{\frac{2}{3}}}$
(b) $\dfrac{81^{\frac{1}{4}} + 81^{\frac{1}{4}}}{81^{\frac{3}{4}} - 81^{\frac{1}{4}}}$
(c) $\dfrac{125^{\frac{2}{3}} + \left(\frac{1}{5}\right)^{-2}}{2^{0}(\sqrt[4]{625})^{2}}$

18 A parabola has its vertex at $(0, 0)$ and passes through the point $(\frac{1}{2}, -2)$.
If the axis is vertical, find
(a) the defining equation and the equation of the directrix.
(b) the co-ordinates of the focus.

19 Solve for $-2\pi \leq \theta \leq 2\pi$

(a) $\cos 2\theta = -\dfrac{1}{2}$
(b) $4 \cos^2 \theta = 1$
(c) $\cos^2 \theta + \cos \theta = 0$

(d) $(1 - \tan \theta)(1 - \cos \theta) = 0$

20 (a) Solve the system, $x + y = 3$ and $y + 13 = x^2 - 8x$.
(b) Illustrate your answer in (a) graphically.

21 In isosceles triangle PRT, $RP = TP$. If a line, drawn parallel to the base,
meets PT and PR at Q and S respectively, prove $\triangle PQS$ is also isosceles.

22 Evergreens sampled from a field had the following heights, in centimetres.
        169   167   171   175   171   172   173   167   169   163
(a) Calculate the standard deviation.
(b) How many trees are within one standard deviation of the mean?

23 From R, a point outside the circle, a line RPQ, meets the circle at P and
Q, respectively. Another line, REC, meets the circle at E and C,
respectively. If RST is tangent to the circle at S and CS $\|$ RQ, then prove
that $\angle RSE = \angle CRQ$.

24 Prove that
(a) $(1 - \sec \theta)(1 + \sec \theta) = -\tan^2 \theta$
(b) $\csc \theta(1 + \sin \theta) = 1 + \csc \theta$

25 Sketch the locus of all points, $P(x, y)$, such that the sum of the distances
from P to $F_1(-3, 0)$ and $F_2(3, 0)$ is 7.

26 What is the equation of each conic for the given properties?
(a) The foci are at $(0, 2)$, $(0, -6)$; the sum of the focal radii is 12.
(b) The foci are at $(0, -4)$, $(0, 6)$; the difference of the focal radii is 8.

27 A small tree is supported at a point on its trunk by two lengths of wood.
The lengths are in the ratio $7:6$. The angle of elevation of the shorter
length is $70°$. Calculate the angle of elevation of the other length of wood.

Exponential Table: $y = 10^x$

x	0	1	2	3	4	5	6	7	8	9
0.00	1.000	1.002	1.005	1.007	1.009	1.012	1.014	1.016	1.019	1.021
0.01	1.023	1.026	1.028	1.030	1.033	1.035	1.038	1.040	1.042	1.045
0.02	1.047	1.050	1.052	1.054	1.057	1.059	1.062	1.064	1.067	1.069
0.03	1.072	1.074	1.076	1.079	1.081	1.084	1.086	1.089	1.091	1.094
0.04	1.096	1.099	1.102	1.104	1.107	1.109	1.112	1.114	1.117	1.119
0.05	1.122	1.125	1.127	1.130	1.132	1.135	1.138	1.140	1.143	1.146
0.06	1.148	1.151	1.153	1.156	1.159	1.161	1.164	1.167	1.169	1.172
0.07	1.175	1.178	1.180	1.183	1.186	1.189	1.191	1.194	1.197	1.199
0.08	1.202	1.205	1.208	1.211	1.213	1.216	1.219	1.222	1.225	1.227
0.09	1.230	1.233	1.236	1.239	1.242	1.245	1.247	1.250	1.253	1.256
0.10	1.259	1.262	1.265	1.268	1.271	1.274	1.276	1.279	1.282	1.285
0.11	1.288	1.291	1.294	1.297	1.300	1.303	1.306	1.309	1.312	1.315
0.12	1.318	1.321	1.324	1.327	1.330	1.334	1.337	1.340	1.343	1.346
0.13	1.349	1.352	1.355	1.358	1.361	1.365	1.368	1.371	1.374	1.377
0.14	1.380	1.384	1.387	1.390	1.393	1.396	1.400	1.403	1.406	1.409
0.15	1.413	1.416	1.419	1.422	1.426	1.429	1.432	1.435	1.439	1.442
0.16	1.445	1.449	1.452	1.455	1.459	1.462	1.466	1.469	1.472	1.476
0.17	1.479	1.483	1.486	1.489	1.493	1.496	1.500	1.503	1.507	1.510
0.18	1.514	1.517	1.521	1.524	1.528	1.531	1.535	1.538	1.542	1.545
0.19	1.549	1.552	1.556	1.560	1.563	1.567	1.570	1.574	1.578	1.581
0.20	1.585	1.589	1.592	1.596	1.600	1.603	1.607	1.611	1.614	1.618
0.21	1.622	1.626	1.629	1.633	1.637	1.641	1.644	1.648	1.652	1.656
0.22	1.660	1.663	1.667	1.671	1.675	1.679	1.683	1.687	1.690	1.694
0.23	1.698	1.702	1.706	1.710	1.714	1.718	1.722	1.726	1.730	1.734
0.24	1.738	1.742	1.746	1.750	1.754	1.758	1.762	1.766	1.770	1.774
0.25	1.778	1.782	1.786	1.791	1.795	1.799	1.803	1.807	1.811	1.816
0.26	1.820	1.824	1.828	1.832	1.837	1.841	1.845	1.849	1.854	1.858
0.27	1.862	1.866	1.871	1.875	1.879	1.884	1.888	1.892	1.897	1.901
0.28	1.905	1.910	1.914	1.919	1.923	1.928	1.932	1.936	1.941	1.945
0.29	1.950	1.954	1.959	1.963	1.968	1.972	1.977	1.982	1.986	1.991
0.30	1.995	2.000	2.004	2.009	2.014	2.018	2.023	2.028	2.032	2.037
0.31	2.042	2.046	2.051	2.056	2.061	2.065	2.070	2.075	2.080	2.084
0.32	2.089	2.094	2.099	2.104	2.109	2.113	2.118	2.123	2.128	2.133
0.33	2.138	2.143	2.148	2.153	2.158	2.163	2.168	2.173	2.178	2.183
0.34	2.188	2.193	2.198	2.203	2.208	2.213	2.218	2.223	2.228	2.234
0.35	2.239	2.244	2.249	2.254	2.259	2.265	2.270	2.275	2.280	2.286
0.36	2.291	2.296	2.301	2.307	2.312	2.317	2.323	2.328	2.333	2.339
0.37	2.344	2.350	2.355	2.360	2.366	2.371	2.377	2.382	2.388	2.393
0.38	2.399	2.404	2.410	4.415	2.421	2.427	2.432	2.438	2.443	2.449
0.39	2.455	2.460	2.466	2.472	2.477	2.483	2.489	2.495	2.500	2.506
0.40	2.512	2.518	2.523	2.529	2.535	2.541	2.547	2.553	2.559	2.564
0.41	2.570	2.576	2.582	2.588	2.594	2.600	2.606	2.612	2.618	2.624
0.42	2.630	2.636	2.642	2.649	2.655	2.661	2.667	2.673	2.679	2.685
0.43	2.692	2.698	2.704	2.710	2.716	2.723	2.729	2.735	2.742	2.748
0.44	2.754	2.761	2.767	2.773	2.780	2.786	2.793	2.799	2.805	2.812
0.45	2.818	2.825	2.831	2.838	2.844	2.851	2.858	2.864	2.871	2.877
0.46	2.884	2.891	2.897	2.904	2.911	2.917	2.924	2.931	2.938	2.944
0.47	2.951	2.958	2.965	2.972	2.979	2.985	2.992	2.999	3.006	3.013
0.48	3.020	3.027	3.034	3.041	3.048	3.055	3.062	3.069	3.076	3.083
0.49	3.090	3.097	3.105	3.112	3.119	3.126	3.133	3.141	3.148	3.155

Exponential Table: $y = 10^x$ continued . . .

x	0	1	2	3	4	5	6	7	8	9
0.50	3.162	3.170	3.177	3.184	3.192	3.199	3.206	3.214	3.221	3.228
0.51	3.236	3.243	3.251	3.258	3.266	3.273	3.281	3.289	3.296	3.304
0.52	3.311	3.319	3.327	3.334	3.342	3.350	3.357	3.365	3.373	3.381
0.53	3.388	3.396	3.404	3.412	3.420	3.428	3.436	3.443	3.451	3.459
0.54	3.467	3.475	3.483	3.491	3.499	3.508	3.516	3.524	3.532	3.540
0.55	3.548	3.556	3.565	3.573	3.581	3.589	3.597	3.606	3.613	3.622
0.56	3.631	3.639	3.648	3.656	3.664	3.673	3.681	3.690	3.698	3.707
0.57	3.715	3.724	3.733	3.741	3.750	3.758	3.767	3.776	3.784	3.793
0.58	3.802	3.811	3.819	3.828	3.837	3.846	3.855	3.864	3.873	3.882
0.59	3.890	3.899	3.908	3.917	3.926	3.936	3.945	3.954	3.963	3.972
0.60	3.981	3.990	3.999	4.009	4.018	4.027	4.036	4.046	4.055	4.064
0.61	4.074	4.083	4.093	4.102	4.111	4.121	4.130	4.140	4.150	4.159
0.62	4.169	4.178	4.188	4.198	4.207	4.217	4.227	4.236	4.246	4.256
0.63	4.266	4.276	4.285	4.295	4.305	4.315	4.325	4.335	4.345	4.355
0.64	4.365	4.375	4.385	4.395	4.406	4.416	4.426	4.436	4.446	4.457
0.65	4.467	4.477	4.487	4.498	4.508	4.519	4.529	4.539	4.550	4.560
0.66	4.571	4.581	4.592	4.603	4.613	4.624	4.634	4.645	4.656	4.667
0.67	4.677	4.688	4.699	4.710	4.721	4.732	4.742	4.753	4.764	4.775
0.68	4.786	4.797	4.808	4.819	4.831	4.842	4.853	4.864	4.875	4.887
0.69	4.898	4.909	4.920	4.932	4.943	4.955	4.966	4.977	4.989	5.000
0.70	5.012	5.023	5.035	5.047	5.058	5.070	5.082	5.093	5.105	5.117
0.71	5.129	5.140	5.152	5.164	5.176	5.188	5.200	5.212	5.224	5.236
0.72	5.248	5.260	5.272	5.284	5.297	5.309	5.321	5.333	5.346	5.358
0.73	5.370	5.383	5.395	5.408	5.420	5.433	5.445	5.458	5.470	5.483
0.74	5.495	5.508	5.521	5.534	5.546	5.559	5.572	5.585	5.598	5.610
0.75	5.623	5.636	5.649	5.662	5.675	5.689	5.702	5.715	5.728	5.741
0.76	5.754	5.768	5.781	5.794	5.808	5.821	5.834	5.848	5.861	5.875
0.77	5.888	5.902	5.916	5.929	5.943	5.957	5.970	5.984	5.998	6.012
0.78	6.026	6.039	6.053	6.067	6.081	6.095	6.109	6.124	6.138	6.152
0.79	6.166	6.180	6.194	6.209	6.223	6.236	6.252	6.266	6.281	6.295
0.80	6.310	6.324	6.339	6.353	6.368	6.383	6.397	6.412	6.427	6.442
0.81	6.457	6.471	6.486	6.501	6.516	6.531	6.546	6.561	6.577	6.592
0.82	6.607	6.622	6.637	6.653	6.668	6.683	6.699	6.714	6.730	6.745
0.83	6.761	6.776	6.792	6.808	6.823	6.839	6.855	6.871	6.887	6.902
0.84	6.918	6.934	6.950	6.966	6.982	6.998	7.015	7.031	7.047	7.063
0.85	7.079	7.096	7.112	7.129	7.145	7.161	7.178	7.194	7.211	7.228
0.86	7.244	7.261	7.278	7.295	7.311	7.328	7.345	7.362	7.379	7.396
0.87	7.413	7.430	7.447	7.464	7.482	7.499	7.516	7.534	7.551	7.568
0.88	7.586	7.603	7.621	7.638	7.656	7.674	7.691	7.709	7.727	7.745
0.89	7.762	7.780	7.798	7.816	7.834	7.852	7.870	7.889	7.907	7.925
0.90	7.943	7.962	7.980	7.998	8.017	8.035	8.054	8.072	8.091	8.110
0.91	8.128	8.147	8.166	8.185	8.204	8.222	8.241	8.260	8.279	8.299
0.92	8.318	8.337	8.356	8.375	8.395	8.414	8.433	8.453	8.472	8.492
0.93	8.511	8.531	8.551	8.570	8.590	8.610	8.630	8.650	8.670	8.690
0.94	8.710	8.730	8.750	8.770	8.790	8.810	8.831	8.851	8.872	8.892
0.95	8.913	8.933	8.954	8.974	8.995	9.016	9.036	9.057	9.078	9.099
0.96	9.120	9.141	9.162	9.183	9.204	9.226	9.247	9.268	9.290	9.311
0.97	9.333	9.354	9.376	9.397	9.419	9.441	9.462	9.484	9.506	9.528
0.98	9.550	9.572	9.594	9.616	9.638	9.661	9.683	9.705	9.727	9.750
0.99	9.772	9.795	9.817	9.840	9.863	9.886	9.908	9.931	9.954	9.977

Table of Trigonometric Functions

ANGLE deg	sin	cos	tan	csc	sec	cot	ANGLE deg	sin	cos	tan	csc	sec	cot
0	0.0000	1.0000	0.0000	undefined	1.0000	undefined	45	0.7071	0.7071	1.0000	1.4142	1.4142	1.0000
1	0.0175	0.9998	0.0175	57.299	1.0002	57.2900	46	0.7193	0.6947	1.0355	1.3901	1.4396	0.9657
2	0.0349	0.9994	0.0349	28.654	1.0006	28.6363	47	0.7314	0.6820	1.0724	1.3673	1.4663	0.9325
3	0.0523	0.9986	0.0524	19.107	1.0014	19.0811	48	0.7431	0.6691	1.1106	1.3456	1.4945	0.9004
4	0.0698	0.9976	0.0699	14.336	1.0024	14.3007	49	0.7547	0.6561	1.1504	1.3250	1.5243	0.8693
5	0.0872	0.9962	0.0875	11.474	1.0038	11.4301	50	0.7660	0.6428	1.1918	1.3054	1.5557	0.8391
6	0.1045	0.9945	0.1051	9.5668	1.0055	9.5144	51	0.7771	0.6293	1.2349	1.2868	1.5890	0.8098
7	0.1219	0.9925	0.1228	8.2055	1.0075	8.1443	52	0.7880	0.6157	1.2799	1.2690	1.6243	0.7813
8	0.1392	0.9903	0.1405	7.1853	1.0098	7.1154	53	0.7986	0.6018	1.3270	1.2521	1.6616	0.7536
9	0.1564	0.9877	0.1584	6.3925	1.0125	6.3138	54	0.8090	0.5878	1.3764	1.2361	1.7013	0.7265
10	0.1736	0.9848	0.1763	5.7588	1.0154	5.6713	55	0.8192	0.5736	1.4281	1.2208	1.7435	0.7002
11	0.1908	0.9816	0.1944	5.2408	1.0187	5.1446	56	0.8290	0.5592	1.4826	1.2062	1.7883	0.6745
12	0.2079	0.9781	0.2126	4.8097	1.0223	4.7046	57	0.8387	0.5446	1.5399	1.1924	1.8361	0.6494
13	0.2250	0.9744	0.2309	4.4454	1.0263	4.3315	58	0.8480	0.5299	1.6003	1.1792	1.8871	0.6249
14	0.2419	0.9703	0.2493	4.1336	1.0306	4.0108	59	0.8572	0.5150	1.6643	1.1666	1.9416	0.6009
15	0.2588	0.9659	0.2679	3.8637	1.0353	3.7321	60	0.8660	0.5000	1.7321	1.1547	2.0000	0.5774
16	0.2756	0.9613	0.2867	3.6280	1.0403	3.4874	61	0.8746	0.4848	1.8040	1.1434	2.0627	0.5543
17	0.2924	0.9563	0.3057	3.4208	1.0457	3.2709	62	0.8829	0.4695	1.8807	1.1326	2.1301	0.5317
18	0.3090	0.9511	0.3249	3.2361	1.0515	3.0777	63	0.8910	0.4540	1.9626	1.1223	2.2027	0.5095
19	0.3256	0.9455	0.3443	3.0716	1.0576	2.9042	64	0.8988	0.4384	2.0503	1.1126	2.2812	0.4877
20	0.3420	0.9397	0.3640	2.9238	1.0642	2.7475	65	0.9063	0.4226	2.1445	1.1034	2.3662	0.4663
21	0.3584	0.9336	0.3839	2.7904	1.0711	2.6051	66	0.9135	0.4067	2.2460	1.0946	2.4586	0.4452
22	0.3746	0.9272	0.4040	2.6695	1.0785	2.4751	67	0.9205	0.3907	2.3559	1.0864	2.5593	0.4245
23	0.3907	0.9205	0.4245	2.5593	1.0864	2.3559	68	0.9272	0.3746	2.4751	1.0785	2.6695	0.4040
24	0.4067	0.9135	0.4452	2.4586	1.0946	2.2460	69	0.9336	0.3584	2.6051	1.0712	2.7904	0.3839
25	0.4226	0.9063	0.4663	2.3662	1.1034	2.1445	70	0.9397	0.3420	2.7475	1.0642	2.9238	0.3640
26	0.4384	0.8988	0.4877	2.2812	1.1126	2.0503	71	0.9455	0.3256	2.9042	1.0576	3.0716	0.3443
27	0.4540	0.8910	0.5095	2.2027	1.1223	1.9626	72	0.9511	0.3090	3.0777	1.0515	3.2361	0.3249
28	0.4695	0.8829	0.5317	2.1301	1.1326	1.8807	73	0.9563	0.2924	3.2709	1.0457	3.4203	0.3057
29	0.4848	0.8746	0.5543	2.0627	1.1434	1.8040	74	0.9613	0.2756	3.4874	1.0403	3.6280	0.2867
30	0.5000	0.8660	0.5774	2.0000	1.1547	1.7321	75	0.9659	0.2588	3.7321	1.0353	3.8637	0.2679
31	0.5150	0.8572	0.6009	1.9416	1.1667	1.6643	76	0.9703	0.2419	4.0108	1.0306	4.1336	0.2493
32	0.5299	0.8480	0.6249	1.8871	1.1792	1.6003	77	0.9744	0.2250	4.3315	1.0263	4.4454	0.2309
33	0.5446	0.8387	0.6494	1.8361	1.1924	1.5399	78	0.9781	0.2079	4.7046	1.0223	4.8097	0.2126
34	0.5592	0.8290	0.6745	1.7883	1.2062	1.4826	79	0.9816	0.1908	5.1446	1.0187	5.2408	0.1944
35	0.5736	0.8192	0.7002	1.7435	1.2208	1.4281	80	0.9848	0.1736	5.6713	1.0154	5.7588	0.1763
36	0.5878	0.8090	0.7265	1.7013	1.2361	1.3764	81	0.9877	0.1564	6.3138	1.0125	6.3925	0.1584
37	0.6018	0.7986	0.7536	1.6616	1.2521	1.3270	82	0.9903	0.1392	7.1154	1.0098	7.1853	0.1405
38	0.6157	0.7880	0.7813	1.6243	1.2690	1.2799	83	0.9925	0.1219	8.1443	1.0075	8.2055	0.1228
39	0.6293	0.7771	0.8098	1.5890	1.2868	1.2349	84	0.9945	0.1045	9.5144	1.0055	9.5668	0.1051
40	0.6428	0.7660	0.8391	1.5557	1.3054	1.1918	85	0.9962	0.0872	11.4301	1.0038	11.474	0.0875
41	0.6561	0.7547	0.8693	1.5243	1.3250	1.1504	86	0.9976	0.0698	14.3007	1.0024	14.336	0.0699
42	0.6691	0.7431	0.9004	1.4945	1.3456	1.1106	87	0.9986	0.0523	19.0811	1.0014	19.107	0.0524
43	0.6820	0.7314	0.9325	1.4663	1.3673	1.0724	88	0.9994	0.0349	28.6363	1.0006	28.654	0.0349
44	0.6947	0.7193	0.9657	1.4396	1.3902	1.0355	89	0.9998	0.0175	57.2900	1.0002	57.299	0.0175
45	0.7071	0.7071	1.0000	1.4142	1.4142	1.0000	90	1.0000	0.0000	undefined	1.0000	undefined	0.0000

nswers to Chapter Reviews, Chapter Tests,
1aintaining Skills, Cumulative Reviews and Year End
.eview can be found in the Teachers Edition of this text.

nventory of Essential Skills, pages 9–16
kills with Integers, page 9
.a)-16 b)34 c)390 d)-126 e)1512 2.a)2 b) $-\dfrac{15}{32}$ c)$\dfrac{16}{9}$

)-1 e)-27 f)$-\dfrac{3}{8}$ 3.a)$\dfrac{7}{6}$ b)$-\dfrac{9}{10}$ c)$\dfrac{4}{3}$ d)$-\dfrac{1}{64}$ e)$-\dfrac{1}{18}$ f)$\dfrac{3}{8}$

.a)-3 b)14 c)7 d)29 e)-6 f)-19 g)-7 h)-1 i)-1
.a)70 b)60

'olynomials, page 10
.a)$8y - 26$ b)$52x - 4$ c)$y^2 + 4y$ d)$6x - 2x^2$
)$4x^2 + 5xy - 3y^2$ f)$-5a^2 + 4ab - 98b^2$
.a)$-x^2 + 2x + 48$ b)$-3 + 16x - 2x^2$ 3.a)$-x^2 - 6$
)$y^2 + 28y + 52$ c)$14 - 85y - 51y^2 - 12y^3$
.a)$x^2(4x + 1)$ b)$(3x^2 + 4y)(3x^2 - 4y)$ c)$(x + 4)(x + 5)$
)$(y - 6)(y - 7)$ e)$(1 + x)(5 + x)$
)$(x + 2\sqrt{2})(x - 2\sqrt{2})(x^2 + 8)$ g)$(x + 2)(x - 2)(x^2 + 1)$
)$(y^2 + 3)^2$ i)$(x + \sqrt{3})(x - \sqrt{3})(x^2 + 5)$
)$(x^2 + 15y)(x^2 - 15y)$
)$(3x - y)(x - 3y)$ 5.a)$-$ b)$-$

)$+$ d)$-$ e)$+$ f)$+$ 6.a)$\dfrac{(x - 2)}{(2x + 1)}$ b)$\dfrac{1}{(a - 1)(a + 3)}$

)$(x - 4)^2 + 16$ d)$\dfrac{-2(a + 8)}{(a + 3)(a - 3)}$ 7.a)1 b)$-\dfrac{91}{90}$

iolving Equations, page 11
.a)41 b)44 2.a)2 b)-2 c)11 d)5 e)6 3.a)1 b)0

.a)$y = \dfrac{5}{2}$ b)$m = 9$ 5.a)$y = 1 - 2x$ b)$y = \dfrac{2}{3}x + \dfrac{1}{3}$

)$x = 3 - 2y$ d)$x = \dfrac{5}{2}y + 3$ e)$s = \dfrac{t}{4} - p$ f)$n = \dfrac{PV}{RT}$

)$t = \dfrac{u^2 - v^2}{-2v}$ h)$a = \dfrac{2A}{t} - b$

nequalities and Inequations, page 11
.a)F b)T c)T d)F 2.(a), (b), (d), (e), (f)
.a)$y > -1$ b)$x > 2$ c)$x \geq 10$ d)$m < 0$ e)$x \leq 8$

Ratio and Proportion, page 12
.a)4 b)30 c)5 d)10 e)$m = 7, n = 21$ f)$x = 15,$
$= 35$ 2.a)$a = 1, b = 0$ b)$a = 16, b = -15$

.a)$3{:}2$ b)$1{:} -1$ c)$15{:}8$ d)$10{:} -11$ 4.a)$\dfrac{19}{2}$ b)$\dfrac{17}{8}$

Analytic Geometry, page 13
.a)$\sqrt{146}$ b)$4\sqrt{10}$ c)$\sqrt{65}$ d)20 2.a)$\sqrt{13}$ b)$\sqrt{13}$
)$\sqrt{13}$ d)same 3.$2\sqrt{10} + \sqrt{13} + 5$ 4.a)1 b)1 c)$\dfrac{1}{5}$
)$-\dfrac{1}{3}$ 5.a)$\dfrac{1}{2}, \dfrac{1}{2}, \dfrac{1}{2}$ c)$k = 0$ 6.c)A, B 7.a)$3, -1$
)$2, -6$ c)$16, -4$ d)$3, -3$ e)$k = -2$ 8.a)A and G,
3 and F b)$k = 1$ 9.a)-1 b)1 c)-2 d)$\dfrac{1}{2}$ e)$-\dfrac{3}{2}$ f)2

g)$\dfrac{2}{3}$ h)$\dfrac{1}{3}$ 10.a)-3 b)$\dfrac{3}{2}$ c)-5 d)6 11.a)y-intercept of 3

b)slope of 3 12.a)$y = \dfrac{2}{3}x + 6$ b)$y = -2x + 3$ c)$y = b,$

$y = 8$ d)$y = mx, y = -\dfrac{5}{2}x$ 13.a)$-\dfrac{1}{4}$ b)$\dfrac{1}{2}$ c)-3

d)undefined 14.a)$k = \dfrac{1}{8}$ b)$k = \dfrac{3}{2}$ 15.$\triangle$DEF, $\triangle$GHI

16.a)25 square units b)21.5 square units 17.a)$\dfrac{21\sqrt{13}}{13}$ units

b)$\dfrac{21\sqrt{26}}{26}$ units c)$\dfrac{9\sqrt{2}}{4}$ units d)0, point on the line

Solving Systems of Equations, page 16
1.a)-2 b)3 2.a)-2 b)-1 3.a)$x = 2, y = 1$ b)$x = -1,$
$y = 3$ c)$x = 1, y = 0$ d)$x = 4, y = 0$ e)$x = 6,$
$y = 2$ f)$x = -\dfrac{4}{3}, y = \dfrac{1}{3}$ 4.a)$(-1, 3)$ b)$(1, -1)$
c)$\left(2, \dfrac{3}{4}\right)$ 5.a)$(-1, 2, 3)$ b)$\left(\dfrac{5}{7}, \dfrac{1}{2}, \dfrac{27}{14}\right)$ c)$(1, 2, 3)$
6.a)$(1, 1, 1)$ b)$\left(-\dfrac{71}{35}, \dfrac{144}{35}, \dfrac{28}{5}\right)$ c)$\left(\dfrac{11}{19}, -\dfrac{22}{19}, \dfrac{31}{19}\right)$

CHAPTER 1

1.1 Exercise, page 22
1.a)$f(x) = 2x + 1$ b)$g(x) = x^2 - 3$ c)$h(x) = \dfrac{2}{x + 1}$
d)$k(x) = \sqrt{x^2 + 1}$ 2.a)$f(x) = 3x - 2$ b)$h(x) = \dfrac{1}{2x - 1}$
c)$g(x) = \sqrt{9 - x^2}$ 3.a)2 b)3 c)4 d)$-1, -2$ 4.(b)
5.a)-9 b)-1 c)2 d)1 e)-2 f)-11 6.a)0 b)6 c)1
d)$2m^2 - 3m + 1$ e)$\dfrac{2}{m^2} - \dfrac{3}{m} + 1$ f)$18m^2 - 9m + 1$

7.a)21 b)-27 c)-3 d)$3(2h - 1)$ e)$3\left(\dfrac{2}{k} - 1\right)$
f)$3(4k - 1)$ 8.a)$x^2 - 2$ b)$x^2 - 4$ c)$x^2 - 4x + 2$
d)$-2x^2 + 4$ e)$4x^2 - 2$ 9.a)undefined b)2 c)$1 - 2x$
d)$\dfrac{2x}{x - 2}$ e)$\dfrac{-x - 2}{x - 2}$ f)$\dfrac{-x + 2}{x + 2}$ 10.a)$m^2 - 7m + 12$
b)$\dfrac{2(a + 2)}{-3a - 5}$ c)$\dfrac{1 + 3x}{1 - 2x}$ 11. 6 12. ± 3 13. $1, 4$ 14. -5

1.2 Exercise, page 25
1.a)Yes, $\{-2, -1, 0, 1, 2\}, \{1, 2, 3\}$ b)Yes, $\{-3, -2,$
$-1, 0, 1, 2\}, \{0, 1, 2, 3, 4\}$ c)No, $\{-3, -2, -1, 0, 1,$
$2, 3\}, \{0, 1, 2, 3\}$ d)Yes, $\{-3, -2, -1, 0, 1, 2, 3\},$
$\{-2, -1, 0\}$ e)No, $\{-2, -1, 0, 1, 2\}, \{0, 1, 2, 3\}$
f)No, $\{-4, -3, -2, -1, 0, 1, 2, 3\}, \{-2, -1, 0, 1, 2\}$
2.b)$\{1, 2, 4, 5\}, \{-2, -1, 3\}$ c)Yes 3.a)Not a function
b)Function c)Not a function 4.a)Yes, $x \in R, y \in R$ b)Yes,
$x \in R, \{y|y \geq -2, y \in R\}$ c)No, $\{x|x \geq -2, x \in R\}, y \in R$
d)No, $\{x|-3 \leq x \leq 3, x \in R\}, \{y|-2 \leq y \leq 2, y \in R\}$
e)Yes, $\{x|-4 \leq x \leq 4, x \in R\}; \{y|-2 \leq y \leq 2, y \in R\}$
f)No, $\{x|-3 \leq x \leq 3, x \in R\},$
$\{y|-3 \leq y \leq 3, y \in R\}$ 5.a)$\{x|x \geq 0, x \in R\},$
$\{y|y \geq 0, y \in R\}$ b)(i)4 (ii)13 (iii)3 c)$A = s^2$ d)Yes
6.a)$\{x|-6 \leq x \leq 6, x \in R\}, \{y|-6 \leq y \leq 6, y \in R\}$

b)(i)0 (ii)0 (iii)6, -6 **c)**No **7.a)**$\{x|x \geq 0, x \in R\}$,
$\{y|y \geq 0, y \in R\}$ **b)**(i)6.5 (ii)6.7 (iii)12 **c)**Yes

1.3 Exercise, page 28
3.b)moves up, moves down **c)**$g: f(x) \rightarrow f(x) + q$
8.b)moves left, moves right **c)**$g: f(x) \rightarrow f(x + p)$
13.b)vertical stretch, compressed vertically
c)$g: f(x) \rightarrow af(x)$ **d)**Yes **17.b)**compressed horizontally,
horizontal stretch **c)**$g: f(x) \rightarrow f(bx)$ **d)**Yes **21.a)**reflection
through x-axis, reflection through y-axis
b)$g: f(x) \rightarrow -f(x)$ **c)**$g: f(x) \rightarrow f(-x)$ **d)**Yes

1.4 Exercise, page 33
1.b)$f^{-1}(-1) = \dfrac{5}{3}, 5; f^{-1}(0) = \dfrac{4}{3}, -4; f^{-1}(1) = \dfrac{-7}{2}, 1$
c)No **2.a)**D:{0, 1, 2, 3};
R:{-2, 0, 2, 4} **b)**D:{0, 1, 3}, R:{-3, -2, 0, 2, 3} **c)**(a)
3.a){(-2, 4), (-1, 1), (0, 0), (1, 1), (2, 4)} **b)**{(0, 0),
(1, -1), (1, 1), (4, -2), (4, 2)} **d)**No **5.a)**-1, 1, $-\dfrac{1}{2}$
b)(-1, -1), (1, 0), $\left(0, -\dfrac{1}{2}\right)$ **c)**$f^{-1}(x) = \dfrac{1}{2}x - \dfrac{1}{2}$
6.c) No **7.e)**(a), (c) **8.b)**$f^{-1}(x) = \dfrac{1}{2}x + \dfrac{3}{2}$
9.b)$x = y^2$ **10.a)**$2x - 3y = -6$ **b)**$y^2 = x + 6$

1.5 Exercise page 36
1.a)$y = \dfrac{1}{2x}$ **b)**$y = \dfrac{1}{x^2}$ **c)**$y = \dfrac{1}{3x - 2}$ **d)**$y = 2x$
e)$y = \dfrac{4x + 1}{2}$ **f)**$y = \dfrac{1}{2x^2 + x - 3}$ **3.b)**Yes **c)**$x = -1$
4.b)$x = -1, 1$ **5.b)**$g : f(x) \rightarrow \dfrac{1}{f(x)}$ **c)**Yes **6.b)**$\dfrac{3}{2}$ **7.b)**$x - 2 = 0$,
$x + 2 = 0$ **8.a)**$x + 2] x - 3] x + 5$ **b)**$2x] x - 3] x + 6] x - 3$
c)$3.1x + 2.5] x - 6.5] x - 3.6$
d)$9.8x - 3.6]x] x + 6.5] x + 9.3$ **9.a)**12, 18, 72,
201.312 **b)**19.5, 52, 120.5 **c)**63.6175, 209.8266,
739.6825 **10.a)**-57, -174.375, 85.659 **b)**16.0768,
67.4368 **11.a)**0.0114, 0.0025 **b)**13.2372, 11.0615

1.6 Exercise, page 41
3.a)$f + g = \{(-4, 6), (-2, 5), (1, 5), (4, 10)\}$
$D = \{-4, -2, 1, 4\}$ **b)**$f - g = \{(-4, 2), (-2, 3),$
$(1, 1), (4, 2)\}$, $D = \{-4, -2, 1, 4\}$
c)$g - f = \{(-4, -2), (-2, -3), (1, -1), (4, -2)\}$,
$D = \{-4, -2, 1, 4\}$ **4.a)**4 **b)**8 **c)**-2 **d)**2 **e)**0
5.a)-18 **b)**2 **c)**10 **d)**0 **6.a)**$5x - 5$, $x \in R$ **b)**$x + 1$,
$x \in R$ **c)**$-x - 1$, $x \in R$ **7.a)**$x^2 - 3x - 1$, $0 \leq x \leq 4$
b)$-x^2 + 9x - 1$, $0 \leq x \leq 4$ **c)**$x^2 - 9x + 1$,
$0 \leq x \leq 4$ **9.a)**$x \in R$ **b)**$x \in R$ **10.a)**$x \in R$
13.b)$-2 \leq x \leq 3$, $-2 \leq x \leq 3$
14.a)$\{x|x \neq \pm 1, x \in R\}$, $x \in R$ **b)**$\{x|x \neq \pm 1, x \in R\}$

1.7 Exercise, page 45
2.a)-3 **b)**-1 **c)**-2 **d)**12 **e)**15 **f)**15 **3.a)**1 **b)**4 **c)**5
d)3 **4.a)**-5 **b)**-13 **c)**1 **d)**11 **e)**10 **5.a)**$4x^2$, $16x^2$
b)$3x^2 - 3$, $3x^2 - 18x + 27$ **c)**$x + 3$, $x + 3$
d)$2x^2 + 12x + 13$, $4x^2 - 8x + 4$ **e)**2, 3
6.a)$6x - 9$, $6x - 3$ **b)**$12x^2 - 36x + 27$, $6x^2 - 3$
c)$6x^2 - 9$, $18x^2 - 3$ **d)**$12x^4 - 36x^2 + 27$, $18x^4 - 3$

e)$\dfrac{3}{2x - 3}$, $\dfrac{1}{6x - 3}$ **f)**$\dfrac{3}{4x^4 - 12x^2 + 9}$, $\dfrac{1}{18x^4 - 3}$

7.a)$g(x) = x$ **b)**$g(x) = x$ **8.a)**$2x + 1$, $2x + 1$
b)$3x^2 - 5$, $3x^2 - 5$ **9.a)**$x^2 - 3$ **b)**$x^2 + 2x - 3$
c)$x \in R$ **d)**$x \in R$ **e)**$\{y|y \geq -3, y \in R\}$ **f)**$\{y|y \geq -4, y \in R\}$
10.a)$6x^2 - 2$, $12x^2 - 1$ **b)**$x \in R$, $\{y|y \geq -2, y \in R\}$,
$x \in R$, $\{y|y \geq -1, y \in R\}$ **11.a)**$fog(x) = \sqrt{x^2 - 3}$,
$gof(x) = x - 3$ **b)**$\{x|x \leq -\sqrt{3}$ or $x \geq \sqrt{3}, x \in R\}$,
$\{y|y \geq 0, y \in R\}$; $x \in R$, $y \in R$, **12.a)**$fog(x) = 3\sqrt{9 - x^2}$,
$\{x|-3 \leq x \leq 3, x \in R\}$, $\{y|0 \leq y \leq 9, y \in R\}$,
$gof(x) = 3\sqrt{1 - x^2}$ $\{x|-1 \leq x \leq 1, x \in R\}$, $\{y|0 \leq y \leq 3,$
$y \in R\}$; **b)**$fog(x) = \sqrt{3x + 1}$ $\left\{x|x \geq -\dfrac{1}{3}, x \in R\right\}$, $\{y|y \geq 0,$
$y \in R\}$; $gof(x) = 3\sqrt{x} + 1$ $\{x|x \geq 0, x \in R\}$, $\{y|y \geq 1, y \in R\}$
c)$fog(x) = \sqrt{9 - x^4}$, $\{x|-\sqrt{3} \leq x \leq \sqrt{3}, x \in R\}$,
$\{y|0 \leq y \leq 3, y \in R\}$; $gof(x) = 9 - x^2, x \in R$,
$\{y|y \leq 9, y \in R\}$ **13.a)**$g(x) = \dfrac{1}{2}x - \dfrac{3}{2}$ **b)**$g(x) = x^2 + 2x + 3$

14.a)$acx + ad + b$, $acx + bc + d$ **b)**$3x^2 + 2$
15.a)$2x - 3$ **b)**$2x - 3$ **c)**$2x - 6$ **d)**$2x - 6$ **e)**$2x - 6$
f)$2x - 3$ **16.a)**$27k - 14$ **b)**$2\sqrt{9k - 16} + 5$

17.$T = 2\pi\sqrt{\dfrac{1 + 0.0035C}{980}}$ **18.a)**2.0071 s **b)**2.0078 s
c)increased **19.a)**90 s **b)**38 s

CHAPTER 2

2.1 Exercise, page 53
1.a)$2x^2$ **b)**$3b^2$ **c)**$4xyz^2$ **d)**$5pq$ **e)**$-x^4$ **f)**$-5rs$ **g)**$-6d^2$
h)$-xyz$ **2.a)**5 **b)**x **c)**$(r - 3s)$ **d)**dx **e)**$(b + 3a - 9ab)$
f)$4mn$ **3.a)**$2p$ **b)**$9xy$ **c)**$-5vw^2$ **d)**a^3 **e)**-6 **f)**$3y$ **g)**9
h)$-13mn$ **i)**5 **j)**$11s^3t$ **4.a)**$(k + p)$, $(b + a)$ **b)**$(a + b)$,
$(k + p)$ **5.a)**$(x + y)(a + b)$ **b)**$(a + b)(x + y)$
6.a)$(a - b)(y - 2x)$ **7.a)**$(x - 1)$, $(x - y)$ **b)**$(a - c)$,
$(a + b)$ **8.a)**$(3v + 2w)(5c + 4d)$ **b)**$(8x - 7y)(5z - 2w)$
c)$(11x - 3y)(2v + w)$ **d)**$(5b - 3c)(4p + q)$
9.a)$(p + q)(a + b)$ **b)**$(r - 2s)(t - v)$ **c)**$(c - 4d)(3m + n)$
d)$(4g - h)(5k + 2m)$ **10.a)**$2(2g + 3h)(4b - p)$
b)$x(x - 2y)(2v - w)$ **c)**$ab(a + 3)(b + 2)$
d)$5n^3(n - m)(m + 3)$ **e)**$(x^2 + 1)(y^2 + 2)$
f)$(a + 2b - c)(3p + q)$ **g)**$(3x + y - 4z)(7v - 3w)$
h)$(x - 1)(x - y)$ **i)**$(a - c)(a + b)$
j)$3x(m - n)(y - 2)$ **k)**$(x^2 + 2y^2)(x^2 + 1)$

2.2 Exercise, page 57
1.a)-3, 2 **b)**-9, -7 **c)**4, 5 **d)**1, 6 **e)**-11, -12
f)13, -3 **g)**-13, 3 **2.a)**$y + 1$ **b)**$b + 4$ **c)**$x - 7$
d)$p + 7$ **e)**$z - 9y$ **f)**$m - 3n$ **3.a)**$s + 9t$ **b)**$q - 5r$
c)$2p - 9q$ **d)**$7a + 3b$ **e)**$5m - 2n$ **f)**$x + 3y$ **4.a)**3
b)p **c)**m^2 **d)**$-7a$ **e)**25 **f)**$16k^4mn$ **5.a)**3, 7
b)$(v + 3w)(v + 7w)$ **6.a)**21, -4 **b)**$(3a - 2b)(2a + 7b)$
7.a)$(p - 7)(p + 1)$ **b)**$(s - 2)(s + 6)$ **c)**$(x - 3)(x - 11)$
d)$(y + 5)(y + 8)$ **8.a)**$(a - 1)(a + 9)$ **b)**$(m - 4)(m + 13)$
c)$(k - 2p)(k - 7p)$ **d)**$(a + 3b)(a + 5b)$ **e)**$(y - 3z)(y - 6z)$
f)$(g - 3h)(g + 4h)$ **9.a)**$3(x - 2)(x + 11)$
b)$y(y - 7)(y + 1)$ **c)**$4p(p + 5)(p + 3)$

d)$m(q - 2)(q - 8)$ **e)**$r^3(r - 2s)(r + s)$
f)$7v^2(t - 4v)(t - 5v)$ **10.a)**$(5x - 2)(x - 3)$
b)$(3p + 1)(2p + 7)$ **c)**$(5 - 3q)(2 + 5q)$
d)$(7a - 6)(a + 5)$ **e)**$(5 - 3b)(2 - 9b)$
f)$(6m + n)(5m + 11n)$ **g)**$(4g + 3h)(3g - 7h)$
h)$(7x - 3y)(2x + 9y)$ **i)**$(13s + 2t)(4s - 5t)$
j)$(11p + 7q)(p + 3q)$ **11.a)**$2m(2m + 7)(m - 5)$
b)$9n^2(11 - 5n)(1 + 2n)$ **c)**$-5x(7p + 2q)(p + 5q)$
d)$x^3(9x - 2y)(5x - 3y)$ **e)**$-4ab(4a - 3b)(6a - 5b)$
f)$s^2t^2(2s - 5t)(s + 13t)$ **12.a)**$(7 - a)(11 - a)$
b)$5(2p + 3)(p - 5)$ **c)**$(11x + 3y)(3x + 7y)$
d)$-(4s - 5r)(3s + 7r)$ **e)**$m^2n^2(5m + 6p)(2m + 3p)$
f)$-3w(5w + 3x)(5w - 6x)$ **g)**$(4v - 5w)(2v - w)$
h)$-m^2(3c + 5d)(12c - 7d)$ **i)**$3(5x^2 + 11)(x^2 - 4)$
j)$2gh(9gh + 7)(2gh - 7)$ **k)**$(4m + 13n)(4m + 3n)$
l)$3r(p^2q - 3r)(p^2q - 9r)$

2.3 Exercise, page 60

1.a)$(m + 2)$ **b)**$(3x + 1)$ **c)**$(a - 2b)$ **d)**$(5p + 3q)$
e)$(x^2 + 2)$ **f)**$(3s^2 - 4t^2)$ **g)**$(mn - 4)$ **h)** $\left(z - \dfrac{1}{2} \right)$
i)$(0.75n + 3m)$ **j)**$(c + d - e)$ **k)**$(3g - h + 2k)$
2.a)$(b - 4)$ **b)**$(x + 3y)$ **c)**$(2g - 3h)$ **d)**$(11 + 4k)$
e)$(p + q - r)$ or $(p - q - r)$ **f)**$(a - 2b - 3c)$ or
$(a + 2b - 3c)$ **g)**$(x^2 + 7)$ **h)**$(c^2d^2 - g)$ **i)** $\left(3n - \dfrac{1}{3} \right)$
j) $\left(m - \dfrac{1}{4} \right)$ **3.a)**$5y$ **b)**$3q$ **c)**$7a$ **d)**$4m^2$ **e)**x^4; $8y^2$ **f)**$2t$
g)$9cd$ **h)**$12x^2yz^3$, $10vw$ **4.a)**1 **b)**$4x$ **c)**$3p$ **d)**$20mn$ **e)**$\dfrac{1}{4}d^2$
f)$5x^2$ **5.a)**$(a + 4)^2$ **b)**$(2b + 3)^2$ **c)**$(3y - 2)^2$ **d)**$(7p - 4)^2$
e)$(3a + 2b)^2$ **f)** $\left(\dfrac{x}{2} - 5y \right)^2$ **6.a)** $\left(2m - \dfrac{n}{3} \right)^2$ **b)**$(xy + 3)^2$
c)$(2ab + 3c)^2$ **d)**$(g^2h - 4)^2$ **e)**$(x - 2yz)^2$ **f)**$(p^2qr + 3t)^2$
7.a)$(b - 4)(b + 4)$ **b)**$(2x - 1)(2x + 1)$ **c)**$(x - 3y)(x + 3y)$
d)$(2g - 3h)(2g + 3h)$ **e)**$(11 - 4k)(11 + 4k)$
f)$\left(m - \dfrac{1}{4} \right)\left(m + \dfrac{1}{4} \right)$ **g)**$\left(3n - \dfrac{1}{3} \right)\left(3n + \dfrac{1}{3} \right)$ **h)**$(x^2 - 7)(x^2 + 7)$
i)$(xy - 1)(xy + 1)$ **8.b)**$(1 + 3x - 2y)(1 + 3x + 2y)$
c)$(7a - 1 + 4b)(7a + 1 - 4b)$
d)$(p^2 + 2 - q^2)(p^2 + 2 + q^2)$
e)$\left(xy - x + \dfrac{1}{4} \right)\left(xy + x - \dfrac{1}{4} \right)$ **f)**$(6m - 2n + 6)(6m + 2n + 4)$
g)$(5g - 7h - 3k - 2p)(5g - 7h + 3k + 2p)$
9.a)$7(x - 2)(x + 2)$ **b)**$50(2 - y^2)(2 + y^2)$
c)$(y - 7z)(y + 7z)$ **d)**$(1 - 2pq)(1 + 2pq)$
e)$3(m - 4n)(m + 4n)$ **f)**$a(a - 1)(a + 1)$
g)$k(gh - 2)(gh + 2)$ **h)**$(4p^2 - 5q)(4p^2 + 5q)$
i)$4(3c - 2d^2)(3c + 2d^2)$ **j)**$m^3(m - 1)(m + 1)$
k)$13(b^2 - 2c^2)(b^2 + 2c^2)$ **l)**$\dfrac{1}{8}(x^2 + 4)(x - 2)(x + 2)$
10.a)$(y^4 - 2)(y^4 + 2)$
b)$(2w^2 - v^2)(2w^2 + v^2)(4w^4 + v^4)(16w^8 + v^8)$
c) $\left(\dfrac{p}{2} - \dfrac{q}{3} \right)\left(\dfrac{p}{2} + \dfrac{q}{3} \right)$ **d)**$(a + 5)(a - 1)$
e)$(2c + 1 - b)(2c + 1 + b)$ **f)**$-3h(2g - h)$
g)$(0.2m - 2.5n)(0.2m + 2.5n)$

h)$3(d^2 - 2g^2)(d^2 + 2g^2)$ **i)**$3 \left(w - z - \dfrac{1}{3}v \right) \left(w - z + \dfrac{1}{3}v \right)$
j)$(b - 3a)(b + 5a)$ **k)**$8k(g - h)$ **l)**$5(-x + 2y)(11x + 2y)$
11.a)$(a + 3 - b)(a + 3 + b)$ **b)**$(x - y - 3z)(x - y + 3z)$
c)$(d^2 - e^2 - 4g^2)(d^2 - e^2 + 4g^2)$
d)$4(m - n - p)(m + n + p)$ **e)**$(x^2 - y + z)(x^2 + y - z)$
f)$2(p - 2r + q)(p + 2r - q)$
g)$(3a + 3b - 2c - 2d)(3a + 3b + 2c + 2d)$

12.a)$(13 - x)(7 + x)$ **b)**cannot factor
c) $\left(\dfrac{a}{8} - \dfrac{b}{7} \right)\left(\dfrac{a}{8} + \dfrac{b}{7} \right)$ **d)** $\left(\dfrac{c}{2} - \dfrac{d}{3} \right)\left(\dfrac{c}{2} + \dfrac{d}{3} \right)\left(\dfrac{c^2}{4} + \dfrac{d^2}{9} \right)$
e)$(5x^2y - 2z^2)(5x^2y + 2z^2)(25x^4y^2 + 4z^4)$
f)$6(g - 3h)(g + 3h)$ **g)**cannot factor **h)**cannot factor
i)$2(2x - 5)(2x + 5)$ **j)**$3(3w - 4)(3w + 4)$
k)$(4m - 4n - r - s)(4m - 4n + r + s)$

2.4 Exercise, page 64

1.a)$4x^2 - 2x + 9$ **b)**$-5x^3 + 2x^2 - x + 11$
c)$x^4 + 7x^3 + 12x^2 - 2x + 5$
d)$x^2 - x + 1 + \dfrac{2}{x} - \dfrac{1}{x^2}$ **2.a)**$4p^2 + 0p - 5$
b)$3x^3 + 0x^2 + x - 2$ **c)**$y^5 + 0y^4 + 0y^3 + 0y^2 + y + 7$
d)$4w^4 + 3w^3 + 0w^2 + 0w - 75$ **3.a)**$q - 3$
b)$2a - 7$ **c)**$x + 3y$ **d)**$2c - d$ **e)**$b - c$ **f)**$2(3x - y)$
g)$2gh + 1$ **h)**$mn + p$ **i)**$2v + 7w$ **j)**$7p + 4q$
k)$(x - y)(x^2 + y^2)$ **l)**$3u^2 - z$ **4.a)**-6 **b)**6 **c)**-12
d)15 **5.a)**$4n + 3$ R5 **b)**$a - 4$ R-11 **c)**$3d - 10$ R2 **d)**$2k + 5$ R-6
6.a)$2xy + 5$ **b)**$(m^3 - 3m^2 + 12m - 38)$ R 119
c)$6q + 1$ **d)**$2s - 1$ **e)**$-a^2 - a + 1$
f)$(d^3 + d^2 - d - 1)$ R(d + 1) **7.a)**$x^4 - x^3 + x^2 - x + 1$
b)$z^5 + z^4 + z^3 + z^2 + z + 1$ **c)**$(y^2 - 1)$ R2
d)$2q + 6$ **e)**$(a - 3)$ R(-a - 5) **f)**$d^4 = 3d^2 + 1$
g)$(m^2 + 3m)$ R(-2m^2 - 6m + 6)$ **h)**$w^4 - 2w^2 + 3$
8.a)7 **b)**3 **9.a)**$x^3 - 4x^2 + x + 9$ **b)**$x^2 - 5x + 6$
c)3 **10.a)**$x^2 + 4x + 3$ **b)**$x^2 - 2x - 3$ **c)**$x^2 + x - 6$
d)$x^2 - 9$ **11.a)**$x \neq -5$, $(x + 10)$ **b)**$m \neq 15$, $(m + 6)$
c)$x \neq 8y$, $(x + 4y)$ **d)**$t \neq -4$, $(t + 3)$ **e)**$a \neq -3b$,
$(a - 8b)$ **f)**$x \neq -2y$, $(x - 4y)$ **g)**$x \neq 3$, $(2x + 1)$
h)$k \neq -6$, $(5k - 8)$ **i)**$3m \neq 7n$, $(2m + 5n)$ **j)**$x \neq \dfrac{1}{3}$,
$(2x - 35)$ **k)**$2x \neq 3y$, $(x + 4y)$ **l)**$x \neq \dfrac{y}{3}$, $(6x + 5y)$

2.5 Exercise, page 68

2. (a) and (c) **3.a)**$(x - 1)$, $(3x + 1)$, $(x + 2)$
b)$(x + 1)$, $(2x - 1)$, $(2x + 3)$ **c)**$(x + 1)$, $(2x + 3)$,
$(x + 2)$ **d)**$(x - 1)$, $(x + 1)$, $(2x + 1)$ **4.a)**8 **b)**-5
5.a)$(x - 1)$, $(2x + 3)$, $(x + 7)$ **b)**2, $(3x - 1)$, $(x^2 - 2)$
c)$(x - 1)$, $(x + 3)$, $(2x + 3)$ **d)**$(x + 1)$, $(2x + 1)$,
$(3x - 2)$ **6.b)**$(x + 1)$, $(2x - 1)$, $(2x + 3)$ **7.a)**1 **b)**7
8. -2, $\dfrac{1}{2}$ **9.a)** -4, 3 **b)** -1, -3 **c)**3, -3, -1 **d)**2
10.a)$x^3 - 3x^2 - 4x + 12$ **b)**$x^3 + 2x^2 - 5x - 6$
11.a)3 **b)**$(3x - 5)$, $(x + 1)$ **12.b)**$(x + y)$,
$(x^2 - xy + y^2)$ and $(x - y)$, $(x^2 + xy + y^2)$

13.a)8 b)−1 **14.**a)3 b)11 **16.** $2x^3 - 5x^2 - 5x + 3$
17.a)$-x^2 + 2x - 4$ b)$-y^2 - 3y - 9$ **19.**a)0 b)$(x + y)$,
$(x^2 - xy + y^2)$ **20.**a)$(y + 1)$, $(y^2 - y + 1)$
b)$(x + 2)$, $(x^2 - 2x + 4)$ c)$(y - 1)$, $(y^2 + y + 1)$
d)$(x - 2)$, $(x^2 + 2x + 4)$ e)$(2x + 1)$, $(4x^2 - 2x + 1)$
f)$(2y - 1)$, $(4y^2 + 2y + 1)$ g)$(x + 4)$, $(x^2 - 4x + 16)$
h)$(x - a)$, $(x^2 + ax + a^2)$ i)$(x + y)$, $(x^2 - xy + y^2)$
j)$(3x + y)$, $(9x^2 - 3xy + y^2)$ k)$(1 - 3y)$, $(1 + 3y + 9y^2)$
l) $\left(\frac{y}{2} + 1\right)$, $\left(\frac{y^2}{4} - \frac{y}{2} + 1\right)$ **21.**a)$(a + 4b)(a^2 - 4ab + 16b^2)$
b)$2(x - 1)(x^2 + x + 1)$ c)$2(1 - 3x)(1 + 3x + 9x^2)$
d)$y(y + 1)(y^2 - y + 1)$ e)$(x + y)(x - y)(x^4 + x^2y^2 + y^4)$
f)$(3a - 5)(9a^2 + 15a + 25)$ g)$27(2m - 1)(4m^2 + 2m + 1)$
h)$(10m - y)(100m^2 + 10my + y^2)$

i) $\left(ab + \frac{1}{2}c^2d^2\right)\left(a^2b^2 - \frac{1}{2}abc^2d^2 + \frac{1}{4}c^4d^4\right)$

2.6 Exercise, page 72

1.a)-5 b)$2, -5$ c)$-\frac{1}{2}$ d)$\frac{2}{3}, -\frac{1}{2}$ e)$3, -\frac{2}{3}$ f)$\frac{1}{2}, \frac{2}{3}$ **2.**a)$3, -2$
b)$5, -5$ c)$\frac{2}{3}, -3$ d)$0, 5$ e)$3, -\frac{1}{2}$ f)$0, 5$ **3.**a)$-3, -5$ b)-10,
3 c)$5, -5$ d) $-\frac{2}{5}, -3$ e)$\frac{1}{4}, -\frac{1}{4}$ f)$-\frac{1}{2}, -4$ g) $-\frac{1}{3}, -3$
h) $-\frac{3}{5}, 2$ **4.**b)$3, 4$ **5.**a)$8, -6$ b)$\frac{5}{3}, -6$ c) $-\frac{1}{3}, \frac{3}{2}$ d)$\frac{3}{2}, 2$
e) $-\frac{5}{2}, 3$ f)$-10, 3$ **6.**b) $-\frac{11}{2}, 3$ **7.**a)$-\frac{2}{3}, 1$ **8.**a)$\{7, -3\}$
9.b)$4, -3$ **10.**a)$-2, -4$ b)$9, -8$ c)$\frac{7}{3}, \frac{3}{2}$ d)$-2, -3$ e) $-\frac{5}{2}$,
3 f)$\frac{5}{3}, 2$ **11.**a)$3, -2$ b)$-3, 5$ c)$-4, -5$ d)$6, 7$ e) $-\frac{1}{3}, 5$
f)$-7, 3$ g) $-\frac{7}{5}, -1$ h)$\frac{5}{4}, -\frac{5}{4}$ **12.**a)$7, -3$ b)$2, -9$
c)$\frac{5}{3}, -3$ d)$\frac{7}{2}, 1$ e)$7, -6$ f)$\frac{6}{5}, 4$ **13.**a) $-\frac{1}{5}, 4$ b)$\frac{5}{3}, -3$
c)$4, -4$ d)-1 e)$\frac{3}{4}, -\frac{5}{2}$ f) -5 g)$\frac{1}{6}, 1$ h)$\frac{10}{3}, -3$ i) -2
j)$\frac{3}{2}, -1$ k)$\frac{3}{2}, -\frac{5}{2}$ l)$\frac{2}{3}, -\frac{1}{2}$ **14.**a)$x^2 - x - 6 = 0$
b)$x^2 + 10x + 24 = 0$ c)$2x^2 - 7x + 3 = 0$
d)$x^2 - 4x = 0$ e)$6x^2 + x - 1 = 0$ **15.**a)$p = -2$,
$x = \frac{1}{3}$ b)$p = -5$, $x = -\frac{1}{2}$ c)$p = 25$, $x = \frac{1}{4}$ d)$p = 5$,
$x = -5$ **16.**b)$3, 2$ **17.**a)$4, -2$ b)$\frac{3}{2}, -\frac{5}{2}$ c)$1, -4$
18.a)$-7, 5$ b)$-5, 3$ c)$\frac{1}{3}, 3$ d)$\frac{1}{2}, -3$ **19.**a)$\frac{2 \pm \sqrt{6}}{2}$
b)$\frac{3 \pm 2\sqrt{6}}{3}$ c)$\frac{5 \pm \sqrt{33}}{4}$ d)$\frac{3 \pm \sqrt{57}}{6}$ **20.** $\frac{1 \pm \sqrt{61}}{10}$
21.a)$1.8, -0.2$ b)$1.9, -0.4$ **22.**a)$2.51, -0.99$ b)3.52,
1.93 c)$-1.75, 0.67$

2.7 Exercise, page 78

1. 20, 22 **2.** 9, 7 **3.**a)5 b)6 **4.** Highway 70 km/h, or
40 km/h **5.** 9 cm, 12 cm **6.** 11, 12 **7.** 30 cm × 70 cm
8. 2, 4, 6, 8 **9.** 1.2 m **10.** 7.2 m × 7.2 m **11.** 10 m
12. 4.9 cm × 4.9 cm, 7.9 cm × 7.9 cm **13.**Alice:
8.4 km/h, 5h, Bob: 8.0 km/h, 5.25 h **14.** Truck
60 km/h, Jeep 80 km/h **15.** $18

2.8 Exercise, page 82

1.a)$m^2 - 17m + 16 = 0$ b)$\pm4, \pm1$ **2.**a)$\pm5, \pm1$
3.a)$k^2 - 8k + 12$ b)$\pm2, 1, -3$ **4.**a)$1, 3, -2, -4$
5.a)$m = \pm1, \pm2$ b)$x = 2, 3$ c)$y = \pm2, \pm5$ d)$x = \pm4, \pm2$
e)$x = 0, 1$ f)$x = \pm\sqrt{3}, \pm\sqrt{2}$ g)$x = \pm1, \pm3$ h)$x = 1, 3$
i)$x = \pm2, \pm3$ j)$x = \pm1, \pm2$ **6.**a)$2, 8$ b)$1, 3$ **7.**a)$1, 3$
8.a)$1, 2, -3, -4$ b)$\pm1, 2, 4$ **9.**a)$\{-3, -2, 1, 6\}$
b)$\{1, 2, 4\}$ **10.**a)$1, 2$ b)$1, 3$ c)$1, 2$ d)$0, 1$ **11.**a)$\pm\sqrt{6}$,
$\pm\sqrt{10}$ b)$\pm1, \pm\sqrt{35}$ c)$\pm3, 4, -2$ d)$1, 2$ e)$1 \pm \sqrt{2}$
f)$1, 4$ g)$\pm1, 3$ h)±1 **12.**a)$-2 \pm \sqrt{3}, \frac{3 \pm \sqrt{5}}{2}$
13. $-\frac{2}{3}, \pm1, \frac{4}{3}$ **14.** $-\frac{5}{2}, \frac{1}{2}, -1, 2$ **15.** $-\frac{1}{2}, \pm1, -\frac{5}{2}$
16. $\cdot 2 \pm \sqrt{2}, 1, 2$ **17.** $\pm1, \pm2$ **18.** 4, $\pm1, 2$

2.9 Exercise, page 86

1.a)$3, 4$ b)$-5, 1, 7$ c)$-3, 5, -\frac{5}{2}, \frac{1}{3}$ d)$\pm3, \pm2$
2.a)2 b)$3, -2$ **3.**a)$1, 2, 3$ b)$0, 1, -3$ c)$0, -2, -3$
d)$0, \frac{3}{2}, -5$ e)$0, \frac{5}{3}, -3$ f)$0, 7, -3$ **4.**a)$\pm3, \pm1$
b)$\{-3, -2, 2, 3\}$ **5.**a)$(x + 2), (x - 3), (x + 3)$ b)-2,
$3, -3$ **6.**b)$1, -2$ **7.** $-\frac{1}{2}, 1, 2$ **8.**b)$3, \pm\frac{1}{2}$ **9.**a)$\pm1, 2$
b)$\pm1, -2$ c)$\pm2, -3$ d)$1, 2 \pm \sqrt{6}$ e)$-1, -\frac{1}{2}, 2$
10.a)$1, 2, -3$ b)$5, \pm1$ **11.**a)$\pm3, -2$ b)$-2, 1, 3$
12.a)$-2, -1, \frac{1}{2}, 3$ b)$-3, \pm2, \pm1$ **13.**b)$-2.4, 0.4, 2$
14.a)$4, \pm3$ b)$-1.3, 0.8, 1$ **15.**a)$-1.9, 0.4, 1.5$ b)2.4
16.a)1.3 b)3.7 **17.**b)$0.9, 4.98$ **18.**a)$-\frac{1}{3}, -\frac{5}{2}$ b)$0, 4$
c)$1, \frac{2}{3}$ d)$-4, 1$ e)$2.06, -0.35$ f)$0.41, -1.75$ **19.**a)-5,
$-3, 4$ b)$-3, \frac{-1}{2}, 1$ **20.**a)$1, 2, 3, 4$ b)$-1, 1$

CHAPTER 3

3.1 Exercise, page 93

1.a)3 b)6 c)36 d)-16 e)-16 f)16 g)-6 h)1 i)-12
j)-8 k)-33 l)32 **2.**a)7 b)-7 **3.**a)-11 b)-11 **4.**a)-12
b)7 c)-7 d)0 **5.**a)36 b)36 **6.**a)162 b)162 **7.**a)2 b)6
c)6 d)2 e)6 f)-14 g)7 h)7 i)2 j)14 k)-12 l)23
m)-160 n)-160 o)32 p)32 **8.**a)$6|y|$ b)$6y^2$ c)$6|y^3|$
d)$5|x||y|$ e)$5x^2y\sqrt{y}$ f)$5x^2y^2$ g)$6x^2|y|\sqrt{x}$ h)$10|xy^3|$ i)$9x^2y^2\sqrt{x}$
j)$7|x|y^2\sqrt{y}$ k)$4|x^3|y^2\sqrt{xy}$ l)$12|x|y^2\sqrt{xy}$ **9.**a)$9|y|$ b)$9y^2$
c)$5x^2$ d)$5|x|$ e)$2x^2$ f)$20|y|$ g)$20x\sqrt{x}$ h)$8|x|y\sqrt{y}$ **10.**a)$\sqrt{a^3}$
b)$\sqrt{xy^3}$ c)$\sqrt{a^3b^2}$ d)$\sqrt{a^4b^3}$ e)$\sqrt{3x^3y^6}$ **11.**a)$7|x|$ b)$5|x|$
c)no **12.**a)12, 12, 12 b)12, 12, 12 c)same **13.**a)2, 2, 2
b)2, 2, 2 c)same **14.**a)9, 1, 9 b)9, 9, 9 c)no
15.a)no b)no c)no d)no e)yes f)no

3.2 Exercise, page 97

1.a)$6, -6$ **b)**$3, -3$ **c)**$4, -4$ **d)**$20, -20$ **e)**$2, -2$
f)$3, -3$ **g)**$3, -3$ **h)**none **i)**$2, -2$ **j)**$5, -5$ **k)**$9, -9$
l)$4, -4$ **m)**$5, -5$ **n)**$3, -3$ **2.a)**$-1, -7$ **b)**$-3, 9$
c)$-4, 5$ **3.b)**$0, -6$ **4.b)**$-\frac{5}{4}, \frac{5}{4}$ **5.b)**$-11, 13$

6.a)$2, -14$ **b)**$-4, 10$ **c)**$6, 4$ **d)**$\frac{5}{2}, -\frac{7}{2}$ **e)**$-3, \frac{13}{3}$

f)$3, -\frac{11}{3}$ **7.a)**$-1, 4$ **b)**none **c)**$-\frac{3}{2}$ **d)**none **e)**none

f) $-\frac{5}{4}, \frac{1}{4}$ **8.a)** $-\frac{5}{4}, -\frac{11}{4}$ **b)**$\frac{3}{2}, \frac{9}{2}$ **9.a)**$3, -\frac{17}{3}$

b)$4, -3$ **c)**$-3, \frac{11}{3}$ **d)**$-3, \frac{17}{5}$ **e)** $-\frac{8}{3}, 2$ **f)**$2, -\frac{6}{5}$ **10.a)**$\frac{2}{3}$

b)-5 **c)**$-\frac{3}{16}$ **d)**$-\frac{4}{3}$ **e)**$-\frac{6}{11}$ **f)**$\frac{5}{6}$ **11.a)**$2, -2$ **b)**$\frac{5}{4}, -\frac{5}{4}$

c)$8, -8$ **d)**$9, -9$ **e)**$10, -10$ **f)**$12, -12$ **12.a)**$\frac{5}{4}, -\frac{5}{4}$

b)$\frac{1}{4}, -\frac{9}{4}$ **13.a)**$9, -11$ **b)**$15, -9$ **c)**$-4, 8$ **14.a)**$4, -2$

b)$3, \frac{1}{5}$ **c)**$2, 4$ **15.** $1, \frac{21}{5}$

3.3 Exercise, page 100

1.a)$|x| \leq 3$ **b)**$|x| \leq 6$ **c)**$|x| \geq 2$ **d)**$|x| \leq 4$ **e)**$|x| \leq 5$
f)$|x| \geq 3$ **g)**$|x| \geq 3$ **h)**$|x| \leq 3$ **4.a)**$y \geq 6$ or $y \leq -6$
b)$-3 < y < 3$ **c)**$y > 4$ or $y < -4$ **d)**$-2 \leq y \leq 2$
e)$-2 \leq m \leq 2$ **f)**$k < -2$ or $k > 2$ **g)**$p < -\frac{5}{2}$ or $p > \frac{5}{2}$
h)$x \leq -6$ or $x \geq 6$ **5.a)**$-2 \leq y \leq 2$ **b)**$y \leq -11$ or
$y \geq 11$ **c)**$-4 \leq y \leq 4$ **d)**$-1 < y < 1$ **e)**$-2 < y < 2$
f)$-3 \leq y \leq 3$ **6.a)**$-2 \leq x \leq 6$ **b)**$x \leq -2$ or $x \geq 6$
c)$-3 < x < 9$ **d)**$x < -3$ or $x > 9$ **7.b)**$0 < x < 4$
8.b)$0 < y < 4$ **9.a)**$-5 < x < 11$ **b)**$x < -5$ or $x > 1$
c) $-\frac{5}{2} < x < \frac{3}{2}$ **d)**$x \leq -1$ or $x \geq 2$ **e)** $-\frac{9}{2} < x < 2$
f)$x \in R$ **10.a)**$-9 \leq x \leq 3$ **b)**$x \leq -7$ or $x \geq -1$
c) $-\frac{7}{2} < x < \frac{15}{2}$ **d)**$-7 < x < 5$ **11.a)**$y \geq -\frac{1}{3}$ **b)**$-1 < m < 3$
c)$x \leq \frac{1}{3}$ or $x \geq 1$ **d)**$\frac{5}{4} \leq k \leq \frac{5}{2}$ **e)**$-1 < y < 3$ **f)**$a < -1$

3.5 Exercise, page 106

1.a)9 **b)**$\frac{1}{64}$ **c)**625 **d)**0.008 **e)**-0.008 **f)**-0.008 **g)**$\frac{1}{25}$

h)$\frac{1}{64}$ **i)**1 **j)**$\frac{9}{4}$ **k)**$\frac{16}{81}$ **l)**1 **2.a)**$\sqrt{7}$ **b)**$\sqrt[5]{x}$ **c)**$\sqrt[4]{a}$

d)$\sqrt[5]{8^2}$ **e)**$\frac{1}{2}$ **f)**$\sqrt[3]{b^3}$ **g)**$5\sqrt[3]{x^2}$ **h)**$\sqrt{\frac{1}{x}}$ **i)**$\sqrt[5]{\frac{1}{a^2}}$

j)0.1 **3.a)**$6^{\frac{1}{2}}$ **b)**$10^{\frac{1}{2}}$ **c)**$7^{\frac{1}{3}}$ **d)**$\left(\frac{1}{4}\right)^{\frac{1}{2}}$ **e)**$3^{\frac{2}{3}}$ **f)**$12^{\frac{3}{7}}$

g)$7^{\frac{4}{3}}$ **h)**$8^{\frac{9}{7}}$ **i)**$-0.027^{\frac{1}{3}}$ **j)**$7^{-\frac{1}{2}}$ **4.a)**$x^{\frac{1}{5}}$ **b)**$x^{\frac{1}{2}}$ **c)**$x^{-\frac{1}{3}}$

d)$x^{-\frac{3}{2}}$ **e)**$a^{\frac{5}{6}}$ **f)**$p^{\frac{7}{8}}$ **g)**$a^{-\frac{1}{4}}$ **h)**$a^{\frac{4}{3}}$ **i)**$a^{\frac{2}{3}}$ **j)**b **5.a)**3

b)4 **c)**27 **d)**0.2 **e)**2 **f)**8 **g)**32 **h)**0.1 **i)**27 **j)**4 **6.a)**8

b)$\frac{1}{9}$ **c)**$\frac{1}{32}$ **d)**$\frac{1}{27}$ **e)**1000 **f)**0 **g)**$\frac{1}{5}$ **g)**$\frac{3125}{32}$ **i)**$\frac{1}{3}$ **j)**5

7.a)$\frac{1}{2}y^{-\frac{1}{2}}$ **b)**y^{-1} **c)**1 **d)**$\frac{27}{2}x^{-\frac{1}{4}}$ **e)**xy **f)**$|x|\sqrt{3}$ **8.a)**$3^{\frac{3}{4}}$ **b)**$5^{\frac{4}{9}}$

c)$a^{\frac{1}{2}}b$ **d)**a^2b^3 **e)**xy^2z^3 **f)**$y + 2y^{\frac{3}{2}}$ **g)**$3x^3y^2$ **h)**$2x^3y^2$

9.a)15 **b)**$\frac{1}{2}$ **c)**$\frac{431}{27}$ **d)**53.064 **10.a)**$4x^4y^6z^8$ **b)**$x^{\frac{2}{3}}y^{\frac{1}{6}}$ **c)**x^n

11.a)$x \geq -3$ **b)**$x \geq 4$ **c)**$x \geq -\frac{5}{2}$ **d)**$x \geq -4$ **e)**$x \geq \frac{4}{3}$

f)$|x| \geq 3$ **g)**$x \geq \frac{2}{3}a$ **h)**$x \geq -\frac{3b}{a}$ **12.a)**18 m **b)**54 m

13.a)24 m **b)**8 m **14.a)**21.4 m **b)**16.6 m **15.a)**No **b)**Yes

3.6 Exercise, page 111

1.a)$4\sqrt{2}$ **b)**$-12\sqrt{2}$ **c)**$2\sqrt{2}$ **d)**$-\sqrt{5}$ **e)**$-12\sqrt{3}$
f)$10\sqrt{3}$ **g)**$2\sqrt[3]{2}$ **h)**$-4\sqrt[3]{3}$ **i)**$2\sqrt[4]{3}$ **j)**$-3\sqrt{5}$ **k)**$2\sqrt[3]{3}$
l)$-6\sqrt{2}$ **2.a)**$\sqrt{18}$ **b)**$-\sqrt{48}$ **c)**$\sqrt{675}$ **d)**$\sqrt{288}$
e)$-\sqrt{108}$ **f)**$\sqrt[3]{24}$ **g)**$-\sqrt[3]{54}$ **h)**$\sqrt[4]{432}$ **3.a)**$2\sqrt{2}$
b)$2\sqrt[4]{2}$ **c)**$-5z\sqrt[3]{z^2}$ **d)**$2\sqrt[3]{2y^3}$ **e)**$2x\sqrt[6]{2x}$ **f)**27
g)$2y^2\sqrt[9]{y}$ **h)**$-2\sqrt[3]{9x}$ **i)**$3a^3$ **j)**$2p^2q\sqrt[4]{q}$ **4.a)**$\frac{\sqrt{3}}{3}$
b)$\frac{\sqrt{12}}{3}$ **c)**$\frac{2\sqrt{3}}{3}$ **d)**$\frac{2}{5}$ **e)**$\frac{|a|\sqrt{b}}{b^2}$ **f)**$\frac{-a}{3b}\sqrt[3]{9ab}$ **g)**$\frac{|q|\sqrt[3]{2p}}{2r^2}$
h)$\frac{-2x}{y^2x}\sqrt[5]{w^3y^3}$ **i)**$\frac{\sqrt[3]{2}}{2}$ **j)**$\frac{x\sqrt[6]{75y^4z^2}}{7z^2}$ **5.a)**$\frac{1}{3}$
b)$\frac{-5}{a^2}$ **c)**$\frac{\sqrt{ab}}{b^2}$ **d)**$\frac{q^2\sqrt[3]{p}}{p}$ **e)**$\frac{-5n}{m^2}\sqrt[3]{m^2n^2}$ **f)**$\frac{-3d}{ce^2}\sqrt[3]{c^2}$
g)$\frac{-2y^2}{x^2}\sqrt[5]{x^4}$ **h)**$\frac{\sqrt{3}}{|v|w^2}$ **i)**$\frac{\sqrt[4]{b^3}}{3|a|}$ **j)**$\frac{\sqrt[3]{2}\sqrt[6]{xy^3}}{y}$ **6.a)**$2\sqrt[4]{8}$ **b)**$\sqrt{2|b|}$
c)$2a|b^3|\sqrt{2a}$ **d)**$a^2b^2\sqrt{|b|}$ **e)**$\sqrt[3]{4a^2}$ **f)**$\frac{2a^2}{b^4}\sqrt{2ab}$ **g)**$|w|2vw$
h)$a|b^3|\sqrt{a}$ **7.a)**$4a\sqrt[6]{4a}$ **b)**$w^2\sqrt[4]{w}$ **c)**$\frac{2p^3}{q^2}$ **d)**$-\frac{a^2b^3}{3}$ **e)**$-\frac{a^2b^3}{3}$
f)$\frac{\sqrt[4]{b^4-a^4}}{ab}$ **8.**B and C **9.a)**$\sqrt{50}$ **b)**$\sqrt{48}$ **10.a)**$\sqrt{72}, \sqrt{75}$ **b)**B

3.7 Exercise, page 113

1.a)$\sqrt{2}$ **b)**$2\sqrt{3} - \sqrt{2}$ **c)**$2\sqrt{3}$ **d)**$\sqrt{3} + \sqrt{7}$ **2.a)**$6 - 6\sqrt{3}$
b)$9\sqrt{2} - 15$ **c)**$8\sqrt{2} - 4\sqrt{3}$ **d)**$15\sqrt{5} + 25\sqrt{6}$
e)$-\sqrt{3} - 2\sqrt[3]{2}$ **f)**$6\sqrt[3]{2} - 9\sqrt{2}$ **3.a)**$12\sqrt{2} + 6\sqrt{2} - 6\sqrt{2}$
b)$12\sqrt{2}$ **4.a)**$\sqrt{3}$ **b)**$-23\sqrt{5}$ **c)**$18\sqrt{2}$ **d)**$5\sqrt[3]{2}$ **e)**$-37\sqrt{2}$
f)$31\sqrt{2}$ **5.a)**$3\sqrt{2} - 4\sqrt{3}$ **b)**$-54\sqrt{2} - 2\sqrt{7}$ **c)**$14\sqrt[3]{2} - 3\sqrt[3]{3}$
6.a)$-5\sqrt{3}$ **b)**$18\sqrt{2}$ **c)**$-16\sqrt{5}$ **d)**$15\sqrt[3]{5}$ **7.a)**$12\sqrt{2} - 45\sqrt{5}$
b)$12\sqrt{2} - 45\sqrt{5}$ **8.a)**$20 - 25\sqrt{2}$ **b)**$\sqrt{2} - 5\sqrt{3}$ **c)**$17\sqrt[3]{5}$
d)$33\sqrt{2} - 4\sqrt{3}$ **e)**$\sqrt[3]{3} - \sqrt[3]{5}$ **9.a)**$-3\sqrt{2}$ **b)**$-4\sqrt{3}$
c)$20\sqrt{3} - 24\sqrt{2}$ **d)**$12\sqrt{2} - 10\sqrt{3}$ **10.**A **11.**B **12.**(a)
14.a)$(7b - 6a)\sqrt{2}$ **b)**$(-6n - 13m)\sqrt{3}$ **c)**$(8y - 3x)\sqrt{5}$
15.a)$36, 36, -36$ **16.a)**$2\sqrt{2}|x|$ **b)**$7x\sqrt{2x}$ **c)**$6|a|b\sqrt{b}$
d)$3a^2b\sqrt{3b}$ **e)**$4m^2n\sqrt{5n}$ **f)**$8|x|z^2\sqrt{y}$ **g)**$3|x|\sqrt{y}$
h)$2pq\sqrt[3]{3q^2}$ **17.a)**$3|x| - 2x\sqrt{x} - 5x^2$ **b)**$2b\sqrt{b} - 3b^2\sqrt{b}$
c)$(3|x| - 2x^2y)\sqrt{y}$ **d)**$(3y - 6x + 2xy^2)xy\sqrt{xy}$
18.$10\sqrt{3} + 9\sqrt{2}$ **19.a)**$(|a| - 3)\sqrt{b} - 2|b|\sqrt{a}$

b)$(1 - b - 3ab - a)\sqrt{ab}$ c)$(9 - 10|a| - 3|a||b|)\sqrt{b}$
20.a)$(7b + 2|a| - 2)\sqrt{b}$ b)$(a + 2)\sqrt{a} + 7(8 - 3b)\sqrt{b}$
c)$(7a + 26ab + 4)\sqrt{ab}$

3.8 Exercise, page 119
1.a)$-6\sqrt{6}$, $6\sqrt{18}$ b)$-72\sqrt{6}$, $-72\sqrt{6}$ c)360 2.a)$6\sqrt{6}$
b)$-8\sqrt{30}$ c)$15\sqrt{15}$ d)$-\sqrt{15}$ e)$15\sqrt{15}$ f)$4\sqrt{5}$ g)$15\sqrt{30}$
h)-6 i)$-6\sqrt{6}$ 3.a)$-108\sqrt{14}$ b)$-96\sqrt{3}$ c)$-180\sqrt{3}$
d)$-240\sqrt{30}$ 4.a)$6\sqrt{10} - 9\sqrt{15}$ b)$9\sqrt{6} - 27\sqrt{2}$
c)$8\sqrt{15} - 24\sqrt{2}$ d)$6\sqrt{6} - 12\sqrt{3}$ e)0 f)$18\sqrt{6}$
g)$12 - 6\sqrt{6}$ h)$18 - 18\sqrt{6}$ 5.a)$3a^2 + 5ab - 2b^2$
b)$5\sqrt{6}$ c)$6\sqrt{6} + 29$, $66 - 36\sqrt{2}$ 6.a)$-1 - \sqrt{6}$
b)$21 - 9\sqrt{6}$ c)$42 - 12\sqrt{10}$ d)$21\sqrt{15} - 77$
e)$100 - 40\sqrt{6}$ f)$180 - 72\sqrt{6}$ 7.a)$40\sqrt{2} - 6\sqrt{10}$
b)$10 + 3\sqrt{6}$ c)$12\sqrt{6} - 18$ 8.a)$8 + 2\sqrt{15}$
b)$20 + 8\sqrt{6}$ c)$38 - 12\sqrt{10}$ d)$265 - 30\sqrt{70}$
e)$18 - 12\sqrt{2}$ f)$100 + 50\sqrt{3}$ 9.a)$15\sqrt{15} - 53$
b)$42 - 81\sqrt{2}$ c)$10 - 22\sqrt{15}$ d)$224\sqrt{3} - 1016$
10.b)-66 11.a)-8 b)75 c)$375 - 96\sqrt{15}$
12.a)$594\sqrt{6} - 1768$ b)$-68\sqrt{6}$ 13.a)$-6\sqrt{6}$ b)$18\sqrt{6}$
c)$-3\sqrt{6}$ d)30 e)6 f)-6 g)$18 - 12\sqrt{6}$ h)$12 + 12\sqrt{6}$
14.a)$40 + 22\sqrt{5}$ b)$-60 - 33\sqrt{5}$ c)$4 - 32\sqrt{5}$

3.9 Exercise, page 123
1.a)-2 b)10 c)39 d)-20 2.a)3 b)37 c)135 d)-2
e)-52 f)61 g)5 h)46 i)-9 j)-270 k)-20 l)1163
3.a)$3\sqrt{3} - 2$ b)$6\sqrt{5} - 2\sqrt{2}$ c)$46 + 16\sqrt{3}$
d)$403 - 108\sqrt{3}$ e)$75\sqrt{6} - 30\sqrt{3}$ f)$72\sqrt{6} - 237$

3.10 Exercise, page 125
1.(b), (c), (e), (f), (g), (i), (j) 2.a)$\sqrt{30}$, $2\sqrt{15}$
3.(c) and (g) 4.a)$6\sqrt{2}$ b)$\frac{2\sqrt{15}}{5}$ c)$\frac{\sqrt{6}}{3}$ d)$4\sqrt{3}$ e)$-3\sqrt{2}$
f)$-\frac{\sqrt{3}}{3}$ g)$-\frac{\sqrt{10}}{3}$ h)$-\frac{\sqrt{2}}{2}$ i)$6\sqrt{5}$ j)$-2\sqrt{3}$ k)$\frac{-2\sqrt{15}}{15}$
l)$-\frac{\sqrt{5}}{10}$ 5.a)$4\sqrt[3]{4}$ b)$12\sqrt[3]{3}$ c)$\frac{\sqrt[3]{28}}{2}$ d)$20\sqrt{2}$ e)$\sqrt[4]{8}$ f)$\frac{1}{2}$
6.a)$\frac{\sqrt{6}}{12}$ b)$16\sqrt{2}$ c)$-2\sqrt{7}$ d)$2\sqrt{5}$ e)27 f)$\frac{3\sqrt{2}}{2}$ g)$\frac{\sqrt{15}}{3}$
h)$4\sqrt{3}$ i)$-\frac{1}{6}$ j)$-\frac{4\sqrt{5}}{3}$ k)10 l)$\frac{\sqrt[4]{5}}{2}$ 7.a)$-3\sqrt{2}$ b)$-\frac{3}{2}$
c)$-\frac{5\sqrt{2}}{2}$ d)$-\frac{\sqrt{2}}{2}$ e)$-\frac{2\sqrt{30}}{9}$ f)$\frac{1}{4}$ g)$\frac{\sqrt{6}}{6}$ h)$-10\sqrt{2}$
i)$\frac{\sqrt[3]{20}}{5}$ j)-30 8.a)$-4\sqrt{2} - 2\sqrt{6}$ b)$\frac{2 - \sqrt{2}}{4}$
c)$6\sqrt{6} - 1$ d)$\frac{3}{2} - \frac{3\sqrt{6}}{4}$ e)$\frac{3 - \sqrt{10}}{3}$ f)$\frac{5\sqrt{10}}{16} - \frac{1}{4}$

3.11 Exercise, page 127
1.a)$\frac{\sqrt{5}}{5}$ b)$\frac{\sqrt{10}}{5}$ c)$\frac{3\sqrt{2}}{2}$ d)$2\sqrt{6}$ e)$\frac{5\sqrt{6}}{9}$ f)$3\sqrt{10}$

2.a)$\frac{\sqrt{6} + \sqrt{10}}{2}$ b)$\sqrt{6} - 3$ c)$2 + \frac{\sqrt{6}}{2}$ d)$\frac{3\sqrt{10} - 2}{4}$
3.a)$\sqrt{5} + \sqrt{2}$ b)$10 - 3\sqrt{10}$ c)$5 - 2\sqrt{6}$ d)$4 - 2\sqrt{5}$
e)$\frac{11\sqrt{6} - 16}{47}$ f)$\frac{35 - 12\sqrt{6}}{19}$ 4.a)$8\sqrt{10} + 24$ b)$8\sqrt{10} + 24$
5.a)$\sqrt{6} + 2$ b)$\frac{8\sqrt{3} + 12\sqrt{2}}{-15}$ c)$\frac{9\sqrt{2} + 2\sqrt{3}}{25}$ d)$2\sqrt{2} + \sqrt{6}$
e)$\frac{3\sqrt{5}}{25}$ f)$6 + \frac{5\sqrt{6}}{2}$ g)$\frac{-12\sqrt{15} - 15\sqrt{10}}{2}$ h)$4 - \frac{3\sqrt{6}}{2}$
i)$\frac{9 + 4\sqrt{6}}{-15}$ 6.a)$4\sqrt{3}$ b)$2\sqrt{5}$ 7.a)$7\sqrt{2}$ b)$\frac{14\sqrt{6} - 21}{5}$
c)$3\sqrt{2} - 2\sqrt{3}$ d)$\frac{38 + 7\sqrt{6}}{25}$ 8.a)$\frac{1}{4}$ b)$\frac{2\sqrt{6} + \sqrt{3}}{7}$
9.a)$\frac{90 + 6\sqrt{10}}{43}$ b)$\frac{49 - 9\sqrt{10}}{43}$ 10.a)3 b)$\sqrt{\frac{19}{3} - 2\sqrt{6}}$

3.12 Exercise, page 131
1.a)$2x - 1$ b)$9x - 6\sqrt{x} + 1$ c)$4m - 32\sqrt{m} + 64$
d)$x + 4\sqrt{x - 5} - 1$ e)$9m - 12\sqrt{m - 2} - 14$
f)$9y - 48\sqrt{y - 1} + 55$ 2.a)6 b)2 c)3 d)7 e)4 f)3
3.a)17 b)3 c)3 d)5 e)5 f)3 4.a)16 b)none c)both
d)both 5.a)36 b)9 c)4 d)63 e)10 f)8 6.a)25 b)17 c)5
d)8 e)32 f)$\frac{49}{9}$ g)36 h)49 7.6 8.a)$\frac{8}{9}$, 2 b)4 c)$\frac{21}{4}$, 1 d)$\frac{3}{4}$,
1 9.a)0, 3 b)4 c)$\frac{117}{4}$ d)2, 38 10.a)-1, 3 b)144 c)$\frac{1681}{144}$
11.a)0, 8 b)12 12.a)13 b)-21 c)12, -12

3.13 Exercise, page 134
1.a)$\sqrt{5x + 4}$ b)$\sqrt{3x}$ c)$\sqrt{5x + 4} = \sqrt{3x} + 2$ d)$x = 12$
e)A:8 units, B:6 units 2.b)$\sqrt{2x + 6} - \sqrt{x + 37} = 4$
c)$\sqrt{26\,243}$; $5\sqrt{171}$ 3.4, 1 or 5, 8 4.3, 5 5.$8\sqrt{3}$, $2\sqrt{205}$
6.4 7.6 8.5 9.6, 3

3.14 Exercise, page 137
1.a)-7 b)-7 2.a)1 3.a)$\frac{225}{16}$ b)$\frac{1}{2}$ 4.a)1 b)3 c)-1
5.a)-1 b)5 6.a)1 b)-1 7.a)9 b)-2 c)3 d)$\frac{11}{7}$

CHAPTER 4

4.1 Exercise, page 144
1.a)$\sin\theta = \frac{12}{13}$, $\cos\theta = \frac{5}{13}$, $\tan\theta = \frac{12}{5}$ b)$\sin\theta = \frac{3}{5}$,
$\cos\theta = \frac{4}{5}$, $\tan\theta = \frac{3}{4}$ c)$\sin\theta = \frac{3}{5}$, $\cos\theta = -\frac{4}{5}$,
$\tan\theta = -\frac{3}{4}$ d)$\sin\theta = -\frac{3}{5}$, $\cos\theta = -\frac{4}{5}$, $\tan\theta = \frac{3}{4}$
e)$\sin\theta = \frac{5}{13}$, $\cos\theta = \frac{12}{13}$, $\tan\theta = \frac{5}{12}$ f)$\sin\theta = -\frac{5}{13}$,

cos $\theta = \frac{12}{13}$, tan $\theta = -\frac{5}{12}$ g)sin $\theta = -\frac{5}{13}$,

cos $\theta = -\frac{12}{13}$, tan $\theta = \frac{5}{12}$ h)sin $\theta = \frac{5}{13}$,

cos $\theta = -\frac{12}{13}$, tan $\theta = -\frac{5}{12}$ 2.a)csc $\alpha = \frac{5}{3}$,

sec $\alpha = \frac{5}{4}$, cot $\alpha = \frac{4}{3}$ b)csc $\theta = -\frac{13}{5}$, sec $\theta = -\frac{13}{12}$,

cot $\theta = \frac{12}{5}$ c)csc $\theta = \frac{25}{24}$, sec $\theta = -\frac{25}{7}$, cot $\theta = -\frac{7}{24}$

d)csc $\theta = -\frac{25}{24}$, sec $\theta = \frac{25}{7}$, cot $\theta = -\frac{7}{24}$ 3.a)$\frac{7}{\sqrt{53}}$

b)$\frac{2}{\sqrt{53}}$ c)$\frac{7}{2}$ 4.a)csc $\alpha = \frac{\sqrt{202}}{11}$, b)sec $\alpha = -\frac{\sqrt{202}}{9}$,

c)cot $\alpha = -\frac{9}{11}$ 5.b)sin $\theta = \frac{\sqrt{7}}{4}$, tan $\theta = -\frac{\sqrt{7}}{3}$

6.a)csc $\theta = -\frac{5}{3}$, sec $\theta = -\frac{5}{4}$, cot $\theta = \frac{4}{3}$

b)sin $\alpha = \frac{-\sqrt{161}}{15}$, tan $\alpha = \frac{-\sqrt{161}}{8}$ 7.a)sin $\theta = \frac{2}{\sqrt{5}}$,

sec $\theta = \sqrt{5}$ b)cos $\beta = -\frac{\sqrt{65}}{\sqrt{114}}$, csc $\beta = \frac{\sqrt{114}}{7}$

8.a)sin $\theta = -\frac{5}{13}$, cos $\theta = -\frac{12}{13}$, tan $\theta = \frac{5}{12}$,

csc $\theta = -\frac{13}{5}$, cot $\theta = \frac{12}{5}$ b)cos $\alpha = \frac{24}{25}$, tan $\alpha = \frac{7}{24}$,

csc $\alpha = \frac{25}{7}$, sec $\alpha = \frac{25}{24}$, cot $\alpha = \frac{24}{7}$ c)sin $\theta = -\frac{8}{17}$,

cos $\theta = \frac{15}{17}$, tan $\theta = -\frac{8}{15}$, sec $\theta = \frac{17}{15}$, cot $\theta = -\frac{15}{8}$

9.a)(i)± 4 (ii)± 6 (iii)± 2 b)(i)sin $\theta = \pm\frac{4}{5}$, cos $\theta = \frac{3}{5}$

tan $\theta = \pm\frac{4}{3}$ (ii)sin $\theta = \frac{4}{5}$, cos $\theta = \pm\frac{3}{5}$, tan $\theta = \pm\frac{4}{3}$

(iii)sin $\theta = \pm\frac{2}{\sqrt{13}}$, cos $\theta = \frac{3}{\sqrt{13}}$, tan $\theta = \pm\frac{2}{3}$ 10.1 11.$\frac{6}{17}$

4.2 Exercise, page 147

2.a)sin $\theta = \frac{24}{25}$, cos $\theta = \frac{7}{25}$, tan $\theta = \frac{24}{7}$ b)sin $\theta = \frac{4}{5}$,

cos $\theta = -\frac{3}{5}$, tan $\theta = -\frac{4}{3}$ c)sin $\theta = -\frac{12}{13}$, cos $\theta = -\frac{5}{13}$,

tan $\theta = \frac{12}{5}$ d)sin $\theta = -\frac{3}{5}$, cos $\theta = \frac{4}{5}$, tan $\theta = -\frac{3}{4}$

3.a)csc $\theta = \frac{\sqrt{34}}{5}$, sec $\theta = -\frac{\sqrt{34}}{3}$, cot $\theta = -\frac{3}{5}$ b)csc $\theta = -\frac{\sqrt{13}}{2}$,

sec $\theta = -\frac{\sqrt{13}}{2}$, cot $\theta = \frac{2}{3}$ c)csc $\theta = -\frac{\sqrt{13}}{2}$,

sec $\theta = -\frac{\sqrt{13}}{3}$, cot $\theta = \frac{3}{2}$ d)csc $\theta = -\frac{\sqrt{34}}{3}$, sec $\theta = \frac{\sqrt{34}}{5}$,

cot $\theta = -\frac{5}{3}$ 4.a)$(-\sqrt{3}, -1)$ b)sin $\alpha = -\frac{1}{2}$, tan $\alpha = \frac{1}{\sqrt{3}}$

5.a)$(-8, 15)$ b)cos $\theta = -\frac{8}{17}$, sec $\theta = -\frac{17}{8}$,

cot $\theta = -\frac{8}{15}$ 6.a)2nd, 3rd c)sin $\theta = \pm\frac{24}{25}$ 7.a)1st,

2nd c)cos $\beta = \pm\frac{3}{5}$, tan $\beta = \pm\frac{4}{3}$ 8.a)2nd, 4th

c)sin $\alpha = \pm\frac{7}{25}$, cos $\alpha = \pm\frac{24}{25}$ 9.a)cos $\theta = \pm\frac{15}{17}$

b)sin $\alpha = \pm\frac{5}{13}$ c)tan $\beta = \pm\frac{24}{7}$ d)cot $\theta = \pm\sqrt{3}$

10.a)sin $\theta = \pm\frac{24}{25}$ b)$\pm\frac{168}{625}$ c)± 1 12.c)$\frac{\sin \theta}{\cos \theta} = \tan \theta$

13.a)+ b)- c)- d)+ e)- f)- 14.a)- b)+ c)-
d)- e)+ f)- 15.a)- b)+ c)+ d)- e)- f)- 16.a)I
b)III c)II d)IV e)II f)I 17.a)I, II b)II, III c)I, III d)III,
IV e)I, III f)I, IV 18.a)I+, II+, III-,IV- b)I+, II-,
III-, IV+ c)I+, II-, III+, IV- d)I+, II+, III-, IV-
e)I+, II-, III-, IV+ f)I+, II-, III+, IV-

4.3 Exercise, page 151

1.a)480° b)-405° c)390° d)-480° 2.a)585°,
945°, -135°, ... b)225° 3.a)-30°, -390° b)330°
4.a)-315°, -675° b)-240°, -600° c)-340°,
-700° 5.a)315°, 675° b)240°, 600° c)300°, 660°
6.b)-290° c)70° d)70° 7.b)210° 8.a)160° b)135°
c)330° d)300° e)60° f)240° 9.a)30° b)45° c)55°
d)255° e)240° f)60° g)10° h)20° i)220° j)260°
10.(c), (d) 11.(a), (c), (d) 12.(a), (b), (c) 13.(b), (c), (d)

4.4 Exercise, page 155

1.sin: $\frac{1}{2}$, $\frac{1}{\sqrt{2}}$, $\frac{\sqrt{3}}{2}$; cos: $\frac{\sqrt{3}}{2}$, $\frac{1}{\sqrt{2}}$, $\frac{1}{2}$; tan: $\frac{1}{\sqrt{3}}$, 1, $\sqrt{3}$

2.a)$\frac{1}{\sqrt{2}}$ b)$\frac{1}{2}$ c)$\frac{\sqrt{3}}{2}$ d)$\frac{\sqrt{3}}{2}$ e)$\sqrt{2}$ f)$\sqrt{3}$ g)$\frac{1}{\sqrt{3}}$ h)$\frac{2}{\sqrt{3}}$

i)1 j)$\frac{2}{\sqrt{3}}$ k)$\sqrt{2}$ l)$\frac{1}{\sqrt{3}}$ 3.a)sin $\theta = \frac{1}{\sqrt{2}}$, cos $\theta = -\frac{1}{\sqrt{2}}$,

tan $\theta = -1$, csc $\theta = \sqrt{2}$, sec $\theta = -\sqrt{2}$, cot $\theta = -1$

b)sin $\theta = -\frac{1}{2}$, cos $\theta = -\frac{\sqrt{3}}{2}$, tan $\theta = \frac{1}{\sqrt{3}}$, csc $\theta = -2$,

sec $\theta = -\frac{2}{\sqrt{3}}$, cot $\theta = \sqrt{3}$ 4.a)sin $\theta = -\frac{1}{\sqrt{2}}$, cos $\theta = -\frac{1}{\sqrt{2}}$,

tan $\theta = 1$ b)sin $\theta = \frac{\sqrt{3}}{2}$, cos $\theta = \frac{1}{2}$, tan $\theta = \sqrt{3}$

c)sin $\theta = -\frac{1}{2}$, cos $\theta = \frac{\sqrt{3}}{2}$, tan $\theta = -\frac{1}{\sqrt{3}}$ d)sin $\theta = \frac{\sqrt{3}}{2}$,

cos $\theta = -\frac{1}{2}$, tan $\theta = -\sqrt{3}$ 5.b)sin 300° $= -\frac{\sqrt{3}}{2}$,

cos 300° $= \frac{1}{2}$, tan 300° $= -\sqrt{3}$ 6.b)csc $(-225°) = \sqrt{2}$,

sec $(-225°) = -\sqrt{2}$, cot $(-225°) = -1$ 7.a)$-\frac{\sqrt{3}}{2}$

b)$-\frac{\sqrt{3}}{2}$ c)Answers same because the angles are coterminal.

8.a) $-\dfrac{1}{\sqrt{2}}$ **b)** $\dfrac{\sqrt{3}}{2}$ **c)** $\dfrac{1}{\sqrt{3}}$ **d)** -1 **e)** $\dfrac{2}{\sqrt{3}}$ **f)** $-\sqrt{3}$ **g)** $\sqrt{2}$ **h)** -1

i) $-\dfrac{2}{\sqrt{3}}$ **9.a)** $\dfrac{1}{\sqrt{3}}$ **b)** $-\dfrac{1}{2}$ **c)** $-\dfrac{1}{\sqrt{3}}$ **d)** $-\sqrt{2}$ **e)** 1 **f)** $\dfrac{1}{2}$

10.a) $\dfrac{-1+\sqrt{3}}{2}$ **b)** $\dfrac{\sqrt{6}}{2}$ **c)** $\dfrac{1}{4}$ **11.a)** $-\sqrt{3}$ **b)** $120°$ **12.a)** $\pm\dfrac{\sqrt{3}}{2}$

b) $120°$ or $240°$ **13.a)** $30°,\ 330°$ **b)** $210°,\ 330°$ **c)** $45°,\ 225°$

d) $225°,\ 315°$ **e)** $120°,\ 240°$ **f)** $225°,\ 315°$ **14.** $\dfrac{2\sqrt{3}-3}{4}$

4.5 Exercise, page 158

1.a) 0.7314 **b)** 1.1106 **c)** 1.4142 **d)** 0.6428 **e)** 1.4663
f) 1.0000 **g)** 1.2361 **h)** 0.7431 **i)** 0.6157 **j)** 0.7986
k) 0.7813 **l)** 1.2868 **2.a)** I **b)** D **c)** I **d)** D **e)** I **f)** D
3.a) 49° **b)** 49° **c)** 50° **d)** 48° **e)** 52° **f)** 54° **4.a)** 48°
b) 47° **c)** 48° **d)** 47° **e)** 52° **f)** 51° **5.a)** 0.9205
b) 0.7193 **c)** 5.2408 **d)** 1.0187 **e)** 1.1223 **f)** 0.3249
g) 6.3138 **h)** 0.8192 **6.a)** 0.7986 **b)** 0.0349 **c)** 0.7071
d) 1.1504 **e)** -0.7071 **f)** 1.8361 **g)** 2.9238 **h)** 7.1154
7.a) 59° **b)** $-304°$ **8.a)** $\sin\theta = 0.7431,\ \cos\theta = -0.6691,$
$\tan\theta = -1.1106$ **b)** $\sin\theta = -0.4695,\ \cos\theta = -0.8829,$
$\tan\theta = 0.5317$ **c)** $\sin\theta = -0.5299,\ \cos\theta = 0.8480,$
$\tan\theta = -0.6249$ **d)** $\sin\theta = 0.9703,\ \cos\theta = -0.2419,$
$\tan\theta = -4.0108$ **9.a)** -0.9336 **b)** 0.4245 **c)** -3.8637
10.a) 27° **b)** 34° **c)** 37° **d)** 74° **11.a)** 30° **b)** 48° **c)** 56°
d) 37° **12.a)** 10° **b)** 18° **c)** 36° **d)** 78° **e)** 65° **f)** 25°
g) 20° **h)** 26° **i)** 73° **13.a)** 155° **b)** 209° **c)** 240° **d)** 275°
e) 55° **f)** 211° **14.b)** 245°, 295° **15.a)** 110°, 250°
b) 155°, 335° **16.a)** 243°, 297° **b)** 22°, 338° **c)** 103°,
283° **d)** 116°, 244° **e)** 15°, 195° **f)** 58°, 122° **17.a)** I
b) D **c)** I **d)** D **e)** I **f)** D **18.** (a), (e); For $0° \leq \theta \leq 90°$,
$0 \leq \sin\theta,\ \cos\theta \leq 1$ **20.a)** 10° **b)** 74° **c)** 15°

4.6 Exercise, page 162

1.a) 60° **b)** 90° **c)** 45° **d)** 30° **e)** 135° **f)** $-90°$ **g)** $-120°$
h) 150° **i)** 240° **j)** $-135°$ **k)** $-300°$ **l)** 270° **m)** 360°

n) $-720°$ **o)** 540° **2.a)** π **b)** 2π **c)** $\dfrac{\pi}{2}$ **d)** $\dfrac{\pi}{4}$ **e)** $-\dfrac{\pi}{3}$ **f)** $-\dfrac{5}{6}\pi$

g) $\dfrac{\pi}{6}$ **h)** $\dfrac{4}{3}\pi$ **i)** $-\dfrac{11}{6}\pi$ **j)** $\dfrac{3\pi}{2}$ **k)** $-\dfrac{\pi}{2}$ **l)** $\dfrac{2}{3}\pi$ **3.a)** π **b)** $-\pi$ **c)** $\dfrac{\pi}{2}$

d) $-\dfrac{5}{4}\pi$ **e)** $\dfrac{7}{6}\pi$ **f)** $\dfrac{5}{3}\pi$ **4.a)** II **b)** III **c)** III **d)** I **e)** IV **f)** IV

6.a) π **b)** $\dfrac{4}{3}\pi$ **c)** 4π **7.a)** 63° **b)** 52° **c)** $-17°$ **d)** 86° **e)** $-46°$

f) 23° **8.a)** $\dfrac{2}{3}\pi$ **b)** $-\dfrac{3}{4}\pi$ **c)** $\dfrac{11}{6}\pi$ **d)** $\dfrac{13}{6}\pi$ **9.a)** $\dfrac{\pi}{2}$ **b)** $\dfrac{\pi}{6}$ **c)** 1 **d)** $\dfrac{\pi}{2}$

e) 2.6 **10.a)** 1.0 rad **b)** 2.8 rad **c)** 4.3 rad **d)** 5.3 rad

11.a) $-\dfrac{5}{3}\pi$ **b)** $\dfrac{7}{4}\pi$ **c)** $-\dfrac{11}{6}\pi$ **d)** π **e)** $-\dfrac{5}{4}\pi$ **f)** $\dfrac{\pi}{2}$ **12.a)** 0.0175

b) 0.1396 **c)** 0.4363 **d)** 1.1345 **e)** 2.9671 **f)** -4.1015

13. $\dfrac{5}{4}\pi$ **14.a)** -0.4161 **b)** 1.5574 **c)** -1.0101 **d)** -0.6421

e) -0.9093 **15.a)** 2 rad/s **b)** 132 cm **16.** 30 rad/s

17. 65.4 rad/s **18.** $\dfrac{\pi}{11}$ rad/s **19.** 7.3 km/s

20.a) 7.4 revolutions **b)** 46.3 rad/s

4.7 Exercise, page 167

1.a) $-2\pi,\ -\pi,\ 0,\ \pi,\ 2\pi$ **b)** $-\dfrac{3\pi}{2},\ \dfrac{\pi}{2}$ **c)** $-\dfrac{\pi}{2},\ \dfrac{3\pi}{2}$ **2.a)** 0
b) 1 **c)** -1 **d)** 0 **3.a)** 0.7071 **b)** 0.7071 **c)** -0.7071
d) -0.7071 **4.a)** $-330°,\ -210°,\ 30°,\ 150°$ **b)** $-150°,$
$-30°,\ 210°,\ 330°$ **5.b)** 0 **c)** $-4\pi,\ -3\pi,\ -2\pi,\ -\pi,\ 0,$
$\pi,\ 2\pi,\ 3\pi,\ 4\pi$ **d)** 1 **e)** $-\dfrac{7}{2}\pi,\ -\dfrac{3}{2}\pi,\ \dfrac{\pi}{2},\ \dfrac{5}{2}\pi$ **f)** -1

g) $-\dfrac{5}{2}\pi,\ -\dfrac{\pi}{2},\ \dfrac{3}{2}\pi,\ \dfrac{7}{2}\pi$ **h)** The values of y repeat at equal

intervals **6.a)** radians: $0,\ \dfrac{\pi}{6},\ \dfrac{\pi}{3},\ \dfrac{\pi}{2},\ \dfrac{2}{3}\pi,\ \dfrac{5}{6}\pi,\ \pi;\ \cos\theta:\ 1,$
$0.866,\ 0.5,\ 0,\ -0.5,\ -0.866,\ -1$ **b)** degrees: $180°,$
$210°,\ 240°,\ 270°,\ 300°,\ 330°,\ 360°;\ \cos\theta:\ -1,$
$-0.866,\ -0.5,\ 0,\ 0.5,\ 0.866,\ 1$ **8.a)** $\{\theta|\ -2\pi \leq \theta \leq 2\pi\}$
b) $\{\cos\theta|\ -1 \leq \cos\theta \leq 1\}$ **c)** 1 **d)** $-\dfrac{3}{2}\pi,\ -\dfrac{\pi}{2},\ \dfrac{\pi}{2},\ \dfrac{3}{2}\pi$
e) 1 **f)** -1 **g)** $-2\pi,\ 0,\ 2\pi$ **h)** $-\pi,\ \pi$ **9.c)** the values of y
repeat at equal intervals **10.a)** the values of the function are
repeated at equal intervals **b)** 2π **12.a)** $\left\{\theta|\ -2\pi \leq \theta \leq 2\pi,\right.$
$\left.\theta \neq -\dfrac{3\pi}{2},\ -\dfrac{\pi}{2},\ \dfrac{\pi}{2},\ \dfrac{3\pi}{2}\right\}$ **b)** $-2\pi,\ -\pi,\ 0,\ \pi,\ 2\pi$ **c)** 0 **d)** For
each θ there is a unique value of $f(\theta)$ **15.a)** 0 m **b)** 4 m **c)** 0 m
d) -3.0 m **e)** 3.0 m **f)** -3.0 m **16.a)** 12 h **b)** 0 h, 24 h **c)** 6 h,
18 h **d)** 5 h, 19 h **e)** 11 h, 13 h **f)** 7 h, 17 h **17.b)** 12 h, 36 h
c) 0 h, 24 h, 48 h **18.b)** 23 h

4.8 Exercise, page 171

1.b) same period, same horizontal intercepts
c) different maximum and minimum values **2.a)** 2, -2
b) 3, -3 **c)** (i) 2π (ii) 2π **3.a)** 4 **b)** 6 **4.b)** same period,
same horizontal intercepts **c)** different max. and min.
values **5.a)** 2, -2 **b)** 3, -3 **c)** (i) 2π (ii) 2π **6.a)** 5 **b)** 8
7.b) same max. and min. values, same y-intercepts
c) different periods, different θ-intercepts **8.a)** 1 **b)** -1

c) (i) 2π (ii) π (iii) 4π **9.c)** $\dfrac{2}{3}\pi$ **10.a)** $\dfrac{\pi}{2}$ **b)** 6π **11.b)** same

max. and min. values, same y-intercepts **c)** different
periods, different θ-intercepts **12.a)** 1 **b)** -1 **c)** (i) 2π (ii) π
(iii) 4π **13.c)** $\dfrac{2}{3}\pi$ **14.a)** $\dfrac{\pi}{2}$ **b)** 6π **15.d)** (i) π (ii) $\dfrac{\pi}{2}$ (iii) 2π **16.c)** $\dfrac{\pi}{3}$

17.a) $\dfrac{\pi}{4}$ **b)** 3π **18.b)** same periods, same max. and min. values
c) different intercepts **19.b)** same periods, same max. and
min. values **c)** different intercepts **20.b)** same periods
c) different intercepts

4.9 Exercise, page 175

1.a) 1 **b)** 4 **c)** $\dfrac{1}{2}$ **d)** 2 **2.a)** π **b)** $\dfrac{2\pi}{3}$ **c)** $\dfrac{\pi}{2}$ **d)** 4π **e)** 2π **f)** 2

3.a) $-\dfrac{\pi}{4}$ **b)** $\dfrac{\pi}{3}$ **c)** $-\dfrac{\pi}{2}$ **d)** $-\dfrac{\pi}{3}$ **e)** $\dfrac{\pi}{3}$ **f)** $\dfrac{\pi}{8}$ **g)** $\dfrac{\pi}{3}$ **h)** 2π **4.a)** 0

b) $\dfrac{\pi}{2}$ **c)** 0, π, 2π **d)** 1.4 **5.a)** 0, π, 2π **b)** -1 **c)** 1 d)$\frac{\pi}{2}, \frac{3\pi}{2}$ **6.a)**1 **b)**$\frac{3}{4}\pi, \frac{7}{4}\pi$ **c)**$\frac{\sqrt{2}}{2}$ **d)**$\frac{5}{4}\pi$ **7.a)**1; 0, 2π

b)$\frac{\pi}{8}, \frac{5\pi}{8}, \frac{9\pi}{8}, \frac{13\pi}{8}$; 0; 0 **c)**3; π; 0 **d)**0.7; $\frac{\pi}{4}$; 1

12.a)$y = \sin 2\theta$ **b)**$y = 3 \sin 2\theta$ **c)**$y = \sin \left(\theta + \frac{n}{3}\right)$

13.a)$y = \cos \frac{1}{2}\theta$ **b)**$y = 2 \cos \theta$ **c)**$y = \cos \left(\theta - \frac{\pi}{2}\right)$

4.10 Exercise, page 180
1.a)0 **b)**3 **c)**0 **d)**0 **e)**3 **2.a)**0 **b)**0 **c)**1 **d)**−1 **e)**0 **3.a)**1, π, 0 **b)**2, 2π, 0 **c)**2, $\frac{2\pi}{3}$, 0 **d)**3, π, 0 **e)**1, 2π, $\frac{\pi}{3}$ **f)**1, 2π, $-\frac{\pi}{4}$ **4.a)**1, $\frac{2\pi}{3}$, 0 **b)**3, 2π, 0 **c)**2, $\frac{2\pi}{3}$, 0 **d)**1, 2π, $-\frac{\pi}{3}$ **11.b)**same amplitudes, same periods; different phase shifts **15.a)**$\{y| -a \le y \le a, y \in R\}$ **b)**$\frac{2\pi}{b}$ **c)**a **d)**$-\frac{c}{b}$ **e)**$-a$ **f)**a **16.a)**50 h **b)**5 m **c)**(12.5 + 50k) hours, k ∈ W **d)**−5 m **e)**(37.5 + 50k) hours, k ∈ W **17.a)**(i)(7 + 50k) hours, k ∈ W (ii)no values **b)**(0 + 25k), k ∈ W **c)**0 **18.b)**(9.5 + 50k) hours, k ∈ W **c)**(34.5 + 50k) hours, k ∈ W **19.b)**2k, k ∈ W **c)**0.4 rad

4.11 Exercise, page 184
1.a)$\{\theta| -4\pi \le \theta \le 4\pi, \theta \ne n\pi, n \in I, \theta \in R\}$ **b)**$\{y| |y| \ge 1\}$ **c)**$\frac{-7\pi}{2}, -\frac{3\pi}{2}, \frac{\pi}{2}, \frac{5\pi}{2}$ **d)**$-\frac{5\pi}{2}, -\frac{\pi}{2}, \frac{3\pi}{2}, \frac{7\pi}{2}$ **2.a)**1 **b)**not applicable **3.c)**$\{\theta|\theta \in R, \theta \ne \frac{\pi}{2} + \pi n, n \in I\}$, $\{y|y \in R, |y| \ge 1\}$ **4.c)**$\{\theta|\theta \in R, \theta \ne \pi + \pi n, n \in I\}$, $\{y \in R\}$

5.c)$\left\{\theta|\theta \in R, \theta \ne \frac{\pi}{2} + \frac{\pi n}{2}, n \in I\right\}$, $y||y| \ge 1, y \in R$ R}

6.b)$\{\theta|\theta \in R, \theta \ne 2\pi + 2\pi n, n \in I\}$, $\{y| |y| \ge 1, y \in R\}$

7.b)$\left\{\theta|\theta \in R, \theta \ne \frac{\pi}{4} + \frac{\pi}{2}n, n \in I\right\}$, $\{y| |y| \ge 1, y \in R\}$

4.12 Exercise, page 186
1.a)$y_1 = 3 \sin (2\theta)$ **b)**$y_2 = 2 \sin \left(\frac{2\theta}{3}\right)$

c)$y_3 = \frac{1}{2} \sin \left(\theta - \frac{\pi}{2}\right)$ **d)**$y_4 = 4 \sin \left(\theta - \frac{\pi}{3}\right)$

e)$\{y_1| -3 \le y_1 \le 3, y_1 \in R\}$, $\{y_2| -2 \le y_2 \le 2, y_2 \in R\}$

$\left\{y_3| -\frac{1}{2} \le y_3 \le \frac{1}{2}, y_3 \in R\right\}$, $\{y_4| -4 \le y_4 \le 4, y_4 \in R\}$

2.a)$y = 2 \sin \left(\frac{\theta}{2}\right)$ **b)**$y = 2 \sin \left(\theta + \frac{\pi}{2}\right)$

c)$y = \frac{1}{2} \sin \left(\theta + \frac{\pi}{2}\right)$ **d)**$y = 4 \sin \left(\theta + \frac{3\pi}{4}\right)$

e)$y = 3 \sin\left(\theta + \frac{\pi}{3}\right)$ **f)**$y = 2 \sin \left(\theta + \frac{\pi}{6}\right)$

3.a)$y = 3 \sin\left(\theta + \pi\right)$ **b)**$y = \sin(2\theta + \pi)$ **c)**$y = \frac{1}{2} \sin\left(3\theta - \frac{\pi}{2}\right)$

CHAPTER 5

5.1 Exercise, page 192
6.a)$(1 - \cos \theta)(1 + \cos \theta)$ **b)**$(1 - \sin \theta)(1 + \sin \theta)$
c)$(\sin \theta + \cos \theta)(\sin \theta - \cos \theta)$ **d)**$\sin \alpha(1 - \sin \alpha)$
e)$(\tan \alpha + \cot \alpha)(\tan \alpha - \cot \alpha)$
f)$(\sec \theta + 1)(\sec \theta - 1)$ **7.a)**$\cos \theta$ **b)**$\sin^2 \theta + \cos^2 \theta = 1$
c)$\cos^2 \theta$ **d)**$\sin \theta$ **e)**$-\frac{\cos^2 \theta}{\sin^2 \theta}$ **f)**$\frac{1}{\cos^2 \theta}$ **8.a)**$\frac{1 - \sin^2 \alpha}{\sin \alpha}$
9.a)$\cos^2 \theta = 1 - \sin^2 \theta$ **b)**$(1 - \sin \theta)(1 + \sin \theta)$

5.2 Exercise, page 195
6.a)$\sin \theta$ **b)**$-\sec \theta$ **c)**$-\tan \theta$ **d)**$-\csc \theta$ **e)**$-\sec \theta$
f)$\cot \theta$ **g)**$-\cos \theta$ **h)**$-\sin \theta$ **i)**$\csc \theta$ **j)**$\sec \theta$ **7.a)**$\sin \theta$
b)$\tan \theta$ **c)**$-\sec \theta$ **d)**$-\csc \theta$ **e)**$\cot \theta$ **f)**$-\cos \theta$
g)$-\cot \theta$ **h)**$-\tan \theta$ **i)**$-\cot \theta$ **j)**$-\csc \theta$ **8.a)**$-\frac{1}{2}$
b)$-\frac{1}{\sqrt{3}}$ **c)**$-\sqrt{2}$ **d)**2 **e)**$\sqrt{3}$ **f)**$-\frac{1}{\sqrt{2}}$ **9.a)**$\frac{-2\sqrt{3}}{3}$ **b)**$-\frac{1}{2}$
c)1 **d)**$-\frac{1}{2}$ **e)**$\sqrt{3}$ **f)**$\sqrt{2}$ **10.a)**$-\frac{1}{\sqrt{2}}$ **b)**$-\frac{1}{2}$ **c)**$-\frac{\sqrt{3}}{2}$
11.a)$-\frac{\sqrt{3}}{2}$ **b)**$-\frac{1}{2}$ **c)**$-\frac{1}{\sqrt{2}}$ **12.a)**$-\sqrt{3}$ **b)**$-\frac{1}{\sqrt{3}}$ **c)**-1
13.a)$-\frac{1}{2}$ **b)**$-\sqrt{3}$ **c)**-2 **d)**$\frac{1}{\sqrt{3}}$ **e)**$-\frac{1}{\sqrt{2}}$ **f)**$\frac{2}{\sqrt{3}}$ **g)**-2
h)$\frac{1}{2}$ **i)**1 **j)**2 **k)**-1 **l)**$-\frac{1}{\sqrt{2}}$ **14.a)**$\sqrt{2}$ **b)**$-\frac{1}{\sqrt{3}}$ **c)**$-\frac{1}{\sqrt{2}}$
d)$\frac{1}{\sqrt{2}}$ **e)**$\sqrt{3}$ **f)**2 **g)**$-\frac{2}{\sqrt{3}}$ **h)**-1

5.4 Exercise, page 199
1.a)0, 2π **b)**π **c)**$\frac{\pi}{3}, \frac{5\pi}{3}$ **2.a)**$-\frac{3\pi}{2}, \frac{\pi}{2}$ **b)**$-\frac{\pi}{2}, \frac{3\pi}{2}$
c)$-\frac{5\pi}{6}, -\frac{\pi}{6}, \frac{7\pi}{6}, \frac{11\pi}{6}$ **3.a)**$\frac{\pi}{2}$ **b)**$\frac{3\pi}{2}$ **c)**no root **d)**$\frac{\pi}{6}, \frac{5\pi}{6}$
e)no root **f)**$\frac{\pi}{4}, \frac{3\pi}{4}$ **4.a)**$\frac{5\pi}{6}, \frac{7\pi}{6}$ **b)**$-\pi$, π **c)**$\frac{\pi}{4}, \frac{7\pi}{4}$
5.a)30°, 150° **b)**150°, 210° **c)**120°, 240° **6.a)**3.7, 5.8
b)0.7, 5.6 **c)**0.8, 5.5 **7.a)**$\frac{\pi}{6}, \frac{11\pi}{6}$ **b)**$\frac{\pi + 12k\pi}{6}$,
$\frac{11\pi + 12k\pi}{6}$, k ∈ I **8.a)**$\frac{3\pi}{2}$ **b)**$-\frac{\pi}{2}, \frac{3\pi}{2}$ **9.a)**$\frac{\pi}{4}, \frac{5\pi}{4}$
b)$\frac{5\pi}{4}, \frac{\pi}{4}, -\frac{3\pi}{4}, -\frac{7\pi}{4}$ **10.a)**$1 - \cos^2 \theta$ **b)**$\pm\frac{\pi}{3}, \pm\pi$
11.a)$(2 \sin \theta - 1)(\sin \theta + 1)$ **b)**$-\frac{11\pi}{6}, -\frac{7\pi}{6}, -\frac{\pi}{2}, \frac{\pi}{6}$,
$\frac{5\pi}{6}, \frac{3\pi}{2}$ **12.a)**$\frac{\pi}{4}, \frac{5\pi}{4}$ **b)**$\frac{3\pi}{4}, \frac{7\pi}{4}$ **c)**$\frac{\pi}{12}, \frac{5\pi}{12}$ **d)**$\frac{\pi}{6}, \frac{5\pi}{6}$ **e)**0, 4π
f)3π **13.a)**$-\frac{\pi}{2}, \frac{\pi}{2}$ **b)**$\pm\frac{\pi}{4}, \pm\frac{3\pi}{4}$ **c)**$\pm\frac{\pi}{3}, \pm\frac{2\pi}{3}$ **14.a)**0,
$\frac{\pi}{2}, \frac{3\pi}{2}$, 2π **b)**0, 2π **c)**0, $\frac{\pi}{6}, \frac{11\pi}{6}$, 2π **d)**0, $\frac{\pi}{2}$, 2π
15.a)30°, 150°, 210°, 330° **b)**±60°, ±120°, ±240°,
±300° **c)**$\frac{7\pi + 12k\pi}{6}$, $\frac{11\pi + 12k\pi}{6}$, k ∈ I **d)**$\frac{\pi + 4k\pi}{2}$,

$\dfrac{3\pi + 4k\pi}{2}$, $\dfrac{\pi + 6k\pi}{3}$, $\dfrac{5\pi + 6k\pi}{3}$, $k \in I$ **16.a)**210°,
330° **b)**15°, 75°, 195°, 255° **c)**90°, 210°, 330°

d)53°, 127°, 210°, 330° **17.a)**30°, 330°

b)60°, 150°, 240°, 330° **c)**45°, 135°, 225°, 315°

d)105°, 165°, 285°, 345° **e)**45°, 90°, 135°, 225°, 270°, 315°

f)60°, 300° **g)**60°, 300° **18.a)**$\dfrac{\pi}{6}$, $\dfrac{5\pi}{6}$ or 30°, 150°

b)π or 180° **c)**$-\dfrac{\pi}{2}$, $\dfrac{\pi}{2}$ or $-90°$, 90° **d)**$\dfrac{\pi}{4}$, $-\dfrac{3\pi}{4}$, $\pm\pi$,

45°, $-135°$, $\pm180°$ **e)**0, π, $-\dfrac{\pi}{2}$, $-\pi$ or 0, 180°, $-90°$,

$-180°$ **f)**$-\pi$, $-\dfrac{\pi}{3}$, $\dfrac{\pi}{3}$, π or $-180°$, $-60°$, 60°, 180°

19.a)$-\dfrac{3\pi}{4}$, $-\dfrac{\pi}{4}$, $\dfrac{5\pi}{4}$, $\dfrac{7\pi}{4}$ **b)**$\pm\dfrac{\pi}{4}$, $\pm\dfrac{3\pi}{4}$, $\pm\dfrac{5\pi}{4}$, $\pm\dfrac{7\pi}{4}$

c)$-\pi$, π **d)**$-\dfrac{11\pi}{6}$, $-\dfrac{7\pi}{6}$, $-\dfrac{\pi}{2}$, $\dfrac{\pi}{6}$, $\dfrac{5\pi}{6}$, $\dfrac{3\pi}{2}$ **e)**$-\dfrac{11\pi}{6}$,

$-\dfrac{7\pi}{6}$, $\dfrac{\pi}{6}$, $\dfrac{5\pi}{6}$ **20.a)(i)**$12 + 24k$, $k \in W$ **(ii)**$6 + 24k$,

$k \in W$ **b)(i)**2 s and 10 s **(ii)**14 s or 22 s **21.a)**$24k$, $k \in W$
b)$12 + 24k$, $k \in W$ **22.**30°, 150°, 210°, 330°

5.5 Exercise, page 205
1.a)sec ρ **b)**csc ρ **c)**cot ρ **d)**cos ρ **e)**tan ρ **f)**sin ρ

2.a)$\dfrac{25}{7}$ **b)**$\dfrac{24}{25}$ **c)**$\dfrac{24}{7}$ **3.a)**14.8 **b)**23.3 **c)**96.6 **d)**58.9

4.a)48° **b)**65° **c)**41° **d)**52° **5.a)**6.1 **b)**67° **c)**19°
d)29.8 **6.a)**$\angle R = 23°$, $r = 6.6$, $p = 15.6$
b)$\angle R = 58°$, $p = 6.0$, $r = 5.1$ **c)**$\angle P = 32°$,
$p = 74.2$, $q = 118.7$ **7.a)**$\angle C = 45°$, $\angle B = 45°$
b)$\angle A = 27°$, $\angle B = 63°$ **c)**$\angle A = 55°$, $\angle C = 35°$
8.a)$\angle P = 35°$, $\angle R = 55°$, $q = 12.2$ **b)**$\angle C = 18°$,
$a = 23.1$, $c = 7.1$ **c)**$\angle A = 58°$, AB = 2.8,
BC = 4.4 **9.a)**97.4 m **b)**3.5 m **c)**1.6 m

5.6 Exercise, page 208
1.84 m **2.**20.9 m **3.b)**12° **4.**60.7 m **5.**39.2 m **6.**Yes, at 73°
7.7.449 m **8.a)**30.9 m **b)**29.9 m **9.**27° **10.**52.8 m **11.**446 m
12.192.5 m **13.a)**147.7 m **b)**1.1 m **14.a)**298.9 m **b)**324.6 m

5.7 Exercise, page 212
1.824.3 m **2.b)**103.7 m **3.**68.5 m **4.**897.8 m
5.71.2 m **6.**64.0 m **7.**175.5 m **8.**15.7 m **9.**7.3 m

5.8 Exercise, page 216
1.a)37.9 **b)**21.0 **c)**39.8 **2.a)**61° **b)**49° **c)**10° **3.a)**$a = b =$
26.1 **b)**$h = 40.3$ $g = 73.9$ **c)** $q = 16.6$, $p = 41.2$

d)$s = 11.5$, $u = 13.9$ **4.**$\angle S = 75°$, UN = 8.4,
SN = 6.7 **5.**$\angle Q = 70°$, $\angle P = 63°$, $p = 8.5$
6.12.7 **7.**$\angle A = 84°$, $b = 17.7$, $c = 15.1$
8.$\angle B = 50°$ **9.**$\angle P = 41°$ **10.a)**$\angle P = 56°$,
MP = 6.8, MN = 19.2 **b)**$\angle B = 67°$, AC = 41.9,
AB = 44.9 **c)**$\angle B = 125°$, AB = 14.2, AC = 44.9
d)$\angle D = 58°$, ED = 10.6, EF = 11.5 **11.a)**$\sqrt{2}$ **b)**$\dfrac{\sqrt{6}}{2}$ **12.**49

5.9 Exercise, page 219
1.341.3 km **2.**8.3 m **3.**10.0 m **4.**566.1 m **5.**37.8 m
6. 25.9 m from A, 1121.8 m from B **7.**34.8°
8.16.0 km **9.**273.4 m **10.**62.0 m **11.**1.53×10^8 km
12.2.50×10^8 km **13.**4200 km

5.10 Exercise, page 224
4.a)6.2 **b)**18.7 **c)**8.0 **5.a)**35° **b)**104° **c)**40°
6.a) $\angle A = 65°$, $\angle B = 43°$, $c = 8.4$ **b)**$\angle D = 48°$,
$\angle E = 35°$, $\angle F = 97°$ **c)**$\angle I = 28°$, $\angle L = 44°$,
$j = 15.5$ **d)**$\angle M = 62°$, $\angle N = 40°$, $\angle Q = 78°$
7.a)$p = 18.0$ **b)**cos $A = 0.0355$ **8.a)**$a = 70.2$ m,
$\angle B = 51°$, $\angle C = 67°$ **b)**$\angle D = 49°$, $\angle E = 60°$,
$\angle F = 71°$ **c)**$\angle X = 75°$, $\angle Y = 58°$, $\angle Z = 47°$
d)$p = 120.7$ m, $\angle Q = 39°$, $\angle R = 29°$ **9.**22°
10.40.7 **11.**19.4 cm **12.**26°

5.11 Exercise, page 227
1.11° **2.**1621 m **3.**$\angle A = 31°$, $\angle B = \angle C = 74.5°$
4.1.0 km **5.**7.0 m **6.**35° **7.**5.5 m **8.**4.2 m
9.459.1 m **10.**244.0 m **11.**134° **12.**246.9 m **13.**391.2 m

5.12 Exercise, page 232
2.a)1.4 **b)**none **3.a)**2.4 **b)**one **4.a)**4.3 **b)**two
5.a)2 solutions **b)**1 solution **c)**no solutions
d)2 solutions **6.a)** $\angle B = 62°$, $\angle C = 47°$, $c = 9.4$
b)no solution **c)** $\angle B = 67°$, $\angle C = 69°$, $c = 12.5$ or
$\angle B = 113°$, $\angle C = 23°$, $c = 5.2$ **d)**$\angle E = 103°$,
$e = 12.4$, $\angle F = 35°$ **e)**$\angle D = 50°$, $\angle F = 92°$,
$f = 21.8$; or $\angle D = 130°$, $\angle F = 12°$, $f = 4.5$
f)no solution **7.a)**1193.1 m **8.**32.0 m **9.**147.5 km or
64.4 km **10.**31.8 km or 5.5 km **11.**125° **12.**10.3 km
or 1.0 km **13.a)**$32.1 < a < 50.0$ **b)**$a < 104.2$
c)$a = 61.8$ or $a > 73.7$

5.13 Exercise 235
1.34.9 m **2.**35.6 m **3.**30.0 m **4.**119.5 m **5.**520.0 m
6.15.4 m **7.**38.7 km **8.**25.4 km **9.**0.4 km **10.**736.3 m

5.14 Exercise, page 237
1.a)27.1 **b)**60° **c)**6.3 **d)**37.9 **e)**52° **f)**23° **g)**4.1
h)18.8 **2.a)**30.7 **b)**0.8 **c)**0.8928 **d)**4.5 **e)**177.2 **f)**28°
3.4.7 m, 5.6 m **4.**1494 m **5.**68 m **6.**27.2 m **7.**4.1 m
8.23.3 km **9.**80.7 km **10.**S 32° W

CHAPTER 6

6.1 Exercise, page 245

1.a) $0° < x° < 90°$ **b)** $90° < x° < 180°$
c) $x° = 90°$ **d)** $x° + y° = 180°$ **e)** $x° = 180°$

6.2 Exercise, page 249

1. $\triangle DEF \cong \triangle VTU(ASA)$, $\triangle MNP \cong \triangle RQS(ASA)$,
$\triangle VUL \cong \triangle HGI(ASA)$, $\triangle ACT \cong \triangle STM(SAS)$,
$\triangle JKL \cong \triangle XWN(SSS)$, $\triangle VMT \cong \triangle PAR(HS)$,
$\triangle ABC \cong \triangle RQP (ASA)$
2.a) $AC = EF(HS)$ or $\angle C = \angle E(ASA)$ or $AB = DF(SAS)$
b) $UT = PQ(SAS)$ or $\angle R = \angle S(ASA)$ or $\angle U = \angle P(ASA)$
c) $\angle U = \angle V(ASA)$ or $\angle S = \angle X(ASA)$
d) nothing(ASA) **e)** $\angle J = \angle N(ASA)$ or $\angle L = \angle M(ASA)$

6.4 Exercise, page 257

5.a) $x° = 27.5°$ **b)** $x° = 54°$ **c)** $x = 20°$ **d)** $x° = 19°$
e) $x° = 60°$ **f)** $x° = 80°$ **6.a)** $x° = 100°$, $y° = 160°$
b) $x° = 60$, $y° = 20°$ **c)** $x° = 120°$, $y° = 30°$

6.6 Exercise, page 264

1.a) T **b)** T **c)** F **d)** F **e)** T **f)** F **g)** F **h)** F **i)** F

6.7 Exercise, page 268

1.a) 1, 5; 2, 6; 3, 7; 4, 8; 9, 13; 11, 15; 10, 14;
12, 16 **b)** 3, 6; 4, 5; 11, 14; 13, 12 **c)** 3, 5; 4, 6;
11, 13; 12, 14 **2.a)** 3, 4; 1, 2; 1, 3; 2, 4; 9, 10; 10, 12; 9, 11; 11,
12; 5, 6; 6, 8; 7, 8; 5, 7; 13, 14; 14, 16; 15, 16; 13, 15
b) 1, 4; 2, 3; 5, 8; 6, 7; 9, 12;
10, 11; 13, 16; 14, 15 **4.a)** $\angle BAC = \angle ACD$,
$\angle B = \angle D$, $\angle CAD = \angle ACB$ **b)** $\angle P = \angle R$,
$\angle PSQ = \angle RQS$, $\angle PQS = \angle RSQ$ **5.a)** $m° = 96°$, $n° = 36°$
b) $m° = 28°$, $n° = 47°$ **6.** $m° = 36°$, $n° = 144°$

6.8 Exercise, page 272

1.a) AB, CD; EG and FH **b)** AB, DC; AD, BC

6.10 Exercise, page 279

1. (a), (e); (b), (c); (g), (i); (d), (k); (h), (l); (f), (j)
13.a) (i) A′(−1, 5), B′(−4, 4) (ii) P′(4, −2), Q′(−1, −6), R′(5, −7)

CHAPTER 7

7.1 Exercise, page 293

1.a) $\angle J = \angle S$, $\angle K = \angle T$, $\angle L = \angle U$,
$\dfrac{JK}{ST} = \dfrac{KL}{TU} = \dfrac{JL}{SU}$ **b)** $\angle A = \angle B$, $\angle C = \angle D$,
$\angle E = \angle F$, $\dfrac{AC}{BD} = \dfrac{CE}{DF} = \dfrac{EA}{FB}$ **c)** $\angle G = \angle M$,
$\angle H = \angle N$, $\angle J = \angle P$, $\angle K = \angle Q$,
$\dfrac{GH}{MN} = \dfrac{HJ}{NP} = \dfrac{JK}{PQ} = \dfrac{KG}{QM}$ **2.a)** $\triangle ABC \sim \triangle DEF$
b) $\triangle PQR \sim \triangle SUT$ **c)** not similar

d) quad WXYZ $\sim$ quad PSRQ **3.a)** $x = \dfrac{7}{3}$, $y = \dfrac{28}{3}$, $z = \dfrac{30}{7}$
b) $x = 4.8$, $y = 13$ **c)** $x = 10$, $y = 12$ **d)** $a = 7.5$
$b = \dfrac{21}{2}$, $y = \dfrac{14}{3}$, $z = 8$ **4.b)** $\triangle ABE \sim \triangle ACD$
c) $x = 8$, $y = 5$ **5.** $AB = 4.0$, $DF = 7.0$ **6.** $DC = 4.5$,
$NP = 10.0$, $MQ = 12.0$ **7.** $\angle E = 150°$

7.2 Exercise, page 297

1.a) $\dfrac{AB}{QR} = \dfrac{BC}{RP} = \dfrac{CA}{PQ}$ **b)** $\dfrac{UV}{LK} = \dfrac{VW}{KJ} = \dfrac{WU}{JL}$
3.a) $\triangle RSV \sim \triangle TPV$ **b)** $\dfrac{RS}{TP} = \dfrac{SV}{PV} = \dfrac{VR}{VT}$ **4.a)** 9 **b)** 16
c) $x = 10$, $y = 6$ **d)** $x = 4$, $y = 3$ **e)** $x = 20$, $y = 12$
f) $x = 17.5$, $y = 14$ **5.** $BC = \dfrac{44}{7}$ **6.** $SQ = 7.8$
12. $\dfrac{48}{13}$ m, $\dfrac{72}{13}$ m **13.** 26.2 m

7.3 Exercise, page 300

1.a) similar **b)** similar **c)** similar **d)** similar **e)** not similar
2.a) $\triangle MNQ \sim \triangle DFE$ **b)** $\triangle RST \sim \triangle YWX$ **c)** $\triangle XYZ$
not $\sim \triangle ABC$ **d)** $\triangle PTO \sim \triangle NCA$ **3.a)** $\angle L = \angle Q$,
$\angle K = \angle R$, $\angle M = \angle P$ **b)** $\angle D = \angle H$,
$\angle E = \angle G$, $\angle F = \angle I$ **5.** 100.0 cm, 120.0 cm,
140.0 cm **6.** 13.6 cm, 14.3 cm, 17.0 cm

7.4 Exercise, page 303

1.a) $\triangle PQR \sim \triangle STU$ **b)** $\triangle XYZ \sim \triangle MNP$ **c)** $\triangle ABC$
not $\sim \triangle DEF$ **2.a)** $x = 4.0$, $y = 18.0$ **b)** $x = 6.0$, $y = 2.0$
c) $x° = 71°$ **d)** $x = 4.5$, $y = 6.0$ **e)** $x = 9.0$, $y = 4.8$

7.5 Exercise, page 306

1.a) $x = 40.0$, $y = 18.0$ **b)** $x = 3.8$, $y = 3.0$ **c)** $x = 1.3$, $y = 1.6$
2.a) 2.0 cm **b)** 15.0 cm
c) 18.4 cm **4.a)** 13 cm **b)** 13 cm

7.6 Exercise, page 308

1.a) 9.0 **b)** 12.0 **c)** 12.0 **d)** 6.0 **e)** 12.0

7.7 Exercise, page 310

1. 70 m **2.a)** 202 m **b)** 4.9 min **3.** 110.2 cm **4.** 3.6 m
5. 10.5 m **6.** 10.5 m **7.** 63.6 m **10.** 52.8 m **11.** 147 m

7.8 Exercise, page 315

1.a) 9:16 **b)** 25:4 **c)** $9:\dfrac{1}{4}$ **d)** 3:16 **e)** 36:5 **2.a)** 25 cm²
b) 21 cm² **3.** 5 cm **4.** 36:25 **5.** 256:169 **6.** 8 **7.** 10 cm
10.a) 98 cm² **b)** 80 cm² **11.** 6 cm **12.** 367.5 cm² **13.** 4:7
14.a) 1 : 4 **b)** 1 : 16 **16.a)** $x° = 75°$, $y° = 60°$ **b)** 33 cm
c) 3.6 cm **d)** 3.5 cm **e)** 2.25 cm² **f)** 8.4 cm **g)** 3.5 cm
h) 12 cm **i)** $\dfrac{7}{3}$ **j)** 5 **k)** 12 cm² **l)** $\dfrac{1}{5}$ **m)** 5.4 cm

7.9 Exercise, page 318

1.a) 1:27 **b)** 1:64 **c)** 8:27 **d)** 64:27 **2.a)** $\dfrac{3}{2}:1$ **b)** 4:5 **c)** $2:\dfrac{3}{4}$

d)$\sqrt[3]{2}$:2 **3.**$\dfrac{512}{125}$ times greater

CHAPTER 8

8.1 Exercise, page 324
2.a)$x = 24$ **b)**$x = 10$ **c)**$x = 2\sqrt{21}$ **3.**48.0 cm **4.**20 cm
5.14.0 cm **6.**5.0 cm

8.2 Exercise, page 329
3.a)$x° = 140°$ **b)**$x° = 45°$, $y° = 30°$ **c)**$x° = 90°$
d)$x° = 40°$, $y° = 110°$ **4.a)**40° **b)**100° **c)**90° **5.a)**25°
b)45° **c)**60° **6.a)**120° **b)**50° **c)**35° **d)**25°
7.a)$x° = 70°$, $y° = 20°$ **b)**$x° = 40°$, $y° = 40°$
c)$x° = 30°$, $y° = 60°$ **d)**$x° = 260°$, $y° = 100°$,
$z° = 50°$ **10.a)**43° **b)**109° **11.a)**160° **b)**3.05 cm

8.3 Exercise, page 333
3.a)$\angle R = 110°$, $\angle S = 95°$ **b)**$\angle P = 75°$,
$\angle S = 153°$ **4.a)**$y = 55°$ **b)**$x = 125°$ **5.a)**60°
b)120° **c)**10° **6.a)**$x° = 30°$, $y° = 70°$, $z° = 30°$
b)$x° = 40°$, $y° = 70°$, $z° = 70°$ **c)**$x° = 50°$,
$y° = 70°$, $z° = 70°$ **d)**$y° = 125°$

8.5 Exercise, page 341
1.a)(i)$\sqrt{2}$ cm (ii)90° **b)**50° **c)**16.0 cm **d)**45° **2.a)**5
b)$x° = y° = 20°$ **c)**$x° = 140°$ **3.** 16.0 cm **4.b)**8.0 cm
10. 3.0 cm

8.6 Exercise, page 344
1.a)70° **b)**40° **c)**$x° = 65°$, $y° = 42°$ **d)**$x° = 85°$,
$y° = 23°$, $z° = 23°$ **2.a)**50° **b)**100° **c)**30° **3.a)**50°
b)72° **c)**140° **4.a)**50° **b)**60°

8.7 Exercise, page 349
1.a)9 **b)**2 **c)**2 **d)**3.53 **e)**1.25 **f)**2.5 **g)**$\dfrac{1}{3}$ **2.** 1.0 cm

3. 12.0 cm **4.** 19.5 cm **5.** 24.0 cm **6.** BC = 4.8 cm,
BD = 6.0 cm **7.**PT = 9.8 cm, RQ = 4.0 cm

8.8 Exercise, page 352
1.a)3.7 cm **b)**18.8 cm **c)**18.4 cm **2.a)**3.9 cm
b)6.4 cm **c)**11.2 cm **3.a)**0.5 rad **b)**1.9 rad **c)**5.0 rad
4.a)22.7 cm² **b)**14.1 cm² **c)**1028.8 cm² **d)**163.3 cm²
e)81.1 cm² **f)**395.1 cm² **g)**148.8 cm² **h)**20.3 cm²
5.a)5.2 cm **b)**12.5 cm **c)**17.4 cm **d)**2.7 cm **6.a)**11.5 cm
b)6.3 cm **c)**22.5 cm **d)**3.9 cm **7.a)**30.0° **b)**3.7 rad
c)154.9° **d)**2.5 rad **8.**4.0 m **9.**0.1 rad **10.a)**0.4 rad
b)4.2 m² **11.**175° **12.a)**37.3 cm² **b)**6.4 cm **13.**4.6 cm
14.a)6.4 cm² **b)**3.7 cm²

CHAPTER 9

9.1 Exercise, page 359
2.a)2^6 **b)**4^6 **c)**3^{16} **d)**9^6 **3.a)**3^{15} **b)**3^{12} **c)**3^9 **d)**3^8 **e)**e^6
f)$3^{m + n}$ **g)**3^{mn} **h)**3^{3m} **i)**3^{2n} **j)**$3^{3 + m}$ **4.a)**32 **b)**6 **c)**-64

d)1 **e)**.09 **f)**32 **g)**-729 **h)**8 **i)**$\dfrac{1}{9}$ **j)**2 **5.a)**125 **b)**$\dfrac{1}{2}$
c)243 **d)**25 **6.a)**2^{12} **b)**2^{10} **c)**3^{14} **d)**2^3 **e)**$(-3)^9$ **f)**$(-4)^6$
g)3^{16} **h)**π^8 **7.a)**$16x^2$ **b)**$-27x^3$ **c)**x^{12} **d)**$3x^8$ **e)**$20x^5$
f)$18x^5$ **g)**$5x^4$ **h)**$-64x^7$ **i)**x^{6p} **j)**$\dfrac{3^{3p}x^{2p}}{2^{2p}}$ **k)**$(-1)^{5p - 3}x^{7p}$
l)x **m)**x^{-5} **n)**x^8 **8.a)**$6x^{10}$ **b)**$5x^5$ **c)**$\dfrac{3}{2}$ **d)**x^5 **9.a)**x^6
b)$y^{a + b - c}$ **c)**a^2 **d)**x^{2b} **e)**$a^{x + 2y + 7}$ **f)**$a^{x + y}b^{2x}$ **10.a)**B
b)A **11.a)**$\dfrac{a^5c}{b^6}$ **b)**1 **c)**$x^{4a}y^{4b}$ **12.a)**36 **b)**4 **c)**$\dfrac{1}{4}$ **14.b)**0, 1

9.2 Exercise, page 363
1.a)1 **b)**$\dfrac{1}{49}$ **c)**$\dfrac{10\,000}{9}$ **d)**$-\dfrac{1}{27}$ **e)**$-\dfrac{1}{27}$ **f)**-27 **g)**$\dfrac{1}{27}$
h)-1 **i)**$-100\,000$ **j)**1 **2.a)**$\dfrac{2}{3}$ **b)**$\dfrac{5}{6}$ **c)**$\dfrac{12}{7}$ **d)**$\dfrac{1}{3}$ **e)**$\dfrac{1}{4096}$
f)$\dfrac{1}{729}$ **g)**-8 **h)**1 **i)**1 **j)**2 **k)**1 **l)**4 **m)**1 **n)**1 **3.a)**x^2 **b)**y^2
c)$\dfrac{1}{x^8}$ **d)**z^6 **e)**1 **f)**$\dfrac{32}{a^5b^5}$ **g)**$\dfrac{1}{a}$ **4.a)**$\dfrac{x^3}{y^2}$ **b)**$\dfrac{3py}{x}$ **c)**$\dfrac{3x}{4}$ **d)**$\dfrac{p^2s^2}{3b}$
e)$\dfrac{4^2s}{5p}$ **f)**$\dfrac{n}{x^3y^2}$ **5.a)**$\dfrac{q^2}{p^3}$ **b)**$\dfrac{3}{x^2}$ **c)**$\dfrac{n^4}{16m^4}$ **d)**$\dfrac{1}{9x^2}$ **e)**$\dfrac{a^4}{9}$**f)**$2187w^7$
g)$7y^2$ **h)**$\dfrac{w^3}{5}$ **6.a)**1 **b)**$\dfrac{1}{v^6}$ **c)**$\dfrac{b^2}{243}$ **d)**k^8 **e)**$\dfrac{144}{d^{12}}$ **f)**$\dfrac{2}{z^5}$ **g)**$\dfrac{xy^{12}}{3}$
h)$\dfrac{7}{m^5}$ **i)**$\dfrac{4}{9s^2}$ **7.a)**$m^{2x + y}$ **b)**$s^{3x - 2p}$ **c)**t^{5x} **d)**$q^{x - 1}$ **e)**$y^{2n - m}$
f)$k^{5m + 3n}$ **g)**$8^{2a + 2b}$ **h)**$q^{a - b - 2c}$ **8.a)**$3 + \dfrac{3}{q^4}$ **b)**$\dfrac{(m + n)^2}{mn}$
c)$\dfrac{p - 2}{p + 2}$ **9.a)**-1024 **b)**-4 **c)**-1024 **10.a)**$\dfrac{7}{10}$ **b)**$\dfrac{3}{56}$
c)-32 **d)**$\dfrac{507}{64}$ **e)**$-\dfrac{2}{3}$ **f)**0 **11.a)**$\sqrt{7}$ **b)**$\sqrt[5]{x}$ **c)**$\sqrt[4]{a}$ **d)**$\sqrt[5]{8^2}$
e)$\sqrt[7]{b^3}$ **f)**$5\sqrt[3]{x^2}$ **g)**$\dfrac{1}{\sqrt[3]{x}}$ **h)**$\dfrac{1}{\sqrt[5]{a^2}}$ **12.a)**$6^{\frac{1}{2}}$ **b)**$10^{\frac{1}{2}}$ **c)**$7^{\frac{1}{2}}$ **d)**$12^{\frac{3}{7}}$
e)$8^{\frac{9}{7}}$ **f)**7^{-1} **13.a)**$x^{\frac{1}{3}}$ **b)**$x^{-\frac{1}{3}}$ **c)**$x^{-\frac{1}{2}}$ **d)**$a^{\frac{5}{6}}$ **e)**$a^{-\frac{1}{4}}$ **f)**$a^{\frac{2}{3}}$ **14.a)**$\sqrt[4]{3^3}$
b)$\sqrt[9]{5^4}$ **c)**$b\sqrt{a}$ **d)**a^2b^3 **e)**xy^2z^3 **f)**$y + 2\sqrt{y^3}$ **g)**$2x^3y^2$ **h)**$\sqrt{8}$
15.a)3 **b)**4 **c)**27 **d)**8 **e)**32 **f)**2 **g)**27 **h)**4 **16.a)**8 **b)**$\dfrac{1}{9}$
c)$\dfrac{1}{32}$ **d)**$\dfrac{1}{27}$ **e)**1000 **f)**0 **g)**$\dfrac{1}{5}$ **h)**$\dfrac{3125}{32}$ **i)**5 **j)**2 **17.**C $\dfrac{1}{32}$,
A $\dfrac{1}{9}$, D $\dfrac{1}{2}$, B 36

9.3 Exercise, page 368
(Last digit of answers may vary)
1.a)2.5 **b)**22.6 **c)**13.9 **d)**6.5 **e)**9.9 **f)**34.3 **2.a)**10.0
b)52.1 **c)**5.2 **3.a)**20.0 **b)**63.1 **c)**31.6 **4.a)**2.6 **b)**1.8
c)1.3 **d)**1.7 **e)**3.4 **5.a)**15.8 **b)**1.7 **c)**31.0 **d)**1.8 **e)**50.1
7.c)$(x, y) \rightarrow (-x, y)$ **8.c)**$(x, y)(-x, y)$ **16.b)**decay
17.a)2.5 **b)**4.7 **c)**14.5 **18.a)**95 **b)**53 **c)**13 **19.b)**growth
c)(i)$430 (ii)$495 (iii)$600 **d)**(i)2.5 years (ii) 3.6 years

9.4 Exercise, page 372
Answers may vary for questions 1 to 7.
1.a)10.1 **b)**10.1 **c)**Approximately the same. **2.a)**2.3
b)2.2 **c)**Approximately the same. **3.a)**21.8 **b)**2.6 **c)**24.1
d)17.5 **e)**8.0 **4.a)**21.8 **b)**2.7 **c)**24.5 **d)**17.7 **e)**8.1

Approximately the same. g)B **5.a)**7.6 **b)**48.4 **6.a)**7.9, 50.2
Approximately the same. c)A **7.a)**5.4 **b)**28.5 **c)**5.0.

.5 Exercise, page 375

.a)6.026 b)6.067 c)6.152 d)4.102 e)1.778 f)1.782
)1.799 h)1.945 **2.a)**$10^{0.650}$ b)$10^{0.191}$ c)$10^{0.949}$
)$10^{0.836}$ e)$10^{0.254}$ f)$10^{0.704}$ g)$10^{0.491}$ h)$10^{0.406}$
.a)$10^{0.7375}$ b)$10^{0.5674}$ c)$10^{0.2062}$ d)$10^{0.4579}$ e)$10^{0.3528}$
)$10^{0.8017}$ g)$10^{0.8895}$ h)$10^{0.9876}$ **4.a)**$10^{1.676}$ b)$10^{2.3862}$
)$10^{-0.396}$ d)$10^{1.926}$ e)$10^{2.499}$ f)$10^{-0.203}$ g)$10^{1.802}$
)$10^{2.831}$ **5.a)**30.4 b)490.9 c)668.1 d)1.11 e)1.17
.a)28.58 b)8.59 c)30.37 d)55.72 e)666.8 f)5480.4
.a)7.023 b)4.276 c)13.76 d)1.0193 **8.a)**38.55
)8.384 c)3.603 d)158.2 e)120.0 f)29.04
.a)6.03 b)0.895 c)339.0 **10.a)**$1.56 b)$3.28
1.$633.72 **12.**B is better by $110.46.

.6 Exercise, page 378

1.a)3.948 b)128.000 c)10.882 d)0.479 e)0.227
)4 014.274 g)0.949 h)16.809 i)1.071 j)0.245
2.a)5.0 b)3.0 c)2.0 d)4.0 e)9.0 f)15.9 g)512.0
)17.9 **3.a)**5.76 b)1.78 c)9.58 d)725.43 e)0.62
)0.70 g)1.50 h)708.22 i)0.20 j)0.00 k)6.16 l)1.55
4.a)1.39 b)0.49 c)0.23 d)-1.00 e)-0.69 **5.a)**1.28
)2.25 c)2.59 d)2.98 e)15.80 f)2.30 **6.a)**38.04
)47.96 c)0.28 d)1.47 e)0.19 f)160 720.91
7.6.5 km/h **8.**502 377 286 **9.**$192.54 **10.**$4.80 \times 10^{-13}$

.7 Exercise, page 380

1.a)100 b)200 c)3200 d)102 400 **2.a)**3200
)51 200 c)200(2^n) **3.a)**2500 b)160 000 **4.**$\frac{1}{2}$ h
5.a)32 000 b)2 048 000 c)500(2^{2n}) **6.**12 min
7.a)10 000 b)2 560 000 **8.a)**16 000 b)64 000
)1 024 000 **9.**179 200 **10.**6400% **11.a)**$986.91
)$1659.52 c)$1853.13 **12.a)**$2585.03 b)$1085.03
13.a)$1665.56 b)$865.56 c)$144.26

.8 Exercise, page 384

1. 25 mg **2.** 5 d **3.a)**280 d b)980 d **4.** 62.5 mg **5.** 138 d
6. 48 d **7.** 5.0 km/h **8.** 32.0 km/h **9.** 11 500 years
10. 8600 years **11.** 17 300 years **12.** 8600 years
13. 20 200 years

CHAPTER 10

10.1 Exercise, page 391

1.c) Yes d)$x = 10^y$, $y = \log_{10}x$ **2.c)**$y = \log_2 x$,
$y = \log_{\frac{1}{2}}x$ **4.a)**0.30 b)0.70 c)0.90 d)1.00 **5.a)**0.60 b)0.85
c)0.93 d)0.54 **6.a)**1.0 b)10.0 c)5.0 d)2.5 e)4.5 f)2.2

10.2 Exercise, page 394

1.a)$\log_2 8 = 3$ b)$\log_3 81 = 4$ c)$\log_5 125 = 3$
d)$\log_7 343 = 3$ e)$\log_m p = n$ f)$\log_{10} 1 = 0$
g)$\log_{16} 4 = \frac{1}{2}$ h)$\log_{81} 27 = \frac{3}{4}$ **2.a)**$3^3 = 27$

---

b)$10^3 = 1000$ c)$2^7 = 128$ d)$x^z = y$ e)$5^0 = 1$
f)$27^{\frac{1}{3}} = 3$ g)$2^{-3} = \frac{1}{8}$ h)$49^{\frac{1}{2}} = 7$ i)$10^{-3} = 0.001$

3.a)$2^4 = 16$ b)$6^1 = 6$ c)$\log_2 \frac{1}{8} = -3$ d)$\log_{49} 7 = \frac{1}{2}$
e)$27^{\frac{1}{2}} = 3$ f)$2^6 = 64$ g)$\log_{12} 1 = 0$
h)$\log_{10} 0.0001 = -4$ i)$7^0 = 1$ **4.a)**5 b)7 c)4 d)3 e)5
f)$\frac{2}{3}$ g)$\sqrt{3}$ h)-3 **5.a)**1 b)2 c)0 d)3 e)0.134 f)1.611
g)2.164 h)-1.520 **6.a)**All equal to 1 b)3, 2, -1,
-3, 2; x **7.a)**3 b)4 c)5 d)-3 e)-5 f)0 g)$\frac{1}{2}$ h)6 i)$\frac{1}{3}$
j)x k)1 l)0 **8.a)**8 b)5 c)3 d)-3 e)$\frac{1}{125}$ f)20 g)2
h)$x > 0$, $x \neq 1$ i)$\frac{7}{2}$ **9.b)**1000 c)316.2278 d)6.3096
e)4073.8028 f)3.4834 g)1.2179 **10.a)**4 b)$\frac{1}{4}$ c)27
d)4 **11.a)**6 b)7 c)13 **12.a)**7 b)5 c)$\frac{21}{20}$ d)$\frac{7}{5}$ **13.a)**4
b)100 c)25 d)$\frac{1}{27}$ e)64 f)m **14.a)**$\frac{1}{3}$ b)1 c)27 d)729
e)512 f)4 **15.a)**1027 b)-27 **16.a)**1000 b)3980
17.10 000 **18.**4 **19.**40 **20.**16

10.3 Exercise, page 398

1.a)A : 1, B : 1.58 b)2.58 c)2.58 d)approximately the
same e)$2 \times 3 = 6$ **2.a)**A : 2, B : 1.4 b)0.7 c)0.6
d)$9 \div 4.5 = 2$ **3.a)**A:0.30, B:0.90 b)0.90 c)same
d)8 e)0.90 f)same **4.a)**(i)0.9 (ii)0.6 b)0.6 c)4 d)0.6
e)same f)yes

10.4 Exercise, page 401

1.a)$\log_3 13 + \log_3 47$ b)$\log_2 3.2 + \log_2 78$
c)$\log_{10} 15.2 + \log_{10} 33.8$ d)$\log_x p + \log_x g$
e)$\log_z x + \log_z y$ f)$\log_b m + \log_b n$ **2.a)**$\log_7(13 \times 41)$
b)$\log_2(28 \times 36)$ c)$\log_7(43 \times 81)$
d)$\log_{10}(22.7 \times 36.3)$ e)$\log_a(mn)$ f)$\log_x(a^3 b^3)$
3.a)$\log_3\left(\frac{37}{22}\right)$ b)$\log_2\left(\frac{85}{74}\right)$ c)$\log_{10}\left(\frac{222}{75}\right)$ d)$\log_x\left(\frac{71}{17}\right)$
e)$\log_b\left(\frac{33}{11}\right)$ f)$\log_a\left(\frac{x^2 y}{sy}\right)$ **4.a)**$\log_2 72 - \log_2 35$
b)$\log_7 352 - \log_7 19.3$ c)$\log_{10} 751 - \log_{10} 82$
d)$\log_a a - \log_a b$ e)$\log_x 52.5 - \log_x 131$
f)$\log_a 741 - \log_a 337$ g)$\log_b 842 - \log_b 61.3$
h)$\log_k 73.2 - \log_k 13.7$ **5.a)**$\log_4 7 + \log_4 6$
b)$\log_6 0.28 + \log_6 536$ c)$\log_7 421 - \log_7 237$
d)$\log_x 22.3 - \log_x 481$ e)$\log_{10} 7 + \log_{10} 27 + \log_{10} 361$
f)$\log_x p + \log_x q - \log_x m - \log_x n$ **6.a)**2 b)2 c)3 d)3
7.a)5 b)3 c)4 d)3 **8.a)**7 b)2 c)3 d)1 e)6 f)2 **9.**B
10.(a), (c) **11.a)**$\frac{3}{2}$ b)9 c)$\frac{14}{3}$ d)$-\frac{1}{2}$ **12.a)**$\log_2\left(\frac{xyz}{a}\right)$
b)$\log_3\left(\frac{xy^2}{a^2}\right)$ c)$\log_3\left(\frac{x^2 - y^2}{xy}\right)$

10.5 Exercise, page 405

1.a)$2 \log_7 15$ b)$8 \log_6 57$ c)$4 \log_x a$ d)$-2 \log_a y$

e)$\frac{1}{4}\log_2 15$ f)$\frac{4}{5}\log_4 25$ **2.a)**$\log_8 7^3$ b)$\log_6 5^{10}$ c)$\log_x a^4$

d)$\log_5 z^{\frac{4}{3}}$ e)$\log_4 7^{-\frac{2}{3}}$ f)$\log_x a^{-\frac{6}{5}}$ **3.a)**$5\log_{10}(1.68)$

b)$\frac{1}{2}\log_{10}(3.81)$ c)$\frac{1}{2}\log_{10}(4.86)$ d)$\frac{2}{3}\log_{10}(4.86)$

e)$\frac{2}{5}\log_{10}(4.26)$ f)$\frac{3}{2}\log_{10}(9.86)$ **4.a)**50 b)10 c)$\frac{3}{4}$ d)$\frac{4}{3}$

e)$\frac{1}{5}$ f)$\frac{19}{3}$ g)$\frac{6}{5}$ h)$\frac{4}{5}$ i)-2 **5.a)**1 b)1 c)1.3979

d)0.8433 e)0.1174 f)2.5054 **6.**A, C **7.**D, C, A, B

8.a)$\log_2 180$ b)$\log_6 5$ c)$\log_4 72$ d)$\log_5 216$ **9.a)**$m - n$

b)$2m + 2n$ c)$2m - n$ d)$3m + 3n$ e)$\dfrac{m + n}{2}$

f)$\dfrac{-2(m + n)}{3}$ g)$\dfrac{m + n}{3}$ h)$-2m - n$ **10.a)**$\dfrac{31}{4}$ b)$\dfrac{3}{2}$

11.a)$\log_5\left(\dfrac{\sqrt{x} \times \sqrt[3]{y}}{\sqrt[4]{z}}\right)$ b)$\log_4\left(\dfrac{\sqrt{xy^3}}{a^2b^2}\right)$ c)$\log_2\left(\dfrac{\sqrt[3]{x^2}}{yz^4}\right)$

10.6 Exercise, page 408

1.a)$x = 6$ **2.a)**54 b)15 c)11 **3.a)**5 b)48 c)7 **4.a)**2

5.a)64 b)16 **6.a)**5 b)12 **7.a)**64 b)4 **8.a)**3 **9.a)**2 b)3

c)$\frac{8}{7}$ **10.a)**5 b)$\frac{13}{9}$ c)5 **11.a)**3 **12.**10.35 km

13.200.5 kPa **14.b)**$19\ 200$ m **15.**15.11 kPa **16.**273.6 kPa

10.8 Exercise, page 412

1.a)0.2663 b)-0.3255 c)2.0103 d)2.0561

2.a)2.3502 b)1.3678 c)3.6824 d)0.6227 **3.a)**0.8420

b)-0.1920 c)4.2031 d)-0.3151 **4.a)**0.3652 b)3.5237

5.a)23.43 b)4920.40 c)17.73 d)1.07 **6.a)**3.1623

b)6.3760 **7.a)**2.902 b)1.9078 c)148.16 d)3166.78

8.a)2.6 b)1.5 c)1.6 d)0.9 e)2.6 **9.a)**-0.7 b)-0.52

c)1.5 **10.a)**-7.7 b)-11.5 c)0.7 d)-0.5

10.9 Exercise, page 414

1.5.39 m **2.a)**4.47 years b)7.64 years **3.**7.19 years

4.a)2.6 cm b) 1.5 m **5.**3400 km **6.**23.6 s **7.**8.02 days

8.15.3 s **9.a)**(i)1.37 m/s (ii)3.01 m/s (iii)5.65 m/s

b)59.1 m **10.a)**158 b)501

10.10 Exercise, page 417

1.a)10% b)6 years c)9% **2.**10% **3.a)**6 years

CHAPTER 11

11.1 Exercise, page 423

1.a)a circle b)a straight line (drawn diagonally) c)a
vertical line d)a straight line which is the bisector of the
angle formed by the adjacent sides **2.a)**a circle b)a spiral
c)a straight line d)a circle e)a cylinder **3.a)**a circle whose
radius is 5 cm b)a line parallel to the given line and 4 cm
away from the line c)the bisectors of the angles of
intersection d)the right bisector of the line joining these 2
points e)a cube with dimensions 2 cm shorter than those of
the original cube **13.a)**$x = 4$ b)$y = -6$ c)$x = -\dfrac{1}{2}$ d)$y = \dfrac{-5}{2}$

14.$x - 2y + 1 = 0$ **15.**$2x - 3y = 5 + 8\sqrt{13}$ or
$2x - 3y = 5 - 8\sqrt{13}$ **16.**$x + y = 3$
17.a)$x + 3y - 8 = 0$ b)$3x - y - 4 = 0$ c)$(2, 2)$

11.2 Exercise, page 429

1.a)$(x, y) \rightarrow (x + 4, y - 3)$ b)$(x, y) \rightarrow (x - 3, y - 3)$
c)$(x, y) \rightarrow (x + 3, y + 2)$ **2.a)**$(x, y) \rightarrow (x - 6, y)$
b)$(x, y) \rightarrow (x, y - 2)$ c)$(x, y) \rightarrow (x + 5, y + 7)$
d)$(x, y) \rightarrow (2x, y)$ e)$(x, y) \rightarrow (x, 4y)$ f)$(x, y) \rightarrow (3x, 2y)$
3.$(x, y) \rightarrow (x + 3, y - 1)$ **4.a)**$(x, y) \rightarrow (x - 4, y + 3)$
b)$(x, y) \rightarrow (x - 5, y - 2)$ c)$(x, y) \rightarrow (x + 6, y + 4)$
5.a)$x^2 + 4y^2 = 16$ b)$16x^2 + y^2 = 32$
6.a)$x^2 + y^2 = 25$ b)$x^2 + y^2 = 100$ c)$x^2 + y^2 = 36$

d)$x^2 + y^2 = \dfrac{1}{36}$ e)$x^2 + y^2 = 4$ **7.b)**centre is $(2, -3)$,

radius is 10 c)$(-8, 6)$ **8.**$(x, y) \rightarrow \left(\dfrac{x}{2}, 3y\right)$

9.$(x, y) \rightarrow (4x, 2y)$ **10.**$(x, y) \rightarrow \left(\dfrac{1}{3}x, \dfrac{2}{3}y\right)$

11.3 Exercise, page 433

1.a)$y^2 = 16x$ b)no **2.a)**$x^2 = -12y$ b)yes **3.a)**$y^2 = 8x$
b)$x^2 = -12y$ c)$y^2 = -16x$ d)$x^2 = 20y$ **4.a)**$(4, 0)$
b)$x = -4$ **5.a)**focus $(0, 4)$, vertex $(0, 0)$, directrix
$y = -4$ b)focus $(-6, 0)$, vertex $(0, 0)$, directrix $x = 6$
c)focus $(8, 0)$, vertex $(0, 0)$, directrix $x = -8$ d)focus
$(0, -5)$, vertex $(0, 0)$, directrix $y = 5$ **6.a)**$y^2 = 100x$
b)$(25, 0)$ c)$y = 0$ d)$D = \{x | x \geq 0, x \in R\}$, $R = \{y | y \in R\}$,

x-intercept 0, y-intercept 0 **7.a)**$x^2 = \dfrac{1}{8}y$ b) $\left(0, \dfrac{1}{32}\right)$

c)$x = 0$ d)$D = \{x | x \in R\}$, $R = \{y | y \geq 0, y \in R\}$, x-intercept
0, y-intercept 0 **8.c)**$(0, 0)$ **9.a)**$x^2 = 12(y - 1)$
b)$y^2 = 16(x + 1)$ c)(a)$(0, 1)$, (b)$(-1, 0)$
10.a)$D = \{x | x \geq -1, x \in R\}$, $R = \{y | y \in R\}$,
$y^2 = 12(x + 1)$ b)$D = \{x | x \in R\}$, $R = \{y | y \geq -1, y \in R\}$,
$x^2 = 12(y + 1)$ c)$D = \{x | x \leq -2, x \in R\}$, $R = \{y | y \in R\}$,
$y^2 = -20(x + 2)$ d)$D = \{x | x \geq 2, x \in R\}$, $R = \{y | y \in R\}$,
$y^2 = 20(x - 2)$ **11.a)**$(x - 2)^2 = 12(y - 1)$
d)$(y + 3)^2 = -12(x + 4)$ **12.a)**$x^2 = -12.5$ b)$(0, -3.125)$, $x = $
0 **13.a)**(i)$(3, -4)$ (ii)$x = 3$ b)(i)$(2, -12.5)$ (ii)$y = 1$ c)(i)$(4, $
1) (ii)$x = 4$ **14.a)**$y = 2(x - 1)^2 - 3$ b)$(1, -3)$ c)$x = 1$
16.a)$y^2 = -12.5x$ b)17.3 cm **17.**1.4 cm
b)$(1, -3)$ c)$x = 1$ **16.a)**$y^2 = -12.5x$ b)$10\sqrt{3}$ **17.**1.4
18.a)$y^2 = 223.2x$ b)$(55.8, 0)$

11.4 Exercise, page 439

1.	Major Axis	Minor Axis	Vertices	Foci	x-Intercepts	y-Intercepts
a)	20	2	$(10, 0), (-10, 0)$	$(\sqrt{99}, 0), (-\sqrt{99}, 0)$	$10, -10$	$1, -1$
b)	10	2	$(0, 5), (0, -5)$	$(0, 2\sqrt{6}), (0, -2\sqrt{6})$	$1, -1$	$5, -5$
c)	20	10	$(10, 0), (-10, 0)$	$(5\sqrt{3}, 0), (-5\sqrt{3}, 0)$	$10, -10$	$5, -5$
d)	16	12	$(8, 0), (-8, 0)$	$(2\sqrt{7}, 0), (-2\sqrt{7}, 0)$	$8, -8$	$6, -6$
e)	10	6	$(5, 0), (-5, 0)$	$(4, 0), (-4, 0)$	$5, -5$	$3, -3$
f)	10	4	$(0, 5), (0, -5)$	$(0, \sqrt{21}), (0, -\sqrt{21})$	$2, -2$	$5, -5$
g)	20	16	$(10, 0), (-10, 0)$	$(6, 0), (-6, 0)$	$10, -10$	$8, -8$
h)	20	16	$(0, 10), (0, -10)$	$(0, 6), (0, -6)$	$8, -8$	$10, -10$

Vertices	Foci	Major axis	Minor axis	Equation
(0, 5), (0, −5)	(0, 4), (0, −4)	10	6	$\frac{x^2}{9} + \frac{y^2}{25} = 1$
(4, 0), (−4, 0)	(2√3, 0), (−2√3, 0)	8	4	$\frac{x^2}{16} + \frac{y^2}{4} = 1$
(0, 3), (0, −3)	(0, √5), (0, −√5)	6	4	$\frac{x^2}{4} + \frac{y^2}{9} = 1$

4. $\frac{x^2}{36} + \frac{y^2}{20} = 1$ **5.** $\frac{x^2}{36} + \frac{y^2}{100} = 1$

6.a) $\frac{x^2}{31.25} + \frac{y^2}{56.25} = 1$ or $36x^2 + 20y^2 = 1125$
b) $\frac{x^2}{2.25} + \frac{y^2}{1.25} = 1$ or $20x^2 + 36y^2 = 45$
c) $\frac{x^2}{k^2 - m^2} + \frac{y^2}{k^2} = 1$ **d)** $\frac{x^2}{p^2} + \frac{y^2}{p^2 - k^2} = 1$

7.a) $\frac{x^2}{16} + \frac{y^2}{9} = 1$ **b)** $\frac{x^2}{16} + \frac{y^2}{4} = 1$ **c)** $\frac{x^2}{16} + \frac{y^2}{1} = 1$
d) $\frac{x^2}{9} + \frac{y^2}{25} = 1$ **e)** $\frac{x^2}{36} + \frac{y^2}{4} = 1$ **f)** $\frac{x^2}{9} + \frac{y^2}{25} = 1$

8. $\frac{x^2}{324} + \frac{y^2}{144} = 1$ **10.** $\frac{x^2}{1600} + \frac{y^2}{900} = 1$

11.a)(i)(0, 2√2), (0, −2√2) (ii)(0, 2√3), (0, −2√3)
b)(i) $\left(0, \sqrt{\frac{10}{3}}\right)$, $\left(0, -\sqrt{\frac{10}{3}}\right)$ (ii)(0, √10), (0, −√10)
c)(i) $\left(\frac{\sqrt{5}}{2}, 0\right)$, $\left(-\frac{\sqrt{5}}{2}, 0\right)$ (ii)(√2.5, 0), (−√2.5, 0)

12.b)x-intercepts 2, −2, y-intercepts 2√2, −2√2
c)(0, 2), (0, −2) **14.** $\frac{x^2}{7.3 \times 10^{18}} + \frac{y^2}{4.7 \times 10^{17}} = 1$

15. $\frac{x^2}{5.19 \times 10^7} + \frac{y^2}{5.18 \times 10^7} = 1$

16. $\frac{x^2}{2.24 \times 10^{16}} + \frac{y^2}{2.23 \times 10^6} = 1$

17.a) $\frac{x^2}{3.36 \times 10^{15}} + \frac{y^2}{3.22 \times 10^{15}} = 1$

11.5 Exercise, page 445
1.a)$a = 2\sqrt{5}$, $b = \sqrt{5}$, $c = 5$ **b)**$a = \sqrt{10}$, $b = 6$, $c = \sqrt{46}$ **c)**$a = 2\sqrt{2}$, $b = \sqrt{3}$, $c = \sqrt{11}$

2.

	Major axis	Vertices	x-intercepts	y-intercepts	Foci
a)	10, horizontal	(5, 0), (−5, 0)	5, −5	none	(√41, 0), (−√41, 0)
b)	4, horizontal	(2, 0), (−2, 0)	2, −2	none	(√13, 0), (−√13, 0)
c)	10, horizontal	(5, 0), (−5, 0)	5, −5	none	(√34, 0), (−√34, 0)
d)	4, horizontal	(2, 0), (−2, 0)	2, −2	none	(√5, 0), (−√5, 0)
e)	16, vertical	(0, 8), (0, −8)	none	8, −8	(0, 2√41), (0, −2√41)
f)	16, vertical	(0, 8), (0, −8)	none	8, −8	(0, 10), (0, −10)
g)	10, vertical	(0, 5), (0, −5)	none	5, −5	(0, √34), (0, −√34)
h)	12, horizontal	(6, 0), (−6, 0)	6, −6	none	(10, 0), (−10, 0)
i)	10, horizontal	(5, 0), (−5, 0)	5, −5	none	(13, 0), (−13, 0)

3.a)(2, 0), (−2, 0) **b)**(3, 0), (−3, 0) **c)**4 **d)** $\frac{x^2}{4} - \frac{y^2}{5} = 1$
4.a) $\frac{x^2}{9} - \frac{y^2}{16} = 1$ **b)** $\frac{x^2}{12} - \frac{y^2}{4} = -1$ **5.a)**$y = \pm\frac{1}{2}x$
b)$y = \pm\frac{4}{3}x$ **c)**$y = \pm\frac{4}{3}x$ **6.a)**$\pm\frac{1}{10}$ **b)**±5 **c)**$\pm\frac{4}{3}$

8.a)$x : 4, -4$; $y :$ none **b)**$x : 3, -3$; $y :$ none **c)**$x :$ none; $y : \sqrt{10}, -\sqrt{10}$ **d)**$x : 6, -6$; $y :$ none **e)**$x :$ none; $y : 1, -1$
f)$x : 2, -2$; $y :$ none

10.$11x^2 - 25y^2 = 275$ **11.a)**$\frac{x^2}{9} - \frac{y^2}{7} = 1$ **b)**$\frac{x^2}{16} - \frac{y^2}{20} = 1$
c)$\frac{x^2}{39} - \frac{y^2}{25} = -1$ **d)**$\frac{x^2}{64} - \frac{y^2}{1024} = 1$ **e)**$\frac{x^2}{9} - \frac{y^2}{55} = -1$
13.$24x^2 - y^2 = 5400$ **14.**$60x^2 - 4y^2 = 9375$

11.6 Exercise, page 449
3.a)transverse 12, conjugate 12 **b)**$y = \pm x$

11.7 Exercise, page 451
1.7.5 **2.a)**$B \times S = k$ **b)**800 **c)**3.2 **d)**80 **3.**8 **4.a)**30
b)4 **c)**8 **5.**39 **6.a)**1.6 **b)**2.5 **c)**10 **d)**1000 **7.a)**15 **b)**3
c)10 **8.a)**2 days **b)**8 **9.**100 cm **10.a)**17.5 cm
b)87.5 cm² **11.a)**4.3 min **b)**5.6 cm

11.8 Exercise, page 454
2.a)parabola **b)**hyperbola **c)**circle **d)**ellipse **3.a)**$\frac{\sqrt{5}}{3}$
b)0 **c)**1.5 **d)**0.75 **e)**1 **f)**2.0 **4.a)**0.6 **b)**0.6 **c)**0.8
5.a)$\frac{\sqrt{13}}{3}$ **b)**$\frac{\sqrt{34}}{5}$ **c)**$\frac{\sqrt{41}}{5}$ **6.** $\frac{x^2}{36} + \frac{y^2}{27} = 1$
7.$\frac{x^2}{45} - \frac{y^2}{36} = -1$ **8.a)**$\frac{x^2}{36} + \frac{y^2}{16} = 1$ **b)**$\frac{x^2}{16} - \frac{y^2}{20} = 1$
c)$\frac{x^2}{25} + \frac{y^2}{9} = 1$ **d)**$\frac{x^2}{4} - \frac{y^2}{12} = 1$ **e)**$\frac{x^2}{4} + \frac{y^2}{16} = 1$
f)$\frac{x^2}{25} + \frac{y^2}{9} = 1$ **9.b)**ellipse **10.b)**hyperbola

11.2.4, 4.4, 0.62, 0.53, 0.38, 0.32, 0.13

11.9 Exercise, page 459
1.a){(−3, 4), (4, −3)} **b)**{(−4, 0), (0, 2)} **c)**{(−4, 8), (2, 2)}
3.b)(−4, 7), (7, −4) **c)**{(−4, 7), (7, −4)} **4.b)**{(−5, 0), (5, 0)}
5.a)(−6, 8), (8, 6) **6.a)**(2, 4), (4, 2)
7.a)(0, −3), (6, 0) **8.a)**(1, 3) **b)** $\left(\frac{8 + \sqrt{29}}{5}, \frac{4 - 2\sqrt{29}}{5}\right)$, $\left(\frac{8 - \sqrt{29}}{5}, \frac{4 + 2\sqrt{29}}{5}\right)$ **c)**(−4, 14), (1, −1)
9.a){(−3, −1), (1, 3)} **b)**{(2, 3), (6, 1)} **c)**$\left\{\left(\frac{\sqrt{2} - 2}{2}, \frac{1}{\sqrt{2}}\right),\right.$

$\left(\dfrac{-\sqrt{2} - 2}{2}, -\dfrac{1}{\sqrt{2}} \right) \right\}$ **d)**$\{(\sqrt{2}, 1), (\sqrt{2}, -1),$

$(-\sqrt{2}, 1), (-\sqrt{2}, -1)\}$ **e)**$\{(-3, -4),\ (-4, -3), (4, 3),$

$(3, 4)\}$ **f)**$\left\{ \left(-\dfrac{12}{5}, -\dfrac{12}{5} \right), \left(-\dfrac{12}{5}, \dfrac{12}{5} \right), \left(\dfrac{12}{5}, -\dfrac{12}{5} \right) \right.$

$\left. \left(\dfrac{12}{5}, \dfrac{12}{5} \right) \right\}$ **10.b)**$\{(-2, -1), (8, -1)\}$ **12.**$(3, 2), (4, -1)$

13. $\left(\dfrac{95 - 2\sqrt{1306}}{21}, \dfrac{38 - 5\sqrt{1306}}{21} \right),$

$\left(\dfrac{95 + 2\sqrt{1306}}{21}, \dfrac{38 + 5\sqrt{1306}}{21} \right)$ **14.a)**$(a^2 + 1)m^2 > b^2$

b)$(a^2 + 1)m^2 = b^2$ **c)**$(a^2 + 1)m^2 < b^2$

15.a) $\left(\dfrac{-2 + \sqrt{46}}{2}, \dfrac{-2 - \sqrt{46}}{2} \right), \left(\dfrac{-2 - \sqrt{46}}{2}, \dfrac{-2 + \sqrt{46}}{2} \right)$

11.10 Exercise, page 464
1.a)parabola, $(5, 3)$ **b)**ellipse, $(2, -1)$ **c)**hyperbola, $(-3, 1)$ **d)**parabola, $(-2, 3)$ **e)**circle, $(1, 2)$ **f)**ellipse, $(-4, 1)$ **2.a)**ellipse, $(1, -1)$, y-axis **b)**hyperbola, $(2, -3)$, x-axis **c)**ellipse, $(-1, 2)$, x-axis **d)**hyperbola, $(-3, 3)$, x-axis **3.a)**1 **b)**$\dfrac{\sqrt{13}}{3}$ **c)**0.8 **d)**1

e)$\dfrac{\sqrt{7}}{3}$ **f)**$\dfrac{\sqrt{3}}{2}$ **4.a)**$(x - 2)^2 = -4(y + 1)$

b)$\dfrac{(x - 1)^2}{9} - \dfrac{(y - 2)^2}{4} = 1$ **c)**$(y + 3)^2 = 12(x + 4)$

d)$\dfrac{(x - 2)^2}{9} + \dfrac{(y - 1)^2}{4} = 1$ **5.a)**parabola

b)$F\left(3, \dfrac{1}{2} \right)$, $V(3, -1)$ **c)**$(x, y) \to (x - 3, y + 1)$,

$x^2 = 6y$ **6.a)**ellipse **b)**$F_1(3, 2\sqrt{7} - 1)$,

$F_2(3, -2\sqrt{7} - 1)$, $V_1(3, 7)$, $V_2(3, -9)$ **c)**$(3, -1)$

d)$(x, y) \to (x - 3, y + 1)$, $\dfrac{x^2}{36} + \dfrac{y^2}{64} = 1$ **7.a)**ellipse **a)**ellipse

b)$F_1(1 - \sqrt{7}, -2)$, $F_2(1 + \sqrt{7}, -2)$, $V_1(-3, -2)$,

V_2 **c)**$(1, -2)$ **d)**$(x, y) \to (x - 1, y + 2)$,

$\dfrac{x^2}{16} + \dfrac{y^2}{9} = 1$ **8.a)**hyperbola **b)**$F_1(-4, -3)$,

$F_2(6, -3)$, $V_1(-3, -3)$, $V_2(5, -3)$ **c)**$(1, -3)$

d) $(x, y) \to (x - 1, y + 3)$, $\dfrac{x^2}{16} - \dfrac{y^2}{9}$

9.a) $(x, y) \to (x - 1, y - 2)$
b) $(x, y) \to (x + 3, y - 4)$ **c)** $(x, y) \to (x - 1, y - 1)$

d) $(x, y) \to \left(x - 3, y + \dfrac{9}{2} \right)$ **10.a)**$\dfrac{(x - 1)^2}{25} + \dfrac{y^2}{16} = 1$

b)$\dfrac{(x + 1)^2}{9} - \dfrac{y^2}{7} = 1$ **11.a)**$\dfrac{(x + 2)^2}{9} - \dfrac{(y - 3)^2}{7} = 1$

b)$\dfrac{(x + 1)^2}{36} - \dfrac{(y - 4)^2}{64} = 1$ **12.a)**ellipse

b)$(-2, 2)$ **c)**$F_1(-2, 2 - \sqrt{21})$, $F_2(-2, 2 + \sqrt{21})$,

$V_1(-2, -3)$, $V_2(-2, 7)$ **d)**$10, 4$ **e)**$\dfrac{\sqrt{21}}{5}$

CHAPTER 12

12.2 Exercise, page 477
1.a)10 **b)**12%, 19%, 4% **2.a)**0.6 **b)**0.6 **3.a)**3, 3

b)$\dfrac{2}{3}, \dfrac{2}{3}$ **5.a)**7 **b)**5 **c)**1 **d)**2 **e)**8 **f)**3 **6.a)**$x \in \{0, 1, \dots, 16\}$

b)$f(x) \in \{0, 1, \dots, 8\}$ **c)**8 **8.a)**44 **b)**4 **c)**48 **d)**16 **e)**27
f)34 **9.a)**3% **b)**1% **c)**44% **d)**40% **10.a)**7 **11.a)**120
b)113 **c)**160 **d)**27 **e)**13 **f)**0 **12.b)**no **13.c)**10%
d)good **14.b)**4, 3 **c)**73 **d)**11, 14

12.3 Exercise, page 483
1.a)11 **c)**70 **2.a)**6 **b)**11 **c)**15 **d)**2 **e)**5 **f)**1 **3.a)**125
b)55 **4.a)**9% **b)**11% **c)**16% **d)**20% **5.**$378 571.43
6.c)62% **d)**yes **7.c)**55-60 **d)**75-80 **10.a)**47% **b)**8%
12.a)10% **b)**31% **14.a)**61 **b)**80 **c)**89 **15.**95 **16.**39
17. 200 **18.a)**(i)60 (ii)90 (iii)30

12.4 Exercise, page 489
1.a)20.9, 20, 24 and 20 **b)**7.2, 7, 12 **2.a)**12.4 **b)**12.4
c)12.2 **4.a)**5.9 **b)**22.4 **5.a)**72 **b)**360 **c)**5 **6.a)**12.7
b)13 **7.b)**13.5 **c)**13 **11.**0.276 **12.a)**26.8 **b)**27.5 **c)**28
13.a)28 **b)**6.5 **c)**7 **14.b)**yes

12.5 Exercise, page 494
1.b)24.3 **2.b)**8.8 **3.**15.5 **4.b)**29.3 **5.b)**29.7 **6.d)**75.9
7.a)76.5 **9.b)**75.9 **10.**$2.18 **11.a)**$2.78 **12.a)**$15.13
b)$14.50 or $16.25; $56.50 **c)**$16.58

12.6 Exercise, page 498
1.a)52.4 **b)**6.6 **2.a)**4.8 **b)**1.3 **3.a)**411.2 **b)**157.8 **4.a)**25.3
5.a)35.7 **6.a)**Myrna: 28.9, 10.2; Eileen: 28.9, 6.4 **b)**Eileen
7.a) David: 79.6, 13.7; Lesley: 75, 10.8; Kevin 60, 9.3;
Andrew: 40.2, 7.4; Joelle: 45.5, 21.1 **b)**Andrew

12.7 Exercise, page 503
1.a)4 **b)**7 **c)**14 **2.a)**15 **b)**14.4 **c)**16.9 **3.a)**1.2 **b)**2.2 **c)**4.3
4.a)C **b)**A **6.a)**0 **b)**5.2 **7.a)**0 **b)**4.2 **8.a)**19 **b)**0 **c)**0
d)64 **9.a)**11 **b)**0 **c)**120 **10.d)**3.0 **e)**4.1 **11.a)**10 **c)**3.16
12.a)20 **c)**6.8 **14.a)**74 **b)**6.3 **c)**no **15.a)**14 **b)**11 **c)**4.6 **d)**no

12.8 Exercise, page 507
1.a)2.4, 1.3 **b)**12.5, 1.4 **c)**42.2, 1.2 **2.b)**20, 275
c)13.8 **d)**5.0 **3.a)**38.3, 13.3 **b)**5.6, 0.7 **4.b)**6.9, 2.7, 9
5.b)4.8, 3.0, 10 **6.**No **7.**Yes **8.**No **9.b)**0.38 **c)**good

12.10 Exercise, page 514
1.b)0.878 **c)**$y = 0.57x + 1.92$ **2.a)**0.949, high
b)$y = 0.733x + 3$ **c)**(i)28.7 (ii)12.3 (iii)41.9 **3.c)**0.604
d)$y = 4.04x + 346.4$ **e)**A:891.8, B: 140.7 C:770.6

Looking Ahead, page 519
1.b)68 **c)**97.5 **2.a)**68 **b)**14 g **d)**yes **3.a)**34 **b)**47.5
c)68 **d)**95 **e)**99.7 **f)**81.5 **g)**13.5 **h)**49.85 **i)**81.5
6.a)16.5 **c)**4.1 **7.a)**110.7 **b)**40.4 **c)**950 **8.a)**6800
b)500 **9.**about 8 **10.a)**950 **b)**135